WONDERS AND RARITIES

WONDERS and RARITIES

THE MARVELOUS BOOK THAT TRAVELED THE WORLD
AND MAPPED THE COSMOS

Travis Zadeh

HARVARD UNIVERSITY PRESS
CAMBRIDGE, MASSACHUSETTS
LONDON, ENGLAND
2023

Printed in the United States of America

Publication of this book has been supported through the generous provisions of the Maurice and Lula Bradley Smith Memorial Fund

First printing

Cataloging-in-Publication Data is available from the Library of Congress
ISBN: 978-0-674-25845-7 (alk. paper)

Contents

Note on Conventions vii

Introduction: Wonder's End 1

I. A WORLD WITHIN WORDS 25

1 Stranger Lands 27

2 Measures of Authority 45

3 Astral Power 73

II. WONDERS TO BEHOLD 95

4 Cosmic Order 97

5 Terrestrial Designs 128

6 Alchemical Bodies 166

III. DISTANT SHORES 197

7 Long Divided 199

8 Across the Globe 223

9 On the Edge 261

Coda: Acts of Enchantment 294

Note on Sources and Method 333

Sigla and Abbreviations 337

Notes 343

Acknowledgments 407

Illustration Credits 413

Index 417

Note on Conventions

This book references terms, names, and phrases originally recorded in different alphabets. This includes materials in Arabic, New Persian, Ottoman Turkish, and Urdu. While these languages were written using the Arabic script, their pronunciation and orthography varied in significant ways. An effort has been made to follow, with some degree of consistency, established conventions for transliteration. Persian, Turkish, and Urdu pronunciations of Arabic loanwords and names are rendered when appropriate (e.g., *fazl, masnavī, mezheb*), save for publication information in the notes originally written in the Arabic script, where Arabic transliteration is generally used for the purposes of consistency. Toponyms and dynasties are rendered with modern English spellings (e.g., Qazvin, Mecca; Ilkhanids, Timurids). Proper names and words common to English are not transliterated. Unless otherwise stated, all translations are my own. Many of the dates for individuals and events found in the original historical sources are only estimates. When relevant, the corresponding Common Era year is noted after the date in the Islamic lunar calendar, or, for modern publications from Iran, the Islamic solar calendar. Without the month or day for a given event, it is often only possible to offer a range of two years when converting into the Common Era. In such situations, a symbol of approximation (~) has been used.

For Supriya and Anoush—*meri duniya*

WONDERS AND RARITIES

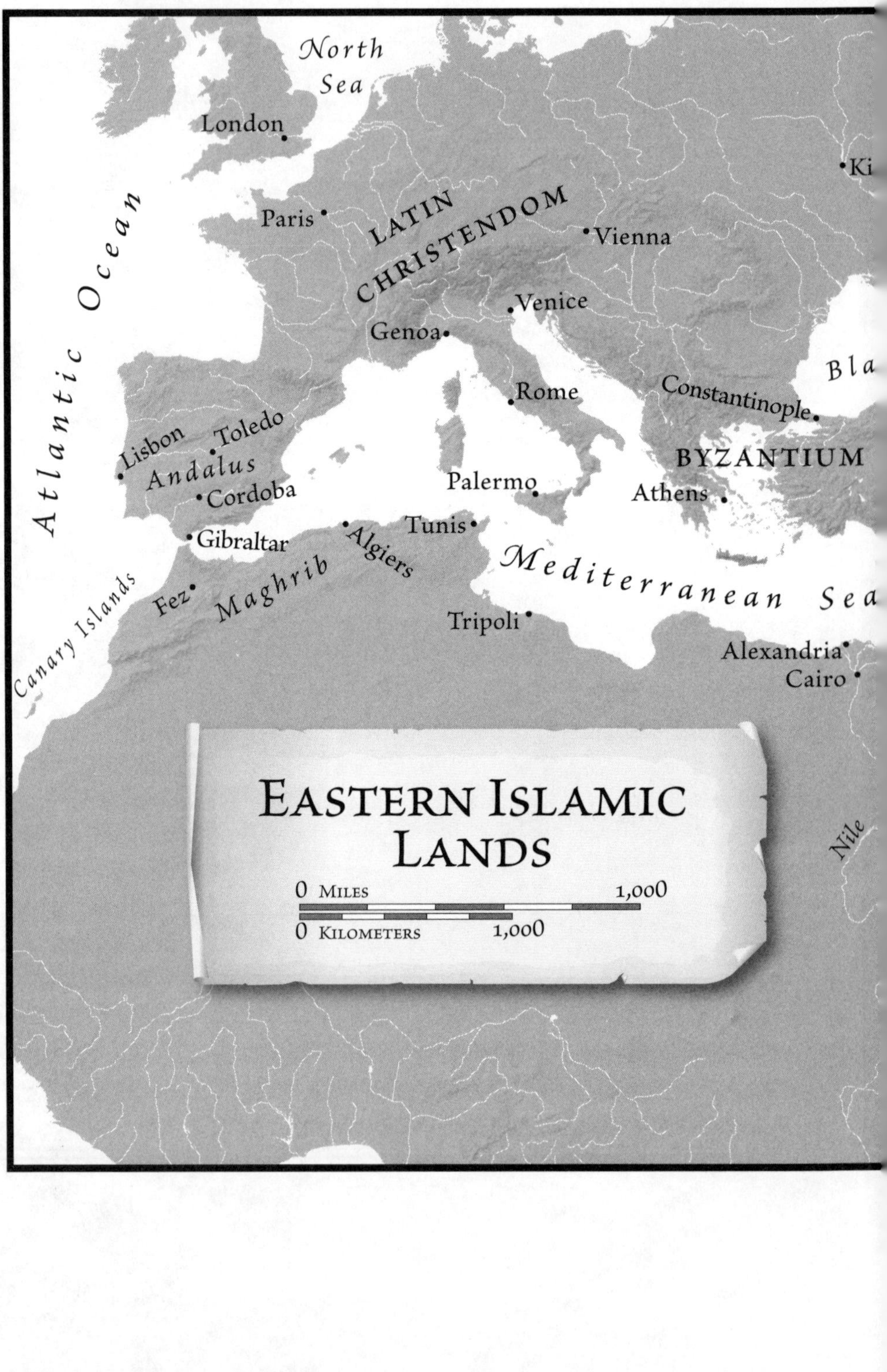
North Sea
London
Paris
LATIN CHRISTENDOM
Vienna
Atlantic Ocean
Venice
Genoa
Rome
Constantinople
Bla
BYZANTIUM
Lisbon
Toledo
Andalus
Cordoba
Palermo
Athens
Tunis
Gibraltar
Algiers
Mediterranean Sea
Fez
Maghrib
Canary Islands
Tripoli
Alexandria
Cairo
Ki
EASTERN ISLAMIC LANDS
0 Miles 1,000
0 Kilometers 1,000
Nile

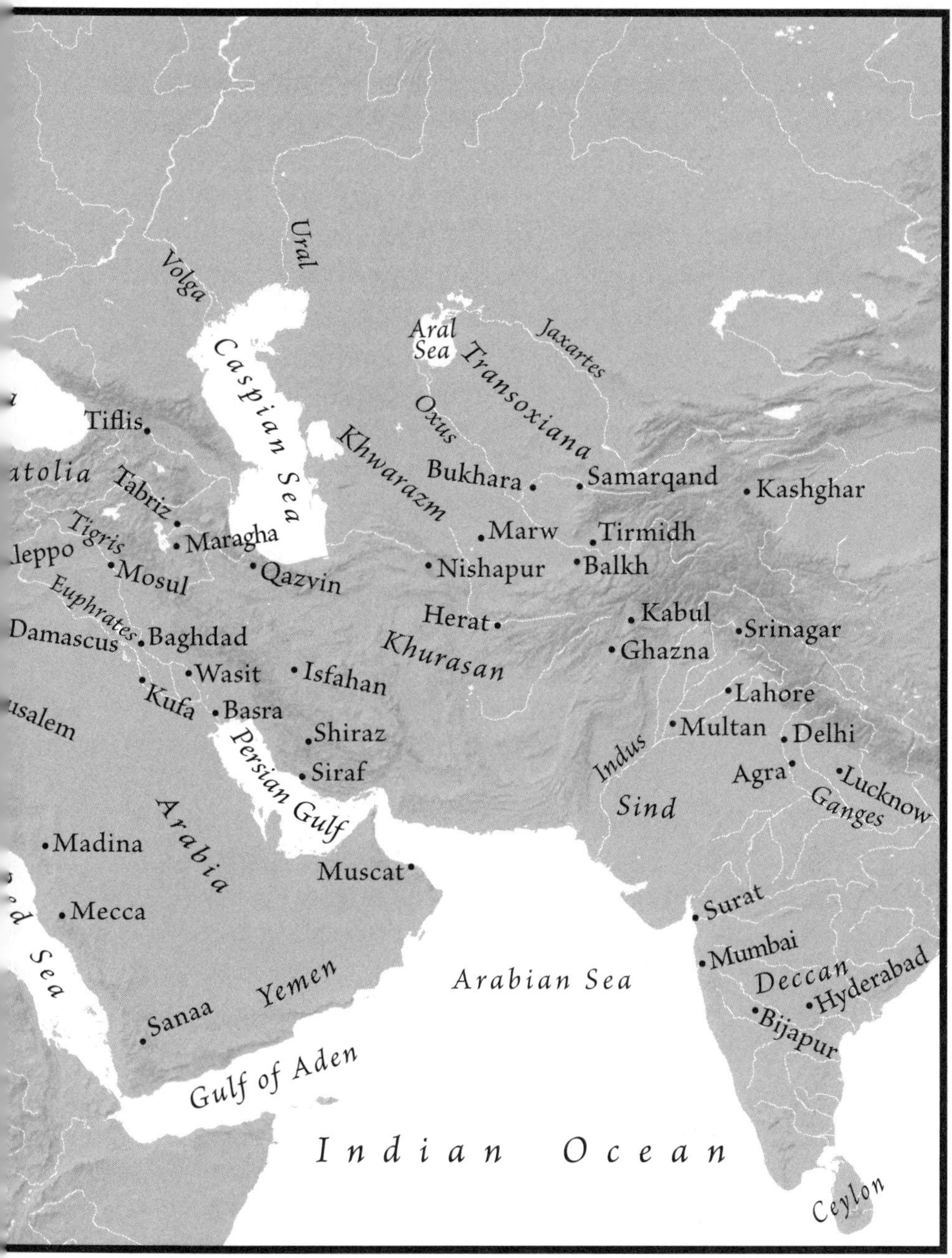

Ural
Volga
Aral Sea
Jaxartes
Caspian Sea
Transoxiana
Oxus
Tiflis
atolia
Khwarazm
Tabriz
Bukhara
Samarqand
Kashghar
Tigris
Maragha
Marw
Tirmidh
leppo
Mosul
Qazvin
Nishapur
Balkh
Euphrates
Herat
Kabul
Damascus
Baghdad
Khurasan
Srinagar
Ghazna
Wasit
Isfahan
Lahore
Kufa
Basra
usalem
Multan
Delhi
Shiraz
Persian Gulf
Indus
Siraf
Agra
Lucknow
Sind
Ganges
Arabia
Madina
Muscat
Surat
Mecca
Mumbai
Deccan
Hyderabad
Arabian Sea
Sea
Yemen
Bijapur
Sanaa
Gulf of Aden
Indian Ocean
Ceylon

INTRODUCTION

Wonder's End

YES, THERE IS A MERMAID. But is that really so farfetched? There is, after all, that singular animal found in the Malayan seas, the dugong, which could easily be mistaken for some strange mixture of human and fish. And yes, if you look over there you will certainly see griffins and other monsters of old. But before casting aspersions at the author of this wondrous book, what about those strange creatures now extinct, whose fossilized skeletons have only recently been unearthed? The same goes for that story from Arabia of the woman with a single torso, but with two heads and two pairs of arms. Why raise objections when practically the whole world has heard of the twins of Siam who have toured London and Boston?

These were thoughts that came to Captain Thomas John Newbold, who had a knack for geology and a penchant for exploring old caves. Some valuable truths, he reasoned, may well be found in this curious compilation of Eastern wonders illustrated with pictures. Newbold was an officer and a scholar who rose in the ranks of the Honorable East India Company and in the estimation of the Royal Asiatic Society. Stationed in south India, the captain had uncovered a manuscript brimming with ancient erudition. It had recently belonged to the collection of the late nawab—the one in Kurnool along the Deccan plain, who had lost his lands and then his life in 1840 to the Company.

Private stockholders in London were always looking for ways to turn a profit, and if it took an army to open up new markets or libraries of rare manuscripts, so be it. After all, the spolia of conquest and the countless souls subjugated only added to the air of righteousness that guided the entire venture. But that was another matter altogether.

For the captain was after minerals, precious stones, buried fossils, and ancient tombs. Like many trained Orientalists and Company soldiers, Newbold was on the hunt for valuable specimens from the East: old manuscripts, soil ripe for cotton production, diamond deposits, the bones of prehistoric creatures, undeciphered scripts. This strange Persian collection might just prove to be of some help.

As Company officials saw it, rarities were after all just commodities. And this book was filled with them. Yes, there were mermaids, conjoined twins, and all variety of monstrous creatures, but these pages, lined with ancient reports of minerals, plants, and animals along with accounts of distant lands, could well contain clues for uncovering even further riches in the East. Or at least that was what Newbold hoped, as he leafed through a copy of Qazwīnī's *Wonders and Rarities,* one of the most influential works of natural history ever written, if measured solely in copies spread across the globe.[1]

The illustrated manuscript in Newbold's hands was an exemplar of a work originally composed in Iraq during the middle of the thirteenth century. Qazwīnī, its author, was no Aristotle, Galen, or Ptolemy, but he made accessible a world of philosophy and science to a broad audience. His collection of natural history captivated readers for centuries with tales of remote regions, accounts of powerful forces in the heavens, and descriptions of rare substances and seldom seen animals.

A naturalist like Newbold would delight in the pages dedicated to minerology, botany, and zoology. With its encyclopedic ambitions, Qazwīnī's compendium, written long ago on the eve of the Mongol conquests, would certainly be an erudite companion through distant worlds of learning. But what to do with all those strange creatures and questionable tales? It is a simple problem with no easy solution. Wonders and rarities confound by design.

Qazwīnī's natural history had journeyed quite far by the time it reached Captain Newbold in India. All the places it has seen could never have been anticipated by its author—from Iraq across Asia, Europe, Africa, and the Americas.

This was not the first such compendium written in Arabic or Persian, nor would it be the last. But over time, Qazwīnī's *Wonders and Rarities* emerged as one of the most enduring examples of an entire field of learning dedicated to cataloguing the wonders of nature.

The collection enjoyed remarkable longevity. But the circumstances of its birth are no less arresting. In his long life, Qazwīnī witnessed unfathomable tumult. He survived two waves of Mongol invasions. The first, led by Genghis Khan, devastated all of the East, including Qazwīnī's hometown in northern Iran. The second came a generation later under the banners of Hülegü, the grandson of Genghis Khan, who in 1258 brought an end to the Abbasid caliphate, which had ruled over Iraq for centuries and had served as a symbol of authority for Muslims well beyond the capital of Baghdad. Qazwīnī not only survived, he continued to flourish as a judge and professor. Like many other notables, he found patronage and support with the Ilkhanids, a successor state to the vast Mongol empire.

With so much upended, Qazwīnī drew inspiration and comfort in the orderly design of nature. The pages of *Wonders and Rarities* aspire to states of felicity. Humor, jest, and clever witticisms meet ornate poetry and unrivaled anecdotes. In this universe of mysteries and unresolved enigmas, play and diversion offer a tested means for both confronting uncertainty and cultivating ambiguity—to face the puzzle of a divine design that is both transcendent and yet manifest, to contemplate the perfection of a creation that is nonetheless riddled with suffering, to entertain the plausible and outlandish through realms of possibility.[2]

Qazwīnī wed geography, astrology, talismans, and alchemical transmutations with accounts of angels, jinn, and savage beasts at the edges of the known world. Sex magic and demon possession can leap out of one page only to be met on another with pious invocations to the majestic power of God. These were not marginal musings at the edge of orthodoxy. Qazwīnī's collection, and the many other similar titles like it, occupied positions of lasting authority and continued appeal. Part of Qazwīnī's particular success lay in his efforts to make accessible sophisticated natural philosophy in a single well-organized book.

Undoubtedly, tantalizing stories of strange creatures, rare gems, and powerful spells also played a role. The panoply of illustrations opens up a world of delectation. As the famed philosopher Ibn Sīnā (d. 428/1037) contends, paintings and sculptures, even of hideous objects, if done with artistry and accuracy could inspire awe through their uncanny capacity to capture reality. Qazwīnī, who followed Ibn Sīnā in countless ways, understood that feelings of wonder and perplexity served as sources of pleasure through intellection and aesthetic appreciation.[3]

Qazwīnī begins his compendium with one of the most significant discussions of wonder in Islamic history. In the vocabulary of wonder, rarity is never far from sight. The two concepts are so tightly bound together in the phrase *ʿajāʾib wa-gharāʾib,* wonders and rarities, that they are nearly inseparable. Yet Qazwīnī also teaches that in order to truly fathom the wondrous, we must understand rarity in very separate terms. They are related, but also distinct.

To write of wonder and rarity was to enumerate, to fascinate, and to numb through the authoritative language of mastery and endless inventory. Faced with the expanse of creation, the abiding illusion—that the wonders of the world could be captured within the pages of a single book—was irresistible. Such books, after all, held the secrets of nature, which when unleashed could produce marvelous results.

The vocabulary of wonder in classical Arabic and Persian spoke for the entire world: animals, plants, minerals, stars, planets, comets, customs, beliefs, poetry and paintings, languages and emotive states, regions near and far, towering buildings and puzzling machinery, fine jewelry and rare gifts, spells and incantations. Yet the enumeration of all things rare and strange was not the end of wonder, nor did wonder cease in the encyclopedic impulse felt among powerful patrons and aspiring scholars to possess and consume everything in sight—divided into clear taxonomies of existence. Wonder and its corollaries of astonishment, awe, curiosity, perplexity, and bemusement conditioned the horizons of experience and the limits of what could be known with certainty in the face of the sublime power of a transcendent and yet immanent God.[4]

The Quran frequently counsels that contemplating the wondrous signs of creation leads those with insight to an appreciation of the divine Creator. It is an ancient idea. In comparative terms, the emotive valences evoked by sensations of awe and curiosity stretch across time and place. By all measures these feelings are quite primal, and, like a range of other sensorial responses shared by animals, do not even appear to be unique to humans.[5]

The ancient Greeks turned to wonder at the cosmos as the first steps in philosophy, pausing in perplexity at the hidden causes behind natural phenomena. Sanskrit poetics cultivated an entire register dedicated to producing wonder. Marvels and miracles occupy early Jewish and Christian teleologies of divine design. Monsters and strange creatures populate maps and bestiaries in Latin letters and Asian cosmologies. Many similar sensibilities organized around appeals to the astonishing and the uncanny guide an array of modern scientific and religious discourses, as well as all manner of consumption and entertainment in the circulation and accumulation of capital.

But pointing out the numerous commonalities may border on the banal. For what makes these sensations and emotions intelligible are the specific practices and ideological values that seek to elicit, contain, and explain them. Wonder talk often flirts ambiguously with problems of being and knowing: the marvel of the real, of what could be, or what is all but impossible. In this sense, the particular vocabularies of wonder and the meanings that they evoke shed light on formations of ideology and authority, offering insight into the diverse and changing configurations of feeling and being in the world. Despite the countless points of connectivity and areas of overlap, so many of the regions that wonder has traveled in the course of Islamic history remain largely uncharted.

Before you, esteemed reader, in the here and now, is a book that holds up a mirror to Qazwīnī's *Wonders and Rarities,* which for hundreds of years has sought nothing more than to capture the entire world. The chapters that follow proceed as a kind of response, in the Persian tradition of the *javāb,* a poetic reply made centuries later to a classic. Such replies generally develop a similar sensibility, structure, or set of motifs, but often for distinct purposes. I have drawn many details directly from Qazwīnī's writings and those of his contemporaries. Through imagery, digressions, and storytelling, I have also attempted to evoke in some measure the intricate conventions that enliven classical Arabic and Persian belles lettres, Qazwīnī included. Long lists, anecdotes, philosophical inquiry, plaintive cries, talismans, strange creatures, exquisite maps, and fine paintings—these are all tried techniques to produce particular feelings and ideas. So too does the hard spine of a taxonomical order lend structure to what could otherwise be an endless array of configurations.

Qazwīnī developed a tripartite division for his compendium, moving in three parts from the heavens, to the sublunar realm of the Earth, and finally to the mineral, vegetal, and animal kingdoms. My book too unfolds in three main parts, though to different ends. The first starts with the broad historical contexts in which Qazwīnī traveled, studied, taught, and wrote. This sets the stage for the second part, focused on *Wonders and Rarities,* roughly following the topics that Qazwīnī developed—from his influential discussion of wonder and rarity, to the heavens and then to the realms of earthly diversity. The final part, in turn, traces the long reception Qazwīnī's collection enjoyed and the momentous changes in learning that ensued. Qazwīnī offered much more than a mere dictionary or a litany of lemmas. Painters, readers, and patrons alike found innumerable inspirations in the illustrations of ingredients, remedies, and stories.

In addition to the life of a book and the worlds in which it traveled, what unfolds here is a road map for reading Islamic history through the prism of emotions and sensibilities. In this way, my book offers a sustained reflection on the singular importance of the phrase *ʿajāʾib wa-gharāʾib,* wonders and rarities, in the development of Arabic, Persian, Turkish, and Urdu letters. The concept offers an aesthetic means for appreciating the limits of the world and a philosophical disposition for traversing them. Wonder is a handy compass along this terrain, both as an object of inquiry and as a method of historical analysis. It points to philosophy and ethics, to the pursuit of pleasure and spiritual states of gnosis, to magic and science. In this way, to wonder at what is different or strange may help us better attend to the innumerable ruptures that separate the past from the present.[6]

As an emotion, wonder can evoke sensations in the body and expressions across the face, to pause in astonishment, admiration, or even horror before something unexpected or seldom seen. This corporeal reaction is captured in Persian with the emblematic phrase *angusht-i taʿajjub,* the finger of astonishment, or the finger of *ḥayrat,* perplexity, a gesture that over time became a common feature in paintings and book illumination (figure I.1). To feel amazement, Qazwīnī teaches, is to be perplexed and even entertained and to thereby view incongruities with delight and bemusement. Wonder, perplexity, confusion, and curiosity are, however, quite distinct from all the objects, sights, and phe-

FIGURE I.1: Dragon devouring cattle on Dragon Island, as bystanders look on, one gesturing with a "finger of astonishment." From a manuscript of Qazwīnī's *Wonders,* copied in 973/1565 for an emir of upper Egypt and then acquired in 1831 by the chief surgeon of the French army during the conquest of Algeria.

nomena that can be fashioned with adjectives of the rare, the marvelous, or the extraordinary. To say that a shooting star or a rainbow is wondrous is to make a claim about its value and to assert that an appropriate response is indeed to wonder or to contemplate, to feel awe or astonishment.[7]

Blanketing physical existence in the grammar of wonder was a practice that conditioned particular ways of seeing nature and of understanding human capacity. Qazwīnī's world within a book could journey so far for it spoke of wonder and rarity in terms of science and religion that were immediately legible. When executed with a fine hand, his natural history was also a masterpiece of elegance designed to convey specific lessons about how to appreciate the world and our place in it.

A book may be the best companion. But it can be quite mercurial, in that alchemical sense, always ready to take on new forms, from manuscript copies to lithograph editions, and movable-typeface prints. A good deal has changed since Qazwīnī put the finishing touches on his collection, with stories of hybrid creatures, such as that two-headed woman from Yemen who was brought to the attention of renowned jurists and centuries later caught the attention of Captain Newbold (figure I.2).[8]

FIGURE I.2: Conjoined twins from Yemen in a Persian translation of Qazwīnī's *Wonders* copied in the late eighteenth century in the Deccan, India.

The early peregrinations of *Wonders and Rarities,* first in both Arabic and Persian, provide a concise index for tracking a range of ideals and values organized around the emotive and cognitive power of wonder. Often illuminated with fine paintings, detailed diagrams, and colorful maps, Qazwīnī's natural history also came to enjoy later Persian, Turkish, and Urdu translations and adaptations. Throughout the nineteenth century, *Wonders and Rarities* continued to be copied, read, and imitated. With the wide adoption of printing press technology, Qazwīnī and countless others like him reached even greater audiences.

But in time, his world would come to represent a medieval vision of a cosmos long forgotten, with the Earth at its center (figure I.3), and with no knowledge of what the Western Hemisphere contained, beyond the great encircling ocean once thought to encompass all inhabitable lands on Earth. Despite their beauty—often celebrated in museums—these natural histories have remained for some time rather poorly understood, so incongruent as they are to the demands and values that animate both secular modernity and religious reform.[9]

Faced with a world that can only conceive of itself as having two hemispheres, orbiting a distant sun, and divested of occult properties, Qazwīnī does not fare well. Following the rise of European colonialism, such collections would

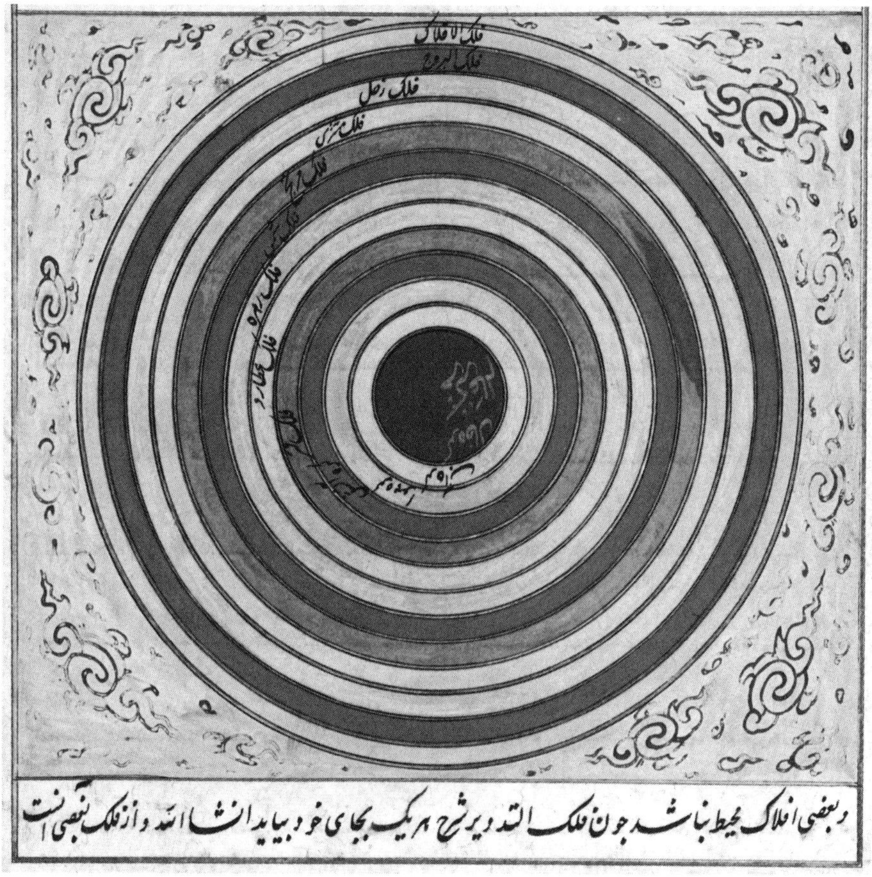

FIGURE I.3: Diagram of the cosmos, with the Earth at its center followed by the spheres of water, air, fire, the Moon, Mercury, Venus, the Sun, Mars, Jupiter, Saturn, the fixed stars, and the sphere which moves all the spheres, from a Persian translation of Qazwīnī's *Wonders*, copied in 974/1566.

take on new meanings with Orientalists, such as Captain Newbold out in the field, documenting, collecting, and shipping them off to imperial archives and university libraries as objects of curiosity. Here they would feature in scholarly publications and museum exhibitions, catalogued as exemplars of Oriental learning and artistry, stagnant yet charming. While some, such as Newbold, hoped to find utility in their pages, others would see emblems of superstition to be either ridiculed or forgotten.[10]

As for Newbold, there is no telling how useful Qazwīnī proved to be. By all measures it appeared promising. While passing through Egypt on Company leave, the captain turned to Qazwīnī for ancient lore as he made an impromptu survey of a route for a canal in Suez to connect the Mediterranean and the Red Sea. However, in the summer of 1850, back in India, not long after publishing his reflections on the late nawab's manuscript with the Royal Asiatic Society, Newbold died of consumption in Mahabaleshwar—a verdant hill station at the far edge of the Bombay presidency renowned for its fresh air.

By then Newbold had already made a tour that had taken him through Damascus, Jerusalem, the recently unearthed ruins of Nineveh abounding with cuneiform tablets outside Mosul, the markets of Baghdad, the port of Basra, and the ancient Achaemenid columns of Persepolis in the heart of Iran—many of the same pathways Qazwīnī had traversed centuries before. But now along these routes hurried European merchants, British Residents, Company regiments, followed by explorers, archeologists, naturalists, and Orientalists. Heralded in the London press as "one of the most celebrated geographers in the East," the captain, who had traveled far in his short life, left behind articles surveying the geology and people of the Orient, numerous specimens and curiosities collected for museums, as well as a study of local customs encountered in the British settlements of the Strait of Malacca, where he presented sweeping assessments of superstitious beliefs held by the natives.[11]

In his brief account of Qazwīnī's collection, Newbold made a good effort at taking seriously what more skeptical eyes would have dismissed as mere fables. The barrier against those savage creatures, Gog and Magog, built by the world conqueror Alexander the Great—that was probably just some exaggerated account from travelers who passed the Great Wall of China, he reasoned. Sure, there were incredible absurdities, talismans, and charms. But why reject the whole when even Herodotus, the very father of history, was known to spin a few yarns? And though Qazwīnī originally wrote in Arabic, he was also fully versed in the venerable authorities of Greek antiquity.

A good number of Qazwīnī's marvels were indeed old, such as that magical ring first made famous by Plato in the *Republic*—whoever wore it would immediately become invisible. Many of the arguments were also familiar to Newbold. In Qazwīnī's celebration of natural wonders as proof of divine design, Newbold heard echoes of leading scientists from Oxford and Cambridge who saw God's higher purpose in the fields of astronomy, anatomy, geology, mineralogy, meteorology, and chemistry.[12]

But there was something overly simplistic in Newbold's efforts to square a natural history written in the middle of the thirteenth century with the science and sensibilities that had taken hold of Europe some six hundred years later. What if that creature over in the water was just a mermaid after all and that barrier at the end of the earth was actually built to protect humanity from the apocalyptic destruction of the savages bottled up on the other side? Was that far-off island of women growing on trees, who made the sound *wāqwāq* when they fell ripe to the ground, really to be identified with Japan? And those dog-headed men on that other island who were fond of cooking up shipwrecked sailors—was that just a tale told of cannibals meant to keep prized trading routes secret? Certainly, there were always skeptics. But with the world so boundless and life so short, who could claim to enumerate all of God's wonders?

Out in the field, Newbold saw a work of immense utility, in some ways close to how his native informants read, circulated, and admired these very writings in Persian and Hindustani. Yet Newbold, who had Indian servants fetch for him the latest European publications unloaded in the docks of Madras and Calcutta, also surely recognized that almost every page of *Wonders and Rarities* could chafe modern scientific sensibilities. Qazwīnī blends together religion, magic, and science as a harmonious unity; he speaks of high-level philosophy while delving into all forms of lighthearted play; he showcases medicaments to be used for countless occult applications; he is doggedly committed to a geocentric universe, yet entirely content to admit the limits of knowledge in a world that is ignorant of the Western Hemisphere; he trades quite seriously in creatures never seen, all in a paean to God's ultimate providential authority.

Newbold's hope of extracting dazzling gems from the dross required sorting out separate piles of fact and fiction to read past the enchantment as mere confusion or fanciful hyperbole. The result was a world denuded of its wonder and unintelligible in its own terms. But there were those who were much less charitable. Newbold, and the others who followed a similar path through the winding tales of distant regions and remote islands, at least made some effort to take seriously, if with a rather patronizing tone and utilitarian aim, the fruits of Eastern erudition.

The more jaundiced chalked up the entire affair to the childish credulity of Orientals. Colonial officials, imperial agents, and zealous missionaries frequently mouthed off ready-made replies: Cast away these ridiculous books of outdated learning. What the natives needed to uplift them from their abject ignorance was a good schooling in modern European science, which would teach them that the Earth orbited the Sun, and that it had two hemispheres, one in the East and the other in the West. Or so the arguments went.

To trace after wonder through the course of Islamic history, as an emotive reaction, a pious performance, or a philosophical disposition, is to traverse some difficult terrain. Today wonder and the marvelous are not far away from fantasy and fictions of the imagination, which as categories of analysis are not particularly helpful for understanding the diverse realities and emotions evoked by the phrase *ʿajāʾib wa-gharāʾib* in its various appeals to wonder and rarity. Following wonder's wake requires an almost impossible voyage past the Scylla of pure reason and the Charybdis of benighted superstition. In part, it is such treacherous territory because the endless marvels of the Orient became a foil for defining the rationality of Christian Europe.[13]

For this, we only need turn to the origins of fiction. The English literary critic and poet laureate Thomas Warton (d. 1790) takes up the topic in a treatise entitled "Of the Origin of Romantic Fiction in Europe." Warton argued that the medieval chivalric romance, with its affinity for fabulous adventures can be traced to the "fictions of Arabian imagination." Warton hypothesized that "amid the gloom of superstition, in an age of the grossest ignorance and credulity, a taste for the wonders of Oriental fiction was introduced by the Arabians into Europe." Beliefs in giants, fairies, dragons—"a sure mark of Orientalism"—magic stones, and other such flights of fancy were not natural to Europeans, but were entirely Arabian inventions.[14]

The idea would prove attractive. Monsters, marvels, and beliefs "in the arts of divination and enchantment" could all be grouped together by the common wisdom that saw the ultimate Oriental origin behind wondrous tales that strain credulity. This theory, while debated, was repeated in the course of the nineteenth century and gained significance with the Romantic and Gothic recourse to the sublime, yet horrific, Orient.[15]

The radical otherness of the East could also offer a space from which to criticize the rationalizing forces of industrialization. And so the cache of Oriental illusions, for which the *Arabian Nights* was the most famous example, could be celebrated and naturalized as an inexhaustible tableau for "fictitious imagery," in a turn that inverted the argument on its head, to embrace the Orient and repel the cold calculus of disenchanted reason.[16]

For European fiction, the *conte oriental,* or Eastern tale, offered a staging ground for lamenting the ills of modern life. Some pined for the idealized utopianism of harems and unbridled libertinism, often with frightening religious tolerance and ambiguity. Others followed the sharp-witted satire of Voltaire (d. 1778) by donning the mask of Eastern superstition and despotism to criticize the structures of power, wealth, and religion at home. Yet, through it all the displacement of medieval ignorance onto an imagined Orient was necessary groundwork for the constitution of European modernity.

The scholars, translators, administrators, merchants, and soldiers whose business was the Orient could turn to Europe as a natural and coherent civilizational unity. Europe, not Christendom, became the default means for measuring the worth of Muslims. As a civilizational marker, Europe was not just made overseas in geographical opposition to the strangeness of others; it also emerged in rising currents of secularism that sought to extirpate religion as the basis for public order. The very vocabulary in its syntactic structure thus forces an opposition, not between Muslims and Christians, but between the ignorance of religious others and the modern forces of progress.[17]

Ultimately the categorical slippage, ethical in its very form, not only makes the Muslim other incompatible with modernity; it also promotes civilizational superiority as founded on an opposition between superstition and reason. These categories are historical products, invested in particular modes of seeing as well as mastering the world. The grammar of such concepts conditions certain forms of thought, opening up one set of possibilities while closing off another.

To understand why *Wonders and Rarities* enjoyed such enduring appeal requires its own act of enchantment to conjure up a cosmos that is quite foreign to modern dispositions. In myriad ways, the world Qazwīnī sought to capture is lost. This is true not only for the variegated contexts of secular modernity, which proceed with a knowledge of the Western Hemisphere, a mastery of a heliocentric solar system, and a guiding suspicion toward all things magical or miraculous. It also holds for the diverse currents of modern Islamic reform that have sought to ignore large swaths of intellectual history that do not readily conform to a vision of Islam as entirely compatible with modern science. In this regard, the diverse means of curating the past, of celebrating certain aspects and ignoring others, reveal a good deal about the values and aspirations guiding the present.

The machinery of empire that took Newbold and many others like him out and back to the Orient was finely calibrated to quicken the circulation of commodities. Human bodies, strange animals, raw materials, precious artifacts, archaic scripts, entire systems of learning, magnificent columns and statues—it all had to go, packed up as cargo, shipped out from port, marked down and underwritten in ledgers. To each its own place and position, catalogued and appraised, with rare curios over here and all variety of forced labor over there.

Museums, libraries, and universal fairs lent a global stage to imperial appetites. The Orient supplied untold possessions to be consumed and absorbed. So could it serve as a fount for perennial wisdom and a fertile ground for spiritualism, mesmeric influences, and ancient biblical secrets waiting to be uncovered. In countless ways, the East would prove indispensable for defining the contours of modern Western sensibilities, sometimes through opposition, of what the West was not, other times through temptation, of what it might strive to become.

Qazwīnī found a home in national archives, museums, and world exhibitions. In the colonies and in the metropolitan centers came new institutions for cataloguing, recording, and giving order to the world. Qazwīnī's appearance in the museum also coincided with a significant reordering of what the spectacle of Oriental wonders meant. Here the value and significance of Qazwīnī's world of curiosities changed as it was contained and redeployed, feted for its paintings of strange creatures, but rejected for its childish ignorance.

A testament to global modernity, the museum transformed from the cabinet of curiosities of Renaissance learning into a public institution connected often quite closely to expressions of colonial and national power. In these halls, wonder too abides, though often with values quite distinct from those that animated Qazwīnī. That is certainly one way of telling the story. But there are also other ways of imagining what museums have meant and what other purposes the display of rare and wondrous objects and specimens could serve.[18]

The term *mathaf* in modern Arabic is a neologism drawn from a classical repertoire. The novel word suggests a place where one could encounter rarities given to powerful patrons, as one might expect to find, say, in a work entitled *tuḥfat al-ʿajāʾib*, "a gift of wonders," a common title for natural histories in classical Arabic. The modern Arabic word stands in for the French *musée* and the English *museum*, both novel repurposings of the ancient Greek *mouseion*, as in a building dedicated to the Muses for philosophical reflection; namely, an institution for higher learning.

In *müze* and *mūze*, modern Turkish and Persian just accept the edifice for what it is, a sui generis and naturalized foreign loanword. But like *matḥaf* in Arabic, the museum in Urdu brings out a palpable sense of wonder, in the *ʿajāʾib-ghar*, the abode of marvels, as in the *ajaib-gher* of Lahore that Rudyard Kipling (d. 1936) knew well. The Lahore museum, which opened its doors in 1864, was a thoroughly colonial enterprise from the start. Yet the formal name in Urdu, "the house of wonders," which still adorns the latticed portico of the Lahore museum in the Indo-Saracenic Revival style so in vogue then, speaks to a dual inheritance.[19]

As with other colonial museums that proceeded and followed suit in Madras, Calcutta, Bombay, and Jaipur, these enterprises for collecting and display joined parallel institutions in London, with captivating exhibitions in the Royal Asiatic Society, the East Indian Company headquarters, and the British Museum. Natural history had an enduring place in the mineral, vegetal, and animal collections that sought to introduce raw materials, commodities, and manufactured goods to be studied, fashioned, and consumed. While utility had a premium, trophies of conquest, nestled in with the dazzling, bizarre, and even grotesque, also took up a good deal of space.

But these newfangled institutions for the public also had other marvelous tales to tell. The real wonder of the East, the lesson went, was to be found in the primitive beliefs that kept the Orientals from being truly modern. In this way the museum could isolate, freeze in time, and decontextualize the goods and materials of an exotic landscape into an unchanging tableau. For all that such lessons were designed to convey, wonder and the institutions and discourses meant to contain and produce it could never truly be fixed to one single meaning or ideological purpose.

Enter the Salar Jung Museum of Hyderabad to further see this twin inheritance on display. The collection represents, as it were, the personal *ʿajāʾib-ghar* of the nawab Salar Jung Mīr Yūsuf ʿAlī Khān, who died in 1949, two years after Indian Independence. Unlike other nawabs of India, who fell one by one, the Salar Jung family served as high-ranking ministers for the princely Niẓāms, a Muslim dynasty of rulers who remained client kings, first under the Company and then through the entire history of the British Raj. Like his father and grandfather before him, the last nawab spent a good deal of his fabulous wealth—measured partly in Golconda diamonds—by amassing curiosities from across the world for his own private collection.

From Europe the nawabs of Hyderabad acquired sundry goods: Swiss clocks and an Italian statue with a gossamer veil of marble, Renaissance paintings, floriated cabinet work, cherubs of enamel, Grecian urns, ornaments of ormolu, Sèvres porcelain, ornithological adornments, and designs from classical lore. European artistry and technical know-how could be its own source of fascination.

After independence the nawab's private collection became a public museum. Still today the wonders of the West abide quite comfortably in a collection meant to showcase the fabulous wealth of a dynasty that withstood the Raj through all forms of stratagem and negotiation. So too did the family amass a vast library of printed books brought in from Europe, which taught the singular importance of modern learning. All these imports joined a collection of manuscripts and lithographs in Persian, Arabic, Urdu, and Sanskrit.

In the archives of the Salar Jung Museum a reader can peruse several copies of Qazwīnī's *Wonders and Rarities* (see figure I.2). The first word in the title of the compendium is *ʿajāʾib*, an Arabic term that carries the sense of wonders, marvels, or miracles to behold. Here are innumerable manuscripts from the full array of classical Islamic learning: histories, volumes on law, precious copies of the Quran, theologies and philosophy, medical encyclopedias, studies on Indian religions with yogis and sanyasis, and numerous works on talismans and the occult arts. There are illuminated manuscripts of poetry and prose, as well the personal horoscopes cast each year for the last nawab by Hindu astrologers.[20]

A reader with tired eyes may pause over a cup of chai at the librarian's desk to take in the framed masterpieces of pious Arabic calligraphy that adorn the walls of the reading room. One dedicated to the nawab in 1940 begins with a prayer to Imam ʿAlī, the son-in-law and nephew of the Prophet Muhammad, venerated by Sunnis and Shia alike: *nādi ʿaliyyan maẓhara l-ʿajāʾib,* "call on ʿAlī, bearer of wonders!"

In the development of Islamic history, the languages of wonder and astonishment were not bound to a single creed or confession or to only a Muslim noble who consulted Hindu astrologers, drove around in a Rolls-Royce, and collected strange objects from distant lands. The nawab's collection survives as a kind of hybrid creature, born of colonialism and independence and a vision of the museum as a house of wonders, where pious appeals shaped by well-worn religious sensibilities can still be encountered throughout its halls.[21]

Well before entering the museum, Qazwīnī's *Wonders and Rarities* had traveled through private collections and libraries dedicated to learning. As with the archives of other major Muslim dynasties, there are exquisite copies housed in the royal collection of the Ottoman sultans in the Topkapı Palace complex of Istanbul. Several can be found in the libraries of the classical colleges, or madrasas, in the city. Some are in Arabic, others in Persian and Turkish. But these archives are quite different from the public displays meant to teach particular lessons about history and heritage.

Despite all the ancient traces of the past—the Hagia Sophia, the Hippodrome with its pharaonic Obelisk, the towering minarets of the Blue Mosque—much in the imperial city of Istanbul has changed. Today in the rose garden of the Ottoman palace is a modern institution, guided by distinctly pious dispositions. Here a visitor can tour the recently opened Museum of the History of Science and Technology in Islam—the life's project of the late Turkish scholar Dr. Fuat Sezgin (d. 2018).

Home to an impressive array of meticulously crafted instruments and objects, in the form of modern replicas and reconstructed models, the museum was founded with the sole purpose of cultivating a greater public awareness of the intellectual achievements of Islamic civilization. Astrolabes, globes, waterwheels, beakers, vials, maps, and miniature observatories with giant quadrants and sextants for measuring the declination of the stars—they all teach the singular importance of reason in Islam. The realization of Sezgin's museum in Istanbul was the culmination of years of study across an encyclopedic range of disciplines in classical Arabic learning.[22]

Like most scholars of Arabic letters, Sezgin was well acquainted with Qazwīnī's natural history and the numerous works like it. But like other important chapters in Islamic thought, compendiums of natural wonders have been a bit like orphan children, abandoned in the grand narratives that have sought to account for what Europe owes Islam. Not a trace of Qazwīnī's collection or the many other similar compilations appear in the museum's exhibits. For revisionist endeavors, wonder, awe, and all those tales of questionable veracity have not aged well. For many, the writings on geography and cosmography that shape Qazwīnī's vision of the world are valuable only in so far as they conform to or anticipate modern ideals of reason and scientific authority.

Starting in the nineteenth century, Muslim reformists sought to turn the tables in response to the withering critiques leveled by colonial administrators, Orientalists, and missionaries. Europe could not have advanced without the ill-gotten and unacknowledged learning taken from Muslims, the argument went. Copernicus (d. 1543) owes quite a bit to Muslim astronomers, it turns out. Moreover, algebra, chemistry, medicine, philosophy, and an entire range of other scientific fields developed in Latin Christendom through sustained engagements with earlier advances in Arabic learning.

As Sezgin would have it, like many other reformists before him, Muslims also reached the Americas well before Columbus (d. 1506). It is an idea with incredible appeal, though it rests on questionable grounds.[23] Leading Muslim politicians have championed the theory, as have countless others. The claim rings true for many as it readdresses a historic imbalance of power by claiming a mastery of the globe well before the rise of the West.[24]

Among the lessons of influence and dependence advanced within the walls of Sezgin's museum is the argument that "Arabic-Islamic geographers" not only laid the foundations for the transatlantic journey to the Americas but had actually discovered the continent before the subsequent European conquests. The theory can be purchased with Sezgin's book on the topic, which announces its argument unambiguously in the subtitle, *The Pre-Columbian Discovery of the American Continent by Muslim Seafarers*. Illustrated with beautiful, if difficult to decipher, old maps, the book is available in the neighboring Topkapı gift shop, in both Turkish and English.[25]

These endeavors to reinsert Islam and Muslims into the history of science and discovery consciously challenge Eurocentric attitudes that advance the natural superiority of Western civilization. Yet they do so while also absorbing the same values of progress and maturation that have shaped the languages of modernity and the concept of civilization as coherent and self-evident. The novel category of civilization used throughout these discussions was born in the eighteenth century during a period of increasing contacts between Europeans and far-off lands. Civilization is as much a spatial metaphor as it is a marker of religion, morality, and intellectual capacity. The notion is not only oppositional, it is also competitive and hierarchical.[26]

Among Muslim modernists and reformists, the study of Islamic civilization, which came to include classical Arabic works of geography, history, and language, formed a basis for challenging the supremacy of Europe. These concerns shaped the work of the early advocate of Pan-Arabism, the Egyptian philologist and statesman Aḥmad Zakī Pāshā (d. 1934), a prominent figure in what is commonly referred to as the *nahḍa,* the renaissance or awakening of Arabic language, literature, and culture. Aḥmad Zakī is well known for his critical editions of classical Arabic manuscripts and his numerous studies on Arabic history, geography, and language.[27]

Writing in both French and Arabic, Aḥmad Zakī sought to uncover the technological and scientific greatness of Islamic civilization that preceded European achievements. Through publications in the Egyptian press, he popularized various theories including the notion that Muslims undertook maritime expeditions to the New World prior to Columbus. He argued that purely through logical inference Muslim scientists were able to deduce the existence of the American continent. Aḥmad Zakī also drew attention to passages, still then in manuscript, detailing expeditions launched across the western sea, which he advanced as the basis for the pre-Columbian discovery of the Americas by Muslims. After Aḥmad Zakī, a string of reformist publications in Arabic, Persian, Turkish, and Urdu advanced the theory that Muslims discovered the Americas before Columbus.[28]

As with earlier Greek and Latin accounts, there is a good deal of speculation about what lay past the horizon's edge. Yet the classical Arabic geographers preserve no record of contacts with the Western Hemisphere, let alone awareness of what it might contain. When Muslims first came to speak of the Americas, they did so through accounts of Columbus and the cartographical material first fashioned by Iberian mapmakers.

Beyond a vague knowledge of the Canary archipelagoes, steeped in wondrous uncertainty, the chorus of authorities, Qazwīnī included, warns of the western ocean as an unfathomable expanse of darkness and a tremendous sea of terrors, which formed part of a great encircling ocean—one of the most recognizable and durable markers of human limitation. As an outer boundary, it signaled where direct experience ended and marvelous reports began. While Qazwīnī does not speak of the Western Hemisphere, he has tales about the sea of darkness. In this he follows a long tradition that entertained the boundaries of terrestrial knowledge with scintillating stories of strange encounters.

After the Iberian contacts across the Atlantic, the limits of what could be known would change forever. Even though Muslim elites were not the first to map out the Americas, they were much quicker to absorb knowledge of the New World than Christian missionaries and Orientalists were ready to acknowledge. As with many other fields of learning, Qazwīnī's collection proved to be invaluable here, as a source for imagining distant seas and as a repository for speculation about exotic lands.

The global stretch of maritime expansion takes on a new character with the circumnavigation of Africa and the exploration of the Americas, which together offer the launching ground for further conquest. The early European sources were adamant that Orientals played only a minor role in this story: while Muslims may have been of the world, they were not masters of it, ignorant of the true nature of the globe. Over the course of the last century, such arguments have inspired reformists and revisionists, through the allure of the counterfactual, to lay claim to the shores of the New World before Columbus, seeking to break free, as it were, from these historical walls of containment.

There is no escaping how much our language for describing the world before Columbus and Copernicus has been shaped by these historical contingencies. To address in English the Islamic histories of wonder is to evoke an exotic geography of difference, far removed conceptually and ethically from everything that is held to be true and obvious in the immanent frame of Western secularism. The specter of liberty brought by empires and markets has made the prospect of fathoming the difference of the past a vexing endeavor. When writing about the wonders of the Orient, there is no practical way to outrun the penumbra of Orientalism.[29]

The vocabularies of Western modernity are enmeshed in the histories of colonialism and the formal study and mastery of others. The logic of secular reason is designed to question the diverse ways of being and knowing that do not end in any obvious division between science and religion, the medieval and the modern, the liberal and the oppressive. Such oppositional markers are not neutral descriptions or value-free assessments; they are ideological claims on reality and morality.[30]

With its encyclopedic impulse for totality and its lasting appeal, Qazwīnī's compendium offers a ready guide to Islamic history before and after the rise of European hegemony. But to conjure a world requires particular techniques of illusion and imagination. Worldmaking has long been a pastime for the powerful, just as it has served as a source for childhood play, intellectual revelry, and social cohesion—beginning with one version of the cosmos and ending in another. A geography of the imagination may motivate imperial ambitions, even while inspiring self-realization, occasions for reflection, or communal recognition.

What separates these variegated cartographies of ideation are the diverse mechanisms of authority—juridical, scientific, religious, or otherwise—that are designed to imprint themselves in the mind and on the body. It is one thing to fashion a world, it is another to demand that we all inhabit it or measure ourselves through its reflection. The ethical implications of such cosmic orders are always shaped by specific contexts in the ebb and flow of power and persuasion.

To denaturalize the categories that grant modernity its authority, while still speaking through the languages and institutions that animate it, requires a good deal of defamiliarization or alternative worldmaking. The effort to make the quotidian strange or unfamiliar may serve to challenge what we think we know or imagine we see—how science, magic, and religion are inseparable, or how entirely coherent it is for the Sun to rise in the east and set in the west on a daily journey around the Earth. Yet the suspension of disbelief can take us only so far.[31]

The stories and anecdotes that lie ahead lead straight through the irreducible aporia that has made the past both necessary and yet largely unintelligible to the demands of the present. Play and jest cannot resolve these problems, although they can reorient them. Diversion and delight may help us to dissent, to differ, or to think otherwise along the twisting arteries that define the walls of these labyrinths.

As Scheherazade of the *Arabian Nights* is quick to tell, deferral can yield powerful results. What could be more wonderful than the end of one tale that demands the beginning of another? When such stories catch on, with their meaningful digressions, no single volume can possibly contain them. Testaments to the physical labor of our textual conditions, they speak back to the illusion of a singular meaning that can be recovered or predetermined, even if the script is so fully internalized that it all but disappears.

An entire cast of individuals are in the wait, some with leading roles, others with merely a passing notice. Before Newbold and his army of imperial agents reappear at the very end, with the rise of Orientalism and European colonialism, much ground will have to be covered. The history of wonder in Islam witnesses empires rise and fall. Law, theology, ecstatic visions, astrology, alchemy, medicine, geography, magic, and miracle all feature—if not entirely in their own terms, then in ways that try to make them more intelligible than the strictures of disenchanted reason would demand. To summon these worlds of astonishment requires vocabularies and grammars that are at times foreign to modern temperaments. So it is that categories—along with all the stories of scholars and patrons, dynasties and regions—play an immeasurable role in the chapters that follow.

To this point, one might say that wonder and enchantment have suffered a good deal in the course of modernity, as Qazwīnī's cosmos, populated with horrifying creatures and alluring delights, has been stripped of its mysterious workings. In its place the language of reason and empiricism has sought to wrap the globe with calculated precision and predictability. Be that as it may, enchantment never left the world, nor has wonder met its end, they have just come to occupy rather different spaces.[32]

Much of this history has been largely forgotten, if only because it has not been readily found or easily told. Paintings of constellations and angels, strange creatures and devices, may be delightful to contemplate as single folios sliced out of manuscripts for museum exhibitions (see figures 2.3 and

8.2). But reading the texts that once encapsulated them is another matter altogether. In the distance between archive and memory lie all the tomes and objects of the past that have been abandoned or simply misplaced as incongruous with the demands and values of the present. Our archives are fragmentary and widely dispersed, while each page of *Wonders and Rarities* builds on realities that are at times so strange and unfamiliar that they require a fair measure of illumination.

This may be a biography of a single book. Yet the chapters that follow tell a broader story of how science and philosophy have long been entwined with all variety of enchantment and miracle; how pleasure and the pursuit of felicity could be cultivated for harnessing occult forces; how formations of learning and authority endure and transform in unforeseen ways. Far from a static repertoire of unchanging motifs or foregone conclusions, the cognitive and emotive grammars of the wondrous and the rare prove to be protean, supple, and capable of absorbing new knowledge and information across distant shores over a vast expanse of time.

In the contours of Islamic history, wonder and rarity are inescapable. They are a means for fathoming all of creation, as well as for organizing and containing it. Qazwīnī with his curious collection—call it a natural history, a cosmography, a book of scientific marvels—is more than an adequate guide. But to understand his language, which is both beautiful and strange, requires a sympathetic ear and a willingness to entertain a world that is now quite hard to find, of stars seen clear in the night sky before the flood of electric lights, of the rhythmic movement of the seasons before the demands of standardized time. A good deal of caution is also warranted. The path is tortuous and only for the brave of heart. There is an invasion coming that will change everything. And yes, there are dragons.

PART I

A World within Words

The routes of learning that Qazwīnī traveled were defined by the authority of philosophy, science, and prophetic revelation. Qazwīnī represents a class of religious scholars whose careers were formed by the scholastic command of natural philosophy. The teachers and contemporaries who shaped Qazwīnī's understanding of the world open up networks of scholarship and patronage. These associations and connections constitute the parameters of an expansive repertoire of cosmic proportions designed to make meaning of the world and to harness its hidden powers.

As a companion to *Wonders and Rarities,* Qazwīnī also wrote a descriptive geography. When read in tandem these two works provide insight into journeys he made and events he witnessed. Together they unveil a wide canvas of time and place designed to account for the entire world.

He began life in the walled city of Qazvin, at the foot of the Zagros Mountains. Decades later, after a distinguished career, he died in the city of Wasit, along the alluvial plains of the Tigris River in the middle of Iraq. The following chapters introduce the worlds of learning and writing that shaped Qazwīnī's journey as a scholar, professor, judge, and author.

Formalized education, courtly patronage, bureaucratic administration, the function of the judiciary, the study of law, theology, natural philosophy, and medicine, the privileged place of poetry and letters, the power of the heavens, and the strange capacity to work miracles and marvels that shatter customary phenomena—these topics and many more await.

The written word can never capture the world it seeks to contain or evoke. But as the natural philosophers might say, the faculty of the imagination, when finely exercised and disciplined, is a powerful force capable of summoning realms from the unseen, even perhaps something so remote as the distant past.

1

STRANGER LANDS

THE FORMS AND SHAPES flickered like long shadows dancing underfoot in the late light of the setting Sun. The séance had been arranged at the request of Abū Ḥāmid al-Ghazālī (d. 505/1111), the famed jurist and theologian from eastern Iran, who was eager to see with his own eyes what jinn looked like and, if possible, to commune directly with them. Trained in the teachings of the Iraqi theologian Abū l-Ḥasan al-Ashʿarī (d. 324/935~) and his many followers who had spread across the East, Ghazālī knew well that along with angels and demons, jinn inhabited the world of the unseen. He also knew that only a select few, from the ranks of prophets and sages, could actually summon these powerful forces.

In an effort to satisfy his curiosity, Ghazālī had enlisted the aid of a master renowned for his power to draw spirits from the *ghayb,* the invisible realm, into the physical world of human perception. The shadows continued flickering and dancing for some time. However, Ghazālī, who possessed one of the sharpest minds of his generation, grew impatient. "I want to speak directly with them and to hear their speech," he demanded. The master scoffed, "You are not capable of seeing any more than this." Just as quickly as it began, the show came to an end with the shapes retreating from view.[1]

This short account on the limits of human knowledge at the threshold of the unseen features in the opening to Qazwīnī's *Wonders and Rarities.* It is not told to highlight Ghazālī as a skeptic, or to paint him as a dupe. Nor does Qazwīnī use it as an opportunity to expose the quick-handed deceptions of fraudulent rogues and tricksters who prey on the gullible and unsuspecting—a well-worn conceit in Arabic and Persian letters of the day.

FIGURE 1.1: The prophet Solomon enthroned, surrounded by jinn and angels, in an Ilkhanid manuscript of Qazwīnī's *Wonders* produced circa 1300.

Stories of summoning, conversing, and even commanding occult forces were not themselves particularly uncommon, told and retold as entertainment and as edification. Such accounts could be bolstered, after all, with the precedent of Solomon, who commanded armies of jinn (figure 1.1), and the model of the Prophet Muhammad, whose wondrous recitation of the Quran converted many among their ranks. For Qazwīnī, however, this particular anecdote, whose historicity he does not draw into question, plays another role altogether: it appears in a list of other strange phenomena cited as examples of the *gharīb,* a category central not only for Qazwīnī, but also for the high-level metaphysical discussions that had come to dominate Islamic philosophy.[2]

While *gharīb* and its direct variants do not form part of the Quranic lexicon, the term features in numerous sayings ascribed to the Prophet Muhammad, such as the eschatologically portentous hadith that "Islam began as a stranger [*gharīb*] and it will return just as it began, a stranger, so may there be blessings upon strangers." As for its categorical value, *gharīb* serves as a meaningful index for a host of fields in the early formation of Arabic literature, from works on rare or difficult words and expressions to collections dedicated to preserving the passing graffiti of strangers and documenting their hardships. Notably, *gharīb* evokes such cognitive and emotive registers as the remote, extraordinary, weird, obscure, peculiar, uncommon, rare, difficult, and foreign, just as it can also be glossed as both queer and curious. The plural form *gharāʾib*—rarities, curiosities, or oddities—plays a pivotal role in Qazwīnī's conception of the world. It features prominently in the alliterative title to his natural history: *ʿAjāʾib al-makhlūqāt wa-gharāʾib al-mawjūdāt* (Wonders of things created and rarities of matters existent).[3]

Qazwīnī suggests to his readers that he was also *gharīb,* at least in the sense of a stranger who settled in a foreign land. In the opening to his natural history, Qazwīnī addresses his own experience of estrangement as a wayfarer: "Worn out with being far from home and country, separated from family and habitation, I determined to study books, on the opinion of the poet who said, 'The best companion of all times is a book.'" In the escape of bookish knowledge, Qazwīnī became engrossed "contemplating the wonders of God's art in His crafts and the rarities of invention among His creations."[4]

The pages of the book lead invariably to the pages of the world. Both here and in his geographical gazetteer, the *Āthār al-bilād wa-akhbār al-ʿibād* (Remnants of the regions and reports of the righteous), Qazwīnī draws on the authority of experience through direct observations made in the course of his own journeys. All of these sentiments are well-worn devices: not only are we left with an author who is both estranged and yet experienced through years of observing and reading, but we have also been handed the world in a book as a companion and as a salve to the toils of the world. While these ideas draw on the artful book culture characteristic of belletristic prose popular in his day, they also share in Qazwīnī's own life story.[5]

What paths did Qazwīnī travel, who did he find along the way, and what fields of knowledge did he pursue in his own training as a scholar and jurist? Answers to these questions can be gleaned here and there from stray references Qazwīnī makes in *Remnants of the Regions,* as well as from passing accounts in later chronicles and biographical compendiums. But as with much of the past so many years removed, the surviving materials preserve only a fragmentary picture.

A good deal, though, can be said about the broader milieu in which Qazwīnī journeyed and the education he pursued. During a historical period of incredible upheaval and social unrest, Qazwīnī could find consistency in the precise measurements of the heavens and the predictable and uniform movements of the stars. His understanding of creation and the physical mechanisms governing existence emerged through the study of philosophy and science in the frameworks of religious learning, which proceeded with its own measures of uniformity and consistency.

In comparison to the spectacular journeys of more renowned travelers, Qazwīnī's peregrinations were modest, in keeping with a particular class of notables who had the means to travel, often in the pursuit of learning. Born on 10 Rabīʿ II 598/7 January 1202, in the city of Qazvin situated in the high plains of northwestern Iran, as a young man Qazwīnī left the land of his birth, which ultimately would serve as his namesake.

After a circuitous route through Iran, Central Asia, and Syria, Qazwīnī settled in the heartland of Iraq in the southern town of Wasit on the Tigris River. His own itineraries appear not to have taken him beyond the familiar *bilād islām*, the lands of Islam—his preferred term for the territories occupied largely by Muslims. For Qazwīnī, the realm of Islam stretched from the receding frontiers of Iberia across North Africa through the Levant, down along the Arabian Peninsula, spreading over Mesopotamia, Anatolia, the Iranian plateau, up through Central Asia, and further along the borderlands of Sind on the alluvial flood lands of the Gangetic plain. Persia marked the midway of the world, perfectly poised in the middle of the seven climes. Mecca, however, served as the spiritual center and the focal point for daily prayer, with all the regions of the world oriented toward God's sanctuary in Arabia.[6]

Across these diverse realms of faith, dotted with ancient churches and synagogues, a traveler would find mosques, some modest and parochial, others magnificent and congregational with spacious courtyards, towering minarets, and finely tooled pulpits. Above the muezzin's call that marked the daily prayers could be heard the bustle of markets filled with metalwork, glassware, and delicate handicrafts, framed by stalls of imported and local fare: from leather hides and fabrics made of silk, satin, cotton, and wool, to spices, perfumes, and medicaments containing special properties.

Goods came and went across thoroughfares with well-kept inns, humble hostels, hospices for the needy, hospitals for the infirm, and communal baths to escape momentarily the din of daily life. These pathways were also lined with schools, libraries, and institutions of higher learning, where the curious could find rare manuscripts, many of great antiquity and in numerous languages, some copied with the sharp reed of a skilled calligrapher, others in a rough hand for personal study.

In more illustrious repositories, from courts to well-endowed madrasas, one could consult celestial globes and astrolabes for mapping with varying precision the movement of the heavens and distances across the Earth. Along the routes connecting these territories, ancient remains of empires lay next to signs of life: milestones marking the paths of caravans and couriers; granaries and windmills; stone bridges and canals that fed wells, cisterns, and waterwheels; farmlands and walled gardens, with troughs for horses, stone-lined sheepfolds, and sweeping camel pastures. Along these pathways moved aristocrats, landowners, and scholars with means, as well as slaves, servants, and all manner of indentured laborers.[7]

Qazwīnī knew this landscape—united in a faith of many shades and of conflicting interpretations—to be marked with villages, vast cities, and imposing fortifications. Contemplating the beauty, majesty, and wonder of the art and learning that brought together people through divine wisdom, it is not hard to see, as Qazwīnī himself did, that the Islamic domains, *al-bilād al-islāmiyya,* contained the best of all the "lands of God," a catchall for the inhabited world. Here thrived the perfect balance of temperaments, measured in handsome bodies, sound minds, and capacious intellects, so completely unlike the Turkmen of the high steppe, the mountain-dwelling Iranian nomads, or the inhabitants of far-off islands, who collectively all possessed in Qazwīnī's view "savage natures, weak intellects, and disproportionate physical forms."

Such notions, while self-serving, were central to established geographical thought. They built on a broad set of teachings undergirded by beliefs in divine providence and the theory of climatic determinism that had been developed by the likes of Aristotle (d. 322 BCE), Ptolemy (d. ca. 170 CE), and

Galen (d. ca. 216 CE). The argument went that the central geographic locations of the Earth, situated neither too far to the north nor too far to the south, offered the best conditions for human inhabitation and in turn brought prosperity, learning, and stability.

Qazwīnī understood that, by dint of climatic variation, those who lived along the edges of the world suffered extremes of heat or cold. Such people were savages by natural disposition, incapable of thriving in the same ways as those who had settled nearer the center of *al-arḍ al-maʿmūra,* the inhabited Earth. This term, the *oikoumenē* of Greek learning, represented the landmasses generally believed to suit human life. Only a fraction of the globe, the so-called inhabited quarter, *al-rubʿ al-maskūn,* could support the conditions for settlement. There were disputes among the geographical authorities with whom Qazwīnī was most familiar as to whether habitation was even possible south of the equator, just as there was consensus that the tribes who lived along the polar circle—who traded furs and pelts for swords and coins with merchants from the lands of Islam—were brutes.

In contrast to these wild creatures, Qazwīnī saw humans in their fullest sense as social animals who by nature relied on collective organization to flourish. Cooperation required urban settlements. Echoing a long tradition of political theory that stretched back to Plato and circulated throughout Islamic thought, Qazwīnī viewed the city as akin to a human body. All the limbs work in concert with one another forming an interconnected whole, such that if one part fails, the entire group is diminished. Through divine determination, people were endowed with an innate social disposition, which Qazwīnī refers to as *al-hayʾa al-ijtimāʿiyya,* the habitus of collectively working together.

The necessities of food and clothing alone required mutual cooperation. To till the land, the farmer needed the carpenter for the plow. The carpenter turned to the blacksmith for the iron blade. The cotton clothing fashioned in the Iranian highlands required not only planting, ginning, carding, spinning, and weaving, but also specialized tools for each stage of production. Daily bread had touched many hands—planting the seeds, harvesting the wheat, milling the flower, and kneading the dough—before being brought to market. But Qazwīnī also knew that without city walls, ramparts, and battlements all the stability brought by cooperation could be easily wiped away.

While tents and mud huts might protect a settlement from the elements, they did nothing to guard against the assault of a hostile force.[8]

This vision of the city bounded by gates and protected by walls was certainly true to Qazwīnī's own experience of urban life. With his attention to vestiges of the past, Qazwīnī found ample precedent for the encircling walls of cities and citadels in the remnants of bulwarks and defensive formations. The ideal of the city rampart stretches back to the earliest written records, in ancient Sumerian and Akkadian cuneiform tablets that praise King Gilgamesh for tending the city walls of Uruk on the edge of the Euphrates.

Yet despite the measure of security that such structures brought, Qazwīnī would not need the distant past to know that these defenses were also vulnerable to the catapults and battering rams of a well-supplied army able to hold a siege. His native city of Qazvin fell to the Mongol forces in 617/1220, when Qazwīnī was not yet twenty years old. We do not know if Qazwīnī witnessed the legions of Genghis Khan (d. 1227) breach the brick city walls. But we can be certain that he was not among the some forty-thousand inhabitants of the city who were massacred by the Khan's men. As with other notables of means, Qazwīnī likely fled well before the onslaught even began.[9]

The first Mongol invasions into the lands of Islam started in the northeast only a year before. With frightening speed, archers on horseback rose up from the high steppe of the north with banners and battle engines blanketing the horizon. One by one the cities fell: Samarqand, Bukhara, and Tirmidh beyond the Oxus River; Marv, Balkh, Herat, and Ghazna, in the East; Nishapur, Qazvin, Rayy, and Hamadan in the heartland of Iran. Those who capitulated were largely spared; those who resisted were not.

The entire area across the Iranian plateau, through Central Asia and abutting the Indus valley, had only recently been brought under the rule of the Khwārazm-Shāh, a Turkic dynasty that had quickly grown well beyond its ancestral center in northern Transoxiana along the Aral Sea. In the first years of the Mongol invasions, the territories claimed by the Khwārazm-Shāh swiftly fell. Many either fled or joined the conquering forces. Peasants and mercenaries were absorbed into the armies of the infidels, while Muslim princes and state officials sought patronage in Genghis Khan's court. It would take several years of consolidating power over the recently conquered territories in Central Asia, Iran, and the Caucasus before the Khan's descendants, led by his grandson Hülegü (d. 1265), swept across Mesopotamia and into the Levant in a second wave.[10]

The invasions sent refugees west into Iraq, Syria, and Anatolia. Qazwīnī settled for a time in the northern Iraqi city of Mosul, on the banks of the Tigris. Mosul was a major center for learning that attracted leading scholars. Here Qazwīnī met the high-ranking statesman and litterateur Ḍiyāʾ al-Dīn Ibn al-Athīr (d. 637/1239), then in the service of the local Mamluk warlord Badr al-Dīn Luʾluʾ (d. 657/1259) who ruled over the city. Among those who kept track of the number slaughtered in the initial conquests was Ḍiyāʾ al-Dīn's brother the chronicler ʿIzz al-Dīn Ibn al-Athīr (d. 630/1233), who wrote from Mosul. Not only did Qazwīnī read ʿIzz al-Dīn's history with his yearly account of those massacred city by city, but he also wove through his own geography references to the destruction let loose by the first stage of invasions.[11]

Though Qazwīnī does not directly address his own experiences with the conquests, his account of the eastern Islamic lands is filled with reports of cities assaulted by siege machines, large numbers of inhabitants fleeing for safety, and whole populations laid to waste by the sword. He reminds his readers that some of the settlements in Central Asia had never recovered from the earlier conquests of the Khwārazm-Shāh and were thus in no position to face another assault.

He paints a landscape of cities left smoldering, orchards and pleasure gardens cut down, and the countryside abandoned. Stories of escaping one invasion only to succumb to another are greeted with the refrain of how beautiful these lands were before the invaders came. Repeated invocations are made to the Almighty, who through it all remains steadfast for "everything but God changes with the ravages of time!"[12]

Living through such upheaval, Qazwīnī recognized the arbitrary nature of political divisions. The conventional demarcations on maps were themselves imaginary, having been first established by the ancient kings of the past. Afarīdūn the Babylonian, Alexander the Great, and Ardashīr the Persian, all had encircled the inhabited quarter of the Earth, so as "to make clear the boundaries of the realms and the routes."[13] Such observations on the imaginative lines, *al-khuṭūṭ al-wahmiyya,* that constituted territorial divisions appear well suited for the age.

The Mongol conquest of the territories held by the Khwārazm-Shāh unfolded across regions marked by political and religious fragmentation. The Abbasid caliphate of Iraq, which once claimed a vast empire, had long since been reduced to a mere shadow of its former glory. To the east lay the Delhi Sultanate, and to the west the Crusader states in Jerusalem, Tripoli, and Antioch, the Ayyubid rulers of Cairo and Damascus, the Saljuq overlords of Anatolia, and their Zangid governors of Aleppo and Mosul. Then there were the Nizārī Ismailis, who controlled impregnable mountain fortresses throughout Iran and Syria and were infamous for the art of political assassination.

Qazwīnī may well have confirmed for himself this lesson on the imaginative nature of political boundaries from the violent destruction that unfolded all around. Yet he asserts his knowledge, not through any direct experience, but through information obtained from books. Here he quotes as an authority Abū Rayḥān al-Bīrūnī (d. ca. 442/1050), the famed natural philosopher, who more than two centuries earlier had made the very same observation that there is nothing natural to the boundary lines drawn on maps.

For his part, Bīrūnī had also experienced significant upheavals in the world order that brought him far from his native land of Khwarazm by the Aral Sea into the court of the Turkic conqueror Maḥmūd of Ghazna (d. 421/1030) on the eastern frontier with India. From Bīrūnī, as well as from a host of other scientific authorities, Qazwīnī also knew the world to be divided into seven climes, which ran as longitudinal bands from south to north. The width of these bands varied, as their north–south boundaries were determined by the length of the longest day at the midpoint of a given clime. This was a model inherited directly from Ptolemy, the famed geographer and astronomer of Alexandria.

Qazwīnī gained his knowledge of the world through his own travels and studies under prominent teachers, as well as through his close reading of an array of Arabic and Persian writings. Mosul—which Qazwīnī refers to as "a pillar upholding the lands of Islam"—offered not only a measure of peace, but also access to books and to the leading minds of the day who had either passed through or had taken up residence in the city. One of the most eminent figures to have a lasting connection with Mosul was the man of letters known as Yāqūt al-Rūmī (d. 626/1229), "the Byzantine Gem," a manumitted slave from Anatolia who as a child was brought to Baghdad, where he was bought by a wealthy merchant. While we do not know if the two ever met, it is clear that Qazwīnī read Yāqūt's massive geographical dictionary closely as he worked on his own gazetteer and natural history.

In Baghdad, the city of caliphs, Yāqūt pursued a classical training in Arabic letters. Yāqūt's education was at first put to the service of his largely illiterate master, in the coming and going of mercantile ventures. Over time, Yāqūt refined his craft as a bookman, developing into a skilled calligrapher who copied hundreds of manuscripts by hand as a means of subsistence. He also found support through wealthy patrons of arts and letters who cultivated a taste for wide erudition wrapped in artful storytelling. With an eye for the complexities of human life, Yāqūt traveled extensively across the east on the eve of the Mongol conquests. The invasions in central Iran sent him back west, first to Mosul and then to Aleppo.

Throughout his journeys, Yāqūt kept detailed notes that became the basis for his multivolume geographical survey, the *Muʿjam al-buldān* (Dictionary of the regions), a first draft of which he finished in 621/1224 while living in Aleppo. This collection joined Yāqūt's other passion, a massive compendium of biographies dedicated to memorializing the lives of litterateurs. Yāqūt's geographical survey enjoyed renown as it deftly wed the art of biography with the plodding consistency of descriptive geography. Ordered alphabetically by place name, it gives colorful accounts of countless towns, cities, and districts across the lands of God, with a focus on the regions of Islam, noting along the way diverse products, customs, strange sights, and marvels. Many of Yāqūt's observations are drawn from his own travels. The volumes are populated with notices of dignitaries and notables, along with the poetry sung, the books composed, and the wonders encountered, entry by entry.[14]

Earlier biographical compendiums also took an interest in geography. One such collection was the *Kitāb al-Ansāb* (The book of lineages) by Abū l-Karīm al-Samʿānī (d. 562/1166), which focused largely on the lives of notable religious authorities, often with direct reference to the places they were from.

Writing from the city of Marv in central Khurasan, Samʿānī organized his massive collection alphabetically by *nisba*. As an adjectival noun of relation, the *nisba* often functions as a last name used to mark any given manner of group allegiance, be it to land, vocation, or party. Conveying a general sense of belonging, the *nisba* can highlight tribal, familial, dynastic relations, affiliations of country, city, village, or quarter, as well as professional training or ideological commitments. Yāqūt was also familiar with the venerable tradition of local histories, which often open with the remarkable characteristics of a given city or district, while supplying a regional history and topography lined with the lives of notable residents.

Yāqūt passed through libraries and private collections that contained not only memorials to local notables but also biographical dictionaries dedicated to individual classes and professions—hadith scholars, jurists, theologians, saints, physicians, philosophers, grammarians, singers, and poets. Although this collective body of writing was almost entirely dedicated to memorializing a male elite, women also feature in various significant capacities.

However, these earlier memorials of past lives made no sustained attempt to account for geography as a universal field of interest. Yāqūt filled this gap by turning to a well-established body of Arabic geographical literature that focused on routes and realms while also attending to the merits and faults of the places found along the way. These motifs, for example, find artful portrayals in such works as the *Aḥsan al-taqāsīm fī maʿrifat al-aqālīm* (The best divisions on the knowledge of the regions), by the quick-witted traveler Abū ʿAbd Allāh al-Muqaddasī (fl. 375/985), who trains his attention on both specialties and defects, highlighting throughout his journeys the pleasures to be had and the perils to be avoided.[15]

Qazwīnī drew from Yāqūt's geography on numerous occasions. This extends beyond merely the shape of Qazwīnī's geographical compendium with its attention to places and people, wonders and rarities, his general focus on the "lands of Islam," and his abiding interest in the broader "lands of God." In fact, many of the descriptions, accounts, and observations, as well as the sources referenced and the poetry quoted, have direct parallels in Yāqūt's *Dictionary of the Regions*.

The parallels often exhibit the hand of a skilled redactor, as well as the voice of a subtle critic. Moreover, Qazwīnī frequently repopulates cities and towns with stories and authorities drawn from his own experiences. Yāqūt would also prove a useful source when drafting *Wonders and Rarities*. Qazwīnī's quotation of Bīrūnī on the arbitrary divisions of the world, for example, appears in Yāqūt's geographical dictionary. It also features in both Qazwīnī's natural history and geography.

As with the conventional turns of phrase, the reworking of entire blocks of writing was a common feature throughout the library of classical Arabic letters. From sprawling legal manuals and monumental histories, to books of etiquette and poetry, the economies of the written word championed ideals of novelty and innovation, while also inviting clever repurposing and subtle reuses.[16]

For Qazwīnī, the creative and extensive reuse of earlier writings offered a primary method of composition. Despite a sustained engagement with Yāqūt's geography, Qazwīnī only makes a passing reference to the work. This distance itself speaks volumes. While Yāqūt's collection of geographical learning spans tomes, Qazwīnī turns his attention to synthesizing and distilling into a concise form a vast amount of material to be both easily portable and a delight to read.[17]

Qazwīnī may have crossed paths with Yāqūt, but nothing from the surviving record offers any indication that he actually did. The same, however, cannot be said for one of the most leading scholars of the day: Athīr al-Dīn al-Abharī (d. ca. 663/1264), who had also come to settle in Mosul. Of the numerous acquaintances mentioned in his geography, Qazwīnī reserves the title of *ustādhunā*, "our master," for Abharī alone. In light of Qazwīnī's abiding interest in the natural sciences, the affiliation is significant.[18]

The body of writings that Abharī produced and the students who studied with him help us better understand the formalized study of natural philosophy that shaped Qazwīnī's intellectual horizons. Scholastic education, largely through the institutional framework of the madrasa, played a notable role in these developments. So did the sweeping philosophy of Ibn Sīnā, whose writings profoundly influenced Qazwīnī's understanding of the world. By far the most important religious authority to systematize Ibn Sīnā's thought was Fakhr al-Dīn al-Rāzī (d. 606/1210), the famed Ashʿarī theologian, philosopher, and scholar of the occult sciences from Khwarazm. Abharī traveled within a larger circle of scholars and authorities who studied with Rāzī and promoted his teachings. Qazwīnī inherits this philosophical legacy and draws on it, page after page, throughout the course of *Wonders and Rarities*.

Through teachers like Abharī, Qazwīnī would have gained access to centuries of learning across fields of natural philosophy. Like Qazwīnī, Abharī appears to have been a refugee from the highlands of Iran. As his *nisba* suggests, he originally hailed from the town of Abhar, a two-day journey west from Qazvin.

Before migrating, Abharī had traveled throughout Transoxiana, where he joined the teaching circles of prominent scholars. These included Rukn al-Dīn al-ʿAmīdī (d. 615/1218) of Samarqand, whom Qazwīnī calls a wonder of the age. Renowned for his sharp tongue and quick mind, ʿAmīdī won admiration as a skillful debater who could silence an opponent before a crowded assembly through the cutting edge of logical reasoning.

The assembly, or *majlis*, served as a central venue for both instruction and disputation, governed by particular rules of engagement that determined not only how and when to speak, but where to sit and when to take leave. Rulers and judges convened assemblies, as did leading courtiers and prominent scholars. As a form of public performance, the assembly fostered the circulation of ideas and the cultivation of knowledge. It also served as testing grounds for aspiring scholars to prove themselves. A deft delivery before a powerful patron could make a career. By developing methods for both rules of engagement and logical principles of argumentation, ʿAmīdī sought to perfect the art of disputation for juridical and philosophical inquiry.[19]

Qazwīnī tells of a memorable debate that his master, Abharī, had witnessed while studying in Samarqand, around the year 610/1213. On this occasion, ʿAmīdī rebutted word by word the arguments of his opponent, who could not even finish his own sentences before ʿAmīdī would retort with a cutting response. ʿAmīdī's adversary was none other than Zayn al-Dīn al-Kashshī (fl. 622/1225). A prominent authority in his own right, Kashshī was well known as a student of Fakhr al-Dīn al-Rāzī. Like Rāzī, he enjoyed the courtly patronage of the Khwārazm-Shāh. Importantly, Kashshī represents the first generation of students who produced digests of Rāzī's teachings on logic, natural science, and metaphysics.[20]

Abharī is usually counted among the disciples of Fakhr al-Dīn al-Rāzī. Yet it is also likely, given his young age at the time of Rāzī's travels in Bukhara and Samarqand, that Abharī studied more closely with Rāzī's older students, such as Kashshī and Quṭb al-Dīn al-Miṣrī (d. 618/1221), who was killed in the Mongol sack of Nishapur. In the knowledge economies of the day, one could legitimately lay claim to a particular authority merely by attending, even as a child, that scholar's public lectures.[21]

According to Qazwīnī, by the time Rāzī had left Khwarazm for the East, his assemblies were attracting huge crowds, often numbering in the hundreds, with eager onlookers from among both leading intellectuals and humble aspirants. At such occasions, Rāzī could simply ask the audience, "What subject would you like me to address?" He would then answer swiftly any topic posed to him from any field of learning, for hours on end.[22]

There is reason to suspect that Abharī himself sat in on such sessions. Whatever the case may have been, Abharī went on to develop a full command of Rāzī's philosophy. Like Kashshī and Quṭb al-Dīn before him, Abharī also produced significant philosophical primers drawn from Rāzī's teachings. In this regard, Abharī stands among a generation of religious scholars who promoted Rāzī's thought across Iraq, Syria, Egypt, and Anatolia. Like his masters, Abharī saw logic, natural science, and metaphysics as interconnected fields of learning. It is a point that Kashshī stresses in the opening to his digest of philosophy, when he explains that logic, *manṭiq*, is the foundational tool humans use to acquire speculative knowledge. Natural science, *al-ʿilm al-ṭabīʿī*, in turn, investigates the wonders of creation and the rarities of existence, which in teleological fashion leads to knowledge of God's oneness, forming the basis of *al-ʿilm al-ilāhī*, metaphysics or the divine science. This vision of nature as filled with signs of a divine Creator profoundly shaped Qazwīnī's understanding of the world, and it is at the heart of Rāzī's philosophy.[23]

For his part, Rāzī systematized the scientific philosophical teachings advanced a century and half earlier by Ibn Sīnā, the philosopher-physician of Bukhara. Known in the Latin Christendom as Avicenna and referred to in Arabic by the simple honorific *al-shaykh al-raʾīs*, the "head master," Ibn Sīnā had produced a full system of thought to account for the totality of existence. Following the curricular study associated with the reception of classical Greek philosophy, Ibn Sīnā had absorbed and then mastered the various branches of logic, physics, mathematics, metaphysics, politics, management, and ethics. As a physician, Ibn Sīnā also expanded beyond Aristotle to produce an influential encyclopedic medical compendium, the famed *al-Qānūn fī l-ṭibb* (The canon of medicine), which consisted of anatomical principles and therapeutic procedures, a catalogue of medicaments and their properties, methods for the diagnosis and treatment of specific ailments and diseases covering the entire body, and formularies for compound remedies.

The fullest expression of Ibn Sīnā's philosophy appears in the *Shifāʾ* (The cure), a multivolume compendium that seeks to cover the Aristotelian curriculum. The philosophical tradition of Aristotelian teachings had been formalized by scholars from Alexandria, particularly during the sixth century, and was known in Sasanian Iran before the advent of Islam. After a full overview of the various branches of logic, Ibn Sīnā moves to natural or physical phenomena, *ṭabīʿiyyāt,* through the study of physics, the heavens, corruption and decay, mineralogy, meteorology, psychology, botany, and zoology. This is followed by a course on mathematics, *riyāḍiyyāt,* through geometry, astronomy, arithmetic, and music.

Only after these propaedeutic studies does the curriculum turn to theology, the so-called divine sciences, *ilāhiyyāt,* which in the Aristotelian tradition are known as the *metaphysica,* literally in Greek those topics that "come after the physics," expressed in Arabic as *mā baʿd al-ṭabīʿa.* For Ibn Sīnā, this final stage of philosophy starts with the development of the first principles for a universal science as the basis for logic, physics, and mathematics. This proceeds to an argument for the divine Creator and concludes with a study of the rational soul, prophetic revelation, and the pleasures and afflictions the soul suffers and enjoys after its separation from the body. These fields of study and forms of classification were well known to Qazwīnī and informed his own understanding of the world.[24]

In his life and shortly after his death, Ibn Sīnā's teachings came to circulate across the lands of God, copied and taught by Muslims, Jews, and Christians alike. From medicine to metaphysics, Rāzī takes up all of these elements and more in the course of his own engagement with Ibn Sīnā's teachings. The result is the full-scale, systematic absorption of Avicennan philosophy within the normative frameworks of Islamic theology. In many ways, Rāzī helped to further concretize a process that had already been initiated a generation earlier by the likes of Abū Ḥāmid al-Ghazālī, whose theological formulations also drew extensively from Ibn Sīnā's philosophical teachings. By this stage, the broader epistemic boundaries that had once separated *kalām*, Islamic dialectical theology, from the tradition of Hellenistic philosophy, known in Arabic as *falsafa*, had begun to shift.[25]

Ibn Sīnā was the beneficiary of the sustained translation of Greek learning that had been initiated by the early Abbasid caliphs, who claimed descent from the Prophet's uncle ʿAbbās. This imperial initiative spanned several generations and made available in Arabic—often by way of Syriac intermediaries, and generally through the aid of Christian translators—a vast corpus of Greek letters. These translations encompassed nearly the entire scholastic tradition of Aristotelian philosophy, from the logic of Porphyry (d. ca. 305 CE) to the emanationism of Plotinus (d. 270 CE). Included were the mathematics of Euclid, the medical and philosophical teachings of Galen, and the astral and terrestrial science of Ptolemy.

There were important translations sponsored by the Umayyad ruling family, well before the Abbasid revolution brought the caliphate centered in Syria to an end in the middle of the eighth century. But the scope and scale of the translation movement in Iraq was unprecedented. Collectively these translations filled bookshelves not only with Hellenistic learning, but also with works from both Sanskrit and the Middle Persian of the former Sasanian Empire of Iran. This movement of letters in Baghdad inspired such philosophers as Abū Yūsuf al-Kindī (d. ca. 256/870) and Abū Naṣr al-Fārābī (d. 339/950), who drew consciously on this newly accessible learning as an ethical basis for society and as a metaphysical foundation for understanding the cosmos. It is this philosophical tradition, moreover, that guided the collective known as the Ikhwān al-Ṣafāʾ, the "Brethren of Purity," who through the course of the early tenth century produced a massive compendium covering a full curriculum of natural philosophy.

Muslim theologians and jurists debated exactly what to do with all these foreign sciences from a pagan past. Many, including Ghazālī himself, condemned a range of tenets held by the ancient philosophers, such as the eternity of the world or the denial of bodily resurrection in the afterlife. Yet, from developing a logical argument to measuring the declination of the Sun, the sheer utility of this body of learning for conceptualizing and mastering existence was undeniable. The sustained systematization undertaken by Ibn Sīnā offered a coherent scientific model to explain physical reality.

For the intellectually inclined, the sheer expanse of Ibn Sīnā's philosophical system, which sought to cover the entire cosmos and was sponsored by powerful courts in Central Asia and Iran, made it all but inescapable. While in some quarters *falsafa* was a tainted term, Ibn Sīnā promoted another category that transcended the particularities of Greek etymology: *ḥikma,* the Arabic word for wisdom, as in the wisdom of the divine Creator who fashions the world in a wondrous design that is not only manifest and perceptible to human intellects, but also stands as a sublime proof for the singular existence of the Almighty. As a standard translation of *sophia,* the Greek word for wisdom, the Arabic notion of *ḥikma* had long been associated with earlier arguments for divine design, as in Galen's influential appeal—well known in Arabic—to contemplate the mysteries of nature for evidence of the wisdom of the divine Creator.[26]

It is through such arguments on the centrality of natural science for understanding the world of God's creation that Rāzī came to incorporate the various branches of logic, physics, and metaphysics into his own philosophical writings. The diverse efforts of synthesizing Avicennan philosophy formed part of a broader debate in Islamic intellectual history that focused on the relative value of the rational sciences, *al-ʿulūm al-ʿaqliyya,* versus the revealed sciences, *al-ʿulūm al-sharʿiyya,* as a means for obtaining knowledge of the world and our place in it.

The tension between the natural law of reason and the law of divine revelation stretches back throughout the early absorption of and reaction to Greek philosophy by Jews and Christians. For Muslims, the bifurcation often pitted classical philosophy against the revealed authority of scripture, as represented by the Quran and the collected sayings and actions of the Prophet, which were codified within massive hadith collections. These tensions had profound legal as well as theological implications regarding what could be known of the world with certainty.[27]

What made Ibn Sīnā so appealing within theological circles was not simply the totality of the philosophical system that he proposed. Rather, the central place Ibn Sīnā accorded to the psychology of the soul in his metaphysics gave religious authorities, who were not necessarily inclined toward Greek philosophy, a powerful means for describing such theological topics as ethics, prophecy, miracles, magic, angels, astral forces, and the state of the soul after death through the language of science. In this way, Avicennan philosophy offered a basis for exposing the division between reason and revelation as a false dichotomy.

In no small measure, the lasting success of this system of thought was indebted to its wide-scale absorption and reformulation within the curricular frameworks of madrasa education. The diffusion of Ibn Sīnā's writings in Iran and Central Asia overlapped with the rise of the madrasa as the primary institution for educating the religious elite in law and theology. As an institution, the madrasa first began to develop in the tenth century in Iran and Central Asia, before it spread throughout Mesopotamia, the Levant, and Anatolia, as well as west into North Africa and east across India.

Legal endowments sponsored by patrons among religious notables and the ruling elite often granted these colleges a fair measure of independence. While the curriculum of study could vary considerably, several discernible features characterize the early growth of madrasa education. There is the focus on grammar, the Quran, and hadith, as the basis for further studies in law, jurisprudence, and theology. Yet even before the absorption of Avicennan philosophy within the institutional frameworks of higher learning, the religious elite read closely the sciences of the ancients, the *ʿulūm al-awwāʾil*, particularly in the fields of logic and mathematics. From an early period, courses in logic offered a foundation for dialectal argumentation and general reasoning, which were central to jurisprudence and theology. Likewise, the study of mathematics, through the fields of arithmetic, algebra, geometry, and astronomy, was often pursued for its practical utility.[28]

Clothed in the language of *ḥikma*, the emergence of philosophy in the context of madrasa education, however, was a significant development for the absorption of the natural sciences in the curricular formation of the religious elite. As promoted by the likes of Fakhr al-Dīn al-Rāzī, logic, physics, mathematics, and metaphysics formed part of a totalizing, universal system of knowledge that emerged as a powerful currency for religious and scientific authority. Scholars with inclinations toward natural philosophy, such as Abharī and Qazwīnī, were able to use the imprimatur of madrasa education to promote their own careers. After all, these sciences taught nothing less than a mastery of the entire world.[29]

2

MEASURES OF AUTHORITY

IN MOSUL, QAZWĪNĪ FOUND ABHARĪ. Abharī, for his part, had earlier found there Kamāl al-Dīn Ibn Yūnus (d. 639/1242), a jurist and theologian renowned for his knowledge of mathematics, philosophy, medicine, and astronomy. Throughout his professional life, Kamāl al-Dīn taught at numerous madrasas, including the famed Niẓāmiyya madrasa in Baghdad, founded in 459/1067 by the mighty vizier Niẓām al-Mulk (d. 485/1092). While in Mosul, Kamāl al-Dīn lectured at various institutions and ultimately settled in a madrasa that took his name. Abharī joined Kamāl al-Dīn as a *muʿīd,* an assistant who explained and repeated lessons already delivered by the professor to the students so as to aid them in their mastery of the material.[1]

Like Abharī before him, Qazwīnī pursued a career in madrasa education. The networks of teaching and learning through which Qazwīnī traveled exhibit several important features that lend further context to the scholastic character of his natural history. The structure of Qazwīnī's *Wonders and Rarities* with its classificatory divisions of all existence is itself a testament to the enduring influence that the formalized study of natural philosophy would play in the course of his own intellectual development.

The increasingly philosophical shape of theology stands as one of the defining characteristics of religious learning during the period. The religious elite did not limit themselves to merely the study of law or the Quran, as modern caricatures of traditional Islamic learning might have it. Just as the halls of madrasas contained illuminated books cataloguing plants, animals, minerals, and the stars, so too did teachers address recondite topics in philosophy and science.

These interests extend from the foundational study of logical argumentation and classification to the shape of the Earth and the movements of the heavens. In this regard, Qazwīnī and his circle of teachers and classmates represent key figures in the development of a particular form of scholastic education that engaged directly with natural philosophy. The broad body of learning taught by the likes of Abharī could account for the physical forces governing creation, including the power of prophecy. Qazwīnī recognized through the course of his own travels and studies that authority and knowledge came in many different varieties, from the precision to solve complicated mathematical problems to the ability to produce miracles.

The biographer and chief justice Ibn Khallikān (d. 681/1282), who like Qazwīnī studied for a time under Abharī, notes that when the books of Fakhr al-Dīn al-Rāzī were first brought to Mosul, no one was familiar with Rāzī's terminology. However, Kamāl al-Dīn, luminary of city, was able to quickly decipher what the sage of Khwarazm meant in his technical prose. The same went for the introduction of ʿAmīdī's method of disputation, which Kamāl al-Dīn mastered after a single night of reading.

This legendary prowess perhaps overlooks the important role played by intermediaries like Abharī in the introduction and dissemination of prized teachings from the East on the eve of the Mongol invasions. Ibn Khallikān also lets on that Abharī was quick to demur to Kamāl al-Dīn's authority out of a general sense of etiquette and respect for the venerable shaykh, who could lecture on a dizzying array of topics. From his own studies in Mosul, Qazwīnī knew Kamāl al-Dīn to be an autodidact who could command any field of learning that he set his eye upon, whether from the revealed or the rational sciences.[2]

With Kamāl al-Dīn in Mosul, students could pursue a course of study that proceeded from logic to physics to metaphysics. If so inclined, they could also focus on medicine, Euclidean geometry, higher-order mathematics, Ptolemaic astronomy, or musical theory. This is not to mention a full range of jurisprudence, Quranic exegesis, and hadith scholarship. In addition to Arabic grammar, poetry, and history, the master also held sessions on the Torah and the Gospels attended by both Jews and Christians.

Famously, Theodore of Antioch (fl. 1243), before rising to the position of court philosopher and Arabic interpreter to the Holy Roman Emperor Frederick II (d. 1250), had traveled to Mosul to study the writings of Ibn Sīnā and Fārābī, as well as Euclid and Ptolemy, with Kamāl al-Dīn. In due course Theodore and the court of the Crusader emperor Frederick II emerged as important conduits for the introduction of Arabic learning into Latin Christendom. With the Crusades and the conquests in Iberia came the translation and absorption of all forms of scientific and philosophical endeavor that thrived in the lands of Islam. This, in turn, helped to connect Latin scholastic communities to Greek antiquity and to the significant contributions made by Muslim authorities to the universal fields of theoretical and practical science.[3]

Among the amazing feats of Kamāl al-Dīn that Qazwīnī records was a reply to a letter sent from the Franks, the umbrella term for Christians of western Europe, as well as the Crusader states that occupied various slices of the Levant. The letter posed queries on matters of medicine, philosophy, and mathematics to Nāṣir al-Dīn (d. 635/1238), the Ayyubid sultan in Damascus, known by the monarchal title al-Malik al-Kāmil, "the perfect king." While Qazwīnī does not specify the letter's origin, it would appear that it came from the Crusader court of Frederick II, who had developed philosophical interests and had sustained diplomatic relations with the sultan.

Scholars in Syria were able to solve the medical and philosophical problems, but a mathematical enigma exceeded their capacities. This they sent on to Abharī in Mosul. The question touched on the ancient paradox of squaring the circle. Unable to arrive at a solution, Abharī turned to Kamāl al-Dīn, who produced a demonstrative proof that was sent back to Damascus. When Qazwīnī traveled through Damascus sometime later, the scholars he met there still marveled at the elegance of the solution that Kamāl al-Dīn had devised.[4]

Some scattered mathematical treatises survive, but the archival record preserves little from the writings of Kamāl al-Dīn. The same cannot be said, however, of Abharī, the modest famulus whose books came to travel in and beyond the courtyards of madrasa education. Having migrated from Iran,

Abharī had to start his academic life anew. Over time he gained considerable prestige and renown, through both his teachings and his written legacy of pithy summaries on a variety of complicated topics. His short introduction to the field of logic, building on the *Isagoge* of Porphyry, alone had a palpable impact in the teaching curriculum, circulating on its own and as the subject of numerous commentaries.[5]

Abharī's précis gives students an overview of the various classes of logical propositions and the ramifications of the Porphyrian tree as a hierarchical model for classification from genus, species, and differentia, a key characteristic of scholastic learning. Abharī covers in terse prose the difference between the universal and the particular, essence and accident. With the foundations for definitions and parts of speech established, Abharī proceeds to the incisive tool in the classical curriculum of logic: the demonstrable proof or demonstration, *burhān,* as a form of syllogistic reasoning based on sound premises used to generate certain knowledge.

Students would learn the various sources of certainty, *yaqīniyyāt,* that could obtain a demonstration, thereby providing a general logical framework for describing how knowledge was produced and confirmed: (1) primary principles, *awwaliyyāt,* such as half of two is one or the whole is greater than the part; (2) eyewitness observations, *mushāhadāt,* as in the common knowledge that the Sun rises in the east or that fire burns; (3) empirical experiences, *mujarrabāt,* as in the efficacy of a recipe that calls for consuming the scammony plant from the morning glory family to relieve excessive yellow bile; (4) correct guesses or intuition, *ḥadsiyyāt,* of middle terms in a syllogistic argument, such as the scientific claim that the light of the Moon is derived from the Sun, which requires a form of intuitive deduction; and (5) the collective transmission of numerous reports of information from independent and reliable sources, *mutawātirāt,* as in the knowledge passed down for generations, by voice and pen, that Muhammad lay claim to prophecy, and that miracles were made manifest through him.[6]

Abharī also developed succinct distillations of the theological philosophy promoted by Fakhr al-Dīn al-Rāzī. Unquestionably, the most famous of these is Abharī's compendium the *Hidāyat al-ḥikma* (Guidance for philosophy). This popular teaching digest served as a standard textbook in the curriculum of higher education and, like Abharī's introduction to logic, enjoyed numerous commentaries. Here the classificatory strictures of logic govern the treatment of physics and metaphysics, through a hierarchical taxonomy, from the *falakiyyāt,* celestial bodies, to the *ʿunṣuriyyāt,* elementary bodies, followed by an outline of the divisions of all existence, a discussion of the Creator and his attributes, an account of angels as immaterial intellects, *al-ʿuqūl al-mujarrada,* and a discussion of the properties of the soul in life and after death.[7] Along with other teaching abstracts of philosophy that Abharī produced, this compendium tethers the mysteries of the divine sciences to the physical universe of the cosmos. The result is a distillation of Avicennan philosophy through the teachings of Fakhr al-Dīn al-Rāzī, whom Abharī frequently refers to by the honorific "our Master who calls us to God," *mawlānā l-dāʿī ilā llāh,* or simply as the Imam.[8]

A centerpiece to the philosophical system that Abharī taught was the metaphysics of the rational soul, *al-nafs al-nāṭiqa,* as the unique basis of human cognition. This philosophy of the mind was shaped by Ibn Sīnā's reformulation of Aristotelian psychology. In page after page, students are guided through the simple argument that the physical and metaphysical worlds are inseparable. These cosmic connections proceed in stages of emanation descending from the first intellect which moves the heavens, and which ultimately serve as the divine origin for human cognition. Abharī follows the celestial realm of the stars and planets with an examination of the various classes of being in the sublunar world.

This leads from meteorology and mineralogy to a discussion of the vegetal and animal souls. The soul or psyche provides the basis for cognition. In the most basic sense, humans share with animals the perceptions of sight, sound, smell, taste, and touch, as well as a common sense that unifies the sensations received from external objects. Additionally, the cognitive processes of the soul function through various faculties, such as estimation, imagination, and memory, situated physically in different parts of the brain. What separates humans from animals in this scheme is the faculty of reason, which is represented by the rational soul. It is a vision of creation and human capacity that Qazwīnī would come to follow closely.[9]

The physical model presented is entirely hierarchical. This grand chain of being—the so-called *scala naturae* of Latin scholasticism—finds mooring in Platonic emanationism and Aristotelian cosmology. A cosmic order extends throughout existence, linking the highest to the smallest, progressing from God, the prime cause, who sets it all in motion, through the heavens, populated by angels and celestial bodies, down to humans, animals, plants, and minerals. Radiating through it all is the emanating power of the soul that infuses creation with the spark of the divine Creator. The highest orders of the rational soul, as Abharī explains, culminate with the sacred faculty, *al-quwwa al-qudsiyya,* a pivotal term in Ibn Sīnā's epistemology.

It is through the sacred faculty of the soul that humans are capable of *ḥads,* the ability to divine new information. Following Ibn Sīnā, Abharī describes *ḥads* as correctly intuiting the middle terms that predicate the conclusions of the major and minor premises of categorical syllogisms. Such forms of intuitive knowledge are reflected in the capacity of the mind to derive new knowledge when presented with a paucity of information.

Discussion of the sacred faculty leads to the general observation that the world is filled with various ranks of humans bestowed with differing degrees of intellectual capacity. There are those who are prodigiously quick at processing information and then generating from it new knowledge, while there are others who are by no measure so intellectually predisposed. Ibn Sīnā was himself renowned as an autodidact, mastering whole fields on his own swiftly and with ease—a talent that one of his chief disciples described as akin to a prophetic miracle, or *muʿjiza.* The analogy is apt. As Abharī explains, the gradation of intellectual capacity, located in the physical power of the soul, serves as the basis for describing in natural terms the foundations for prophecy, itself one of the primary conclusions of Avicennan psychology.[10]

Upper-level students could read with Abharī not just abstract theoretical topics such as the scientific account of miracles; or logical arguments for the existence of a singular divine Creator, through the rejection of *tasalsul,* an endless chain of causation without an ultimate, original cause; or the defense of the void in space—against both Ibn Sīnā and Aristotle—as a logical necessity born out of the demands of movement alone. For Abharī's teaching also turned to such practical matters as geometry, mathematical geography, and observational astronomy.[11]

With the absorption of Avicennan physics, the discipline of astronomy grew into a formalized feature of science as it was taught within the instructional frameworks of higher learning. In the fields of astral knowledge, Abharī follows the well-established geocentric model, promoted by Aristotle and reformulated by Ptolemy, which holds that the stars and the seven planets, which included the Sun and the Moon, all rotate around the stationary Earth in uniform circular motion.

Abharī read and taught the *Almagest,* Ptolemy's compendium of mathematical astronomy, which provided a theoretical model for predicting the motions of the planets. For his students, Abharī produced a concise teaching summary on the science of the cosmic structure, *ʿilm al-hayʾa,* the common term at this point for both planetary astronomy and mathematical geography, two fields that were profoundly interrelated. Abharī's compendium built directly on the Ptolemaic vision of the universe with its elegant solution of epicycles, eccentrics, equants, and deferents to account for and predict the apparent retrograde motion of Mercury, Venus, Mars, Jupiter, and Saturn. Qazwīnī would draw on this theory of planetary motion in the opening chapter of his natural history (figure 2.1).[12]

To make observations of celestial bodies required specialized instruments for tracking and measuring the movement of the heavens. These tools could range in intricacy from the common quadrant, a graduated quarter circle used to determine the altitude of celestial objects above the horizon, to the highly sophisticated astrolabe, an elegantly tooled measuring instrument that, like Ptolemy's *Almagest,* was inherited from the astronomical learning of late antiquity.

The term *astrolabe* is from the Greek, meaning "taker of stars." The instrument offered a mirror of the cosmos through its stereographic projection of the celestial sphere onto the surface of a circular plane. Measurements were made through the use of a sighting arm that moved over a rotating openwork metal plate, known in Arabic as the *ʿankabūt,* the spider, and in Latin as the *rete,* the web that marked various stars as well as the ecliptic—the circular path in the sky along which the Sun, the Moon, and the planets orbit. This flat, open-faced celestial web, in turn, rotates over the terrestrial plane, represented by a series of interchangeable graduated metal discs marking such points as the horizon, the meridian, and the zenith, for a given latitude (figure 2.2).[13]

As a device, the astrolabe could help solve numerous practical problems. Not only was it used to measure the movement of stars and planets, to calculate latitude, or to determine on land any given geographical location, but it also served to tell time. For his part, Abharī complemented his own theoretical

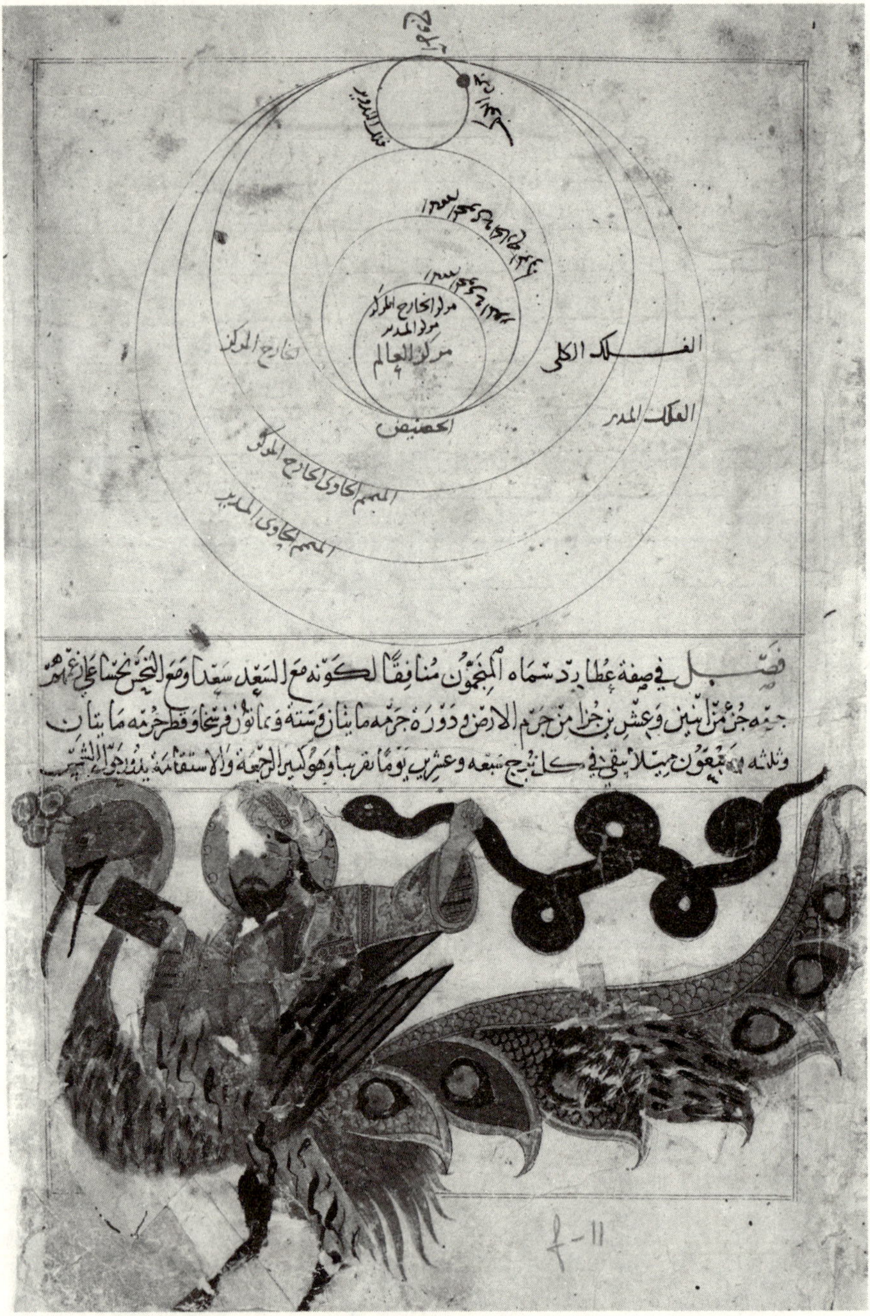

FIGURE 2.1: Diagram explaining the retrograde motion of Mercury (above) and talismanic representation of the planet (below), from an Ilkhanid manuscript of Qazwīnī's *Wonders* produced circa 1300.

FIGURE 2.2: Aristotle taking measurements with an astrolabe as students consult a handbook, in Ibn Fātik, *Mukhtār al-ḥikam* (Selection of wise teachings), copied in the early thirteenth century.

writings on astronomy with a concise manual on how to use the astrolabe, which drew on and refined an earlier treatise on the topic by the Iranian mathematician and astronomer Kūshyār ibn Labbān (fl. ca. 400/1010).[14]

Measurements produced with such astronomical instruments were often recorded in the form of a *zīj*, a specialized handbook of tables for tracking the planets and stars. Along with his guide to the astrolabe, Abharī also finalized one such book of tables, which he entitled *al-Zīj al-shāmil* (The complete astronomical handbook). As a digest of observations, these handbooks offered invaluable data, collected generation after generation, recording, refining, and correcting earlier observations originally preserved in Greek, Sanskrit, and Middle Persian.

The Arabic word *zīj* itself comes from the Middle Persian *zīg*, for a rope or bowstring, likely after the tabular forms of astronomical information contained within such collections. Abharī drew his astral observations by correcting and verifying calculations contained in the *zīj* of the Iranian mathematician and astronomer Abū l-Wafā' al-Būzjānī (d. 388/998), who had settled in Baghdad. As with other such handbooks, Abharī opens his tables with a system for intercalation between various calendric models. Along the way, he provides descriptions for how to use the tabulated data within his handbook for determining various positions of the stars and planets at given times and locations, methods for predicting solar and lunar eclipses, as well as an overview on how to establish in any given location the *qibla*—the direction facing Mecca, which orients daily prayer—itself a complex problem of spherical trigonometry.[15]

The renown of such an array of skills across a full curriculum of study brought students seeking access to higher learning and the prestige that came with it. Through his associations with Rāzī, as well as his own specialized mastery of logic, mathematics, and practical astronomy, Abharī could offer access to the leading philosophical and scientific teachings. Qazwīnī was merely one among many prominent students and intellectuals who sought out Abharī's instruction.

Like Qazwīnī, several of these students came from Iran as part of the waves of migrants who had fled the Mongol sieges. This was the case with Shams al-Dīn al-Iṣfahānī (d. 688/1289), whose father was a high-ranking bureaucrat. His *nisba* marked him from Isfahan, the lush ancient settlement at the foothills of the Zagros Mountains. As a teenager, Shams al-Dīn escaped the city of his birth before the Mongols finally burned it to the ground in 633/1235~.[16]

Under Abharī, Shams al-Dīn pursued a course of study that proceeded along the division of philosophy into logic, physics, and metaphysics. Instruction followed an age-old practice of oral recitation, which served as an ideal for discipleship and as a means of ensuring the proper transmission of knowledge, from prophetic hadith to works of practical science. A teacher would recite from memory or read from a textbook both to individual pupils as well as before larger groups.

In turn, students took notes and repeated back what they had heard. Here the master could verify that the lesson was both properly understood and correctly relayed. Students would also read aloud from a textbook, one that the teacher had either written or mastered. At any moment the session could be paused to verify the correct reading, to test the comprehension of material, or to address tangential questions as they arose. For large works, these audition sessions, or *samāʿāt,* could stretch over the course of several months.

Records of both private lessons and large assemblies are inscribed throughout countless individual manuscripts that document and authorize the generational transmission of learning. Students who fully comprehended the teachings could receive an *ijāza,* a certificate of authorization, which allowed them in turn to teach and transmit to others the material they had mastered. By no measure did all books circulate through the strictures of such direct instruction or authorization. Similarly, there were countless ways to game a knowledge economy that prized access to star teachers and renowned works of scholarship. But the ideal of face-to-face instruction served as a powerful communal marker for the transmission of knowledge and the continuity of authority.

Abharī's disciple Shams al-Dīn went on to have an illustrious career as a chief judge in various towns and cities across the Levant and in Egypt, serving as an official of the recently founded Mamluk sultanate. Along the way, he taught at numerous institutions, including the recently built madrasa attached to the mausoleum of the famed jurist Imam Muḥammad ibn Idrīs al-Shāfiʿī in Cairo, which was founded by Ṣalāḥ al-Dīn al-Ayyūbī (d. 589/1193), the celebrated sultan who had brought the Crusader Kingdom of Jerusalem to its knees.

Shams al-Dīn made a rule of teaching *falsafa,* as peripatetic philosophy was known, only to those who had first fully mastered the religious sciences. Throughout his travels, from the castle of Karak in Jordan through the remote bends of the Nile in southern Egypt, he carried a prized manuscript of Abharī's *Kashf al-ḥaqāʾiq fī taḥrīr al-daqāʾiq* (The revelation of verities through the study of intricacies), a work that follows the philosophical curriculum of logic, physics, and metaphysics. Shams al-Dīn had copied this compendium while studying with Abharī.[17]

In a short autobiographical note added to the end of his personal copy of the *Revelation,* Shams al-Dīn explains that he had finished copying this work directly from Abharī's authorial exemplar in the winter of 646/1248. He did so while residing in a madrasa that had been recently established by the powerful general Sharaf al-Dīn al-Sharābī (d. 653/1255) in Baghdad. Shams al-Dīn further notes that he read the entirety with Abharī, thereby gaining authority in turn to teach the collection to others. We know that Shams al-Dīn did just this, for an authorized manuscript, finalized in the fall of 674/1275 in Cairo and based on his own personal copy, has survived. It contains a note verifying the line of transmission back to Abharī as well as a short account of Shams al-Dīn's peregrinations and professional life.[18]

Shams al-Dīn also relates that once his studies with Abharī had come to an end, he took up a course with the judge Tāj al-Dīn al-Urmawī (d. 653/1255~) in Baghdad, a leading scholar originally from northern Iran. In 628/1231, Urmawī was appointed to lead the Sharābī madrasa next to a bustling market in the city. Qazwīnī knew Urmawī to be unmatched in law, jurisprudence, philosophy, and belles lettres, and notes that "he had an excellent turn of phrase and socializing with him was always a source of pleasure for the soul, with his splendid stories, subtle parables, rare analogies, and amazing expressions." Urmawī shared much with Abharī. He too was a migrant from northern Iran who laid claim to having studied directly with Rāzī.[19]

Unlike Abharī, however, Urmawī is remembered largely for his writings on the religious sciences, *al-ʿulūm al-dīniyya*. As with many other authorities of the day, Urmawī felt that while all branches of science were valuable and illuminating, there were gradations of excellence that separated them. In Urmawī's schema, the religious sciences crown them all. Collectively these sciences consist of both traditional and speculative disciplines. The fields of traditional learning focus on the communal transmission of knowledge over time. These consist of such subjects as hadith and Quranic interpretation, which Urmawī counsels are in their own right relatively simple to master, as they involve correctly copying and relaying the knowledge preserved by earlier authorities. The speculative branches, in turn, encompass dialectical theology, or *kalām*, as the foundation of religion, followed by jurisprudence, which Urmawī classifies as a branch of theology, and then positive law, which focuses on contingent human actions. These fields require precision of mind and thought developed through sustained study and reflection.

Urmawī draws this lesson on the hierarchical order of knowledge in the opening to his commentary on *al-Maḥṣūl fī ʿilm al-uṣūl* (The ascertained in the science of the foundations), Rāzī's early and highly influential handbook of *uṣūl al-fiqh*, or jurisprudence. As a field of writing, commentaries offered an important means for distilling, explaining, and teaching authoritative texts. They also could be used for critiquing, challenging, and reformulating the ideas of past masters. Under Urmawī, Shams al-Dīn read Rāzī's handbook and then went on to produce his own commentary on the *Ascertained*. This commentary not only confirmed Shams al-Dīn's training and development from student to teacher, but it also marked him as part of a generation of students qualified to expound, promote, and refine Rāzī's teachings.[20]

The same holds true for Najm al-Dīn al-Qazwīnī (d. ca. 693/1294), known by the honorific *kātibī*, highlighting a family affiliation with scribes or administrators. The *nisba* suggests that Najm al-Dīn was originally from the city of Qazvin. As with Shams al-Dīn, he too pursued a course of study with Abharī on philosophy, developing in particular a mastery of logic. A unique manuscript also survives that documents Najm al-Dīn's studies with Abharī on the topics of logic, physics, and metaphysics. Najm al-Dīn copied the collection himself from Abharī's master exemplar during the summer of 627/1230. A series of attestations written by Abharī note that Najm al-Dīn read the manuscript directly with him. This thereby authorized him to transmit to others his mastery of Abharī's teaching.

As with several other scholars of his generation, Najm al-Dīn engaged extensively with the writings of Rāzī. However, he focused his attention on the fields of philosophy, producing significant commentaries on Rāzī's *Muḥaṣṣal* (The summary) and *Mulakhkhaṣ* (The epitome), philosophical compendiums that cover logic, physics, metaphysics, and theology. In addition to sophisticated writings on logic, Najm al-Dīn also produced an influential philosophical compendium, the *Ḥikmat al-ʿayn* (The philosophy of the quintessence), which, much like Abharī's own handbook, circulated widely and was the subject of commentaries in its own right.[21]

Najm al-Dīn went on to work under the famed philosopher and astronomer Naṣīr al-Dīn al-Ṭūsī (d. 672/1274), at the observatory of Maragha south of Tabriz in the northern highlands of Iran. As with others associated with the observatory, Ṭūsī also had important connections with Abharī. Foremost, he is said to have read Ibn Sīnā's *al-Ishārāt wa-l-tanbīhāt* (Pointers and reminders) under Abharī's instructions. This highly succinct and enigmatic handbook marked the last major articulation of Ibn Sīnā's philosophy. Abharī knew the *Pointers* well and drew on it in the course of his own writings. As with many other teachings, Abharī came to the *Pointers* through

Rāzī's influential commentary, which divided Ibn Sīnā's short guide into sections on logic, physics, and metaphysics.[22]

In Qazwīnī's telling, Rāzī had charged himself with defending the *Pointers* against doubts that were posed by a scholar in Bukhara. Having got wind of these criticisms during his stay in the city, Rāzī sought out the impudent braggart and succeeded in publicly humiliating him: "How can you come up with such questions, when you haven't even covered the entirety of the work!" The story is an allusion to the famed debates between Rāzī and Sharaf al-Dīn al-Masʿūdī (fl. 575/1180), who had written a short treatise impugning various aspects of Ibn Sīnā's *Pointers*. In a single evening Rāzī digested the treatise, which, Qazwīnī notes, Masʿūdī had attempted to hide when he heard that Rāzī was in town. While Rāzī's presentation of the *Pointers* in many ways offers a defense, it also lays bare several problems he found in Ibn Sīnā's philosophical system. For his part, Ṭūsī also produced a commentary on the *Pointers* that sought to address Rāzī's objections.[23]

Commentaries that built on and responded to earlier commentaries were a lasting feature in the public lives of scholars who strove to absorb, refine, or challenge the authoritative teachings from past masters. This spirit of inquiry follows the epistolary exchange that Ṭūsī carried on with Abharī, which, among other problems, drew into question foundational tenets of Ptolemy's planetary model, following explicitly in the footsteps of the astronomer and mathematician Ibn al-Haytham (d. 430/1039), who had produced his own *shukūk*—puzzles, doubts, or contradictions as in the Greek *aporia*—on numerous aspects of the Ptolemaic picture of the cosmos. This included the problem of the retrograde motion of the five outer planets, which Ptolemy addressed through a model of epicycles attached to epicycles. Among the many arenas where Ṭūsī would gain fame, one of the most significant would be his mathematical critique of Ptolemaic astronomy that sought to better accord with the observed phenomena of planetary motion.[24]

The tension between past precedent and independent judgment is often distilled in Arabic through the pithy and alterative dyad *naql* and *ʿaql*, tradition and reason. As with each generation, the circle of scholars in Qazwīnī's orbit came to speak for the boundaries of what could be said and thought. Populating the provincial villages and capital cities of Qazwīnī's geography in regular intervals are authorities whose expertise was born of both the written and the spoken word. Functionaries, secretaries, copyists, booksellers, jurists, judges, preachers, poets, wonderworkers, alchemists, physicians, philosophers, saints, and sages all jostle along a landscape of reputation and renown. Many were flashes in the pan, whose names are all but forgotten save for Qazwīnī's record. A few were luminaries who could harness such clarity of purpose that others, far and wide, come to see the world anew.

In Qazwīnī's estimation, Rāzī stood out as such a person, a visionary who appeared only once in a lifetime. To accentuate this point, in the course of both his geography and natural history, Qazwīnī returns to the famed hadith of the Prophet that promises continual *tajdīd,* or renewal. The prophecy predicts that every hundred years God will send a *mujaddid,* a renewer, to revivify the religion of the *umma,* the community of the faithful. Competing lists of candidates who could be qualified as the *mujaddid* for a given age formed part of a partisan exercise that sought to affirm the God-given superiority of particular theological and juridical factions.

Drawn from arguments that advanced Shāfiʿī jurisprudence and Ashʿarī theology, Qazwīnī's list reveals a good deal about his own affiliations. In it he names Shāfiʿī, the eponymous founder of the Shāfiʿī school of law; Ibn Surayj (d. 306/918), the lead exponent and teacher of Shāfiʿī jurisprudence in Baghdad; Abū Bakr al-Bāqillānī (d. 403/1013), renowned for systematizing and popularizing Ashʿarī theology; and Ghazālī, who stood out in a generation of authorities from Iran and Iraq trained in Shāfiʿī jurisprudence and Ashʿarī theology. Qazwīnī's list of luminaries culminates with Rāzī, who like Ghazālī before him was both a Shāfiʿī jurist and an Ashʿarī theologian, and who, also like Ghazālī, was steeped in the philosophical writings of Ibn Sīnā.[25]

In addition to the study of *ḥikma* through the prism of Ashʿarī theology and Avicennan philosophy, Rāzī, Abharī, Kamāl al-Dīn, Urmawī, Shams al-Dīn, Kātibī, and Qazwīnī were all trained in the juridical norms, legal maxims, and practical methods associated with the teachings and writings of the Shāfiʿī *madhhab,* or school of law. A primary instrument of juridical authority, the *madhhab* was not a school in the physical sense of a singular institution. Rather the term came to express a communal set of legal principles, guidelines, and writings that were disseminated generation after generation throughout a network of teachers and students. Madrasas, in turn, emerged as one of the primary vehicles for the study of law and were often associated with particular legal affiliations.

The development of diverging legal schools built on the personal relations of masters and disciples, the systemization of teachings that were passed on and commented upon generationally, as well as networks of patronage and prestige that ebbed and flowed across diverse classes of local elites. The Shāfiʿī school coalesced in the tumult of rival juridical affiliations and regional factionalism. In the early development of Sunni legal affiliations, the Shāfiʿīs had to contend with competing claims to authority and conflicting practices of legal interpretation. Already powerful were the Ḥanafīs, who followed the legal authority of Abū Ḥanīfa (d. 150/767) from Kufa and his chief disciples in Baghdad. The early generations of Ḥanafī scholars came to promote a legal tradition that sought to balance *raʾy*, the independent judgment and reasoning of jurists, with the prophetic precedent, or Sunna of the Prophet, and the binding authority of the Quran.

For his part, Shāfiʿī traveled through Arabia and Iraq before settling in Egypt. He knew the legal and logical methods developed by the *ahl al-raʾy*, the partisans of independent judgment or reason. He also was steeped in the traditionalism of hadith scholars, having studied with the leading scholar of Medina, Mālik ibn Anas (d. 179/796), author of one of the earliest legal handbooks based on the Quran, the sayings of the Prophet, and the rulings of the first generation of jurists from the Hijaz of Arabia. Around Mālik's teachings came to coalesce the school of Mālikī law.

But in addition to Mālik and Shāfiʿī, traditionalists could also turn to the circle of Aḥmad ibn Ḥanbal (d. 241/855) of Baghdad, who had a prodigious mastery of hadith and had developed an active resistance to the caliphate as the imprimatur of law and belief. Ibn Ḥanbal stood out as a chief representative of the *ahl al-ḥadīth*, the so-called partisans of prophetic tradition, who championed the Quran and the Sunna as the prime basis for positive law. This strain of traditionalism was in notable contrast to the Ẓāhirīs, who promoted the teachings of Dāwūd ibn Khalaf (d. 270/884) of Baghdad, with his focus on the *ẓāhir*, the literal, external meaning of the Quran as a foundation for juridical norms.

Not to be forgotten in this competing terrain of legal authority are the Karrāmīs, named after the pious reformist Ibn Karrām (d. 255/869), whose followers attracted adherents among rural converts in eastern Iran, and who adopted many of the legal and theological positions promoted by Ḥanafī authorities. Over time, through a complex confluence of forces, the Ẓāhirīs and the Karrāmīs, as well as a host of other schools that had developed around individual jurists, ceased to attract new adherents. This left, by and large, the followers of Abū Ḥanīfa, Mālik, Shāfiʿī, and Aḥmad ibn Ḥanbal as the collective representatives of Sunni law.[26]

In addition to Sunni factions, numerous Shia groups asserted their own authority in matters of both law and theology. In the face of early Umayyad hegemony, the Shia could reasonably claim precedence as the rightful inheritors and custodians of Muhammad's message. Their messianic call not only led to revolts against Umayyad oppression, but ultimately gave voice to the revolution that brought the Abbasids to power. But the Shia also split into factions, largely over the question of the Imam and the debate over who should lead the community of the faithful. In palpable ways, the assertion of juridical independence and the rise of Sunni jurisprudence were responses to the early messianic peals of Abbasid imperialism and the absolute religious authority that the early Abbasid caliphs sought to wield.[27]

Certainly there were historical tensions between Sunnis and Shia. This is felt most notably in Qazwīnī's day with strident Sunni reactions against the Ismailis, who viewed their Imam as the living authority of God on earth. But there was also a good deal of factionalism that came to divide Sunnis. These tensions could have an acrimonious flare, marked not only with public debates, but defamations, insults, and even riots, which, while no means uniform or always present, left a palpable imprint on the fabric of social life. Baghdad had lived with the fear of Ḥanbalīs, who could burn down entire neighborhoods; Nishapur once witnessed internecine violence so intense that it spread into the neighboring town of Bayhaq, uniting the Shāfiʿīs and Ḥanafīs to destroy Karrāmī institutions of learning; and the fall of Isfahan to the Mongols was quickened by the Shāfiʿīs and Ḥanafīs in the city having turned one against other.[28]

Across the East in Iraq, Iran, and Central Asia, Shāfiʿīs had to contend with spaces first occupied by Ḥanafī jurists, who for much of the early history of the Abbasid caliphate enjoyed patronage as the preferred legal tradition of the empire. Part of the success of Shāfiʿī scholarship lay in the development of norms and hermeneutical methods of jurisprudence that sought to harness the Quran, the Sunna of the Prophet—represented in the sayings and actions of Muhammad—and analogical reasoning as probative foundations for positive law. Critique, argumentation, and the development of juridical theory played parts in the emergence of Shāfiʿī authority, and social conditions, familial relations, and local patterns of patronage quickened its rise.

The theological dimensions to these social transformations are also telling. Many Ḥanafīs in Iraq and beyond had also trained in Muʿtazilī theology, once the preferred dogma of the Abbasid caliphate, with its emphasis on divine justice, strict monotheism, free will, rational argumentation, and a general ambivalence concerning the binding authority of the Sunna as a source for law. Abū l-Ḥasan al-Ashʿarī, for his part, leveled a sustained critique aimed at many of the core formulations of Muʿtazilī theology, laying claim to both tradition and reason through the cutting force of logical argumentation. One by one, Shāfiʿī scholars came to embrace Ashʿarī teachings. Similarly, many Ḥanafīs, notably in eastern Iran and Central Asia, turned to the theological lessons of Abū Manṣūr al-Māturīdī (d. 333/944), who promoted a traditionalism in keeping with the broad parameters of what became Ashʿarī theology.

By the time Qazwīnī made his way to Baghdad, toward the end of the reign of the Abbasid caliph al-Mustanṣir billāh (d. 640/1242), the Shāfiʿīs had long enjoyed a prominent place in various institutions throughout the capital. The preeminent Niẓāmiyya madrasa served as a bastion for Shāfiʿī jurisprudence, as stipulated in the endowment of its charter. Several institutions in Baghdad catered to Shāfiʿī scholars, such as the recently founded Sharābī madrasa, where Tāj al-Dīn al-Urmawī taught, endowed by the high-ranking general Sharaf al-Dīn al-Sharābī. But the caliphate, which had significantly diminished in power, if not prestige, also had to balance competing factions to rule over the capital and the surrounding territories.

In the beginning of his reign, the caliph Mustanṣir erected in his name a madrasa, with a towering archway at the entrance, an adjoining hospital for the sick that was supplied with a full pharmacy, quarters overlooking the Tigris, a kitchen for scholars in residence, and a public library that was designed for making copies of unique manuscripts. A senior physician in attendance led the study and practice of medicine. Further in the charter of the endowment, the caliph stipulated posts for head professors to teach Mālikī, Ḥanafī, Shāfiʿī, and Ḥanbalī law, representing the four juridical schools that had long gained the status of Sunni orthodoxy.[29]

In his geography Qazwīnī speaks of the educational complex as the pride of the city, with its wondrous timepiece for telling the hours of daily prayer; it towered over the entrance to the courtyard and was celebrated in verse. Early witnesses describe the large water clock at the portico with hidden levers, cams, weights, and pulleys. Completed in 633/1235~, the ingenious device marked time with an azure dial for the heavens. Across its face ran the Sun and Moon at different phases in measured progression. Two golden falcons signaled the hour with hazelnuts made of stone that dropped through their beaks chiming into two goblets beneath, as one of twelve doors opened.

In the clock at the caliphal madrasa of Baghdad, Qazwīnī would have also recognized the massive clepsydra erected a generation earlier at the entrance to the Great Mosque of Damascus, which he had also visited. That device was maintained and described by Fakhr al-Dīn Riḍwān (fl. 600/1203), a physician and son of the clockmaker who had built it. A parallel design, with a rotating disc for the zodiac tracking the hours and the cosmic movement of the heavens over the course of the year, features in the famed collection of mechanical devices, or *ḥiyal,* by Ibn Razzāz al-Jazarī (d. 602/1206), a renowned engineer, inventor, and courtier from Upper Mesopotamia (figure 2.3). Exciting the interests of caliphs and powerful patrons, the Arabic literature on mechanics and automata included translations of Greek writings from classical antiquity, as well original works with further refinements and ingenious inventions. Jazarī's treatise circulated throughout the period in illuminated manuscripts that made their way into numerous madrasas and royal collections. A vocabulary of wonders inspired an appetite for these intricate devices. Their hidden mechanisms were designed to delight and perplex, but also to be explained, tested, and refined.[30]

The marvelous public clock at the entrance to the Mustanṣiriyya madrasa with its representation of celestial order, the vast library, the attending physicians, jurists, and experts in the Quran and hadith, all lent the complex a lasting authority as a premier institution of learning. Outside the courtyard, the streets bustled. With endowments for professors and students, numerous other madrasas of the city also attracted a steady stream of aspiring talent. Conflicting interpretations, sectarian divisions, and competition over patronage and prestige could consume a good deal of energy, but countervailing forces also contributed a measure of unity and a pride of place.

During Mustanṣir's caliphate, scholars representing various Sunni juridical factions had a hand in running the judiciary, both in Baghdad and in

FIGURE 2.3: A single folio of the castle clock described in Jazarī's book of automata, which parallels the description of the water clock in the caliphal madrasa of Baghdad, from a manuscript copied 755/1354, likely in Cairo.

the provinces. After the death of Mustanṣir, his son al-Mustaʿṣim billāh (d. 656/1258) was installed with the aid of his father's generals, including the powerful Sharaf al-Dīn al-Sharābī, a patron of Shāfiʿī scholars and institutions. Soon afterward Qazwīnī was appointed judge for the provincial township of Hilla, south of Baghdad, known for its population of prominent Shia families.

In his judgeship, Qazwīnī enjoyed the support of the chief judge of Baghdad, Sirāj al-Dīn al-Nahruqallī (d. 654/1256), an influential Shāfiʿī jurist of the day. Two years later, in the spring of 652/1252, after the Ḥanafī judge for the city of Wasit had resigned his post while on pilgrimage to Mecca, Sirāj al-Dīn appointed Qazwīnī to the position. This took place over the initial objections of the powerful chief finance minister, Abū Ṭālib al-Dāmghānī (d. 658/1260), who was a descendant of a family of Ḥanafī judges and administrators. Ḥanafī and Mālikī judges had recently run the judiciary of Wasit, but when Qazwīnī came to the post, it had been stipulated from the head judiciary in Baghdad that Shāfiʿī law would be followed.[31]

Located on the Tigris, in Lower Mesopotamia, Wasit had long been an important administrative center and a source of tax revenue for both the Abbasids and the Umayyads before them. Founded as a garrison city by the Umayyad strongman al-Ḥajjāj (d. 95/714), Wasit took its name, meaning "at the middle," from its location midway between the rival garrison cities of Kufa and Basra.

Mustaʿṣim visited the city with Dāmghānī, his chief finance minister, ostensibly as a pleasure excursion but also to check the local administration and to maintain a caliphal presence in the surrounding provinces. Qazwīnī not only served as judge of Wasit but was also appointed to run the madrasa founded there by Sharābī two decades earlier in 632/1235 on the east side of the Tigris. In this position Qazwīnī was the third in a line of Shāfiʿī jurists to run Sharābī's madrasa in Wasit. Qazwīnī held on to both of these posts until his death in the spring of 682/1283. During his long tenure as a notable of the city, Qazwīnī lived through one of the most turbulent periods of upheaval the region had ever witnessed.[32]

The madrasa and the judiciary sustained their authority through the predictable rhythms of institutional life. Both were strengthened through the force of the pen and the power of the purse. A judge drew a stipend through the central administration of the state treasury, while a madrasa professor collected a salary directly from the pious endowment established for the institution. Qazwīnī received both. He also had a whole coterie of attendants and servants who worked under him.

In addition to the professorship, the charter for the Sharābī madrasa provided income for two *muʿīds*, or teaching assistants, as well as posts for twenty-two *faqīhs*, or jurisconsults. Abutting a congregational mosque, the Sharābī madrasa was also situated next to a recently built *ribāṭ*, which functioned as a kind of lodge for the pious, with space to house mendicants and endowments for a scholar to teach the Quran, another to transmit hadith, and a third to serve as Imam, all of whom were supplied with monthly stipends.

Following the arching passageways of clay brick walls through the turreted gateway of the complex, opened a generous courtyard, flanked by a course of chambers and alcoves, offering ample space for living quarters, rooms for instruction, and a library of books both rare and easily found—all common architectural features of madrasas. The regular teaching curriculum demanded a whole economy of labor—not only in attending to the daily lessons but also in meeting the material conditions of instruction. The city's numerous madrasas produced a steady market not only for farmers, weavers, and carpenters to feed, clothe, and house students, but also for papermakers, bookbinders, and copyists to provide them with the means of instruction and composition.[33]

The judiciary was also a creature of paper. Originally an import introduced into Iraq from Central Asia early in the Abbasid caliphate, paper soon filled all corners of public life. Relatively cheap and easy to produce, paper replaced leather parchment, which could require a herd of sheep for a single tome, and was preferred in quality and price to papyrus, which was imported from the Nile River delta.

As an official organ of the state, a central facet of the judiciary was the *dīwān*, the registry that preserved the written records presented before the court. Judges consulted wills, deeds, endowments, sworn affidavits, lists of evidence, as well as notebooks of local precedent, compendiums of legal rulings, and encyclopedias of positive law. Judges also came to master rules of decorum, both oral and written. These were memorialized in collections that stipulated the procedural guidelines for running the judiciary. Often circulated in the frameworks of differing juridical schools, writings on manners and etiquette, as in the refined comportment or *adab* required of a judge, offered guidance for administering the courts.[34]

The judiciary supported several officials who oversaw that the *majlis al-ḥukm,* or the court assembly, followed measured procedure. On entering the assembly, the judge would recite standard supplications to God, seeking guidance and forbearance in the execution of justice. Overseeing the assembly were various authorities, such as the gatekeeper and chamberlain, who held order and kept an eye out for any rabble-rousers in attendance. There were jurist adjuncts assigned to the judge to assist with gathering and reading materials, notaries to certify documents, and scribes to register written records. Here too were translators who sought to ensure that litigants and defendants not fluent in the language of the court would be both properly debriefed and correctly understood.

From high-ranking nobles to illiterate peasants, the judiciary bore witness to hopes and aspirations as well as petty disputes and family feuds across all walks of life. Christians, Jews, and Zoroastrians often sought recourse to courts run by Muslims, which could, depending on the circumstance, be more favorable than their own communal institutions of law. Likewise, as in other realms of public authority, women's voices can be heard, particularly in such matters as inheritance, property disputes, and domestic quarrels. Women also developed rather novel means of navigating legal frameworks that in obvious ways privileged freeborn Muslim men.

This being said, above all the written record preserves and sanctifies the ideal of a social order dominated and controlled by an educated male elite. The court, like the madrasa, derived its authority through the routine of men in communion with each other. As with all bureaucratic forces, the power of the judiciary was also occulted in arcane layers of jargon and expertise, which sought, through measured reason, to both obtain and produce predictability and facticity.[35]

The terrain of quotidian affairs, of hardship and relief in the normal unfolding of custom and habit, at any minute could be torn asunder with awe-inspiring signs of divine providence. Such is the promise of saintly wonders.

Wandering through the pages of Qazwīnī's geography are countless sages, addressed as *aṣḥāb al-karāmāt al-ẓāhira,* masters of manifest miracles.

As a term of art, the word *karāma* gave theologians a way to distinguish saintly wonders from the inimitable miracles, or *muʿjizāt,* performed by the last Prophet Muhammad and the line of prophets stretching back to Adam who preceded him. In terms of effect, the two could look very similar. Casting out demons, healing incurable illnesses, filling hats with gold, pulling dirhams out of thin air, leaving footprints of dinars, taming wild beasts, producing fruit in the dead of winter, attaining angelic visions, and having dreams with visionary knowledge—are all among the many signs and wonders, *āyāt wa-ʿajāʾib,* that flow through Qazwīnī's geography. From one place to the next run the storied adventures of such masters as Dhū l-Nūn (d. 248/862) of Egypt, the peripatetic Sahl al-Tustarī (d. 283/896), the Baghdadi circle of al-Muḥāsibī (d. 243/857), his disciple al-Junayd (d. 298/910), and the miraculous feats and ecstatic execution of al-Ḥallāj (d. 309/922), famed martyr of divine love.

In addition to the masters of old, Qazwīnī celebrates more recent miracle workers. There is the Persian sage Shihāb al-Dīn al-Suhrawardī (d. 587/1191), whom Qazwīnī refers as a *ṣāḥib al-ʿajāʾib wa-l-umūr al-gharība,* "a master of wonders and extraordinary phenomena." Suhrawardī developed a philosophy of illumination that taught how, through the discipline of the soul, a sage could learn to produce wonders and harness hidden forces in the cosmos. Efforts to revive ancient wisdom occupy Suhrawardī's philosophy, as do references to the prophetic teachings of Hermes, which promise to unlock the secrets of nature. For his part, Qazwīnī dutifully notes how much Fakhr al-Dīn al-Rāzī venerated the writings of Suhrawardī. Qazwīnī also records accounts he heard in his hometown and beyond of those who had witnessed Suhrawardī work wonders.[36]

By this period, belief in saintly miracles had long been a formalized aspect of Ashʿarī dogma, featuring not only in theological discussions but in creeds—often the first teachings that children memorized—outlining the basic tenets of orthodox faith. There were, to be sure, naysayers, such as certain factions among the philosophers as well as among the Muʿtazila, the group of theologians from which Abū l-Ḥasan al-Ashʿarī, and with him his disciples, famously diverged.[37]

Following good Ashʿarī teachings, Qazwīnī is quick to ridicule those who deny the reality of saintly miracles, by citing the story of a village in

Syria. The residents had long mocked such miracles as nonsense, until one day a wild, hungry beast came to town and they were all but helpless against it. Only then did they rush quickly to the local saint, Jābir, a master of wonders, beseeching his aid. He replied coolly, before saving the village, "Where now are the people who deny the miracles of saints?" Such stories feature frequently in the memorials of saints' lives and in handbooks of Sufi piety and practice. Qazwīnī read this material avidly and knew well the multivolume collection by Abū Nuʿaym al-Iṣfahānī (d. 430/1038), with its numerous accounts of saintly adventures and pious teachings.[38]

Not only did Qazwīnī read the stories of the *awliyāʾ*—the helpers, protectors, or friends of God, as sages were generally known—but he also met several such masters, both in Iran and during his travels. There was Nūr al-Dīn al-Jīlī, whom Qazwīnī encountered in his youth, when in 614/1217~ the venerable shaykh moved to Qazvin. With a pure countenance, a thick beard, and a towering presence, Nūr al-Dīn commanded attention. Anyone, even a king, who saw him could not but feel awe at his dignity. In addition to accounts of various miracles, Qazwīnī notes that Nūr al-Dīn wrote about the wonders of his ecstatic states and his otherworldly visions of angels, paradise, and hellfire, the condition of life after death, and the unique powers controlled through the names of God and Quranic verses.

While the renown of Nūr al-Dīn did not extend beyond northern Iran, the same cannot be said of Ibn al-ʿArabī (d. 638/1240). Qazwīnī met the famed mystic from Murcia in 630/1232~ while residing in Damascus. The writings and teachings of Ibn al-ʿArabī would come to wield an incredible influence in the development of piety and philosophy.[39]

Qazwīnī was not merely an observer of this saintly world. He also participated actively in networks of mystical learning, cultivating relations with gnostics and sages. In Qazvin he studied under Muʿīn al-Dīn Ḥasanwayh, a respected Sufi shaykh from among the notables in the city. Through these studies, Qazwīnī received an *ijāza*, which granted him the authority to transmit all the received learning that Muʿīn al-Dīn had acquired. Muʿīn al-Dīn also produced *fawāʾid*, pious teachings in the service of the poor and for mendicant Sufis. In Qazwīnī's estimation, Muʿīn al-Dīn was renowned as one of the learned masters of *tajrīd*, capable of stripping his soul from attachments to the body through ascetic practice. Qazwīnī also confirms that Muʿīn al-Dīn could speak directly with jinn and successfully perform exorcisms.[40]

In addition to fleeting references in his geography, Qazwīnī redacted details of his life story in a personal handbook of all the teachers with whom he had studied, which he then passed on to Kamāl al-Dīn Ibn al-Fuwaṭī (d. 723/1323) of Baghdad. Such personal collections recorded the individual relations and curricular studies that a student developed, often in the fields of the traditional sciences, where oral transmission was at a premium.

These collections could then be used for granting teaching licenses to others, as students developed into instructors. To those with an eye for memorializing notable authorities, such notebooks offered a treasure trove of historical detail. Qazwīnī's handbook of teachers is all but lost, save for a few scattered references that Ibn al-Fuwaṭī makes in the course of producing his own vast biographical compendium. Ibn al-Fuwaṭī's collection focuses largely on the lives of notables who lived right before and after the Mongol invasions; it is organized alphabetically by *laqab*, or honorific, such as Qazwīnī's title *ʿImād al-Dīn*, "pillar of the faith." For his part, Ibn al-Fuwaṭī went on to oversee the library housed in the observatory complex in Maragha, run by Naṣīr al-Dīn al-Ṭūsī. He then later returned to Baghdad to run the library of the Mustanṣiriyya madrasa in the city famed for its wondrous clock.

To produce his memorial of administrators and rulers, scholars and saints, Ibn al-Fuwaṭī reached out to notables throughout the region, gathering personal documents, local histories, and other biographical compendiums. Only a small fraction of Ibn al-Fuwaṭī's collection survives, covering honorific titles from *ʿIzz al-Dīn*, "might of the religion," through *Muwaffaq al-Dīn*, "prosperous of the religion." Among the nearly six thousand entries that do survive is one of the earliest biographical accounts of Qazwīnī, whom Ibn al-Fuwaṭī refers to as *shaykhunā*, "our teacher."[41]

As with most of these lives, Ibn al-Fuwaṭī's entry for Qazwīnī is rather laconic. It registers Qazwīnī's full name with customary titles and honorifics and notes his esteemed ancestry as a descendant of Anas ibn Mālik (d. 93/712), one of the early Companions of the Prophet in Medina. Mention is made of

Qazwīnī's appointment as judge in Hilla and the date when he was transferred to the judiciary of Wasit on the order of the chief judge in Baghdad and over the initial resistance of the finance minister, Dāmghānī.

Ibn al-Fuwaṭī also mentions that Qazwīnī had a *taṣnīf*, or compilation, and that he dispatched from Wasit an *ijāza* for Ibn al-Fuwaṭī. The written collection in question is not specified. The word *taṣnīf*, from *ṣannafa*, "to separate into categories," often carries the sense of a work organized thematically by topic, and this may well have been a reference to *Wonders and Rarities*, whose title page announces that it is one of Qazwīnī's *taṣānīf*, or compilations. Ibn al-Fuwaṭī also cites Qazwīnī's collection of teachers a handful of times in the course of his own biographical dictionary. The entry for Qazwīnī notes that he was born in Rabīʿ II 598/January 1202. After Qazwīnī passed away in Muḥarram 682/April 1283, his body was brought from Wasit to Baghdad for burial. The birthdate, as well as much of the other biographical information, may well have been drawn from Qazwīnī's personal handbook of authorities that he had entrusted to Ibn al-Fuwaṭī. No direct mention is made of a natural history or a geographical compendium of cities or towns, though other biographical collections and historical compendia reference both.[42]

By piecing together these other references, a fuller, if nonetheless still fragmentary, picture emerges. In his youth Qazwīnī studied under a local jurist in Qazvin, who granted him an *ijāza* certificate that allowed him to pass on all of the oral and written reports that his teacher had amassed; this included the canonical hadith collection of Ibn Mājah (d. 273/887), another luminary of the city. As a professor, Qazwīnī also held assemblies of learning where he taught and transmitted his own writings to others, including specifically his natural history, through the conventions of *ijāza* certification. Among his students were not only Sunni authorities, but also prominent Shia. In addition to the judgeship of Wasit, Qazwīnī served as the head professor of the Sharābī madrasa. He remained in these positions until his death. He was buried in the Shūnīzī cemetery in Baghdad, a resting site for many notable scholars and officials, including his colleague Tāj al-Dīn al-Urmawī. Qazwīnī was renowned as chaste, honest, and upright, an excellent scholar and righteous man, who had beautiful penmanship.[43]

Several more details can be gleaned from the early manuscripts of the two works that survive of Qazwīnī's writings. In both, Qazwīnī claims an Arab ancestry that stretches back to Anas ibn Mālik, known as the servant of the

Prophet. As a descendant of Anas ibn Mālik, Qazwīnī publicly took the *nisba* Anṣārī, from the helpers of Medina who embraced the Prophet and the early believers when they fled the persecution that they had faced in Mecca.

Qazwīnī was also proud of his ancestors from Qazvin and notes that he came from a family of jurists who had settled in the city. This connection is affirmed through the additional *nisba* Kammūnī, "the cumin seller," a name that Qazwīnī explains in passing came from his forefather five generations earlier, Abū l-Qāsim Ibn Hibbat Allāh al-Kammūnī, who had first settled in Qazvin and whose descendants were notables in the city. From his numerous references to Persian poetry and his youth living in Iran, it is also clear that he was a Persian speaker, though he makes no direct mention of this.[44]

Very little can be said about what it was like to live a life far from home, though Qazwīnī makes the passing observation that the people of Iraq were unique for their hatred of *ghurabā'*, strangers, which they particularly directed at Persians. And other than relating a comment that Urmawī had shared with Qazwīnī on the successful policies of Mustanṣir that kept the Mongols at bay, Qazwīnī offers very little insight into his own experiences with the invasions.

In this Qazwīnī had much in common with the scattered voices that document the lives of his generation. These writings largely take as accepted knowledge the coming onslaught and the profound aftershocks that followed. A mere four years into Qazwīnī's term as judge and professor, Baghdad had fallen, the caliph and his family had been killed, and the walls of Wasit had been breached.[45]

3

ASTRAL POWER

PIECES OF METAL CAME RAINING DOWN amid bolts of lightning, with fire streaming across the sky. Sharp as arrows, they pierced deep into the ground, in the land of the Turks and along the southern rim of the Caspian Sea. Heavier forms could also be seen falling from the heavens, including a huge chunk of iron, weighing over five hundred pounds, that smashed into the land of Juzjan on the high steppes of the Iranian plateau. Like a ball slamming against a wall, it rebounded twice before finally plowing into the ground, setting off a terrifying boom.

After much effort, a large splinter of the thick mass was finally broken off and sent to Maḥmūd of Ghazna (d. 421/1030), the Turkic warrior king and conqueror of the Gangetic plain. An unsuccessful attempt was made to forge a sword from the ferrous metal, which could not be worked with normal tools. Ibn Sīnā documents these phenomena, which occurred within his own lifetime, in his meteorological section of natural philosophy in the *Cure,* where he treats the formation of metals, minerals, and mountains, as well as other such topics as clouds, lightning, earthquakes, shooting stars, comets, rainbows, halos, and the cataclysmic force of floods.

Ibn Sīnā notes that he attempted to smelt stones that had fallen from the sky, which for him offered insight into how metals could form in the upper atmosphere through extreme heat, following the general framework of Aristotelian meteorology. For Qazwīnī, who cites Ibn Sīnā's authority on these strange events, such rarities offer examples of extraordinary natural phenomena. Qazwīnī includes references to these falling objects in the opening to *Wonders and Rarities,* where he provides an extensive definition of the term *gharīb*, one of the primary categories of the strange, extraordinary, and uncanny that structure his natural history.[1]

In addition to Ibn Sīnā's meteorites, other topics generally treated in the classical Arabic reception of Aristotelian meteorology feature in Qazwīnī's introductory treatment of rare phenomena, such as shooting stars, comets, strange weather patterns, and earthquakes. In Arabic, Aristotle's *Meteorologica* was first known through the translation by the Baghdadi philosopher and convert Yaḥyā ibn al-Biṭrīq (fl. 204/820); it was entitled *al-Āthār al-ʿulwiyya*, literally "upper influences" or "atmospheric signs." As a field of natural philosophy, the discipline of meteorology probed the interaction between the Earth and its climate.[2]

Many of these matters had astrological implications, as treated extensively by Ptolemy in his *Tetrabiblos*, a study in four books on the effects and influences that celestial and climatic forces had on the movement of the Earth and the course of human life. This study was first translated into Arabic by Ibn al-Biṭrīq's father at the request of the Abbasid caliph al-Manṣūr (d. 158/775). As a work of predictive astrology, the *Tetrabiblos* complemented Ptolemy's mathematical geography and his writings on theoretical and mathematical astronomy, represented chiefly in the *Almagest* and the *Handy Tables*—which also enjoyed Arabic translations. Where Aristotle's *Meteorologica* gave physical explanations to climatic and terrestrial phenomena, Ptolemy's *Tetrabiblos* offered astrological models for reading the signs of the heavens and the climatic influences that conditioned human life and society across the globe.

Far from a marginal body of knowledge or set of practices, judicial astrology had long occupied an important place in the halls of higher learning and in the assemblies of mighty rulers. As with other leading authorities, Qazwīnī viewed astronomy and astrology to be intertwined. He also knew, given the complexity of the calculations, that very few individuals could claim true mastery of the field.

The recondite nature of astrological knowledge lent it an enduring allure. In contrast to the strident condemnation of astrology in the course of modern religious reform, scholars, religious authorities, and the ruling elite, spanning regions, dynasties, and sectarian divides, turned to the predictable movement of the heavens as a rational basis for explaining the influence of the stars on health and the course of human affairs. There were certainly philosophical debates over astrology's speculative character, just as there were concerns in theological and traditionalist circles that ascribing predictive forces to the stars might challenge God's ultimate power. By Qazwīnī's day, though, the argument had been so thoroughly turned around that denying the force of the stars risked imputing limitations to divine providence. The heavens proceeded through divine decree and beneficence, all balanced in perfect harmony.[3]

The abiding interest in judicial astrology also intersected with a view that occult forces in the physical world could be harnessed for sundry purposes. Just as the world could be read for its therapeutic and predictive powers, it could also be manipulated by alchemical processes or spells and incantations. The movement to explain in scientific terms how physical forces could produce extraordinary phenomena forms part of broader developments in the authoritative traditions of scholastic philosophy. Qazwīnī's view of the interconnected powers governing the cosmos guides his compendium of wonders and builds on earlier efforts to rationalize magic and miracle.

Along with the study of matter and the movement of bodies in time and space came a concern for the elemental forces that governed and shaped the physical world. These forces and influences could be seen in obvious ways through the power of the Sun to nourish plants and the effect of the Moon on the cycles of the tides. How far such celestial influences extended and the extent to which they could be predicted were matters of considerable inquiry. The relationship of the celestial phenomena of stars, planets, and comets with the terrestrial, sublunar world of the Earth had long occupied Muslim intellectuals, who inherited, reformulated, and extended the astral learning of antiquity.

Aristotelian natural philosophy divided the study of the cosmos between the phenomena of the upper and lower spheres—referred to in Arabic as the *ʿulwiyyāt* and the *sufliyyāt*—treating the celestial and terrestrial as separate and yet interconnected domains. In *Wonders and Rarities,* Qazwīnī follows this division, advanced by the likes of Ibn Sīnā and Rāzī, as he proceeds to account for the entirety of existence from the sublime movements of the heavens above through the various forms of life and matter supported on Earth below.

By Qazwīnī's day the mathematical study of the universe and its structure was referred to as *ʿilm al-hayʾa*, as taught by his teacher Abharī and others, denoting the science of the configuration or form of the cosmos, which included both the upper celestial regions of the heavens and the lower terrestrial realm of the Earth. As a discipline focused on cosmic order, *ʿilm al-hayʾa* combined the fields of mathematical astronomy and geography.[4]

Where judicial astrology, with its focus on the predictive power of the stars, fit into the schema of the rational sciences, however, had long been a delicate matter. Countless caliphs and courtiers, physicians and philosophers promoted and pursued the study of the judgments of the stars, the *aḥkām al-nujūm*, as the discipline of judicial astrology was generally known. Yet the ethical and epistemic status of astrology also received scrutiny. Many Muslim theologians took issue not only with the suggestion that God's will could be predicted, but also with the implication that the stars themselves could produce causal effects in the world, independent, as it were, of God's omnipotence. At a theological level, one of the primary concerns proceeded from the problem of causation and the question of whether there were secondary causes apart from God, a topic that was itself highly debated.[5]

For many, the scientific basis of astrology, its rules, and its results appeared to fall beyond the domain of demonstrable knowledge, which could be verified or replicated. This is the critique developed by Ibn Sīnā, who takes issue with what he saw as the speculative and capricious nature of the field. To be sure, Ibn Sīnā does accord judicial astrology a place among the rational sciences, as one of the applied subfields of *al-ḥikma al-ṭabīʿiyya*, or natural philosophy, situated after medicine. In his view, the study of the judgments of the stars was a conjectural science, *ʿilm takhmīnī*, which he locates alongside physiognomy, dream interpretation, the study of talismans, amulets, and alchemy.

For Ibn Sīnā, astrology seeks to divine the conditions for the cycles of the world, kingship, kingdoms, regions, and nativities, the length of lives, the propitious timings for events, and various questions of the unknown through the movements of the stars, following their correspondence with each other, with regard to the degrees of the zodiac, and with respect to the Earth. In his classification of the sciences, Ibn Sīnā sees astrology as separate from the astronomical study of the cosmos, *ʿilm al-hayʾa,* which he identifies as one of the principal or theoretical fields of mathematics.

As with alchemy, Ibn Sīnā develops a sustained critique of astrology, not only questioning its demonstrative validity, but also doubting whether Ptolemy, the author of the *Almagest,* could have actually produced works on judicial astrology, such as the *Tetrabiblos,* which Ibn Sīnā ridicules as wildly speculative. Given the complex forces at play, Ibn Sīnā concludes, it is impossible to derive with any certainty a scientific basis for predicting how celestial influences will actually unfold.[6]

Qazwīnī, in contrast, follows a tradition much more amenable to judicial astrology. Where Ibn Sīnā derides the identification of particular planets and the signs of the zodiac with good or bad fortune, Qazwīnī dutifully outlines a full array of astrological associations concerning the predictive nature and significance of the stars. In his treatment of the celestial realm, Qazwīnī draws on the mathematical models, measurements, and diagrams in Ptolemy's *Almagest.* But he also turns frequently to the authority of *munajjimūn,* masters of the stars, for their astrological descriptions of the qualities of the planets and constellations. Moreover, astrological and talismanic imagery formed the basis for the illustrations accompanying Qazwīnī's treatment of the heavens (see figure 2.1). While the term *munajjim* often referred to practitioners of judicial astrology, the title could also evoke the learned study of the stars more broadly. So while Qazwīnī turns to *munajjimūn* for astrological interpretations of the planets, he also cites their authority when identifying the names and titles of the various stars drawn from the Hellenistic traditions of astral learning, which he contrasts with the names and constellations used among the Arabs before the rise of Islam.[7]

Although there were meaningful efforts to separate mathematical astronomy from judicial astrology, the lines of demarcation could easily blur. Foremost, the science of the stars, *ʿilm al-nujūm,* as astral knowledge was also known, could encompass the observational and theoretical models of astronomy alongside the practical methods of reading the stars for their astrological significance.

For Rāzī, the predictive utility of astral learning was a question that he pursued throughout his intellectual career. Not only did he champion various astral arts as licit within the frameworks of religious law, but he also produced an influential guide to astral and talismanic magic, *al-Sirr al-maktūm* (The hidden secret). In this Arabic work, produced for a royal patron in the court of the Khwārazm-Shāh, Rāzī taught how to harness the angelic powers associated with the celestial spheres of planets and stars. Rāzī was intimately familiar with Ibn Sīnā's ambivalence toward astrology; he also recognized that, as with knowledge of medicine and drugs, judicial astrology was not based on demonstrative proof. But Rāzī also defended astrology, arguing that like medicine its benefits could be obtained through continual trial and observation.[8]

Similar arguments can be traced throughout the literature in the field, such as the efforts by Kūshyār ibn Labbān to elevate astrology as an applied science in his influential introduction to the discipline, modeled directly on Ptolemy's *Tetrabiblos*. Kūshyār argues that a proper reading of the stars and their movements can help to understand the significance of comets and shooting stars and to foresee earthquakes and floods. By observing and predicting the position of the stars, a master of the discipline can cast horoscopes and calculate the length of a life and the rise and fall of a dynasty, as well as determine the auspicious times to carry out such mundane matters as bathing, tailoring, buying and selling, consecrating buildings, sowing seeds, harvesting crops, ensuring conception, treating illnesses, contracting marriages, and engaging in war.[9]

As for Rāzī, many of his writings express an ambivalence and at times even notable antipathy toward the predictive claims of judicial astrology, particularly in the course of his public disputations and in the field of Quranic exegesis. Yet the emanating power of the soul guiding the philosophical system that Rāzī had inherited from Ibn Sīnā also accorded a natural basis for explaining and harnessing the astral forces of the heavens. As Rāzī knew well, a close reading of Ibn Sīnā's emanationism demanded a recognition of the celestial influences that structured the cosmos.

Unlike Ibn Sīnā, in his own classification of the sciences Rāzī situates the study of astronomy, denoted as the science of the cosmic structure, *ʿilm al-hayʾa*, directly beside judicial astrology, which in turn precedes geomancy or terrestrial astrology, *ʿilm al-raml*, and the art of summoning jinn and demons through incantations, *ʿilm al-ʿazāʾim*, all of which culminate in the divine science of metaphysics, *ʿilm-i ilāhiyyāt*. For Rāzī, it would be foolhardy to ignore the practical power obtained through astral knowledge, for, as he affirms in his own handbook of astral magic, the master of these arts can learn to conquer a foe without ever directly touching him.[10]

The Mongols certainly had no qualms with directly engaging their enemies. But as they marched on Baghdad, their forces were armed not only with swords, bows, battering rams, catapults, and kegs of gunpowder. There also came soothsayers and astrologers who could interpret dreams, converse with the spirit world, heal the sick, change the weather, and foresee events to come.[11]

With their divinations, the horse-riding nomads brought along a belief in *möngke tenggeri*, the everlasting heaven, not only as a source for good fortune and success, but as a preordained basis for divine rule and dominion over the entire world. In the course of their conquests, the Mongol armies swept up local administrators, secretaries, tax collectors, market inspectors, ministers, princes, physicians, ascetics, jurists, judges, linguists, and translators, as well as alchemists and fine craftsmen. Particularly prized were those who, like their own divines, could augur the course of events to come from the heavens and who, through prayers and supplications, could conjure forces of angels and command armies of jinn.[12]

The empire founded by Genghis Khan stretched across Eurasia, from Mongolia, Turkistan, Tibet, the Korean Peninsula, north and west China, down into Vietnam, up through Siberia into Russia. All of Iran fell. With the armies of Hülegü, the grandson of Genghis Khan, came Iraq and Anatolia. Their westward movement through the Levant was checked only by the Mamluk armies of Syria and Egypt. The kingdoms conquered and territories submitted constituted the largest land-based empire the world had ever seen, all acquired with breathtaking speed. Countless Muslim elites joined ranks in the expansive state and the successor dynasties that followed. Often, but not always, those who acquiesced and yielded to Mongol authority were safe from the severity of their wrath.

Genghis Khan and his descendants made it an imperial policy to give protection to those who submitted to them, whether Muslim or Jew, Christian or Buddhist. Many of the conquered came to embrace the nomad invaders and flourish through their patronage. The high-ranking Persian functionary ʿAlāʾ al-Dīn ʿAṭāʾ Malik al-Juwaynī (d. 681/1283), born after the first waves of Mongol invasions across Central Asia, lived his entire life in their entourage. As patron, Juwaynī would play an important role promoting Qazwīnī's *Wonders and Rarities*. In due course Qazwīnī would come to celebrate Juwaynī's magnificence in the opening to his compendium of natural history.

As Juwaynī rose through the ranks to serve as governor over Iraq, he also found time to celebrate in flowing Persian prose the new regime founded by Genghis Khan. Dedicated to Hülegü, Juwaynī's *Tārīkh-i jahān-gushā* (History of the world conqueror) survives as the earliest history of the invasions written by a Muslim from within the Mongol administration. In it Juwaynī offers an ideological justification for the conquests, which in both might and expanse easily surpassed the legendary feats of Alexander the Great (d. 323 BCE). Not only did Juwaynī come to see the Mongols as instruments of divine providence, but he also advanced their claims to universal sovereignty through their policies of broad-minded religious impartiality. As Juwaynī saw it, Genghis Khan was an idol worshiper who followed neither religion nor confession. But the statesman also affirmed that the emperor expressed neither partisanship nor preference toward any one faith. The warlord "honored, respected, and venerated the learned and ascetics from all sects," recognizing, in Juwaynī's view, that such equanimity was a step toward "the divine presence of the Almighty Truth." This ideal of universal sovereignty would have a lasting impact on conceptions of kingship in the following centuries.[13]

While the Mongol elite laid claim to religious impartiality, Juwaynī also knew them to be deeply concerned with the predictive power of the planets, scrupulously seeking propitious timings for their affairs and concluding nothing until their astrologers had given approval. Throughout the conquests of Central Asia and Iran, Juwaynī references the role that astrologers played as they traced the movement of beneficent and maleficent stars to determine auspicious days for battle, to predict insurrections, and to cast natal horoscopes.

Juwaynī identifies these holy men, seers, or healers by the Turkic *qam*, though he also refers to them using the Arabic *munajjim*. Through their knowledge of magic and their power to subjugate demons, they sought to predict the future and heal the ill. Though he expresses a good deal of disdain, Juwaynī does not reject these claims out of hand, for he also acknowledges that it was quite likely that some of these soothsayers were indeed in communion with maleficent spirits. But the Mongols were not alone in their appeal to extraordinary powers, a point Juwaynī makes throughout the course of his history. Many of their Muslim adversaries sought out astrologers and those who could harness the strength of jinn and the aid of angels.[14]

In the course of the conquests, the Mongols were quick to enroll learned Muslim sages who had obtained mastery of the stars and the forces of the unseen. Long before the invasions, Muslim religious scholars had cultivated astrological knowledge and had developed techniques for summoning demons, commanding jinn, and propitiating angels. Such learning was often highly prized.[15]

But these fields also extended well beyond the ruling elite. Qazwīnī was intimately familiar with numerous authorities renowned for occult knowledge; and he was conversant with the body of practical guides that explained how to subjugate unseen forces. In the East, one of the better-known handbooks of spells and incantations was the Arabic collection *al-Shāmil fī l-baḥr al-kāmil* (The comprehensive compendium on the entire ocean), composed by Abū l-Faḍl Muḥammad al-Ṭabasī (d. 482/1089), a resident of Nishapur. Both Rāzī and Qazwīnī read Ṭabasī's recipes for spells, talismans, and incantations; the work survives in multiple manuscript copies and also circulated through Persian translations. Rāzī drew extensively on Ṭabasī as an authority in his own collection of talismanic and astral magic. Qazwīnī knew Ṭabasī's handbook to be a large collection containing specific methods for commanding jinn, with descriptions of various incantations and the conditions for their use.[16]

Often activated through appeals to God's omnipotence, the power of divine and angelic names and prayers, frequently derived from Hebrew and Aramaic, as well as the efficacious recitation of the Quran, and pious supplications, Ṭabasī's talismans and spells open a world of demons, jinn, and angels waiting to be harnessed for countless aims, some mundane, others cosmic. Remembered as a pious ascetic and a Sufi who composed numerous works, Ṭabasī was trained in Shāfiʿī law and Ashʿarī theology. He delivered lectures in the Niẓāmiyya madrasa, established by the powerful Saljuq vizier Niẓām al-Mulk, in the city of Nishapur.

In this regard Ṭabasī was not a marginal figure. His standing can be seen in his patron Abū l-Barakāt al-Furāwī (d. 549/1155), who requested that

Ṭabasī produce a handbook of spells, as well as in his pupil and primary transmitter of his works, Abū l-Qāsim al-Qāyinī (d. 547/1153). Both were notables of Nishapur. Abū l-Barakāt was a Shāfiʿī legal scholar from the prominent Furāwī family, which counted numerous religious authorities among their ranks. Qāyinī was also a trained Shāfiʿī jurist and a venerated scholar of hadith and for forty years he headed a group of Sufis in a monastic lodge outside Herat. As Ṭabasī also makes clear a host of earlier Sufi masters had preceded him in the discipline of summoning forces from the unseen. According to Qazwīnī, it was in Ṭabasī's capacity as a respected authority that Ghazālī sought out his assistance to conjure jinn.[17]

Ṭabasī's incantations draw heavily from pious prayers and the divine speech of the Quran. Hadith collections commonly discuss the legitimacy of Quranic charms under the rubric of medicine. This forms part of a larger set of apotropaic and prophylactic uses of the Quran, in both oral and written forms, many of which can be traced back to the earliest formation of Islamic ritual devotion.

These practices include recited charms as well as written amulets. The consumption of Quranic verses was also a largely normative means of drawing upon the power of the Quran, as the very word of God, into the body. This generally takes the form of writing particular verses known for their special properties, usually chosen for their semiotic content, onto a dish or a piece of paper, often using saffron, and then ingesting the verses after they dissolve in water. In his treatment of protective amulets and the directions for Quran ingestion, referred to here as *nushra,* Ṭabasī gives several recipes for consuming the Quran that were in wide circulation across the region.[18]

The traditions of occult learning and inquiry represented by Ṭabasī continues with Rāzī's handbook of practical magic. They are also fully expressed in the writings of Sirāj al-Dīn al-Sakkākī (d. 626/1229). Like Rāzī, Sakkākī was a leading scholar from Khwarazm who found patronage in service to the Khwārazm-Shāh. Both were also renowned for their handbooks on rhetoric and style drawn from classical Arabic poetry and the Quran.

Sakkākī's famous compendium *Miftāḥ al-ʿulūm* (Key to the sciences), which offers a systematization of Arabic morphology, syntax, and literary rhetoric, shows affinity with Rāzī's *Nihāyat al-ījāz* (The utmost of brevity), a study of *bayān,* eloquence, within the Quran. Both works enjoyed a considerable presence in the curricular development of madrasa education. Rāzī and Sakkākī also developed a reputation for the promotion of occult learning, particularly in the form of incantations and astral talismans. This confluence of interests moves beyond the famous prophetic hadith that "eloquence has its share of enchantment [*inna mina l-bayān la-siḥran*]." The force of *bayān* as divine speech, exposition, and eloquence taught to humanity is itself a theme at play in handbooks of talismanic magic. The affective power of speech fits well with the theophany of the soul developed in Avicennan natural philosophy, with its ties to the logical study of poetics as a means of influencing others through the force of language.[19]

For his treatment of talismanic magic and the science of summoning jinn, Rāzī drew extensively from Ṭabasī's collection. Sakkākī also read Rāzī's manual of practical magic closely. Although in his own Arabic handbook Sakkākī does not appear to cite Rāzī by name, parallels abound. Moreover, the codicological record suggests that Sakkākī transcribed Rāzī's *Hidden Secret* by hand, in a manuscript that served as the basis for further copies. In contrast to Rāzī, however, Sakkākī lived to witness the initial Mongol invasions and the collapse of the Khwārazm-Shāh dynasty that he once served. Like many other high-ranking officials, Sakkākī found a place within the Mongol administration, entering into the service of Chaghatay (d. 642/1244~), the second son of Genghis Khan. Chaghatay came to rule over large sections of the territory his father had conquered from the Khwārazm-Shāh in Central Asia.[20]

According to a famous account, Sakkākī rose in the ranks of the Mongol court through his renown for occult powers and his mastery of the stars. One day while Sakkākī accompanied the royal entourage, Chaghatay saw a flock of cranes fly overhead. As the khan turned for his bow, Sakkākī asked which of the cranes the emperor wanted to strike. Chaghatay indicated, "The first, the last, and one of those in the middle." So Sakkākī drew a circle on the ground, recited a spell, and then merely pointed. Instantly the three cranes fell to the earth. The emperor raised a finger of astonishment to his mouth in awe. So it was that Sakkākī drew close to the emperor's inner circle and succeeded to dislodge for a time Chaghatay's powerful minister, Ḥabash ʿAmīd, with a rather adroit astrological reading. But Sakkākī's favor with the emperor did not last long and through the machinations of the disgraced minister, Sakkākī was imprisoned and left to die.[21]

Like Ṭabasī, Sakkākī was celebrated as a skilled enchanter, referred to as a *muʿazzim*, one who could conjure jinn, demons, and angels through *ʿazāʾim*, spells and incantations. Sakkākī's Arabic collection takes a very similar title as Ṭabasī's earlier manual, *The Comprehensive Compendium,* and their handbooks often circulated within the same manuscripts. Sakkākī offers practical instructions for how to summon planetary forces and details means for subjugating demons, jinn, and angels, through the use of talismans, seals, and magic diagrams. He also includes an array of sources including Arabic Hermetic teachings and incantations drawn from Sanskrit Tantric magic. Like the collections by Ṭabasī and Rāzī, Sakkākī's writings on occult learning also circulated in Persian.

Sakkākī's renown as a master of Arabic letters grew tremendously. He was also remembered during his life and after his death as an authority on the extraordinary sciences and the wondrous arts, *ʿulūm-i gharība wa-funūn-i ʿajība.* The Mamluk physician and alchemist of Cairo, Ibn al-Akfānī (d. 749/1348), hails Sakkākī's handbook of spells in his influential classification of the sciences as "a work of significant standing, whose benefit is manifestly of great utility, but whose methods are of extreme difficulty." Notably, Ibn al-Akfānī catalogues Sakkākī's collection under the science of talismans, located next to the *Ghāyat al-ḥakīm* (The aim of the sage). Produced in al-Andalus, this influential handbook of astral magic circulated in the name of Abū l-Qāsim Maslama al-Majrīṭī l-Qurṭubī.[22]

Alongside magic diagrams, talismanic charms, mysterious cyphers, and seals, Sakkākī draws on sacred phrases taken from the Quran, the names of God, and prophetic hadith. The supplicant is often asked to watch the heavens for auspicious timings, following astral and planetary forces with incantations to bind spirits. Recipes call for the invocation of strange demonic names, the sacrifice of various animals, suffumigation with incense and medicaments, the use and placement of ritual tools, the writing and burning of spells and secret scripts over sacred flames, all while seated in protective circles and diagrams, generally referred to as a *mandal,* the mage's circle (see figure 9.4). Practices of summoning demons and reading stars could be justified through a philosophical system that rationalized extraordinary phenomena as an integral part of creation woven into the fabric of the cosmos. Such beliefs shaped the vocabulary of both Sufis and messiahs, who could mobilize armies and drive revolts.

Guides by the likes of Qurṭubī, Ṭabasī, Rāzī, Sakkākī, and many others circulated widely, traversing courts and occupying scholars with promises to open up a universe of extraordinary forces and hidden powers. In Qazwīnī's day, powerful patrons commissioned illuminated manuscripts with secret characters, diagrams of talismans and magic circles, and paintings of jinn and demons. Across their pages march armies of jinn and legions of angels, called upon to perform wondrous acts and marvelous deeds. Although a good deal separated them, Qazwīnī's natural history, with its interest in talismans, hidden properties, alchemy, and astrology, shared much with these practical sciences for manipulating the physical world.[23]

Some ten years after Sakkākī met his ignoble end, Chaghatay had to contend with a popular insurrection that started in the countryside and soon swept over the city of Bukhara. In the year 636/1238~ astrologers at the Mongol court witnessed an ominous conjunction of Saturn and Mars, the two malefic planets, in the house of Cancer; this led them to predict the outbreak of rebellion and the emergence of a heretic. Despite the prognostication, the Mongol authorities in Bukhara were caught off guard when peasants bearing axes and clubs overran the city.

Led by a charismatic wonderworker, a certain Maḥmūd from the nearby village of Tārāb, the uprising garnered support from the masses as well as a faction of the notables. They were united in their dissatisfaction with Mongol rule. In addition to assuming the messianic title of the Mahdi, the restorer of religion and justice predicted to rule before the end of the world, Maḥmūd claimed to have mastery of jinn and angels, whom he could summon in battle. Detractors called him a charlatan. Supporters saw a visionary who could heal the sick, restore sight to the blind, and fulfill a prophecy with promises to kill the Mongols and restore religion. Attracting thousands of followers, the uprising succeeded in occupying Bukhara and driving out the Mongol forces, only to be crushed after an expeditionary army was sent by the central court to reclaim the city.[24]

Whether joining the Mongols or opposing them, an array of scholars and authorities found careers through appeals to the power of the heavens and the forces of the unseen. By far the most famous and influential was Naṣīr al-Dīn al-Ṭūsī, the philosopher and physician who made a life for himself as a scholar of the stars. Ṭūsī spent much of his formative career in the courts of the Nizārī Ismailis. Through their patronage he produced significant works of ethics and philosophy and pursued esoteric theories of cosmic emanation that advanced Ismaili doctrines of the divine nature of the Imam.[25]

The Nizārīs of Iran and Syria were a Shia dynasty established in the name of the Ismaili Imam Abū Manṣūr Nizār (fl. 488/1095), who led an unsuccessful revolt to reclaim the throne of the Fatimid caliphate in Egypt from his brother al-Mustaʿlī. The political and religious movement that advanced Nizār's claims in Syria and Iran was led by the general and missionary, Ḥasan-i Ṣabbāḥ (d. 518/1124).

Gaining control of several remote mountain fortresses, Ḥasan-i Ṣabbāḥ established the center of his rule in the imposing citadel of Alamut, high in the mountainous peaks of the western Alborz. Situated just north of Qazvin, the stronghold served as a remote outpost for the Nizārī Ismailis to train converts and to perfect techniques of dissimulation and assassination, which they used to often terrifying ends, murdering both political and religious adversaries. Notoriously, the word assassin derives from the term *ḥashīshiyyūn,* a derogatory label for Nizārī militants, colored by elaborate stories told by Muslims and Crusaders in the Levant of fanatics who had been whipped into maniacal devotion through the consumption of hashish.

As with most Arabic and Persian authorities of the day, Qazwīnī does not use the term *ḥashīshiyyūn,* though he knew well the legacy of their assaults. Referring to the followers of Ḥasan-i Ṣabbāḥ as *fidāʾiyyūn,* zealot devotees ready to sacrifice their lives for the spread of the mission, Qazwīnī recounts several of their attacks, including the famous assassination of Niẓām al-Mulk, the powerful Saljuq vizier and bitter opponent of the Ismailis. Qazwīnī describes Ḥasan-i Ṣabbāḥ, the master of Alamut, as a scholar of philosophy, the science of the stars, geometry, and the arts of *siḥr,* magic or enchantment, a description that fits well with the esotericism then associated with the Ismailis. Under Ḥasan-i Ṣabbāḥ, Alamut became the site of a sophisticated tradition of astral study and observation.[26]

Many Ismaili scholars and leaders cultivated astral learning, with esoteric interpretations of scripture and the cosmos that accorded their Imams, as direct descendants of the Prophet, unbridled power. Qazwīnī's description of the Ismailis as arch-heretics reflects the general attitude of Sunni authorities of the day, who took exception to what they viewed to be an antinomianism that invested the Ismaili Imams with powers to transcend the legislative authority of the Prophet. In his world history, Juwaynī famously interpreted the westward campaigns of the Mongols as the *sirr-i ilāhī*, the divine secret of God's design to drive out the scourge of Ismaili heresy, which evidently in his view justified the massacre of Sunni Muslims.[27]

In Alamut, Ṭūsī had access to a vast library of books of ancient science and occult teachings, collected over years. The holdings not only documented secret Ismaili doctrines on the cosmic order, but also recorded astral knowledge passed down for generations. The library was furnished with finely tooled astrolabes and armillary spheres with their rotating discs and rings used to track the movement of celestial bodies as they arc across the equator and pass along the line of the meridian connecting the north and south poles.

Equipped with some of the best instruments for observing the stars, Ṭūsī knew well how to read the heavens and predict when a dynasty would fall. In 654/1256, when Hülegü besieged Alamut with Juwaynī by his side, it was Ṭūsī who negotiated the surrender of the citadel. Abandoning old masters for new ones, Ṭūsī secured Hülegü's favor and rose in the Mongol entourage as a high-ranking advisor, diplomat, and court astrologer.

Juwaynī ensured that the heretical books from the library of Alamut were burned, only after removing from the collection the most valuable materials, which included prized astronomical instruments. Both Juwaynī and Ṭūsī followed Hülegü on the march westward, which brought the Mongol armies to the walls of Baghdad. Later accounts note several ominous signs leading up to the siege. In contrast to judicial astrology, which required considerable training in mathematics and astronomy, almost anyone could find prognostic meaning in the unexpected appearance of comets, shooting stars, or earthquakes. There was that bright burning star that appeared in the heavens. Then in the same year that Alamut fell, Baghdad experienced a series of tremendous floods, while Medina, the city of the Prophet in Arabia, suffered an earthquake, which set off a fire that consumed the roof of the Prophet's mosque and tomb.[28]

A generation later, Rashīd al-Dīn (d. 718/1318), chief minister to the Ilkhanid dynasty founded by Hülegü, explains in his universal history the importance that formal astrological prognostications played in the march on the Abbasid capital. Throughout his campaigns, Hülegü consulted astrologers to determine the most auspicious course of action. When it came to the siege of Baghdad, the court astrologer Ḥusām al-Dīn (d. 661/1262) warned of apocalyptic consequences should the caliph be killed: horses would perish, soldiers would grow ill, the Sun would cease to rise, rain would stop, plants would wither, strong winds and a powerful earthquake would ravage the world. And if that were not enough, the Mongol emperor himself would die.

Confronted with these predictions, Hülegü turned to Ṭūsī, who rejected the dire warnings. Numerous caliphs from the line of ʿAbbās had been murdered and the world kept going, he counseled, adding simply, "In the place of the caliph will be Hülegü." On January 22 of 656/1258, at the rise of Aries foretelling success in war, the siege began. Soon after, the Mongols breached the city walls. Baghdad fell and the caliph surrendered in a negotiation that Ṭūsī helped broker. Out of an abundance of caution, lest spilling royal blood on the earth would enrage the heavens that had so far favored the Mongols, Hülegü had the caliph wrapped in a blanket and beaten to death. After Mustaʿṣim, the Mongol forces put to the sword most of the royal house of ʿAbbās, along with many of their ministers and generals.

In Rashīd al-Dīn's telling, once the city had fallen and the Abbasid forces had surrendered, Ṭūsī was charged with granting amnesty. Brought in shackles was the once-powerful general Shihāb al-Dīn Sulaymān Shāh, who had been in charge of protecting the city walls and was known for his knowledge of the heavens. How could it be, it was inquired, that someone who claimed to be a master of the stars and the selection of auspicious timings could not foresee this coming? Soon after, on Hülegü's orders, the head of the caliphal general joined those of several other Abbasid warriors dangling from the walls of Mosul as trophies and object lessons for anyone who should think of resisting.[29]

The chief finance minister in Baghdad, Abū Ṭālib al-Dāmghānī, who had once opposed Qazwīnī's appointment to judge over Wasit, succeeded in having one of Mustaʿṣim's sons spirited out of the city before the slaughter. But though a hope of a male descendant survived, it was all for naught. With the death of Mustaʿṣim, a dynasty that had ruled from Iraq for over five hundred years came to a close. While the political and emotive ideal of a caliphate by no means ceased to inspire, the fall of the black banners of the Abbasids saw the end of Baghdad as a center of spiritual and political power. The rise of the Mongol successor state of the Ilkhanids established by Hülegü also weakened both Aleppo and Damascus, though ultimately the Mamluks of

Cairo succeeded in halting the westward movement of the waves from the East. The entire order had been upended, now with infidels ruling over vast lands once controlled by Muslim dynasties.[30]

Baghdad witnessed days of rape and pillage. While the carnage may have been selective, it was also unrelenting. Untold numbers were slain, strewn in piles throughout the streets. The mounds of decaying flesh and rotting food quickly turned the drinking water putrid. Soon a plague descended, as swarms of flies, which fed on the remains, rose up through the city.

To staunch the stench, survivors took to smelling onions. Many perished of hunger, hiding in canals, wells, and catacombs. Profiteers from the surrounding towns and cities brought in fresh food for the starving residents, who, parting with brass cookware, expensive vessels, and priceless books, had no choice but to pay exorbitant prices. Children were enslaved, high-ranking women married off. Through the misery, some found new fortunes while others fled. The surrounding towns and cities were also put to the test. Hilla and Kufa quickly capitulated. Wasit unwisely resisted. Once the Mongol forces breached the walls, the slaughter lasted for days. Qazwīnī was likely among the many notables and ruling dignitaries of Wasit who escaped to the neighboring countryside with their families and property before the carnage began.[31]

When the blood had dried, the money flowed. The Mongols were quick to reestablish the administration of Iraq and set themselves up as the new heads of finance, controlling the patronage that came with it. Here they followed the age-old pattern of promise and threat. The entire enterprise of the state had long worked by balancing the fear of the sword with the distribution of wealth.

Qazwīnī left no trace of his direct experiences with or reflections on the conquests, save for passing allusions, such as when he quotes from his colleague Tāj al-Dīn al-Urmawī, who had led the Sharābī madrasa in Baghdad and died two years before the sack of the city. According to Qazwīnī, Urmawī frequently commented that the abundant alms, or *ṣadaqāt*, that had once been freely given by Mustaʿṣim's father throughout the region of Iraq had served to protect the caliphate from the Mongols. Qazwīnī knew this to be the case, for he remarks that during the rule of Mustaʿṣim, however, the alms decreased and then immediately the Mongols came and conquered.

The comment matches the frequent complaint that Mustaʿṣim had neglected to pay the salaries of his soldiers, many of whom had dropped out of the ranks of the army. Ṭūsī puts this critique directly in the mouth of Hülegü, who, before having Mustaʿṣim executed, chides the caliph for hoarding treasures instead of funding his own defense.[32]

Work on rebuilding started almost immediately. The madrasas and monasteries reopened. Jurists and Sufis were confirmed back in their positions and given daily rations and monthly salaries. Like Qazwīnī, many remained in the same offices they had held before the conquests, such as the Shāfiʿī jurist Niẓām al-Dīn al-Bandanījī (d. 667/1268~), who just the year before had been appointed to the position of chief judge of Baghdad.

Others in the Mongol administration took on new positions that now needed filling. In the place of the caliph, Juwaynī was appointed the provincial governor over all of Iraq. For his efforts and loyalty, Ṭūsī was charged with control over the *awqāf,* or legal endowments, which funded various institutions of learning and piety. In the summer of 657/1259 Ṭūsī was also tasked with a project dear to his own intellectual aspirations: the construction of a new observatory in the city of Maragha, near Hülegü's capital, Tabriz.[33]

According to Rashīd al-Dīn, the order for the new observatory came directly from Hülegü's brother Möngke Khān (d. 1259), who had developed an interest in the mathematical sciences. While Ṭūsī may well have set his sights on the creation of such an institution, nothing like it had been undertaken in his lifetime. Of the many changes brought by the Mongol invasions, the observatory stands out for its intellectual ambitions. Built high on the hill just to the west of the city, with the aid of a local architect, Fakhr al-Dīn al-Marāghī, the complex had at its heart a circular tower bisected by a multistory sextant gradated to measure meridian altitudes of the stars and planets. Ibn al-Fuwaṭī, a protégé of Ṭūsī, preserves scattered references to the formative history of the observatory in his biographical dictionary. Enslaved as a child during the conquest of Baghdad, Ibn al-Fuwaṭī was ultimately freed by Ṭūsī, who trained and employed him as the head librarian for the vast collection of books housed in the observatory complex.[34]

During his tenure in Maragha, Ibn al-Fuwaṭī kept a log of all those who visited the observatory in a handbook that he drew on in the course of later biographical writings. At Ṭūsī's request, hundreds of scholars and their families migrated to Maragha to support the work of the institution. They were greeted with handsome stipends that funded their studies.

Renowned Sufis, Sunni authorities, and high-ranking Shia frequented the observatory, as did ministers, jurists, and poets, who waxed eloquent on the "wandering stars," as the seven visible planets, which included the Sun and the Moon, were commonly known for their movement in the sky. Copyists and schoolteachers, skilled artisans and technicians, innkeepers and merchants met the needs of the bustling city below, where Persian, Arabic, and various Turkic languages could be heard. From as far away as Tunis and China, the observatory attracted scholars who sought the prestige of an imperial center of learning. Here Ṭūsī held large assemblies, teaching crowds of students—some who came with years of training, and others who could barely understand the lectures. His classes covered diverse topics that often included the teachings of Fakhr al-Dīn al-Rāzī, with whom Ṭūsī frequently disagreed.[35]

Many masters of the stars ascended the observatory, equipped with knowledge that wedded the theoretical realms of mathematical observation with practical models for astrological prognosis. In his day, Ibn al-Fuwaṭī met many a *munajjim* from the religious elite who took measurements of the heavens to cast horoscopes. Several were prized at the Ilkhanid court, such as ʿImād al-Dīn Abū Ṭāhir (d. 687/1289), the son of the former Ḥanafī minister Abū Ṭālib al-Dāmghānī, who like his father entered into the Mongol service after the fall of Baghdad.

As with the masters of the stars who traverse Ibn al-Fuwaṭī's collection, ʿImād al-Dīn was trained in judicial astrology and had the technical skills to produce *taqāwīm*, astronomical tables. He was also skilled at tracking the interdependent movement of the planets and constellations to calculate nativities and life spans. With a coterie of highly trained scholars and aides, Ṭūsī recorded the observations conducted in Maragha in his *Zīj-i Īlkhānī* (The Ilkhanid tables), dedicated to the dynasty that favored him. The Persian introduction prefacing the royal copy celebrates, in true astrological fashion, the auspicious rise of Genghis Khan and his successors, while the technical tables included in the work juxtapose the celestial coordinates of Islamic learning with Chinese calendric models and terminology.[36]

As a natural philosopher, Ṭūsī saw the forces of nature to be fundamentally interconnected. It is a sentiment that inspired his study of stars. These interests are also on display in the Persian lapidary that Ṭūsī dedicated to Hülegü, to edify and strengthen the soul of the great emperor. Ṭūsī drew copiously from the Hermetic collection the *Kitāb al-Aḥjār* (The book of stones), ascribed to Aristotle. The work is filled with talismanic forces and the strange powers of precious stones and minerals, several taken from the adventures of Alexander the Great. As with many other areas of natural philosophy, Qazwīnī, like Ṭūsī, read closely Aristotle's *Book of Stones,* and although he was not a preeminent scholar of the stars, he recognized their power over human life.[37]

While overseeing the Maragha observatory in northwestern Iran, Ṭūsī made frequent trips to Iraq. At least twice he visited Wasit, once in 662/1263~ and then again ten years later. All the while Qazwīnī lived in the city serving as judge and professor. Ṭūsī traveled throughout the provinces not only to oversee the administration of institutional endowments, but also to collect books for the observatory library, which he did while visiting Wasit. These activities likely brought the two together in some fashion. Although no record survives of any encounter or exchange, Ṭūsī's network of patronage and support of scholars directly intersected with the spheres of learning and teaching that Qazwīnī inhabited.

In addition to the promise of patronage, the presence of imperial authorities also served to instill fear. Juwaynī's brother, Shams al-Dīn, the chief finance minister for the entire Ilkhanate, in 660/1261~ uncovered a vast misappropriation of state funds. After examining the financial records for Iraq, he saw that the local governor of Wasit, Majd al-Dīn Ṣāliḥ ibn al-Hudhayl (d. 680/1281~), had been siphoning off revenue meant for the central administration. Shams al-Dīn had the governor shackled, beaten, and paraded through the streets. For good measure the governor's property was confiscated. Several other corrupt officials were also dealt with before a new agent was installed to manage the city.[38]

Shortly after the Mongol conquest of Wasit, Qazwīnī put the finishing touches on *Wonders and Rarities,* though he makes no direct mention of all the suffering. The earliest redaction was completed in 659/1260~, during tumultuous times. Qazwīnī was drafting the collection at least a year earlier, shortly after the caliphate had collapsed and the fabric of daily life had frayed. By his own account the project of collecting and collating material for the book took years. Through all the changes and upheaval, providential design offered an inspiration that would continue to guide Qazwīnī for some time to come, well after his first recension, as he worked on finely retooling the compendium. Soon ʿAṭāʾ Malik al-Juwaynī, the new governor of Iraq, would be leafing through a lavishly illustrated world of wonders celebrating his name.[39]

PART II

Wonders to Behold

Other collections of wonders circulated before Qazwīnī put ink to the page. What he added that was truly novel, however, was a deep engagement with natural philosophy as developed within the frameworks of scholastic education. His definitions of key terms and his taxonomical hierarchy reflect a broader endeavor at classification and systematization that had taken hold in madrasa education.

The structure of *Wonders and Rarities* is delineated through an introduction and detailed table of contents. The text then proceeds in three parts and a final conclusion. The first part, or *naẓar,* as Qazwīnī describes these three examinations, covers the heavens, the so-called *ʿulwiyyāt,* celestial phenomena, with subdivisions on the seven wandering planets, the fixed stars, the constellations, the stations of the Moon, the angelic inhabitants, and the movement of time, itself a product of heavenly motion. The second part turns to the *sufliyyāt,* the sublunar realm composed of everything beneath the orbit of the Moon, covering the elemental spheres of fire, air, water, and earth. The sphere of water opens up a discussion of the seas and oceans, and then the chapter on the earth makes up the bulk of the compendium. This terrestrial realm, starting with mountains, rivers, and springs, is divided into the last of the three parts, focused on the mineral, vegetal, and animal kingdoms. The collection then culminates with a final short conclusion dedicated to strange creatures of rare shapes and sizes.

The following chapters turn to the contents of Qazwīnī's natural history—from his influential introduction, through the heavens, to the sublunar realm of all earthly diversity. Even though Qazwīnī delineated these topics in hierarchical orders, he also understood them to be interconnected. So it is that psychology, ethics, and metaphysics connect to astral knowledge, geography, botany, and alchemy, along with medicinal forces and talismanic powers. Together they form a coherent understanding of the world and our place in it, governed by physical forces that can be fathomed and deployed for spectacular ends. As Qazwīnī understood it, to craft such a book, in such a clear and lively manner, could even make a career.

4

COSMIC ORDER

"THE GREAT AND JUST PATRON AND GRANDEE, master and chief, the victorious, triumphant, scholar, supporter and might of the faith, pillar of Islam, order of sovereignty, succor of the Imams, sun of the dynasty, backer of religion, ʿAṭāʾ Malik ibn Muḥammad ibn Muḥammad, may God double his greatness and prolong his glory!" Qazwīnī stayed in the good graces of his new masters. Not long after the public flogging of the provincial deputy in Wasit, Qazwīnī presented his *Wonders and Rarities* to ʿAṭāʾ Malik al-Juwaynī, the new governor of Iraq, with fulsome praise. Following a time-honored ritual of boundless adulation before an august ruler, Qazwīnī heralds Juwaynī, the mighty lord, with a litany of titles and honorifics.

Qazwīnī's dedication comes after a short anecdote meant to illustrate the mysterious power of divine justice and providence. In the face of what to mere humans may seem to be senseless murder and the suffering of innocents, God, just and wise, reigns supreme. The appeal to divine justice skillfully raises Juwaynī's sagacity to hallowed heights: "Through the nobility of his station and the stature of his rank, among the people of the age he is famous for generosity and renowned for learning, endowed by God almighty through noble traits, superior discernment, by glory inherited and acquired."

In the humble submission of a servant, Qazwīnī presents *Wonders and Rarities* to Juwaynī's assembly, concluding the ornate offering with the hope that "my name might live on through the everlasting recollection of his nobility." Lest his own labor be forgotten, Qazwīnī petitions those who look through this book to reflect on the toil it took him to gather everything in it: "I have acquired from hearing, seeing, thinking, and examining so many wondrous mysteries and extraordinary properties [*ḥikam ʿajība wa-khawāṣṣ gharība*], that I desired to note them all down and record them, fearing they might slip away and be overlooked."[1]

Juwaynī was known, like his brother in the central court, to lavish piles of gold dinars on those who sang his praise. The statesman was not the first patron to receive a book of natural wonders that opened up the entire world. Writings on geography and cosmography in the language of wonders had long attracted the interest of the powerful. Qazwīnī knew the example of the famed traveler from the West, Abū Ḥāmid (d. 565/1169~) from Granada, who upon reaching Baghdad dedicated a book of wonders to the powerful Abbasid vizier, Abū l-Muẓaffar Ibn Hubayra (d. 560/1165), renowned for his support of preachers and madrasas. Such patronage could change lives.[2]

For his labor, Qazwīnī was undoubtedly rewarded. It is also clear, as Qazwīnī suggests, that he spent a good deal of time composing the collection. There are redactions of *Wonders and Rarities* that do not contain the opening dedication to Juwaynī. Most likely Qazwīnī had already been at work on the compilation before the Mongol conquest of Iraq. The multiple recensions that survive indicate several stages of authorial composition. Yet expansions and abridgments also highlight the forces of transmission and reception that shaped works of renown.

To fashion a book required trained artisans: manufacturing paper from pulp, drying and then smoothing it out, collecting the quires, cutting the edges, rubricating the lines, and then binding the spine. An illuminated book required even more planning and expertise, with skilled painters and illuminators who made their own pigments and knew how to balance the spread of text and image across the page. Qazwīnī sought not only to define wonder and rarity in distinctly scholastic terms; he also aimed to create a book that would inspire awe at the beauty of its folios and the art of its design. The concise interplay between form and content, which aspired to both describe and generate astonishment, contributed in no small part to the lasting success of Qazwīnī's compendium.

In an age of individual, unique manuscripts, composition and publication involved the work of several hands, from the initial ink on the page to the final tooling of the leather binding. Drawing on notes and copious citations from other books, it was not uncommon for an author to employ an assistant to help collect materials or an amanuensis who could aid in collation, redaction, and the actual process of writing from dictation. As a judge and professor, Qazwīnī had untold resources at his disposal, both in terms of material support and in the form of human labor to assist him in the various stages of publication. Qazwīnī was renowned for his own elegant calligraphy, and there are manuscripts of both the geography and cosmography based on versions that he wrote by hand; this includes the formal presentation for Juwaynī, which Qazwīnī appears to have copied himself. Yet the archival record also indicates that Qazwīnī followed an established practice of those with means to employ professional copyists.[3]

A rough draft was commonly referred to as a *musawwada*, so termed because the paper had been blackened or inked up with notes. Draft copies could serve as the basis for lectures and sessions of dictation that could themselves circulate among students and teachers. In contrast, a fair copy, referred to as *mubayyaḍa* for its pristine, unblemished appearance, would be made from earlier drafts produced either by the author or under the author's direct guidance with attention to consistency and accuracy. Formal copies could, in turn, be presented as gifts to wealthy patrons and to various institutions, from humble libraries in a province to royal archives at the heart of an empire.[4]

Redactions, expansions, and abridgments, authorial or otherwise, could circulate at any stage along the line, as a work made its way into private and public collections. Several manuscript copies of *Wonders and Rarities* survive with the record of Qazwīnī's grandiloquent praise of the powerful governor. Evidence internal to the manuscript recensions suggests that the presentation copy for Juwaynī was finished in 661/1262~ or shortly after. Qazwīnī is also known, though, to have held assemblies of learning where he taught the work, and redactions produced in the course of Qazwīnī's life circulate with no mention of the governor.[5]

Exactly when Qazwīnī first composed his natural history and its precise relationship to his geography also remain to be seen. Together the two volumes present an interconnected account of all creation. A recension of his geography, which circulated with the title *ʿAjāʾib al-buldān* (Wonders of the regions), appears to have been completed in the same year as the presentation copy to Juwaynī, though it survives without any formal dedication.

In broad terms, the two works complement each other, though Qazwīnī appears to have made little effort to integrate them through cross-references or direct citations. While there is some degree of overlap, each is distinct in structure and content. They also generally circulated as stand-alone collections. Qazwīnī continued to fine-tune both works over the course of several years, adding sections here, changing wording there. He issued another version of the geography in 674/1276, which he had copied out himself by hand.[6]

Early in its history *Wonders and Rarities* circulated in Persian. The Persian text follows closely its Arabic twin, though with notable divergences. It contains an entire section dedicated to the talismanic sciences, in a detailed treatment of crafts and arts missing from the Arabic text. This version also features an array of Persian verse citations, also absent from the surviving Arabic versions, with references to paragons of Persian poetry, such as Firdawsī (d. ca. 411/1020), Sanāʾī (fl. 525/1131), Niẓāmī Ganjavī (d. ca. 613/1217), and Kamāl al-Dīn Iṣfahānī (d. ca. 635/1237). These divergences may well have been intended to suit Persian readers, who often included administrators and courtiers.[7]

The earliest dated manuscript of the Persian text known to survive was copied in 699/1300. The frontispiece notes 660/1261~ as the year when Qazwīnī produced the work; that is, before he dedicated a copy to Juwaynī, and before the earliest Arabic redaction, which was issued a year later. Nothing in the Persian text announces that the compendium is a translation, and the introduction is spoken directly in Qazwīnī's voice, using honorifics that imply he was still alive when it was produced.

During this period Persian had gained tremendous prestige as a vehicle for scholarship and literature, used increasingly in the fields of astronomy, astrology, philosophy, and natural science. Royal Persian commissions were also common, such as the lapidary by Naṣīr al-Dīn Ṭūsī produced for Hülegü, and the Persian translation of an Arabic bestiary commissioned by the emperor's great-grandson, Ghāzān Khān, the Mongol ruler of Iran. A courtly redaction of this particular translation survives in a lavishly illuminated manuscript produced in Maragha at the end of the century. Much like Qazwīnī's natural history, the collection, entitled in Persian *Manāfiʿ-i ḥayawān* (Beneficial uses of animals), weaves together pedestrian fare with wondrous creatures seldom seen (figure 4.1).[8]

FIGURE 4.1: The Sīmurgh identified as the ʿAnqā' from a Persian translation of the *Beneficial Uses of Animals* commissioned in the Ilkhanid court, circa 690/1291 in Maragha.

Just as there was a sustained appetite for an array of scientific and philosophical writings in Persian, it was not uncommon for the leading scholars of the day to switch languages for their written compositions when addressing different audiences, or to publish both Persian and Arabic redactions of the same work. Ibn Sīnā, Bīrūnī, Ghazālī, Rāzī, and Ṭūsī were all at home writing in both languages, and it is entirely possible that Qazwīnī composed the Persian version of natural history himself. Whatever the case may be, the close relationship between the early Persian and Arabic recensions of *Wonders and Rarities* points to an interconnected corpus of material. Notably the Persian text maintains the same extensive quotations of sophisticated Arabic verse and prose, suggesting a readership conversant in both literary languages. The exact process of composition remains unclear from the manuscript record alone. Nonetheless, Qazwīnī's fame was cemented early on through both the Arabic and the Persian redactions of his natural history.[9]

Yet perhaps most notable of all the early materials to survive is a unique Arabic manuscript of *Wonders and Rarities* that was produced, evidently under Qazwīnī's direct supervision, in Wasit two years before his death. According to the colophon statement at the end of the manuscript, the copyist was a physician from Damascus who had taken up residence in Wasit. The collection was finished on February 26, 678/1280, nearly two decades after Qazwīnī had originally dedicated the work to Juwaynī. This copy lacks the royal dedication but expands in subtle ways on the earlier recensions (figure 4.2).[10]

While there are numerous copies of the Arabic text, the Wasit redaction is the earliest dated manuscript of the work known to survive. As with many of the later witnesses, this particular manuscript is illustrated with some five hundred lavish paintings and detailed diagrams and maps integral to the authorial design. Spread out over more than two hundred folios, the paintings combine together several fields of book art, drawing into a single volume a star catalogue with accompanying planetary iconography, a sequence of angels, heavenly hosts, and meteorological phenomena, a cycle of exotic islands and marine animals, a *materia medica* of plants, trees, and shrubs, a section on jinn and demons, and a bestiary of animals found on land—all of which culminate in a final account of hybrid animals, strange creatures, and prodigious births. While references to and illustrations of handmade marvels are found here and there, Qazwīnī's natural history focuses above all on the natural wonders of creation.[11]

Although executed with a wide palette of colors, the illustrations and the rosette of the opening frontispiece to the Wasit copy lack the expensive gilding of gold leaf and ink often reserved for the most exclusive royal manuscripts. Just as Qazwīnī had cultivated an appreciation for the marvelous capacity of paintings and sculptures to mimic reality, the paintings in the unique Wasit copy demonstrate connections with the vibrant tradition of illuminated manuscripts produced in Iraq and Iran during the period. These include scientific compendiums with significant attention to astronomy, botany, medicine, and zoology, as well as literary and historical works, which collectively intersected with Qazwīnī's intellectual horizons. Several illuminated manuscripts from Iraq demonstrate affinities with the Wasit copy, such as the Arabic translation of Dioscorides (d. ca. 90 CE), *De materia medica*, copied in 621/1224, with intricate paintings of both plants and people, along with an-

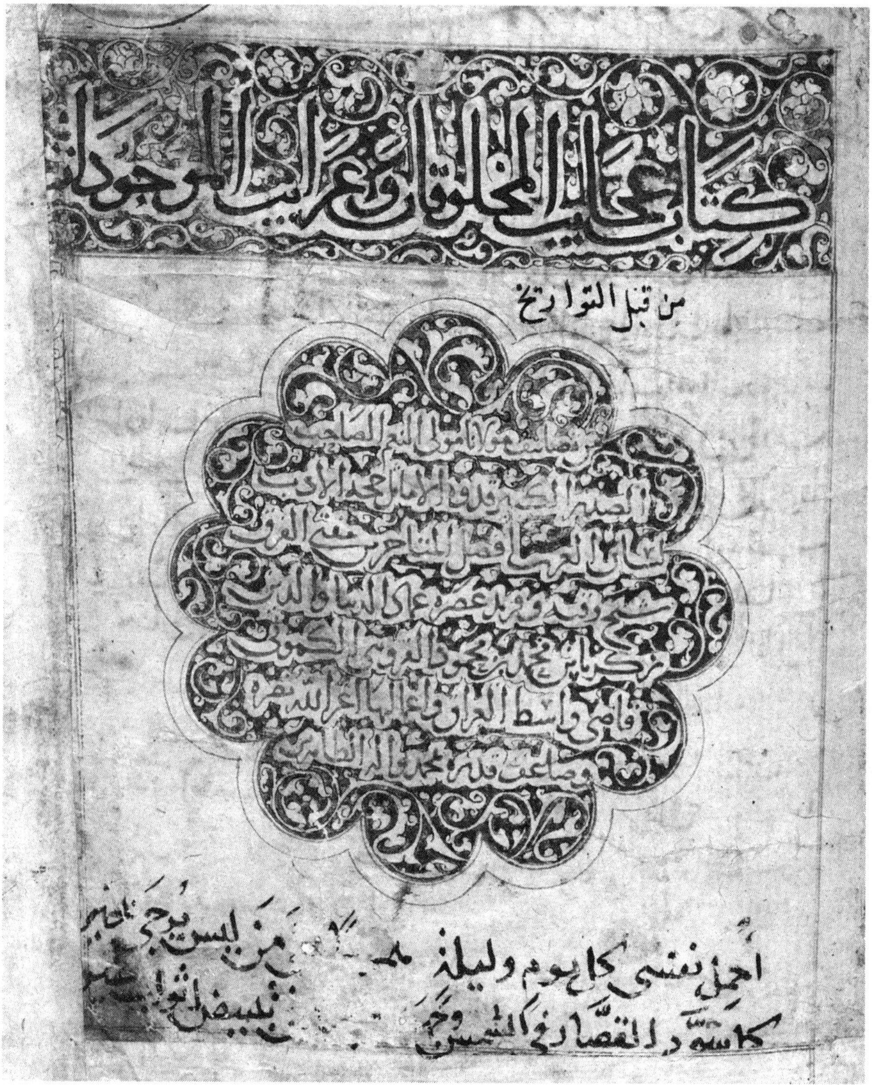

FIGURE 4.2: Frontispiece with title in the header and Qazwīnī's name and honorifics in the central medallion, copied 678/1280 in Wasit during Qazwīnī's lifetime.

other manuscript of the same work finished in 637/1240 in one of the madrasas established in the name of Niẓām al-Mulk.

Also in the field of illustrated scientific collections there survives from the same period a fully illustrated copy of the *Kitāb Naʿt al-ḥayawān* (The description of animals), a well-known bestiary that draws on the authority of both

Aristotle and the physician Ibn Bakhtīshūʿ (d. ca. 450/1058). Works on animal husbandry and veterinary medicine also circulated with illustrated paintings. Technical diagrams and maps frequently accompanied astronomical and geographical writings, by the likes of Bīrūnī and Abū Isḥāq al-Iṣṭakhrī (fl. 324/936), as well as works on the talismanic arts—bodies of writing that Qazwīnī knew well. Numerous literary and historical works enjoyed lavish illuminations and traveled throughout the region, such as the tongue-twisting delights of the *Maqāmāt* (Assemblies), a picaresque series of complex and colorful tales by al-Ḥarīrī (d. 516/1122), a surviving copy of which was illustrated by a painter from Wasit during Qazwīnī's lifetime (see figure 5.4).[12]

The Ilkhanid elite inherited and expanded this thriving tradition of painting. Circulating through a range of institutions were also illuminated works of monumental history by such high-ranking officials as Juwaynī and Rashīd al-Dīn. Drawing skilled artisans into their court, the Ilkhanids cultivated new tastes and styles, which in turn transformed the repertoire of painted books. Many painters were celebrated poets and trained scholars who had mastered the technical skills to produce paintings of great artistry, executed with rich pigments.[13]

In light of the range of illuminated manuscripts that circulated through both courts and madrasas, the honorifics used to describe Qazwīnī in the frontispiece of the Wasit copy are telling. Qazwīnī's formal position is given with full-throated praise: "the master of blessings, the mighty, great chief, model among peers, proof of refinement [*adab*], tongue of the Arabs, most excellent of the moderns, mufti of the sects, shaykh of his time, unique of his age, pillar of the world and of the faith, Zakariyyāʾ ibn Muḥammad ibn Maḥmūd al-Qazwīnī al-Kammūnī, judge of Wasit, the center of Iraq, and its dependencies." The title page of the Wasit manuscript also offers a prayer that God strengthen and double the stature of its author, indicating that Qazwīnī was still alive when the copy was made. This is a point also confirmed in the opening of the text, which further prays for Qazwīnī's longevity and celebrates him as a complete, learned scholar and a leader among followers.[14]

Along with his position as both judge and mufti, Qazwīnī identifies himself with *adab* and with mastery of the Arabic language. As a term of art, *adab* bears ethical implications of civility and comportment associated with educated decorum and refined manners. Although *adab* in Arabic serves today as an equivalent for literature, historically the category was much broader, in many senses closer to the ethical implications of belles lettres, particularly in the performative sense of the full corporeal and intellectual discipline of mind and tongue. In this way the opening honorifics also speak to the manners in which the worlds of science, law, and entertainment could cross on the pathways of education and refinement.

More than mere marvels to behold, the cycle of paintings integral to Qazwīnī's natural history offer insight into this confluence of interests. In addition to the array of phenomena presented as specimens and models for observation and consultation, there are also interspersed here and there paintings of narrative scenes with details drawn from wondrous stories (see figure 5.5). The selection of dramatic tales in the Wasit manuscript, however, is modest when compared to the extensive illustration of anecdotes found in other early copies, including the group derived from the formal presentation copy dedicated to Juwaynī. Qazwīnī may have directly overseen the painting sequences in the Juwaynī commission, which develop more attention to narrative detail.[15]

But it is also clear that painters drew from their own individual repertoires and stylistic preferences. An early exemplar of the royal commission to Juwaynī survives, likely produced at the beginning of the fourteenth century. The stylistic features are characteristic of other Ilkhanid productions, demonstrating a fusion of Far Eastern motifs with earlier book arts in Iraq and Iran. These features are on full display in the illustration for the entry on the *thuʿbān,* a massive serpent painted in the style of a Chinese dragon (figure 4.3). Moving forward, painters would turn to an array of narratives for inspiration, illustrating further subjects in a range of distinctive styles.[16]

As with the traditions of miniature painting, Qazwīnī's natural history was preceded by earlier writings and collections of wonders. There were geographic accounts and the reports of intrepid voyagers. Qazwīnī copiously bookmarked the geography of Ibn al-Faqīh (fl. 289/902), with its penchant for poetry and marvelous tales; he was also steeped in the breathtaking travels of Abū Ḥāmid, and books on the marvels of the seas, with their predilection for morbid tales of castaways and cannibals.

The marvelous design of creation was also a topic that motivated pious, religious writings, such as the *Kitāb al-ʿAẓama* (The book of majesty) by

FIGURE 4.3: Depiction of a *thuʿbān*, a giant serpent or dragon, with Ilkhanid stylistic elements, based on the Juwaynī recension, produced circa 1300.

Abū l-Shaykh al-Iṣfahānī (d. 369/979), which draws on stories of the prophets and ancient biblical lore to track the wonders of the heavens and the terrifying punishments of hell. Such works, with their focus on the eschatological details of the life to come, often served a hortatory purpose to warn of the threat of hellfire, while holding out the promise of celestial salvation. Hosts of angels and armies of demons feature, as do prophetic accounts that disclose the vast expanse of the Earth and the true scope of the heavens. These collections build on repeated Quranic commands to contemplate creation as a means of recognizing the sublime power of the Creator.

Arguments for God's existence through appeals to design feature in an array of early Arabic sources, from the Quran and hadith to works of dialectic theology and scholastic philosophy. Numerous parallels can be drawn with writings from antiquity, reflecting an age-old defense of a Creator God who sets the cosmos in motion. Qazwīnī builds on this argument for God's existence through the signs of ordered creation, which he captures in the poetic maxim "in motion and stillness always for God a witness/with all things there is a sign that He is one divine." The claim that contemplating wonders of creation offers a basis for fathoming divine wisdom features throughout the earliest developments of Islamic theology. Jews and Christians took up similar ideas, as did ancient philosophers, such as Galen in his contemplation of the human body as a study of providential order.[17]

Similar sentiments also play a role in various occult writings, such as the *Sirr al-khalīqa wa-ṣanʿat al-ṭabīʿa* (Secret of creation and the art of nature), ascribed to the Greek philosopher Apollonius of Tyana and translated into Arabic most likely during the reign of the Abbasid caliph al-Maʾmūn (d. 218/833). Qazwīnī knew Apollonius for his command of talismans and his knowledge of *khawāṣṣ,* the hidden properties found in minerals, plants, and animals. With a focus on the powers locked in the physical world, the *Secret of Creation* played a significant role in the development of Arabic and Persian alchemical writings. The collection spans meteorology, astrology, and human psychology, and advances a theophany of causes—referred to in Arabic as *ʿilla,* plural *ʿilal,* related to the Syriac *ʿelātā*—that emanate through existence, all of which can be traced back to God as the first cause of everything.[18]

If God is the ultimate cause of all creation, then what are the origins of evil? It is a problem that Qazwīnī addresses in the opening to his natural history, which begins, after its benediction, with four introductions, or *muqaddimāt,* dedicated to defining the key terms of the title, *ʿajāʾib al-makhlūqāt wa-gharāʾib al-mawjūdāt,* classified as wonder, created beings, rarity, and existent things. The preface to these sections cautions that whoever starts contemplating creation without first mastering the *muqaddimāt,* or propaedeutic principles, will only obtain uncertainty and doubt, and will be left lost, "like one who looks at ailments of children, the short-lived, misfortunes suffered by good people, the rising power of the wicked, and then asks: 'Why is this one wealthy, but the other poor, why did this one live so long, the other so short, why was this one created healthy, the other ill, this one made beautiful, the other misshapen?'"[19]

The dilemma was an ancient one. God's justice and providential order in a world of corruption and decay were long topics of concern. Well before Qazwīnī addressed the matter, Muslim authorities had developed an array of sophisticated responses to why a just and all-powerful God would allow for suffering in the world. In philosophical circles, Ibn Sīnā famously argued that evil itself had no essence, but rather should be defined as the lack or privation of good. Although frequently debated, this solution wielded a lasting influence. For many the issue of theodicy, or divine justice, in the face of evil took on renewed meaning in the course of the Mongol invasions. It is not hard not to see why Qazwīnī's appeal to the wonders of divine order and everlasting stability could provide a salve in tumultuous times.

One common response among Ashʿarī theologians was to argue that divine justice remained uncompromised despite the manifold evils found in existence. Standard Ashʿarī teaching held that evil had no ontological power that stood apart from or was independent of God. Instead both good and evil occurred through divine decree. Ghazālī, for his part, notably took this position to its logical conclusion, arguing that not only did good and evil, benefit and harm, originate in providence, but that God created the best possible world. It is an idea encapsulated in Ghazālī's rhyming maxim, *laysa fī l-imkān abdaʿ min-mā kān*, "in all possibility there is nothing more wondrous than what is."

Similar formulations were also known through the teachings of Galen that had long circulated in Arabic on the marvelous majesty of divine design. To detractors, optimism, in the strict philosophical sense of viewing creation as the optimal or best possible world, appeared to limit God's power to manifest even better worlds. Nonetheless, such expressions of maximal theodicy continued to exert influence in the development of Islamic thought. Through it all, these discussions accorded very little place to evil as an independent metaphysical force that could drive the world in any way apart from God's will.[20]

Qazwīnī sidesteps the intricate philosophical questions raised by the problem of evil with a standard Ashʿarī sensibility on the limits of knowledge. Providence works in mysterious ways, which the human mind cannot possibly hope to fathom. What appears evil or wicked from one perspective, from another can reveal itself to be part of a divine orchestration that transcends the span of human reason. As Qazwīnī stresses, all creation points to God's power and magnificence, filled with wonders of design, only a fraction of which can be mentioned, let alone known, "for everything that humans have ever comprehended is merely a drop in the sea and a grain in the desert of all existence." Found through every aspect of creation from great to small are signs of the complete perfection of God's power and the totality of divine beneficence.

Confronting the majesty of it all, Qazwīnī notes that the proper response is perplexity and astonishment at *badāʾiʿ ḥikmatihi,* the marvels of His wisdom, for "no atom moves in the heavens nor on the Earth save that its movement has a wise purpose [*ḥikma*], or two, or ten, or a thousand, and all this is proof of the oneness of the Creator, His power, greatness, and sublimity." This vision of divine immanence and transcendence does not brook independent evil, just as it dispenses, in its own way, with the problem of suffering, for as Qazwīnī argues, "God creates all things, organizes the whole, and decrees the particulars." The wonders of creation stand as signs of divine goodness emanating through existence. In this sense, there are no malevolent forces in Qazwīnī's cosmos that can outstrip providence, even if strange, uncanny, or horrific, or seen in the baleful acts of demons and jinn with all their mischief and cunning. It is this view that lends the language of natural marvels and wonder, even if misshapen and terrifying, a lasting munificence.[21]

The encyclopedic character of *Wonders and Rarities* blends curious anecdotes of distant peoples, marvelous buildings, strange talismans, and mysterious idols with an abiding concern for the scientific learning of universal philosophy. Through a wide selection of stories and verse, Qazwīnī also tethers traditional religious knowledge to scholastic taxonomy. What made Qazwīnī's work so compelling for his early readers was his capacity to mold earlier inquiries, spanning various genres and fields, into a systematic account of all creation.

The illusion of totality that *Wonders and Rarities* obtains is produced through the hard spine of its taxonomy, which supports countless diversions. Systematic taxonomies for dividing both knowledge and being were staples of madrasa education with its curricular attention to logical models for classification. What Qazwīnī did that was novel, however, was to apply the cutting edge of logic and an abiding concern for pleasure to questions of astonishment.

Before delving into his primary definitions of wonder and the extraordinary, Qazwīnī opens the collection by calling upon God—almighty possessor of *ʿaẓama,* divine majesty, and *kibriyāʾ,* glory, the source of all goodness. The term *ʿaẓama* features in pious collections of prophetic sayings, as well as works on geography and cosmography that reflect on the sublime design of creation. By pious convention, the first words of Islamic manuscripts consisted of praise of God and blessing on the Prophet Muhammad. The wording often conveyed particular importance, both as a way of distinguishing a given work and as means of singling out its general orientation, be it in creedal, literary, or philosophical terms. Qazwini's opening prayer, invoking the majesty and glory of God, through whom proceeds all goodness in creation, also subtly signals his own pedigree, as it closely echoes the benediction that his master, Abharī, used in the opening to his teaching digest on logic and natural philosophy, the *Talkhīṣ al-ḥaqāʾiq* (The epitome of verities). It is also here in the opening words that Qazwīnī celebrates God as *wājib al-wujūd,* the "necessary existent." This category was central to Ibn Sīnā's influential proof of God through the ultimate contingency of all things apart from a unique first cause that sets creation in motion.[22]

In the call to behold the divine through creation, Qazwīnī braids together the authority of the Quran and the hadith with the language of natural philosophy and the rhetorical flair of a skilled litterateur. The Quran frequently critiques the polytheists who fail to appreciate the self-evident signs, *āyāt,* of God's design that surround us all. "Do they not look to the sky above them, how We formed it and adorned it with no rifts within?" God chastises the unbelievers and enjoins, "Look at what is in the heavens and what is on Earth!" As Qazwīnī explains, in these Quranic passages the verb to look, *naẓar,* "does not signify merely casting the pupil of the eye in the direction of the heavens or toward the Earth, for animals share with humans this capacity."

Rather, the lesson to be derived from scripture is that we ought to wonder at the world and our place in it. For Qazwīnī, *naẓar* carries a philosophical meaning of "contemplating intelligibles and investigating sensibles and the divine wisdom, *ḥikma,* spread therein, so that their true realities may become manifest." Examining the perceptible world through the senses and through the intellect leads us to contemplate God's creations and to appreciate the judicious design and principles that govern existence. For Qazwīnī what distinguishes humans from animals, and those who possess reason from those who do not, is the cognitive faculty, called *al-nafs al-nāṭiqa*—the rational soul—which provides the mental capacity for conceptualizing and fathoming intelligibles.[23]

The technical vocabulary here is indebted to the philosophical study of how knowledge is produced and obtained. The Arabic categories *maʿqūlāt* and *maḥsūsāt,* intelligibles and sensibles, the *intelligibilia* and *sensibilia* of Latin scholasticism, point to concepts obtained through the intellect or reason and sensations derived through physical perceptions. Both are key to Aristotle's metaphysics, where knowledge is a product of the various faculties of the soul—the *psychē* in Greek and the *nafs* in Arabic—all working in concert. This is a subject central to the psychological teachings promoted by Ibn Sīnā and those who followed him. In this regard, *naẓar,* which means to look, behold, or speculate, also carries a specific philosophical significance, directly parallel to the Greek *theōria*—speculation, contemplation, and reflection. Aristotle comments, when defining philosophy as the pursuit of truth in the opening to the *Metaphysica,* that whereas the object of practical knowledge is action, the goal of theoretical knowledge is truth. In Arabic letters, the division between the theoretical, *naẓarī,* and the practical, *ʿamalī,* is a defining feature for the classification of knowledge and the various branches of learning.[24]

From the Quranic injunction to behold the world, Qazwīnī turns to prophetic precedent. "Show me everything as it really is," the Prophet implores God, just as he enjoins his followers to "contemplate God's creation!" The kind of reflection called for is not aimless woolgathering, for in it lies the pious practical purpose of *tahdhīb al-nafs* and *taḥsīn al-akhlāq,* disciplining the soul and refining character traits. Qazwīnī stresses that the pursuit of knowledge is itself an ethical endeavor that leads to higher states of awareness and refinement.

While providence and divine decree formed a core dimension of classical Islamic metaphysics, a good deal of space was also accorded to individual agency and moral responsibility. Theologians and philosophers debated the varying means by which individuals are ethical agents who could be held responsible and judged for their acts in cosmic terms. Qazwīnī eschews the problem of how humans could be accountable for their actions when God has power over everything, though his Ashʿarī teachers developed recondite means to account for how individuals "acquired" voluntary actions through a process of secondary causation. The moral argument driving Qazwīnī's cosmology holds that individuals can improve themselves through disciplines of the mind and body, by cultivating a greater awareness of the world and the diverse benefits found therein.[25]

The study of the various branches of knowledge leads to both worldly pleasures and otherworldly bliss. The attention to pleasure, *ladhdha*, and spiritual happiness, *saʿāda*, and their relationship with contemplation has a long history, both in the study of aesthetics and in the ancient philosophies of the good life. Aristotle in the *Ars rhetorica* argues that the act of wondering is, in and of itself, a source of pleasure. This idea appears in the Arabic translation, known as *al-Khaṭāba*, where that which is *ʿajīb*, marvelous or wonderful, or which causes one to marvel, *yataʿajjabu minhu*, is deemed pleasurable.[26]

In his call to contemplate natural wonders as a source of spiritual pleasure and discipline, Qazwīnī turns to the phrase *taḥsīn al-akhlāq*, the refinement of character traits. The expression evokes the language of moral philosophy with its attention to improving the individual and society. The term *akhlāq* is a calque of the Greek plural *ēthē* and with it the field of *ēthika*, just as the word *saʿāda* parallels the goal of *eudaimonia*, in the pursuit of happiness or human flourishing. This ideal of happiness features in the title to Ghazālī's Persian manual of virtue ethics, the *Kīmiyāʾ-i saʿādat* (The alchemy of happiness), itself a condensed restatement of his Arabic magnum opus, the *Iḥyāʾ ʿulūm al-dīn* (The revival of the religious sciences).

Both pleasure and happiness are categories that Ibn Sīnā develops throughout his writings on the soul, which in turn inform later Islamic moral philosophy. The tradition of virtue ethics is also promoted by the likes of Rāzī and Ṭūsī, both of whom approach the metaphysics of the soul and the progressive stages of refinement as part of a grand chain of being that connects the quotidian minutia on Earth to the sublime heavenly orbs above. In this, they shared in a universal field of science and philosophy that sought to account in taxonomical form for the entire order of existence, emanating from God through the tiniest particles of being.[27]

Ibn Sīnā taught that the emotive pleasure of intellection was itself the source of real felicity. The ultimate goal of the rational soul is the attainment of happiness through the pursuit of knowledge. "Learning is pleasurable," he argues, "on account of the wonder that it arouses." Ibn Sīnā frequently refers to the emotive states obtained through contemplation as beyond description, a notion he stresses in the *Pointers and Reminders*, his influential final philosophical primer.[28]

This sentiment motivates Qazwīnī's approach to the marvelous, where pleasure figures in the framing of both text and image. By attaining greater knowledge, higher levels of perception become possible. Qazwīnī punctuates this point when underscoring the ineffable states of awareness obtained through contemplation and spiritual discipline: "The eye of insight opens up as one comes to see wonders in every matter, the smallest part of which is impossible to articulate, for even if someone expressed just a bit of it to others, they would not believe him."

Joining pleasure and the pursuit of truth, Qazwīnī makes a case for beholding the rare and wondrous as a pious and moral endeavor designed to purify the soul. This is precisely the argument that Ghazālī makes throughout the *Revival of Religious Sciences,* in which he advances contemplation of the wonders of creation as a moral pursuit on the path of salvation. The entire world is God's written collection, a *taṣnīf,* an ordered compendium, Ghazālī counsels, containing all the wonders of His creation. And so it is that meditation on God's book of natural wonders, with perplexity and amazement, offers a pious means of awakening to the world. The ancient metaphor of nature as divine scripture was a powerful conceit in the development of Islamic philosophy.[29]

Qazwīnī directly evokes this tradition of wonder as an activity of spiritual discipline and ethical refinement. Reading marvelous stories and beholding paintings of heavenly hosts, strange creatures, and far-off lands offer pleasurable intellection designed to make manifest the divine design of otherwise normal experience. The value that Qazwīnī places on seeing, speculation, and ocular authority builds on long-held associations between wonder and knowledge. The emphasis on bearing witness promotes vision as a privileged mode of knowing through observation.

In Arabic, the power of direct observation is often referred to by such verbal nouns as *muʿāyana* and *mushāhada,* meaning to view, witness, confront with the eye, face, behold, observe ocularly, or experience directly. These associations are frequently voiced when beholding the strange or uncanny, and in this way they echo the expressions *thauma idesthai* in Greek and *mirabile visu* in Latin, "a wonder to behold." In such economies of knowledge and pleasure, wonder seeks its end in truth. One can only marvel, as it were, at something that is actually there. The significance of wonder lies in its veridical status, the "there-ness" of the object witnessed or the event as it unfolds.[30]

Qazwīnī and his readers could deploy numerous categories to differentiate between truth and falsehood and the real and the unreal, as they also had at their disposal a range of terms with a spectrum of meanings to distinguish various narratives and forms of writing. The category of *khabar*—for narrative, account, report—indicated factual value, as in *akhbār Makka,* historical accounts of Mecca. The term *qiṣṣa*—story, or tale, as in the *qiṣaṣ al-anbiyā'*, the stories of the prophets crafted by early preachers or *quṣṣāṣ*—was often associated with pious storytelling, conveying edifying morals, even if considered in certain contexts to be of questionable authority. This is in contrast to *khurāfa*, fable or tall tale, which was most explicitly identified with material of dubious veracity and in time became associated with far-fetched stories and baseless superstitions.

The difficulty is that verisimilitude and the veritable depend in good measure on what is believed to be real or possible. In this sense, the wide-ranging associations evoked by the modern category of fiction, let alone the fantastic or imaginative, do not neatly map onto the classical repertoire of Arabic and Persian storytelling or the writings on wonders and rarities that Qazwīnī had at his disposal. Fables of speaking animals, yarns spun by sailors, and stories of ancient peoples were certainly told, read, and copied solely for the delight they produced, a point accentuated by various Muslim philosophers, such as Fārābī, who promoted a poetics of pleasure for pleasure's sake. The writings on wonders that Qazwīnī inherited and further developed were neither conceived of, nor consumed as, fiction in the modern sense of the word. Nonetheless, these works, which aspire to scientific authority, also demonstrate an awareness of the pleasure of tales whose veracity cannot be fully verified.[31]

Above all, such writings sought to cultivate a sense of awe in the face of the extraordinary. But just as a touchstone of astonishment is truth, the problem of the counterfeit is never far from view. The tension between veracity and mendacity is a theme that Qazwīnī tackles head-on:

> Mentioned in this book are matters that the disposition of a negligent idiot might reject, but that the soul of a rational man would readily accept. Although these affairs may be far from customary phenomena or commonly witnessed experiences, nothing should be deemed too great for the power of the Creator or the cunning of creation and everything therein. They are the marvels of the art of the Creator, which are either perceivable by the senses or intelligible by the intellect, concerning which there can be no doubt or imperfection; or they are elegant tales ascribed to transmitters and I have no truck with them; or they are strange properties for which a lifetime would not be enough to test them, and thus it would make little sense

> to ignore all of them, merely because there is doubt concerning just some of them.[32]

The original Arabic deftly mixes the register of high philosophy with the quick wit of a tongue unleashed in assemblies of letters. When protesting that he bears no responsibility for relating stories of dubious pedigree—rendered here as "I have no truck with them"—Qazwīnī uses a colorful proverb taken from the battlefield of pre-Islamic Bedouin warrior poets, *lā nāqata lī fīhā wa-lā jamala*, literally, "in this, I've neither a female nor male camel," meaning "I've got nothing to do with the matter." Qazwīnī moves his argument along, both in this passage and throughout the preface, with the driving rhythm of rhyming prose, a technique known in classical Arabic as *sajʿ*, which threads a rhetorical force through the work, tying it to the highbrow literary conventions of *adab*. With elegant tales, some of dubious veracity, *Wonders and Rarities* promises to delight both in text and image.[33]

Earlier compendiums of geography and natural history had long traded in the value of astonishment at the strange and extraordinary, mixing stories that strain credulity with quotidian fare and pedestrian facts. In the opening to his geography, Ptolemy famously raised an eyebrow at the hearsay of travelers and merchants who were more interested in boasting about the remote places they had visited than gathering accurate knowledge of the world's edge. Ptolemy's solution, well known to Qazwīnī and emulated in the field of Arabic geography, was to send out scouts across the earth to gather reliable information. But the problem still remained of having to rely on others to verify knowledge drawn from far-flung regions, which a lifetime of globetrotting, even if by horse or by ship, could not solve, given the vast distances of uncharted territory involved.[34]

In his geographical dictionary, Yāqūt approached this problem in terms intimately familiar to Qazwīnī. Indeed, Qazwīnī draws from Yāqūt's language on this particular matter, though without ever stating so. Yāqūt reasons, "I have mentioned many matters that rational minds would reject and from which the natural dispositions of whomever has procured knowledge flee due to their remoteness from customary phenomena or ordinarily witnessed events."

While Yāqūt cautions that "nothing should be deemed as too great for the power of the Creator or the cunning of creation," he is rather "skeptical of such things" and shrinks from them, "discharging responsibility over the truthfulness of these matters to the reader." Yāqūt notes that, solely out of a desire to preserve the benefits of knowledge, he included accounts which may well prove spurious: "If they are true, then we may claim a portion of what is correct; if, however, they prove false, they nonetheless share part in the truth, for I have related them just as I have found them, so that you may know what has been said, whether it be true or false." With this Yāqūt coyly concludes, "For if someone were to say, 'I heard Zayd lie,' certainly you would be interested in knowing the manner in which he lied."[35]

The parallels between how Yāqūt and Qazwīnī addressed the issue of credulity are striking. Qazwīnī's reference to the power of the Creator and the cunning of creation finds an almost natural outlet in the narrative universe of geography and cosmography. However, when reformulating this expression, Qazwīnī retools Yāqūt's justification for the inclusion of material that cannot withstand the strongest scrutiny. Both engage the problem of authenticity by washing their hands of responsibility, with a clearly worded *caveat emptor* in plain sight in their introductions. Yet Qazwīnī adroitly gives a twist to Yāqūt's logic. For while Yāqūt acknowledges that he has mentioned many things that rational minds would question, Qazwīnī folds this statement back onto itself, maintaining that in his own work he has included marvels that only "a negligent idiot might reject, but that the soul of a rational man would readily accept." For a writer so attentive and indebted to Yāqūt's work, the distinction presents a playful reassessment of the argument.

With this reprisal Qazwīnī asserts that only those who contemplate existence and develop the faculty of reason are ready to accept the reality of the wondrous and rare; the negligent, out of ignorance, fail to perceive the marvelous design of creation, and thus renounce that which they do not comprehend. This line of reasoning echoes similar sentiments advanced by Ibn Sīnā at the conclusion of the *Pointers and Reminders*, in his treatment of prophecy and other extraordinary phenomena found in nature.

In a concluding piece of advice, Ibn Sīnā ridicules as fickle and weak those who reject that which they do not see, all the while puffed up with an air of shrewdness and a face of disdainful superiority toward the masses. This skepticism Ibn Sīnā calls idiocy, and he counsels restraint, for *fī l-ṭabīʿa ʿajāʾib*, "in nature are wonders," just as "the upper, active and lower, passive faculties of the soul" can join to produce "extraordinary phenomena [*gharāʾib*]."

Ibn Sīnā's short adage was frequently repeated in the philosophical primers of the day. It is an argument Qazwīnī alludes to on the last folio of *Wonders and Rarities.* When discussing animals with strange or extraordinary forms, *al-ḥayawānāt gharībat al-ṣuwar,* such as a two-headed child born in Khurasan, or chicks found with multiple heads and limbs, Qazwīnī quotes Ibn Sīnā's maxim, "in nature are wonders," citing the great sage by his simple honorific, *al-ḥakīm,* "the Philosopher." And with this last allusion, which points to Ibn Sīnā's final arguments for philosophical open-mindedness in the face of wonders, Qazwīnī draws his collection to a close.[36]

Ibn Sīnā's imprint is felt throughout *Wonders and Rarities.* Not only do references to the philosopher feature in the opening and the conclusion of the work, but his name is repeatedly invoked as an authority on weather, plants, animals, and the power of the rational soul. The teachings of Ibn Sīnā's natural philosophy came to circulate widely in the courtyards of scholastic education, promoted and reformulated by the likes of Rāzī and Qazwīnī's teacher Abharī. In addition to the vast scope of his writings, among the many reasons Ibn Sīnā proves useful to Qazwīnī is that he had developed tools for discussing wonder. He speaks in a language that pulls at the very fabric of reality itself.[37]

One of the most influential pieces to Ibn Sīnā's metaphysics is the view that miracles and extraordinary phenomena can be explained in scientific terms. A close reader of the *Pointers and Reminders,* Qazwīnī draws directly on Ibn Sīnā's rationalization of the strange and marvelous. He does so first by developing the philosophical premise that by contemplating the wonders of creation one ultimately comes to know the existence of the divine Creator.

In the introductory remarks on the key terms in the title of his collection, Qazwīnī offers a definition of *ʿajab,* wonder, that fits into the scholastic study of philosophy. "Wonder," he explains, is "perplexity that occurs in humans due to their lack of knowledge concerning the cause of something, or lack of knowledge concerning the influence of the cause in the effect produced." In the *Cure,* Ibn Sīnā pairs together perplexity or confusion, *taḥayyur,* as a form of astonishment, *taʿajjub.* Similarly, Ibn Sīnā notes when discussing the attractive power of magnetic stone that people do not go searching after the causes of matters that they habitually witness and to which they have grown accustomed: "as astonishment disappears, so too does the desire to search for the underlying cause." Thus, to ponder and reflect upon the world requires a constant sense of perplexity in the search for new knowledge. Ghazālī takes this idea, that through familiarity we lose a sense of amazement, to argue that contemplating the quotidian with awe and perplexity is a spiritual exercise for appreciating God's creation.[38]

Comparable sentiments on the loss of wonder through habituation trace back to the earliest stages of Greek philosophy. It is a problem raised by Plotinus in the *Enneads*. An influential Arabic paraphrase drawn from large portions of the Greek text was produced in ninth-century Baghdad. The Arabic translation came to widely circulate, not as the teachings of Plotinus, but as the *Theology* of Aristotle. Importantly in Arabic the work offered an influential framework for fathoming the power of the soul governing an interconnected cosmos, the reality of magic as a physical force capable of producing wonders, and the significance of contemplating in awe the unity of divine design as a spiritual discipline.[39]

For his part, Aristotle approached wonder as an epistemic problem, in a set of associations that came to inform later developments in Islamic philosophy. In the introduction to the *Metaphysica,* Aristotle famously builds on Plato's earlier association of wonder with the development of philosophy, stating that it is through the act of being astonished that humans began to philosophize. Aristotle explains that wonder arises out of a curiosity to uncover the puzzles of existence, starting with obvious difficulties and progressing toward greater perplexities, such as the changes of the Moon and of the Sun, the movement of the stars, and the origin of the cosmos. We experience wonder, Aristotle concludes, when we are perplexed and ignorant of the cause of a given phenomenon.

Wonder is thus produced by a necessary perplexity about the order of the universe that leads to curiosity and contemplation, and ultimately concludes progressively with knowledge of truth. The emotion of perplexity, referred to as *aporia* in Greek and *ḥayra* in Arabic, stimulates the desire to uncover the hidden causes animating the world. In this sense, perplexity is not the end of philosophy but a primary stimulus. This idea leads to the claim that through wonder at creation one comes to know God.[40]

Just as astonishment is the provenance of the human condition when beholding the majesty of creation, the obverse sentiment is also repeated by countless Muslim authorities: God, omniscient, does not experience wonder in the sense of not knowing the cause behind something. Muslims could turn to the Quran and sayings of the Prophet to affirm that God, majestic, did indeed feel wonder or amazement. But as a divine attribute, this variety of wonderment, *ʿajab,* is not born out of ignorance. Rather, it arises though a sense of contentment or satisfaction, feeling pleased or gratified at the course of events.[41]

Likewise, wonder's corollary as contentment is fondness, as in *ʿajiba ilayhi,* to like or love something. When applied to oneself, such feelings can be rather repugnant, summed up in the expression *uʿjiba bi-nafsihi,* to have self-admiration or to be haughty, connected etymologically to the related term *ʿujb,* meaning pride, vanity, and conceit. Ghazālī dedicates an entire chapter in his *Revival* to pride, or *ʿujb,* as an abhorrent human trait that leads to perdition and stands as an obstacle to purifying the soul before the sublime power of God. In contrast, Ghazālī views contemplation, *tafakkur,* done with humility, to be a key form of self-discipline on the path of salvation. Here the contemplative life is the pinnacle of spiritual exercise. For Ghazālī, among the greatest objects of reflection are the wonders of God's creation, from the human body, produced from a lowly drop of sperm, to the perfectly balanced movement of the heavens. Far from self-admiration, the desired effective state produced by such contemplation is amazement at the complexity of providential design.[42]

These broad associations circulated widely in standard lexical collections, studies of the Quran and hadith, philosophical compendiums, anthologies of stories, and manuals of ethics. For instance, the erudite bookman of Baghdad Abū Ḥayyān al-Tawḥīdī (d. 414/1023) defines *taʿajjub,* the feeling of astonishment or admiration, as the desire to discover the cause and reason of any unknown issue. It is an explanation in line with Aristotelian metaphysics. On the topic, Tawḥīdī relates a question that Plato was said to have posed his students in Athens: "What," the philosopher inquired, "is the most amazing thing in the world?" Some replied the sky or the stars, others daily sustenance or humankind. Finally, Aristotle offered up a concise response: "The most amazing thing of all is that whose cause remains unknown." It is a standard definition shared by the famed ethicist Abū ʿAlī Miskawayh (d. 421/1030), who, in conversation with Tawḥīdī, approaches astonishment largely as perplexity or ignorance before unfamiliar phenomena. In Qazwīnī's day, similar sentiments are taken up by the physician-philosopher and general skeptic ʿAbd al-Laṭīf al-Baghdādī (d. 629/1231). In his own treatment of metaphysics, ʿAbd al-Laṭīf offers a paraphrase of Aristotle's foundational arguments on wonder and contemplation, arguing that progressive stages of knowledge obtained by wondering at the diverse phenomena of creation ultimately lead to knowledge of God.[43]

Look at the beehive, Qazwīnī instructs, at the precision of its form, with symmetric hexagons of equal size; contemplate that mere bees can produce a complexity that skilled engineers and artisans fail to achieve; in preparation for the winter these humble insects completely fill their hive with honey. "This is the meaning of the marvelous," Qazwīnī insists. While children wonder at such sights, with the progression of age and the maturity of the intellect, adults, caught up in the daily din of work and desire, cease to see the world with eyes of curiosity and awe, having forgotten the sensation of what it means to wonder. Only when presented suddenly with some strange animal, rare plant, or miraculous act, "do we cry out to God almighty with words of praise."

For Qazwīnī, wonder does not just end with recognizing that bees created beehives. Perplexity endures, for there is a greater cause behind it all. Schooled in a theology concerned with assaying causal powers, Qazwīnī stresses that God is the ultimate, primary cause for all creation. Thus, wonder does not cease by simply uncovering the immediate cause behind something. Rather, contemplation is merely the first step toward greater levels of astonishment at the sublime wisdom of a transcendent God whose magnificence and intelligence outstrip all human capacity to fully contain or delimit. There is, in this sense, no end to wonder. The failure to feel astonishment or perplexity before creation is not a marker of sophistication, but instead a sign of ignorance.[44]

Affirming the point in terms that parallel how both Ibn Sīnā and Ghazālī approached the loss of wonder through habituation, Qazwīnī notes that throughout life we are confronted with matters that "readily perplex those with rational intellects and souls of the insightful" but that the negligent quickly overlook. Anyone who wants proof of this need only cast an eye of discernment at the perpetual motion of the heavenly bodies. Look at the planets, the Sun and the Moon and their yearly course through the heavens, their variegated hues, the cycles of day and night, the change of seasons. The human body is itself a boundless microcosm of wonder, Qazwīnī observes, echoing Ghazālī, who in turn closely read Galen's arguments for divine design through the wonders of human anatomy. From the celestial spheres above to the minerals below the surface of the Earth, we are called to see the world anew.

Qazwīnī's homiletic treatment of wonder seeks to cultivate a renewed awe in the face of the familiar. His turn to *gharīb,* wonder's primary corollary in Arabic, however, focuses on the rare, strange, weird, and extraordinary. Qazwīnī defines *gharīb* as "every seldom-occurring wondrous matter that diverges from customary phenomena and commonly witnessed experiences [*mukhālif li-l-ʿādāt al-maʿhūda wa-l-mushāhadāt al-maʾlūfa*]." In this definition, Qazwīnī explains that strange divergences occur "through the influence of powerful souls, celestial forces, or elemental bodies, all of which are determined by the power of God almighty and His will."

To further illustrate the point, Qazwīnī lists as examples the miracles of prophets and saints, the prophecies of soothsayers, the power of jinn, the evil eye, and the innate disposition, or *fiṭra,* of certain souls for paranormal mental capacity often associated with Indian sages. Here too are various forms of divination and astrological prognostication, as well as rare astral, meteorological, and geological phenomena, such as comets, meteors, blizzards in the middle of the summer, violent hailstorms, powerful earthquakes, clouds that rain fish and frogs, lands that disappear into the sea, large bodies of water that evaporate, as well as prodigious births, speaking infants, and talking animals.[45]

Once again Ibn Sīnā proves invaluable. Qazwīnī does more than just reference the philosopher's personal experiments with meteorites. Qazwīnī's entire definition of *gharīb* is drawn in the shadow of Ibn Sīnā's metaphysics. As Qazwīnī explains, without directly citing the great master, philosophers have established three categories for extraordinary phenomena, referred to as *al-umūr al-gharība:* (1) psychic influences and subsidiary reactions on the faculties of the imagination that occur without any direct intermediary—when used for good, these produce either prophetic miracles or saintly wonders; when used for ill, they represent the illicit magic of evil souls; (2) talismans that function through their connection by form, shape, and location with special celestial forces and elemental bodies; and (3) *nīranjāt,* potions and amulets, which produce strange phenomena through the power of terrestrial bodies, such as how magnetic stones attract iron.[46]

The category of extraordinary phenomena, *al-umūr al-gharība,* the three-tiered division, and the examples given all directly parallel the treatment of the topic in the final section of the *Pointers and Reminders.* Here Ibn Sīnā presents a concise scientific basis for paranormal phenomena, which includes a general theory of prophecy and miracles. In earlier writings, Ibn Sīnā had developed other divisions of the extraordinary, but in this particular work uses the following classification: (1) a psychic disposition of the soul as the basis for both miracles and magic; (2) the unique properties of elemental bodies referred to as *nīranjāt,* such as the force of magnets to draw iron; and (3) talismans. Ibn Sīnā classifies the art of fashioning talismans and the elemental magic of *nīranjāt* as branches of natural science. Both, he explains, seek to produce extraordinary reactions on Earth. Talismans achieve this by mixing celestial forces with earthly bodies. In contrast, *nīranjāt* are drawn from unique elemental properties, *khawāṣṣ,* as the active basis for spells, amulets, and potions. As for his first category, referred to as a particular psychic disposition, *al-hayʾa al-nafsāniyya,* this spiritual force can be harnessed by saints and prophets, who through the purification of their souls are capable of producing miracles, as well as by evil sorcerers, who wield extraordinary powers even though they have not reached the same degree of perfection. Here what distinguishes the saint from the sorcerer is the individual soul's disposition for good or ill.[47]

Other than slightly transposing the order of the list, Qazwīnī follows the same classification with its tripartite division. Arguably more importantly, Qazwīnī fully accepts the naturalization of magic and miracle, a central tenet of Avicennan metaphysics. Significantly, the *Pointers and Reminders,* a highly concise text, enjoyed several major commentaries over the course of history. This specific taxonomy of the strange and uncanny became quite well known, as the *Pointers* circulated in philosophical circles and in madrasa education.

Qazwīnī then discusses the evil eye as an extraordinary phenomenon. This also intersects with Ibn Sīnā's treatment of the topic in the *Pointers*. Those who possess the evil eye, Qazwīnī explains, can inflict harm through feelings of amazement toward a given object or person. It is the very emotion of admiration, *taʿajjub,* in such people that produces damage through a unique quality of the soul. The explanation again parallels Ibn Sīnā's philosophical handbook, which highlights the psychic power of the soul to influence other bodies without any direct intermediary or physical body. The emotive process at work suggests intersections among wonder, fascination, charm, and admiration. Such feelings, in turn, have measurable influences on the world.[48]

Emotions can exert powerful forces over the body and are themselves central facets to Ibn Sīnā's psychology. As an explanation for the power of the soul to influence other bodies, Ibn Sīnā offers the famous example of how easy it is to run over a tree trunk laid on the ground, but if the same plank of wood were placed across a deep chasm, a person would quickly be seized with fear and trepidation, imagining in the soul the experience of falling. The idea of psychic influence is at the heart of Ibn Sīnā's explanation of the evil eye, and it guides his understanding of the soul's capacity to produce miracles and magic. It also provides the theoretical basis for *The Hidden Secret,* Rāzī's practical guide to astral magic, pneumatic forces, and talismanic recipes.[49]

Qazwīnī's master Abharī read and taught the *Pointers,* and both Rāzī and Ṭūsī published influential commentaries on it. The bond between magic and miracle, presented here as two sides of the same coin, had far-reaching repercussions for the history of Islamic thought. Such ideas, as scientific explanations, lend context to the wide acceptance of occult learning as a licit field of natural science. Rāzī, for one, when discussing this passage in the *Pointers* references his own detailed treatment of the topic in *The Hidden Secret,* a work that, in his own words, delves into *al-ʿulūm al-ghāmiḍa al-sharīfa,* the noble hidden or occult sciences, which are "amazing and terrifying, beneficial and harmful."[50]

The categories for Ibn Sīnā's three-part division of extraordinary phenomena point to distinct registers evoked by the strange and the marvelous. The term *muʿjizāt* was developed as a technical term of Islamic theology to refer to the inimitable miracles of prophets. These were distinguished in name, if not in kind, from *karāmāt,* marvels produced by sages and saints. The other two categories, in contrast, have very different pedigrees. The word *ṭilasm* derives from *telesma* in Greek, via the Syriac *ṭelesmata*, whereas *nīranj* comes from the Middle Persian *nērang* and the universe of Zoroastrian ritual incantations.[51]

Both the talisman and the *nīranj* are associated with harnessing hidden forces in nature, the first through astral powers drawn into an object, usually carved with some form of writing or representation, and the second through the elements binding together the cosmos. Both also build on the efficacious power of language. Talismans generally are activated through prayers, mysterious cyphers, illegible scripts, and sacred symbols designed to capture planetary and astral forces of the heavens within the medium of engraved objects activated through prayers at particular moments of planetary and stellar alignment.

In contrast, *nērang*, at least as incantations, potions, and amulets from Zoroastrian cultic life, are often associated with the efficacious power of ritual recitation drawn from the hymns of the Avesta, which function as a sacred formulary. These spells are activated through elemental influences of water, air, fire, and earth in both the visible and the invisible worlds, and are connected to the astral movements, as well as to the conditions of the various regions of the world and to the humoral and elemental forces governing health and well-being. They also build on dualistic beliefs in the cosmic power of good and evil forces. While *nīranjāt* in early Arabic and New Persian sources echo a good deal of these ideas, the larger Zoroastrian cosmography is largely suppressed and abandoned, as the term takes on a life of its own.[52]

When spoken by early Persian converts, the term *nīranj* could evoke efficacious power as well as the cosmic language of religious outsiders, given its strong associations with Zoroastrian ritual and Avestan formularies. The same rings true for the term talisman, a loanword from Greek. This is not entirely distinct from, say, the etymology of the word "magic," from the Greek *mageia*, the activity of the *magos*, from the Old Persian *magu-*, deployed in Greek often as a derogatory means of referring to Zoroastrian priests and their ritual activities. In this sense the label "magic" has long been a way of infusing the religious beliefs and practices of outsiders with strange or hidden powers.[53]

In Ibn Sīnā's schema, the faculties of the intellect form the natural basis for the capacity of the soul to act directly on other bodies or other souls through a form of paranormal causation. Prophetic miracles are the complete realization of this natural capacity within the form of an individual who, through inherent disposition, has obtained a level of intellectual and spiritual perfection. As a physical force, such extraordinary powers of the soul could be used to achieve both beneficial and harmful results.

Rāzī takes up these arguments to affirm that the study of astral magic is legitimate, even as it draws on sympathetic forces of the stars and the psychic powers of the imagination, with suffumigations, charms, incantations, spells containing incomprehensible words, the sacrifice of animals for summoning forces, and the invocation of planetary spirits. This learning builds on the physical forces in the cosmos. Thus the question is rather the ends to which such practices are deployed and not the actual validity of the science itself. This is not entirely dissimilar from how knowledge of medicine can be used to heal or harm. Indeed, just as Rāzī's handbook is filled with therapeutic recipes for a variety of ailments, it also lists dangerous poisons and harmful spells. Rāzī also stresses that only those with good dispositions who abide by the precepts of religious law and belief should pursue this body of learning.[54]

But even without training, the psychic power of the rational soul could exert all variety of strange influences. Among the examples of extraordinary phenomena, Qazwīnī offers the cases of those with prodigious capacities to read the stars, not from formal study or accurate observations through the use of astrolabes, but merely from a unique influence of the faculty of the soul that guides them, providing insight into the unfolding of events. The point is of note. While Ibn Sīnā criticized astrology as a speculative science, lacking foundation built on demonstrable proof, he also invested the rational soul as it emanates through existence with a divine power to produce marvels and miracles. Ibn Sīnā's objections were not aimed at the physical power of the stars to influence human life. Rather, he argued that the complexity involved in measuring cause and effect is simply beyond human intellectual capacity to master.

Ibn Sīnā emphasizes the necessity of suspending judgment, *tawaqquf,* in the face of extraordinary phenomena, and cautions his students not to quickly reject what they fail to understand. This sentiment—that nature is itself pregnant with the strange and marvelous—proved particularly appealing to theologians who sought a scientific language to account for miracles and magic, both of which are confirmed unambiguously in the Quran and Sunna.

It goes some measure to also explain the wide appeal of Avicennan metaphysics across sectarian divides and religious communities.

This view of natural physical forces gave license to the pursuit of *al-ʿulūm al-gharība,* the strange, extraordinary, or paranormal sciences, as they were frequently known, if rarely mastered. These fields of learning often included various forms of astral divination. Although not its main focus, references to judicial astrology feature throughout *Wonders and Rarities.* A rather telling example is Qazwīnī's discussion of the special properties of the Sun, where he references the ancient theory of the great year, the moment when the Sun had reached its apogee—its furthest distance from the Earth—in each of the twelve signs of the zodiac, after a course of thousands of rotations.

Qazwīnī notes that according to the Brahmins, renowned for their mastery of the stars, this revolution takes 36,000 years to complete. Once the cycle is consummated, the Earth will suffer cataclysmic events: the ocean will dry up while the inhabited Earth will turn into a great sea; the south will become the north and the north will become south. The theory of cosmic upheaval through the movement of the stars was by no means unique to Indian astrology, as the idea also circulated in ancient Babylonian, Greek, and Iranian astral learning.[55]

This vision of the stars and their power also intersects with the field of historical astrology and its focus on dynastic cycles to predict in millenarian terms the coming of new world orders. Qazwīnī took judicial astrology quite seriously. While he acknowledged that Ptolemy's work on astrological teachings was not built on demonstrable proof, he also recognized that it could claim the highest level of probability. The science of predictive astrology represented years of repeated experimentation and calculation based on observing events and their correlation with the movement of the stars and constellations. Qazwīnī comments that in his own day only a very few could actually master the complexity required to read the movement of the heavens.

For Qazwīnī even more impressive than Ptolemy was the *Aḥkām* (Decrees) ascribed to the legendary Jāmāsp, the renowned minister and disciple of Zoroaster. A work of apocalyptic astrology drawn from horoscopes written in the voice of the ancient Persian sage, the collection circulated widely in both Arabic and Persian redactions. Whoever wants to see the validity of astral learning, Qazwīnī advises, need only pick up a copy of Jāmāsp's decrees.

There was never a master of the stars like Jāmāsp, for though he lived before both Moses and Jesus, he accurately foresaw the coming of their prophetic missions. He foretold the end of Zoroastrian rule and the rise of the Arabs and Islam. He predicted the coming of the Turks and the slaughter they would bring. Redactions of this collection of judicial astrology and cycles of dynasties written in the voice of Jāmāsp circulated widely in Qazwīnī's day in both Arabic and Persian. The language of marvels and miracles features throughout. Notably, there are copies in which Jāmāsp further forecasts the coming of Genghis Khan and the impending destruction of the Mongol invasions.[56]

The uncanny power of the stars was well known to Qazwīnī. In the discussion of the apocalyptic chaos that the cosmic revolution of the Sun ultimately would wreak, Qazwīnī notes dryly that the current year is 678/1280. This is also the year that appears in the copyist's colophon at the end of the copy of *Wonders and Rarities* produced in Wasit, the final surviving authorial redaction of the work. For each of his earlier versions, Qazwīnī also supplied the current date, returning several times to this passage and updating it according to the current year. These dates provide some sense of the various revisions Qazwīnī made during the course of his life. The earliest surviving recension takes note of the solar apogee in year 658/1259~, as Mongol armies continued their forward march.

All along, the apogee of the Sun remained in Gemini, the third of the twelve signs in the zodiac, as it had for centuries. In the coming years, scholars in the observatory of Maragha would also record the movement of the solar apogee, and though it had traveled some distance since the days of Ptolemy, it still dwelled in the house of Gemini. There was, it would seem, quite some time yet before the apogee would complete a circuit through the other nine constellations, when the waters of the ocean would once again consume the Earth. But on this, as with so much else, Qazwīnī concludes that God alone knows the truth of such matters.[57]

5

TERRESTRIAL DESIGNS

AS THE HEAVENS MOVED with the rhythm of the seasons, the plaster began to chip and the clay bricks cracked. Over the winter months, snow blanketed the hilltop and covered the mountains that stretched up along the horizon. The long shadows cast by the towers and the circular platforms slowly retreated, as the Sun returned on its yearly climb toward the north during its summer ascent. Once more the buildings tiled with floriated calligraphic inscriptions and the open-air structures built to house finely calibrated instruments of wood and metal faced the midday heat. Then came the rivulets carved out by rain.

Within just a few generations, the observatory at Maragha, funded by Hülegü and led by Naṣīr al-Dīn Ṭūsī, was deserted. The rooms, once bustling with dignitaries, scholars, and Sufis, were empty. The books of the library, many of which had been swept up in the course of the Mongol invasions, were dispersed across the valley, carted off volumes at a time. With the precipitous fragmentation of the Ilkhanid confederation and the successor states that then vied with each other, the observatory was abandoned. Over the years a passerby might comment on substantial ruins that could still be seen. Eventually the numerous buildings receded into the ground, save for the faintest outlines that can still be made out among the rubble.[1]

While almost nothing of the original complex on the bare hilltop remains, numerous traces escaped the ravages of time. There is the actual record of observations made by Ṭūsī and his circle, along with the various theoretical models they devised for correcting Ptolemy and refining earlier calculations. The monumental measuring devices, tall as multistoried buildings, have disappeared. But manuscript copies of a treatise survive on the astronomical instruments designed specifically for the observatory by Mu'ayyad al-Dīn al-ʿUrḍī (d. ca. 665/1266), who traveled from Damascus to join Ṭūsī's team. Directions are given for making numerous devices—such as a double copper quadrant within a circular structure, used to observe the altitudes and azimuths of celestial bodies; an armillary sphere for tracking astral coordinates; even larger armillary spheres for determining the obliquity of the ecliptic and the Sun's entry into the equinoxes; and a giant mural quadrant aligned with the meridian, for measuring the transit altitudes, the highest point a star reaches in the sky.[2]

ʿUrḍī commences his short, illustrated treatise by explaining how these instruments can be used to determine the movement of the stars, their locations in the heavens, their distances from the Earth—the very heart of the cosmos—and their courses as they travel through the sky. The practical discipline of mathematical instrumentation, he explains, serves to complement the theoretical field of metaphysics, and for this reason it is honored for both its subject and its demonstrative precision, "for the heavens are its object of inquiry, the greatest of God almighty's creations and the most wondrous of His designs." The scholars at the observatory drew on these instruments to refine earlier star catalogues, planetary tables, and celestial globes that cast the night sky onto the shape of a sphere. Among ʿUrḍī's many writings, there also survives his treatise on how to use a celestial globe. Perhaps most notable of all the fragmentary material testaments to this intellectual activity is a finely inlaid globe that ʿUrḍī's son, Shams al-Dīn, produced from brass and silver while working at the observatory toward the end of the thirteenth century.[3]

Qazwīnī's illustrations of asterisms, constellations, and stellar movements evoke this highly specialized field of astral learning without ever delving into the details of calculations or theoretical models. The technical language that accompanied Ptolemy's theory of planetary motion at the foundation of classical astronomy can be heard here and there, counting off the apogee and perigee, the deferent and epicycle, with eccentrics and equants. Yet as Qazwīnī makes clear, *Wonders and Rarities* is more concerned with providing a concise digest. While anchored in the higher orders of science, Qazwīnī's compendium does not actually seek to account for these complexities in great detail.

For Qazwīnī the terms of art and allusions to sophisticated discussions in natural philosophy and theology serve a greater purpose. Contemplating creation is a pious activity designed to affirm God's majesty and to appreciate the interconnected beauty and strangeness of existence. The rational explanations of natural phenomena lend the exercise authority. Hot water freezes quicker than cold water. Rain clouds are generated through the evaporation and condensation of the sea. The Moon takes its light from the Sun, which is so powerful that it blocks out all the stars during the course of the day. The spectacular colors of rainbows are formed by the reflection of light through drops of dew. The facticity of these explanations speaks to empirical experience, drawn entirely from a corpus of trusted authorities.

Qazwīnī's collection came to travel across distant lands in part because it pointed to practical means for both explaining and harnessing existence, with entertaining stories of seldom-seen shores. Along the margins of the world could be encountered all variety of singular animals and peculiar creatures, some with bemusing shapes, others with terrifying forms. In this, *Wonders and Rarities* has much in common with earlier writings from antiquity that sought to catalogue strange and rare phenomena. Humor and pleasure produced by beholding diverse possibilities and combinations also lent many of these works a lasting allure.

The illustrated star catalogue that opens Qazwīnī's first chapter, on the heavens, follows the same iconographic sequences of images engraved on Arabic celestial globes. One of the most influential sources for depicting constellations on both flat folios and round spheres was the *Kitāb Ṣuwar al-kawākib al-thābita* (The book of diagrams of the fixed stars) by ʿAbd al-Raḥmān al-Ṣūfī (d. 376/986), which circulated in numerous illustrated manuscript copies in Qazwīnī's day. One of the more notable is a manuscript that passed through Qazvin and drew on the most recent astronomical teachings from the Maragha observatory. Ṭūsī found the manual so valuable that he translated it into Persian. As with a good deal of the scientific and philosophical learning produced in Arabic, Ṣūfī's *Fixed Stars* also enjoyed an audience in Latin Christendom. Like the skilled artisans who fashioned celestial

FIGURE 5.1: Constellation of Centaurus (*qinṭūris*) grasping the constellation of Lupus (*sabʿ*) in the Wasit copy.

globes, Qazwīnī took inspiration from Ṣūfī's iconography and descriptions, which had built upon and refined Ptolemy's list of more than a thousand heavenly bodies connecting forty-eight constellations (figure 5.1).[4]

Celestial globes were generally suspended on stands by three rings crossing each other at right angles representing the horizon, meridian, and the zenith. They were calibrated for various practical purposes, with intersecting circles that marked the celestial equator and the ecliptic path of the Sun through the heavens, along with latitudinal and longitudinal divisions of the night's sky. By taking measurements of the stars on one of the larger celestial globes it was possible to calculate the unequal hour on a given day in any particular town or city, which varied with the seasonal movement of the Sun. In addition to measuring the passage of time, a celestial globe could help determine the azimuth of the direction of prayer, marking the *qibla* for any geographical location.

Furthermore, with attention to the movement of the zodiac, celestial globes could be used to observe the houses of a horoscope at any moment by tracking the ascendant star sign. Measurements could predict both the movement of the Sun as it crossed the vernal equinox, the so-called year transfer, and the nativity-transfer, the juncture when the Sun returns to the position in the zodiac where it stood when an individual was born. All bear significance for casting horoscopes. Such calculations were designed to harness the predictive power associated with the stars and the instruments used to measure them.[5]

Qazwīnī not only records images for all forty-eight constellations that populate celestial globes, citing the authority of Ptolemy's *Almagest*, but he also includes a section on the lunar mansions, known in Arabic as *manāzil*, signifying stations or abodes. Here are diagrams of asterisms, or star groups, for each of the Moon's twenty-eight mansions (figure 5.2). The system of lunar

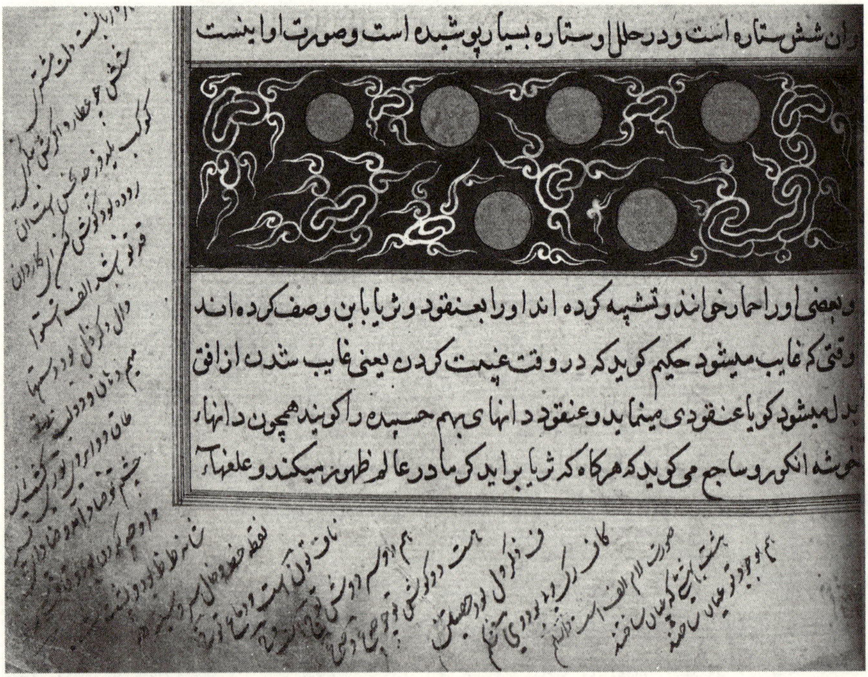

FIGURE 5.2: Dot asterism of the third lunar mansion identified with *Thurayyā*, the Pleiades, from a Persian translation commissioned for Ibrāhīm ʿĀdil Shāh in 954/1547, manuscript copied before 1066/1656 in the Deccan, with marginalia added in the early eighteenth century.

stations was developed in pre-Islamic Arabia as a means for measuring the passage of seasons and for predicting changes in weather. The twenty-eight mansions divide the solar year into periods of approximately thirteen days.

These stations in turn correlate with star groupings from the zodiac as the Moon follows along the general ecliptic of the Sun. Referencing the mnemonic tradition of Arabic poetry and proverbs for tracking the progress of the Moon through the heavens, Qazwīnī also explains when to plant and harvest various crops according to the calendar of the Moon's movement, as well as how to predict the coming of rain. Well before Qazwīnī, the Arabic system of lunar mansions had incorporated elements from Indic and Hellenistic astral learning and had developed into a complex means for reading the night sky.[6]

As with the constellations, in Qazwīnī's day celestial globes could be calibrated to measure the annual progress of the Moon through its twenty-eight stations, starting with Aries and ending with Pisces, thereby passing through all twelve signs of the zodiac. In addition to its importance for agriculture and meteorology, the system of lunar mansions was also connected to techniques for mastering the power of the stars. Early Arabic treatises on talismanic magic, many written in the names of Hermes and Aristotle, link the lunar mansions with specific timings for fashioning talismans and *nīranjāt* recipes for producing potions and amulets.[7]

Both the theory and the iconography of lunar mansions intersect with the development of terrestrial astrology, referred to as *ʿilm al-raml,* the science of sand. This branch of learning came to be known in Latin as *geomantia,* whence the modern English term geomancy. As a divinatory science, the study of geomancy developed intricate calculations, using dots drawn on a tablet or in the sand, to produce prognostications covering all aspects of human life. Although Qazwīnī makes no reference to geomancy in his treatment of the lunar mansions, there is reason to believe that he would have been familiar with the art, which over time became increasingly important in the repertoire of occult learning. The science of sand was studied by Fakhr al-Dīn al-Rāzī as well as by Naṣīr al-Dīn al-Ṭūsī, who was considered a leading authority in the field.[8]

In his travels through the markets of Baghdad, Damascus, and Mosul, Qazwīnī likely came across finely tuned devices used for astrological prognostication. A particularly elegant mechanical instrument for practicing geomancy survives from the period, made in 639/1241~ by a skilled artisan from Mosul for a wealthy market inspector. Fashioned of brass inlaid with silver and gold, the device works through a set of twenty rotating dials and

FIGURE 5.3: Detail of instrument for divination through geomancy dated 639/1241~, with dot asterisms of the lunar mansions, copper inlaid with gold and silver, the work of Muḥammad ibn Khutlukh al-Mawṣilī.

four sliding levers, which when calibrated together are designed to predict the future and anything hidden from immediate perception (figure 5.3).

The largest dial bears the inscription "We devised this dial so that you might learn from it the semblances of the forms of the figures with the forms of the rising and setting mansions." The figures referenced are the dot patterns at the center of geomancy, which here are said to have an explicit correspondence with the asterisms of the lunar mansions. The groups of divinatory dots on the device bear a striking resemblance to the stellar diagrams of the lunar mansions known to Qazwīnī. "I am the revealer of secrets," a poetic description on the face of the instrument announces, "in me are marvels of wisdom, extraordinary occurrences, and hidden matters."[9]

With the fine movement of dials and spheres, time itself could be measured, if not mastered. And it is with the subject of time that Qazwīnī concludes his opening study of celestial phenomena, the first in his three main classifications of creation. The treatment of the heavens opens with the spheres of the seven planets, known as *al-kawākib sayyāra,* a calque of the Greek *planētes asteres,* the wandering or moving stars. Their cycles can be observed in the orbits of the Sun and the Moon and the retrograde movements of Mercury, Venus, Mars, Jupiter, and Saturn.

With accounts of their orbits and special properties, the seven planetary spheres are followed by a description of the fixed stars, so called because they do not appear to change position in relationship to each other as they progress across the heavens. Like the planets beneath them, the fixed stars are situated within their own sphere, the eighth in the cosmic structure. Crossing the northern and southern hemispheres are entries on the forty-eight constellations, twelve of which represent the signs of the zodiac, as well as the twenty-eight lunar mansions, which collectively form part of the fixed stars, whose number is greater, Qazwīnī notes, than all the grains of sand in the desert.

Moving at different speeds, these eight spheres are nestled one within another, from smallest to greatest. Expanding outward, they all are surrounded by the ninth celestial orb, known as the *falak al-aflāk,* the sphere of all spheres, also referred to as the sphere of Atlas, a phrase that by all measures echoes the ancient Greek story of Atlas supporting the world on his back. This great orb encompasses the entire cosmic structure, with the Earth at its center. The outer sphere, the *primum mobile* of Ptolemy's cosmology, sets all the celestial bodies in motion (see figure I.3). From here Qazwīnī turns to the celestial inhabitants, with a cavalcade of angels in an array of forms.

The vantage from the final sphere looking inward toward all the moving parts of the cosmos is the perspective generated on the surface of a celestial globe, whose most practical use was telling time. Qazwīnī saw time as a function of the heavens, either in the sequence of day and night, or, following Aristotle, as a measure of the movement of celestial bodies. In this sense, centuries, years, months, days, and hours could all be used to measure the distance traveled by the wandering stars and the fixed constellations.[10]

On the question of whether anything exists beyond this outer realm, Qazwīnī turned to Fakhr al-Dīn Rāzī, who argued that the cosmos was neither bounded nor finite. This position stands against the doctrines of philosophers such as Aristotle who believed that beyond the celestial sphere was neither void nor plenum, *khalāʾ wa-malāʾ,* meaning that there was nothing else apart from the cosmos itself. On this matter Qazwīnī follows the teachings of both Abharī and Rāzī. He does so, though, without developing any of the intricate arguments used to defend the possibility of a void and with it the plurality of worlds. Instead he simply echoes Rāzī's statement that "the measure of reason is incapable of delimiting the kingdom of God."

This sentiment of boundless cosmic potential accords well with theological sensibilities which often sought to harmonize philosophical teachings with traditional religious learning. This can be seen in the equivalences Qazwīnī draws between scripture and the authority of philosophers, such as the notion that the *kursī*, the pedestal of God mentioned in the Quran, corresponds to the eighth sphere of the fixed stars, or that the *ʿarsh*, God's throne, represents the ninth orb, which encloses all the spheres. "God alone knows whether this interpretation is correct or not," Qazwīnī explains, "but as to the existence of the pedestal and throne, there can be no doubt, for they are established by authoritative Quranic texts." The position accords with standard Ashʿarī theological teachings that place a limit on the capacity of humans to fathom the simultaneous transcendence and immanence of God.[11]

As for the cause and effect driving the movement of the heavens, Qazwīnī upholds the established belief that while the cosmos proceeds from God's decree, angels play vital roles by helping to keep the entire structure in motion. Qazwīnī defines angels as immortal, pure, spiritual substances, not attached to physical bodies, free of desire and anger, and possessing rational intellects. The greatest is known as the Rūḥ, the "soul" or "breath," who keeps the orbs on course and the stars in motion, a force greater and more powerful than the entire celestial sphere and all physical matter of the cosmos, capable, on God's command, of stopping the movement of the heavens and thus bringing time to a standstill. Angelic forces were delegated the role of maintaining the cosmic order. This view echoed philosophical theories, promoted by the likes of Fārābī, Ibn Sīnā, Ghazālī, and Rāzī, according to which angels function as intermediary expressions of the divine intellect animating existence.

According to tradition, angels could come in countless sizes and shapes, such as bulls, lions, horses, eagles, roosters, peacocks, trumpeters, and scribes (see figure C.1). God alone knows how many kinds or how many numbers fill the ranks among the heavenly hosts. Through divine decree, angels keep existence in perpetual motion. Some authorities hold that there is not even a single atom in the cosmos that is not animated by an angel or a group of angels. "So if this is the case with the smallest atoms and particles," Qazwīnī inquires, "then what do you suppose is behind the celestial orbs and the planets, behind the clouds, winds, rains, mountains, deserts, seas, springs, rivers, minerals, plants, and animals?" It is through angels that the cosmos thrives and that existence reaches toward perfection, all by the order of God.[12]

Qazwīnī follows a cosmology where cause and effect emanate from God through the forces of the physical universe, and where angelic mediation moves the realm beneath the Moon. He does so while largely eschewing thorny theological concerns that parsed such problems of predestination and secondary or efficient causation. In the face of God's complete majesty and wisdom, Qazwīnī prefers to focus on the bounds of human reason. The physical mechanisms of causation do not strip power away from God, "the Eternal whose being cannot be described by a beginning, the ever-lasting who never ends and has no limit." To distill sophisticated teachings of natural science into an accessible form, *Wonders and Rarities* does away with the lengthy demonstrative proofs of scholastic learning found in compendiums of philosophy. "Our book," Qazwīnī explains, "is not occupied with such matters."[13]

What matters is a good story. Threaded through technical descriptions drawn from natural philosophy on the heavens and the Earth are countless occasions for diversion. Poetry, proverbs, and pious tales mingle with strange anecdotes and wondrous sights. Traditional authorities of ancient lore, such as Kaʿb al-Aḥbār (d. ca. 32/652) and Wahb ibn Munabbih (d. ca. 110/728), jostle alongside tales of caliphs and quick-witted poetry by the likes of Abū Nuwās (d. ca. 198/814) and al-Mutanabbī (d. 354/965). Mixing utility and pleasure, Qazwīnī seeks to edify and delight. This comes to a head in the concluding discourse on time, the final section on the upper atmosphere.

No discussion of calendric models for reckoning and naming months, days, and years, among the Arabs, Greeks, and Persians, with digressions on the religious festivals of Muslims, Christians, and Zoroastrians, would be complete without a reflection on the passage of time. Time conveys its own bewildering patterns of staggering magnitude. God, for instance, is said to have sent a prophet every thousand years with "strange miracles and manifest marvels" to "lift the flags of His religion and to make manifest His straight path" for the entire world to see.

Through the passage of time, amazing phenomena unfold, in the birth of animals with bizarre forms on account of strange elements mixing together, or the generation of extraordinary minerals and wondrous plants through the change of atmospheric forces. Time's steady movement bears countless transformations. Regions once inhabited become abandoned and areas abandoned are now inhabited, as the sea transforms into the land, the land the sea, the mountain the valley, and the valley the mountain, all through divine decree.

In the face of these vicissitudes, Qazwīnī could hear plaintive cries. "Times have changed for the worse," Abū ʿAlāʾ al-Maʿarrī (d. 449/1058) once complained to the wonder of the age, Badīʿ al-Zamān al-Hamadhānī (d. 398/1008), one renowned litterateur to another. "You say that time has rotted. Well, tell me then, when was it healthy?" retorted Hamadhānī, author of the famed *Maqāmāt* (Assemblies), a collection of tongue-twisting tales filled with the adventures of itinerant rogues. Things have always been a mess, it turns out, from the Abbasids to before Adam. Even when the angels heard that God had decided to install Adam on earth as a successor or vice-regent, they inquired, "Will You put in place there someone who will corrupt it?" It's not that times have changed. They have always been this way, Hamadhānī concludes. Whether in thick or thin, the corruption of time, known as *fasād al-zamān*, was a frequent lament.[14]

While the orbs move in perfect perpetual motion, the Earth and its atmosphere are subject to the forces of generation and corruption, *al-kawn wa-l-fasād*, a central concept in the curriculum of Aristotelian natural philosophy. As the final topic in the chapter on the heavens, the discussion of time serves as a bridge between the celestial spheres and the realm beneath the Moon, the closest celestial orb to the Earth. In the terrestrial world of unceasing elemental flux, change is always constant. Following these forces of fluctuation, Qazwīnī draws the discussion of the heavens to a close with what he promises to be "an amazing tale."

It is the story of a pious youth from the Tribe of Israel who had befriended a saint known simply as Khiḍr, "the Green One." Qazwīnī's audience would have been immediately reminded of the Quran. In the sura of the Cave, an unnamed servant of God leads Moses on a series of adventures righting wrongs in mysterious ways through divine providence. According to established lore, this guide had knowledge of past events, for he had drunk from the fountain of life at "the confluence of the two seas" and thereby obtained immortality. After tasting the waters of perpetual youth, this servant of God is said to have taken on a green hue, or alternatively to have turned the ground he walked on into grass, whence his verdant honorific Khiḍr.

Traveling the world working wonders, Khiḍr had knowledge of the ages. Having witnessed the continual passage of time, Khiḍr features as a record of human history, and as a testament to the Quranic theme of ancient cities laid to waste as divine punishment for profligate ways. In Qazwīnī's story, once news of Khiḍr reached the king of Egypt, the immortal sage was summoned to the court. "What is the most amazing thing you have ever seen?" the king inquired. "As for wonders, I have seen many," Khiḍr replied, "but I will tell you what just happened to me."

"In my travels," Khiḍr continued, "I came across a vast, flourishing city. I asked a man I met, 'When was this city built?'" "Why, this city is ancient, neither my ancestors nor I know how old it is," was the reply. After five hundred years, Khiḍr returns, but not a single trace of the vibrant city remained. There he encountered someone gathering greens and inquired when was it that the city had been destroyed. "What city? Neither my ancestors nor I have ever seen a city here." Another five hundred years pass and Khiḍr returns only to find an ocean and fishermen nearby. "When did this land turn to ocean?" They too reply that for time immemorial it has always been this way. More centuries pass. The sea dries into a desert. But no one has any recollection.

One last time Khiḍr returns again to find a flourishing city, even more beautiful than what he had seen millennia before. But not a soul has any knowledge of what preceded. "Why, this city is ancient, neither my ancestors nor I know how old it is," he is told. On hearing the tale, the king exclaimed, "I want to abandon my kingdom to follow you." "You are not capable of this," the immortal saint replied, "But you would do well to follow this youth here, who can lead you to righteousness." It would not have been much of a leap for Qazwīnī's audience to recognize in this tale of obliteration a foreshadowing of how the pharaoh of Egypt would reject the prophecy of

Moses, Khiḍr's onetime companion, only to see his kingdom plagued by the wrath of God.[15]

It took less than five hundred years after the Mongol siege of Wasit for the city where Qazwīnī served as judge and professor to all but disappear. Had Khiḍr visited on his crisscrossing peregrinations to inquire what had happened to such a thriving settlement, he would have learned that neither hordes of foreign invaders nor internal strife brought the walls of the city to the ground. Over the centuries the riverbed that bisected Wasit and connected it as a vital artery to Baghdad in the north and Basra in the south, began to dry.

Meandering through the wetlands of southern Iraq, the Tigris, not so much fickle as indifferent, slowly began to shift course miles to the east, as it spilled into the marshes. Below, the Euphrates still greeted its northern twin. From their confluence, the ancient rivers continued to flow into the Persian Gulf, linking Iraq with distant trading routes across the Indian Ocean. Once renowned for fine reed pens made from the nearby marshes, and envied for its rich alluvial farmlands and vast tax revenues, Wasit had been abandoned.

Almost nothing of the city survives, save for a vaulted clay-brick gateway still lined with intricate geometric designs incised in terra-cotta, but shorn of the multihued tiles and calligraphic inscriptions that had once adorned its exterior. The gate rises above scattered desert shrubs, towering over the foundations of otherwise unidentified ruins. Some even say, given its location near the dried banks of the Tigris, that the arch, with its spiraling lines and interlacing rhythmic patterns bound together with gypsum plaster, once opened onto the courtyard of the Sharābī madrasa, where Qazwīnī taught and lectured—though there is no way to be sure.[16]

"Travel the earth and see how those who came before met their end!" Variations of this refrain are recorded throughout the Quran. Traces of once-grand buildings rising from the desert line the horizon as reminders of past peoples. Look at all the ancient kingdoms destroyed for their unbelief, goes the warning. So what will be your fate? The tracings of bygone ages not only mark the physical landscape, they shape geographies conjured in verse and prose. Early Arabic poetry born in the desert and recited before potentates often lingers along abandoned campsites of lovers lost and the grand colonnades of empires forsaken. Pausing to contemplate the *āthār* and *aṭlāl*, the physical ruins and remains, is a well-worn motif. One time-tested response was to leave graffiti; another was to preserve the memory of remnants in books.[17]

If the ancient tombs, temples, columns, and lintels were mute, it was only because the inscriptions that covered them remained indecipherable. The crumbling battlements, temples, and pyramids kept their secrets. The elites who had cultivated the divine art of their scripts had long since disappeared. Stories circulated of those who could crack the forgotten cyphers of the ancients, but this was a recondite art of dubious pedigree generally hidden in the pages of occult learning. Centuries after they had been abandoned, the angular scripts of Arabia, the pictorial hieroglyphics of Egypt, and the wedge-shaped cuneiform of Mesopotamia remained silent.

The cursive used to record the Quran, the first book written in Arabic known to history, did not suffer a similar fate. Not only did successive generations dedicate themselves to preserving the words of God and His messenger, but they also traveled with the Arabic script, a relatively novel writing system, refining and transforming it over time. Lining mosques and minarets, palaces, and shrines, Arabic's calligraphic power as the language of divine command filled the horizons. Promoted by empires and refined by scribes, in a few generations Arabic emerged as a language of learning and prestige that stretched across vast portions of the earth, proclaiming in its cursive lines a community of faith. *Wonders and Rarities* traveled along these well-trodden routes that united a republic of letters. While the city of Wasit met a slow demise, Qazwīnī's writings journeyed far beyond the city walls.[18]

Qazwīnī's was by no means the first collection of curiosities to move quickly through book markets or to be feted in royal treasuries. Well before Qazwīnī had taken up the subject, catalogues of wonders occupied shelves in the pantheon of Arabic and Persian letters. So did compilations celebrating the lives of prophets and saints, and the adventures of heroes and kings speak the language of awe in the face of ever-increasing feats of derring-do, cunning intelligence, and miraculous power.

Economies of rarity excited preachers and storytellers and swayed their audiences, in the pursuit of entertainment and edification. The emotive states of bewilderment moved poets and scholars, who developed sophisticated theories of eloquence that focused on marvelous figures of speech and the astonishing effects of language to evoke wonder. By delicately balancing form and content, *lafẓ* and *maʿnā*, eloquent speech could animate the psyche through the faculty of the imagination to conjure in the mind's eye images both real and imaginary. Similarly, the mental capacity to produce wondrous creatures that exist solely in the mind occupied natural philosophers in their pursuit of the realities of existence and the mental powers of conception to fathom God.[19]

Then there were the intriguing vignettes of merchants and sailors spun along with courtly missions and caliphal quests that skirted the edges of the known world and plumbed the depths of towering monuments to uncover ancient mysteries. Rulers and their families consumed luxuries procured from distant lands and remote islands, all the while, they were regaled with extravagant tales, in verse and prose. In the trade of minerals, spices, and animals, were apothecaries and grocers who promised potent mixtures, powerful herbs, and refined tastes. Sailors and merchants imported ambergris, camphor, obsidian, ebony, quicksilver, musk, aloe wood, cinnamon, and clove—all with unique properties that when skillfully applied could cure ailments or excite passions. Recorded in iron gall ink, accounts of the specialties of the world circulated in dazzling tomes presented by courtiers and poets, some well traveled, others merely well spoken, who sought success in the torrents of dirhams and dinars that generous patrons could unleash.[20]

In the pursuit of refinement, the early Abbasid court cultivated a literary taste for the exotic and the unusual. In this the Abbasid elites were preceded by countless other courts with universal ambitions. Earlier centuries had witnessed bustling processions of emissaries from foreign nations and conquered subjects bearing tributes of rare animals, precious objects, and crops by the bushel brought as offerings to the kings of Assyria and Iran, the pharaohs of Egypt, and the emperors of Rome, scenes depicted for time immemorial on alabaster panels, basalt obelisks, triumphant stone arches, gilded temples, and tombs.

Recorded on relief sculptures and preserved in scrolls, the frenetic power of the cosmopolis frequently turned to the peripheries as subjugated sources of wealth, pleasure, and knowledge to be mined, enjoyed, and absorbed. It is this spirit that animated the statesman Pliny the Elder (d. 79 BCE) to write his sprawling Latin encyclopedia, *Naturalis historia,* which he dedicated to the court of the Roman emperor. It is a work characterized by an abiding interest in marvels, monsters, and rarities near and far. Pliny's natural curiosities and amazing physical properties did not find a direct line into Arabic, but countless other imperial models for astonishment from antiquity did. Of these, one of the most influential is the narrative cycle that celebrates the feats of Alexander the Great, in the course of his journeys to conquer the world, documenting the strange creatures, people, and magical forces he encountered. During the Umayyad period these adventures made their way into Arabic, where Alexander features as a pious model of a philosopher king whose conquests bring knowledge and power.[21]

What remains of the Umayyad caliphate, before the Abbasid revolution toppled the Arab state centered in Syria, suggests that they too memorialized their territorial possessions through the wonders witnessed and the spolia gained during the early conquests. Stories of countless marvels accompanied the Arab armies as they spread out from the desert, across North Africa, Syria, Iberia, Anatolia, Mesopotamia, the high steppe of Central Asia, and to the banks of the Indus River.

In addition to anecdotes from later writings, partially preserved paintings survive in Quṣayr ʿAmra, a pleasure palace two days east of Jerusalem that the Umayyad elite would frequent. At the whims of mighty lords and capricious princes, paintings of full-bosomed handmaidens pouring libations set the stage for poetry and celebrations of the hunt. A ceiling fresco of kings conquered or forced into obeisance from faraway lands asserts the reach of Umayyad power, while a vaulted dome above the heated baths depicts the constellations of heaven with the twelve signs of the zodiac, pointing to a cosmic order that maintains it all.[22]

The Abbasids developed many of the intellectual and literary pursuits that had already been set in motion by the Umayyads, though often these earlier forays were left unacknowledged. Nonetheless, the Abbasids supported in an unprecedented fashion the absorption of Greek learning, frequently via Syriac intermediaries. With their seat of power in Iraq, the caliphs of the new regime advanced their own tastes and values, often presented in the vocabulary of ancient Iranian statecraft and courtly refinement. These intellectual pursuits were celebrated with variations of the *translatio imperii et studii* theme—the linear transference of power and knowledge—which in this case advanced the Abbasids as the rightful inheritors of ancient learning and patrons of the arts and sciences.[23]

During this period Arabic began to transform into a language of universal science, spoken and written not just by Arabs or Muslims but by a growing circle of communities and faiths. Among the many prized fields of learning, spanning philosophy and medicine, both geography and cosmography were particularly valued. The science of the stars offered the predictive power to decipher the designs of the cosmos, whereas geographical knowledge proved indispensable for the administration of an empire with spheres of influence that spanned the Mediterranean, Mesopotamia, and Central Asia.

Ptolemy's systematic cosmography, with his focus on mastering the stars and the Earth, enjoyed particular renown. Courtiers, administrators, and scholars in the service of the Abbasids also contributed to the development of terrestrial knowledge and cosmic learning. In these courtly contexts circulated works that mixed practical learning with entertainment.[24]

The first Arabic administrative geographies offer insight into this confluence of forces. The earliest to survive somewhat intact is the influential *al-Masālik wa-l-mamālik* (Routes and realms), composed by the high-ranking Persian administrator and caliphal companion Ibn Khurdādhbih (fl. 270/884). In addition to such practical matters as the rates of taxation, the pilgrimage to Mecca, and the postal and mercantile routes connecting the Abbasid imperium to the provinces, Ibn Khurdādhbih weaves through his geography opportunities for delectation and perplexity.

Poetry lines the way stations and settlements, some obscure, others well known, as various classes of wonders to be found, in edifices and natural dispositions, unfurl across exotic landscapes. Along the islands of the Indian Ocean are tribal kings, cannibals, and pygmies, while past the forests of the north, beyond the Jewish kingdom of the Khazar, the brutish Rus merchants, and the rough Turkic tribes, can be found the savage hordes of Gog and Magog waiting to devour all humanity in an apocalyptic fury long foretold by the Quran.[25]

Many of the interests in the strange and curious that shape Arabic geographical writings have antecedents in classical antiquity. Lists of wondrous buildings circulated in Greek, with the pyramids of Giza, the lighthouse of Alexandria, and the Colossus of Rhodes chief among them. There also developed a Greek encyclopedic tradition dedicated to strange and paradoxical oddities, with accounts of hybrids, monsters, remote peoples, and marvelous regions, as well as omens and portents foreshadowing unusual events. Zoroastrian writings in Middle Persian abound with strange creatures that directly parallel the panoply of monstrous nations known to classical antiquity. Stories of strange places and distant people were often refashioned into Christian apocalyptic literature and collections on apostolic miracles and saintly wonders, which circulated widely in Syriac. As with numerous other arenas, writings in Syriac on natural wonders and descriptive geography over time connected with parallel developments in Arabic.[26]

The language of marvels and curiosity also follows the absorption of Hellenistic writings on geography, history, and philosophy, which developed a taste for wonders, *thaumata,* singularities, *idia,* and antinomies or peculiar phenomena, *paradoxa.* One such example is the treatise on mountains and rivers ascribed to Plutarch (fl. 110 CE), which during the course of the Abbasid translation movement made its way into Arabic under the title *Kitāb al-Anhār wa-khawāṣṣihā wa-mā fīhā min al-ʿajāʾib wa-l-jibāl wa-ghayr dhālik* (On rivers, their properties, the wonders contained therein, mountains, and other matters). The reference to wonders captures the spirit of the collection, with its stories of prodigious oddities drawn from a host of classical authorities on geography and nature. When Qazwīnī turns to the wonders of springs, rivers, and mountains in the sphere of the Earth, he is following this well-established body of writing.[27]

Many of these topics feature in Ibn Khurdādhbih's geographical compendium, where wonderment provides a discrete way of fathoming and classifying existence. Ibn Khurdādhbih's geography is the earliest to survive from the period. But the appearance of *ʿajāʾib,* wonders, as a unifying framework for viewing the world can be traced further back in Arabic letters. For instance, the Iraqi scholar Abū l-Mundhir Ibn al-Kalbī (d. 206/821), who served in the court of al-Maʾmūn, relates an account of the wonders of ancient Babel, which features in one of the manuscript recensions of Ibn Khurdādhbih's geography.

Among the marvels listed in Ibn al-Kalbī's name is an iron mirror through which the entire world could be viewed. The all-seeing mirror is a theme with notable parallels in antiquity, kept alive in Persian letters with the world-revealing goblet of the ancient Iranian king Jamshīd. Ibn al-Kalbī had a prodigious literary output, though the vast majority of his works are lost. References to his titles, however, survive in the *Fihrist* (Catalogue), the list of books and biographies compiled by Ibn al-Nadīm (d. 380/990), a bookman of Baghdad. Here Ibn al-Kalbī is identified with several works connected to wonders. He authored both major and minor compendiums on geography, a study of climes, a work on rivers, a treatise on the four wonders, *al-ʿajāʾib al-arbaʿa,* and another on the wonders of the sea, *ʿajāʾib al-baḥr.* Ibn al-Kalbī also had an abiding concern for history, verse, desert lore, and stories of jinn and the poetry they sang.[28]

Ibn al-Nadīm not only offers a glimpse of early Arabic writings on geography and wonders otherwise lost; he also provides a measure of insight into the social contexts of their production and circulation, both as edification and as entertainment. An illustrative case is the Persian courtier Abū l-ʿAnbas al-Ṣaymarī (d. 275/888), who like Ibn Khurdādhbih rose as a *nadīm,* an intimate or drinking companion in the caliphal entourage. Along with Ibn Khurdādhbih, Ṣaymarī appears in Ibn al-Nadīm's chapter on drinking partners, convivialists, men of letters, singers, buffoons, clowns, and humorists at the court. The chapter also features several court geographers and astrologers. Ṣaymarī served as a judge and jurist and wrote on an array of subjects, including astrology, alchemy, and dream interpretation.[29]

Many of Ṣaymarī's titles point to an interest in humor. An epistle on replies that silence opponents, referred to in Arabic as *al-jawābāt al-muskita,* suggests a master of repartee in assemblies. His treatise *Faḍl al-sullam ʿalā l-daraja* (Superiority of ladders to stairs) bespeaks the cunning wordplay of a sardonic mind, whereas the title *Faḍl al-surm ʿalā l-fam* (Superiority of the anus to the mouth) gives a taste of the libertine obscenity that found a ready audience in ribald drinking sessions of caliphs and courtiers. Ibn Sīnā knew Ṣaymarī as a master of deceptions famed for exposing the tricks of charlatans and rogue astrologers who sought to dupe the unsuspecting. Ṣaymarī's affinity for bawdy entertainment landed him an entire chapter in Hamadhānī's *Assemblies,* where he features as a hero narrating his own colorful tale of just deserts.[30]

Just as sorrow affects the body through crying, Qazwīnī observes, the response to *taʿajjub,* astonishment or bemusement, can express itself through laughter. The idea comes close to the notion of humor as a reaction to something incongruous. But humor is also both play and performance. When facing ambiguity or complexity, it can offer some measure of deferment, if not solace, in matters unresolved, before paradoxes, perplexities, and uncontrollable circumstances. *Wonders and Rarities* is woven through with amusing anecdotes drawn from a broad repertoire of storytelling and poetry. Many of the accounts that find a happy home in the pages of Qazwīnī's natural history can be read with an eye for diversion: a talking bird that speaks through obscene double entendres in drunken couplets, given as a gift to the chief judge of Baghdad; jinn who, after having whisked away a slave, debate the famed battles of wit and verse between Farazdaq (d. 114/732) and Jarīr (d. ca. 110/728), both masters of poetic invective; a massive giant of lore who drowns out his enemies below by urinating all over them.[31]

Like Qazwīnī, Ṣaymarī was also a judge from Iraq with an interest in celestial forces and amusing anecdotes. Ibn al-Nadīm recognized Ṣaymarī to be not only a humorist but also a scholar of the stars. One of Ṣaymarī's astrological treatises, *Kitāb Aṣl al-uṣūl* (Principle of all principles), survives in multiple manuscripts. Among numerous topics, the work probes the auspicious times for conducting various activities, citing a range of ancient Greek, Persian, and Indian authorities. Here are measured nativities for caliphs, the fate of dynasties, and cosmic revolutions, as well as the correspondences of the planets to various countries, and the natural climatic and cosmic forces that influence dispositions.

Planets regulate regions, behavior, and professions. Auspicious Venus is the sign of the Arabs and the ruling state of the Banū Hāshim—the tribe of the Prophet—who make up the Abbasid royal family. The language of Venus is Arabic, its religion Islam, its emotions desire and felicity; the planet abides in gardens of repose with flowing fountains, buildings of joy and perfume, drinking halls, and pleasure palaces; its books are of songs, poetry, and the whisperings of lovers, their sayings, stories, and epistles. Under its sign flourish magic, pressing fruit, burning incense, the arts of women, entertainment and games of backgammon and chess, as well as the composition of tunes and melodies, the music of lutes, cymbals, and drums, and the wonders of crafts and fine workmanship, with sculptures, carving, painting, gilding, ornament, multi-hued dyes, colors, and bejeweled designs.

Other planets are not so propitious. Many of the activities governed by dark Saturn, which rules over India, are occult matters done in darkness through deceptions, acts of enchantment, potions, and sleight of hand trickery. Mars brings out the wonders of dreams, but also signs of calamity and insurrection. With fine attention to the stars and their course over the Earth, Ṣaymarī's manual teaches how to determine the auspicious timings for engaging in battle, undertaking journeys, moving from one abode to another, embarking on the hunt, and commencing trade, as does it reveal methods for finding hidden treasure and objects lost from sight.[32]

Such techniques would come in handy for a raconteur on the move, which is how Hamadhānī depicts Ṣaymarī in the assembly by his name. After falling on hard times, abandoned by friends and patrons, Ṣaymarī takes up a journey to distant lands from Egypt to India, wandering through flourishing cities and abandoned abodes, gathering all the while rarities, stories, witticisms, and traditions, filling up on "the poetry of dandies, the vulgar frivolities of pleasure seekers, the evening tales of the love-struck, the judgments of would-be philosophers, the legerdemain of hucksters, the ruses of illusionists, the anecdotes of boon companions, the flimflam of astrologers, the cunning of quacks, the artifice of the effeminate, the trickery of cheats, and the devilish deeds of fiends." As Ṣaymarī eulogized and satirized, he gained wealth, presents, and curios counted in fine blades, coats of mail, javelins, fleets of horses, and silk brocades. With specialties of the world and stories to spin, he returned to Iraq seeking redress in a masterful turn that ends with inebriation and punch-drunk revenge.[33]

Hamadhānī's artful tale may be just that, but it offers a measure of insight into what a clever wit could leverage through experiences gleaned from far-off lands. The picture is entirely consonant with Ibn al-Nadīm's account of Ṣaymarī's books, which includes not only a treatise on how to uncover the ruses of astrologers, but an otherwise lost collection on the marvels of the sea, *ʿajāʾib al-baḥr*. In a separate chapter of his catalogue Ibn al-Nadīm also has an entry for writings on wonders by sea and land. This section concludes his treatment of tales and fables, referred to as *al-asmār wa-l-khurāfāt*, told generally in courtly settings, often adapted from earlier Persian, Greek, and Indian models.

Some of these collections were written in the voice of animals, like the nestling frame narratives told by two jackals named Kalīla and Dimna, whose stories were translated from Sanskrit into Middle Persian and then into Arabic before circulating in countless other languages. Ibn al-Nadīm comments that the first to take an interest in evening tales, which the word *asmār* connotes, were the ancient Persian kings, a tradition that the Sasanians of Iran continued and expanded, though he also credits Alexander the Great with the courtly custom of telling stories as evening entertainment.

Ibn al-Nadīm opens his chapter on tales with the *Hazār afsān* (The thousand nights), the Persian title for the lost archetype that would form the basis of the Arabic cycle known as *Alf layla wa-layla* (The thousand and one nights). "A tale for your life" is the bargain that keeps alive the princess Shahrāzād, the enchanting heroine. Story after story, night after night, Shahrāzād tames the wrath of a king who is bent on murdering every woman he marries. The result is a wager of ever-increasing astonishment, each tale set to outdo the next, in a labyrinth of building pleasure and horror.

Licentious and magical, titillating and bizarre, tales of dismemberment fold into unbridled orgies where slaves, princesses, and jinn commingle in sundry positions and occasions, aided by talismans, enchanted rings, and ruses to escape from impending doom. *Lā ḥawla wa-lā quwwata illā bi-llāh* can be heard throughout, "there is no power, nor strength, save through God." The refrain saves lives in appeals to divine intervention for the restoration of order and justice, uttered often in the face of danger as a protective incantation, invoking the power of God, who is able to produce wonders beyond human capacity. The invocation features throughout the stories preserved in the latter adaptations of the collection, as well as on the title page of the earliest known Arabic fragment of the *Nights,* a single worn piece of paper copied before 266/879, according to the date of legal testimony scribbled on the margins of the page.[34]

Listed along with the *Nights,* animal fables, and books of wonders, are stories of lovers and sex with jinn. Qazwīnī has his own tales to tell of speaking animals and erotic encounters with jinn as well as besting them in the arts of poetry. For his part, Ibn al-Nadīm classifies some of these evening tales as authentic, *asmār ṣaḥīḥa,* offering examples from the annals and adventures of

Sasanian kings. The Abbasid caliphs had a bottomless appetite for such tales, and the bookmen of Baghdad were more than happy to oblige. The craze reached its heights, Ibn al-Nadīm reports, during the caliphate of al-Muqtadir (d. 320/932), which saw countless awe-inspiring volumes circulating in the markets and the palace.[35]

Qazwīnī could draw from several collections that had preceded him on land and sea. Ibn al-Nadīm gives a sense of these earlier works, most of which are lost. But his indexical endeavor was by no means exhaustive, and with its focus on Baghdad and Iraq much on the peripheries of Abbasid rule and taste is missed. This is the case with the *Kitāb al-ʿAjāʾib al-gharība* (Book of strange marvels)—a collection that also does not appear to have survived the vicissitudes of time—by Abū ʿAbd al-Raḥmān Ibn al-Mundhir (d. 303/915), a Shāfiʿī jurist and hadith scholar from Herat.

Ibn al-Mundhir traveled extensively throughout Khurasan, Iraq, and the Hijaz, both as a merchant and as a scholar studying with hadith authorities. His vita suggests the proverbial *al-riḥla fī ṭalab al-ʿilm*—the travels that aspiring religious authorities with means undertook in the search for knowledge, usually through collecting traditional sayings from the Prophet and the early Companions. In its most ideal form, the circulation of religious learning was a pursuit that transcended material concerns. Yet trade and the movement of merchants also had pious precedent in the early community of believers. While wonders and rarities excited the refined tastes of courtiers, they also addressed the aspirations of audiences well beyond the walls of caliphs and their courtesans, appealing to both the salacious and the devout.[36]

Along mercantile routes, by land and sea, journeyed not only goods but also the belief in a universal call to faith meant for all humanity. As the tales of merchants and travelers frequently tell, the world was populated with infidels, idol worshipers, and savages who knew no religion. Sailors from the Persian Gulf not only packed their hulls with Yemeni frankincense, beads, glassware, ceramics, brass, iron, and tin ingot but also brought along the Arabic script, numbers for counting, and stories of pious miracles and everlasting salvation for winning hearts. In this movement of merchants and mystics is the story of how Muslims journeyed and settled across the Indian Ocean, from the shores of Africa to the archipelagoes of the Far East.

Qazwīnī populates his accounts of savage islands and marvelous seas with merchant vessels and maritime lore. For the intrepid, incredible riches were waiting to be had along the fabled islands of the Indian Ocean and farther east into the encircling sea. Merchants traded for prized camphor from Sumatra and Borneo, teak and black pepper along the Malabar Coast, cardamom from the island of Sarandīb, known also as Ceylon, along with cloves, nutmeg, and sandalwood from the lush rainforest islands of brutes and jinn. Some even made it as far as the Wāqwāq islands, where it was said that women grew on trees.[37]

A shipping vessel promised staggering returns. But where capital investments underwrote adventures driven by monsoon winds, there was always the threat of disaster, blown out by typhoons or wrecked by creatures beneath the deep. On the sandy bottom, across the archipelagoes, the ocean took its share, with sewn-plank ships strewn in pieces, some discovered, but most lost, forgotten relics of the trade that busied the ports of Basra, Muscat, and Siraf. These cargo ships crossed the Red Sea and the Persian Gulf out beyond Sumatra and the east coast of Africa (figure 5.4).

One ill-fated vessel set sail while Ibn al-Nadīm was notating shelves of books in Baghdad. Recently discovered, the ship sank off the coast of Java hauling coral pendants, Tang dynasty ceramics adorned with dragon motifs, parrots, and phoenixes, gold and silver jewelry, pearls, minerals, and aromatics, delicate glassware in emerald-green and aquamarine, chunks of raw cobalt and yellow glass for further smelting, fine swords, axes, and chains, Buddhist statuettes, Arabic prayer beads, and a cast for manufacturing amulets that proclaimed for centuries from the ocean's bed *al-mulk li-llāh al-wāḥid al-qahhār*, "Dominion is God's, the One, the Conqueror!"[38]

But in addition to wealth, those who survived could garner attention with their tales of exotic spices, supple natives, and improbable escapes from certain death. The motif of relief after suffering, known in Arabic as *al-faraj baʿd al-shidda*, informs many of the encounters with cannibals, slavery, and chicanery. Traveling from port towns through courtly assemblies, sailors' lore circulated in several early Arabic collections on the marvels of the sea. These

FIGURE 5.4: Merchant ship in the Indian Ocean, from a manuscript of Ḥarīrī's *Assemblies* copied and painted in 634/1237 by Yaḥyā ibn Maḥmud from the city of Wasit.

works, in turn, left a lasting imprint not only on the languages of wonder and awe but also on ideas about the East, ocean trade, and the edges of the inhabitable world.

Countless stories of high-sea adventures never made it into writing, or if they did, had limited circulation, such as the journeys of the Sufi merchant and hafiz of the Quran Abū l-ʿAbbās al-Qarmīsīnī (d. 599/1202), who returned to Baghdad after more than three decades of traveling. For some twenty years he took up residence in India and the island of Sarandīb, where he became a preacher. In Baghdad he settled in a recently opened Sufi lodge, the *ribāṭ* in the Maʾmūniyya quarter of city, which was run by the mystic Abū Ḥafṣ al-Suhrawardī (d. 632/1234) and supported by the caliphal family. Attracting eager listeners, Qarmīsīnī related stories of the wonders he saw in the course of his many journeys. No written record of his adventures survives, though the memory of his tales from far-off shores still circulated during Qazwīnī's lifetime.[39]

In contrast, Qazwīnī drew extensively from a written repertoire of accounts, much of which had been in circulation for centuries. His treatment of strange islands and frightening creatures builds directly on compendiums of maritime marvels. One of the earliest Arabic collections to survive is the *Akhbār al-Ṣīn wa-l-Hind* (Accounts of China and India) redacted by Abū Zayd al-Sīrāfī (fl. 303/915). This cycle of anecdotes found a lasting readership with its tales of island encounters, curious animals, and luxurious wares.

The famed encyclopedist of Cairo, Abū l-Ḥasan al-Masʿūdī (d. 345/956), met Abū Zayd in Basra during the course of his own sea-bound journeys. Marvelous tales and strange reports feature throughout his surviving writings. Masʿūdī had an ear for stories told by sailors during his travels that took him to India and beyond. He knew that some of the best navigators were captains from the Persian port city Siraf, an entrepôt in the vast expanse that made up Indian Ocean trade.[40]

Collectively the titles on the marvels of the ocean came to speak for a well-defined field. Ṣaymarī's *Wonders of the Sea* is lost, but a collection of merchant tales circulated in the name of his pupil Abū ʿImrān Ibn Rabāḥ al-Awsī (fl. 377/987), who also hailed from the port city Siraf. The Mamluk administrator of Cairo, Shihāb al-Dīn al-ʿUmarī (d. 749/1349), enjoyed Ibn Rabāḥ's book so much that he preserved large portions of it in his own multivolume encyclopedia. ʿUmarī knew the work as *al-Ṣaḥīḥ min akhbār al-biḥār wa-ʿajāʾibihā* (The authentic from accounts on the oceans and their marvels), and it is through ʿUmarī's extensive quotations that fragments in Ibn Rabāḥ's name survive.[41]

The title of Ibn Rabāḥ's collection is noteworthy for its insistence on having selected accurate reports of the ocean and its wonders. In addition to the question of reliability that this material faced, the word *ṣaḥīḥ*, authentic, in the title is reminiscent of Ibn al-Nadīm's classification of authentic stories. It also evokes the vocabulary used in the canonical Sunni hadith collections by the likes of Bukhārī (d. 256/870) and Muslim ibn al-Ḥajjāj (d. 261/875), which took the name *ṣaḥīḥ*, asserting reliable and authenticated material. ʿUmarī notes that Ibn Rabāḥ wrote the collection for the patron of poets and scholars Abū l-Misk Kāfūr al-Ikhshīdī (d. 357/968), a eunuch slave warrior, by all measures originally from Nubia, who succeeded in gaining control over Egypt, reigning as de facto ruler for some twenty years. Before making his way to Egypt and wandering through its book markets, Ibn Rabāḥ had established himself as a Muʿtazilī theologian who frequented assemblies of learning held by Abū l-Fatḥ Ibn al-Furāt (d. 327/938), the powerful vizier of the caliph al-Muqtadir in Baghdad.[42]

In light of the many awe-inspiring tales in Ibn Rabāḥ's collection, the use of "authentic" in the title may be a bit of tongue-in-cheek, in the mode of the satires by his teacher Ṣaymarī. But nothing of what survives suggests anything other than earnest reportage from the mouths of actual captains, pilots, boatswains, fishermen, and merchants who made their way across the ocean, along the coasts of Africa, India, China, and through the Malay Archipelago. The repeated citations of names and dates serve to authenticate spectacular experiences and horrifying customs from strange lands and dangerous seas.

Intriguingly, a good deal of Ibn Rabāḥ's material overlaps with another selection of marvels, the *Kitāb ʿAjāʾib al-Hind* (On the wonders of India), which survives in a single manuscript copied in Mamluk Damascus. Here the author is identified as the ship captain Buzurg ibn Shahriyār, an otherwise unknown figure from the city of Rām-Hurmuz in southern Iran. A comparison suggests that more than mere differences of wording separates the two, as Ibn Rabāḥ's *Authentic Accounts of the Oceans* appears to have excluded several anecdotes that a skeptical eye would view as far-fetched.[43]

Gone is the island, well known to Qazwīnī, populated only with fierce women whose sexual appetite for male captives is insatiable and lethal. On its shores echo the battle cries of Amazonian warriors from Greek lore. In the romance cycle of Alexander the Great's adventures, the Isle of Women makes a memorable appearance. Ibn Rabāḥ seems to have skipped over it, though Buzurg ibn Shahriyār dutifully records the tale. Also left aside are the beautiful islanders who seem just like humans except for their small heads and dorsal fins. When loaded enslaved on ships by unsuspecting merchants, the creatures—offspring of man and fish—quickly escape by jumping into the ocean and swimming away. Such prodigies, it turns out, are quite hard to keep hold of when far from land.[44]

Then there are the giant sea turtles and an account that Buzurg ibn Shahriyār acknowledges strains credulity, but is delightful in the telling. Reaching a barren island in the Indian Ocean, a group of sailors proceeded to celebrate Nowruz, the Persian New Year festival, by lighting a bonfire while preparing a meal, only to discover that rather than having alighted on an island, they had camped on the back of a giant turtle. Agitated by the flames, the creature plunges back into the deep and the sailors barely escape with their lives (see figure 8.4).

With swift currents, the colossal creature, repeatedly mistaken for an island, crossed the seas. In its wake can be seen not only a similar tale in the Alexander romance, but also the entry on the *aspidochelone,* a giant sea monster listed in the *Physiologus,* an early Christian bestiary and lapidary written in Greek and composed in Alexandria, that circulated in numerous languages, including in a late Arabic translation. Qazwīnī draws on a similar story for his discussion of sea turtles. A version appears in the Latin seaborne adventures of Saint Brendan (d. ca. 577) of Ireland, composed in the early tenth century.

Traces of the tale can also be heard in ancient Iranian lore, with the adventure of the hero Karsāsp who lights a fire on the back of a three-horned dragon that dives into the sea, recounted not only in the Avesta, but also in late Middle Persian priestly writings. With his own taste for awe and astonishment, the Muʿtazilī luminary of the Abbasid court, Abū ʿUthmān al-Jāḥiẓ (d. 255/868~), in his book on animals, *Kitāb al-Ḥayawān,* rejects the account out of hand. It is an outlandish report, he concludes, drowning in the absurdity of *al-khurāfāt wa-l-turrahāt,* fables and hoaxes. While many may relate the tale of kindling a fire on a giant creature mistaken for an island, Jāḥiẓ complains that they all ascribe it to sailors, never claiming to have witnessed it with their own eyes.[45]

In ʿUmarī's redaction Ibn Rabāḥ appears to have shorn off some of this more dubious, albeit ancient material. But what makes the grade is no less arresting: man-eating crocodiles charmed by drums; elephants trained to shop the markets for groceries carrying baskets by their trunks; temple prostitutes in India whose wealth adorns idols; giant crabs that toy with sailors by dragging heavy anchors in their claws; kings sold as slaves, paupers made affluent; enormous pearls of exorbitant value; nearly invincible pirates who terrorize islanders, having discovered a plant that protects them from the harm inflicted by iron weapons; self-immolating Indians who prepare their own pyres; idol worshipers who on an annual festival plunge from a mountain peak to their deaths, torn to pieces by the sharp thorns of a copper tree below; a multicolored bird that jumps into flames only to rise reborn from its ashes; Indians who believe that the soul migrates forty days after death, transforming into a dog, donkey, elephant, or some other such creature; a stone from China that aids childbirth; a market for jinn in the upper regions of Kashmir where the clamor of buying and selling can be heard but not a soul can be seen.[46]

Watch out overhead for falling elephants. High above, gargantuan birds carry the beasts by their claws. They toss their lumbering prey to the ground, breaking the hapless creatures to pieces before plunging down for the feast. Though terrifying, these massive birds, it turns out, can prove useful for the brave of heart, when stranded on islands overrun with cannibals or when captured by a naked giant large enough to crush a full-sized ewe in his bare hands. Many a voyager has wagered it all, grabbing hold of their claws to fly away from certain death. Such an ordeal of savagery and salvation features in Ibn Rabāḥ's *Authentic Accounts*. Qazwīnī picks up a similar version of escape by winged flight with the misadventures of a merchant from Isfahan, excerpted from an otherwise anonymous work on the wonders of the sea (figure 5.5).[47]

Several of Qazwīnī's island escapades appear refashioned in the Arabic cycle of stories told in the name of Sindbad the sailor, in a series of adventures driven by dark tales of ever-increasing mayhem, wealth, and pleasure. The French Orientalist Antoine Galland (d. 1715) wove the voyages of Sindbad into his translation of *The Thousand and One Nights*, where they have resided ever since. Included here are accounts of how shipwrecked Sindbad survived a massive sea creature mistaken for an island, deftly fought off a giant cannibal, escaped the death choke of a string-legged beast, and was rescued by grabbing hold of a colossal bird known as the Rukhkh. Qazwīnī reports similar stories, though not attached to Sindbad's name.[48]

FIGURE 5.5: Shipwrecked merchant from Isfahan grasping the talons of a giant bird, from the Wasit manuscript.

Many accounts had traveled quite far by the time they had emerged in early Arabic writings on the sea, adapted and transformed along the way. Much of this has the flair of encounters with temple sex, beliefs of transmigration, and practices of immolation, along with giant creatures of the ocean and terrifying shipwrecks. But tales not only have a way of getting bigger in the telling; they can also be retold and repurposed in seemingly endless reconfigurations.[49]

The misshapen shepherd who terrorizes castaways rings of Polyphemus, the Cyclops, with his flocks in Homer's epic of the clever and cunning Odysseus. In the giant bird that rises from its ashes can be seen the ancient phoenix, a creature that the Greek historian Herodotus (d. ca. 425 BCE) identifies with Egypt and Arabia, but that other authorities from antiquity associate with India. The string-legged savages, known to Pliny as *himantopodes,* appear in the Alexander romance. Birds large enough to carry elephants also look a lot like Garuda, the giant winged mount of the god Vishnu in Indic cosmography who features in the Sanskrit epic the *Mahābhārata*, grabbing a giant elephant and tortoise by its talons, an image that adorns temples and shrines.[50]

Qazwīnī identifies the colossal bird that had the capacity to fly off with water buffalos, rhinoceroses, and elephants as the *ʿanqāʾ*, a creature from ancient Arabic lore. Qazwīnī's Persian recension associates this creature with the Sīmurg, a massive bird with magical powers known throughout ancient Iranian letters. The identification was old and long lasting, found throughout collections of natural history (see figure 4.1 and figure 8.2). Ever skeptical, Jāḥiẓ also equates the Sīmurg with the *ʿanqāʾ* and notes that the majestic bird is frequently woven into the tapestries and carpets of kings. But here again he registers doubts about its reality. So it is that in philosophical circles—represented by the likes of Fārābī and Ibn Sīnā—the *ʿanqāʾ* came to stand for the mental capacity of the soul to conceive of entirely imaginary beings with no basis in reality.[51]

No single expression was as powerful for organizing all the strange creatures of the world as the phrase *ʿajāʾib wa-gharāʾib*, wonders and rarities. It could evoke the monstrous, as in the prodigious and grotesque, while also conveying the general sense of the miraculous and sublime, as in the astonishing design of God's creation, ordered by divine decree. Certainly, demons and jinn could inspire horror and fear, as could the mismatched creatures that jump out from illustrated natural histories in Arabic and Persian, and then in Turkish and Urdu.

The line between beast and monster can blur in the Persian word *jānvar*, meaning generally "animal," from the Middle Persian *gyānwar*, for any animate life. It is a sense that carries over from Persian in the modern Turkish *canavar*, "creature" or "monster." There is also another term in Persian that in time came to be used for monster: *hayūlā*, originally an Arabic philosophical loanword from the Greek *hylē*, for prime matter at the basis of generation, as in the raw unformed elemental stuff of creation. In Arabic there is *maskh*, as in misshapen or deformed, as well as numerous other ways for thinking about wild savages and vicious beasts connected to the word *waḥshī*, beastly, brutish, untamed. Along with the Persian *dīv,* for demon, and *ghūl* and *jinnī* in Arabic—whence *ghoul* and *genie*—Urdu also drew on an Indic repertoire of demonic monstrosities populated with *rākṣas, bhūt,* and *piśāć,* demons, ghosts, and goblins.

In the emotive allure of wonders and rarities, dog-headed people or those whose heads grew on their chests were not merely horrific monstrosities. They also could stand as strange signs of the perfectly ordered harmony of creation. In the higher orders of theology, Muslim authorities were generally quite loath to invest the physical world with forces of evil that could in any way stand apart from or compete with the absolute power of God.

In this there are palpable differences between the face of monstrosity in Latin and what the category of *ʿajāʾib wa-gharāʾib* served to convey. Many of the same creatures and tales that inhabit classical Arabic and Persian letters found their way into the collections of curiosities, bestiaries, isolarios, and secrets of nature that filled the libraries of monasteries, universities, and courts in Latin Christendom. In this Muslims, Christians, and Jews shared a common fount in classical learning that had long cultivated an interest in strange creatures and beasts from faraway regions. With influential translations of works on medicine and natural philosophy came a sustained attention to the hidden properties of minerals, plants, and animals. Appeals to wonder and rarity found in Arabic writings on astrology and talismans circulated in translations produced in Christian scriptoriums. Like their Arabic and Persian counterparts, collections on natural wonders and the secrets of nature found ready audiences in Latin and European vernaculars.[52]

As with *ʿajāʾib,* the Arabic word for wonders, *mirabilia* in Latin stood for marvels and miracles to behold. In contrast, the Latin *monstra* over time took on a distinctly negative sense, as ominous portents, prodigies, and baleful creatures—menacing monstrations of divine will. The term is associated with the verb *monere,* "to warn" and "to remind." As a sign of what God is not, through the power of paradox and negation, the monstrous came to offer a proxy for things evil, demonic, and polluted, as it betokened chaos and disorder.

As an affective state, the wonder evoked in the Latin discourses of *admiratio* was also important for discussing the monstrosities that could be found in an array of phenomena from animals and humans to language and even numerical patterns. But as opposed to merely the portentous or prodigious, monstrosity in medieval Christendom could also evoke physical imbalances that deviated from natural order. Over time the ethical implications would become quite profound, as the monsters on the edges of the world offered a means for classifying and imagining the people encountered across the Western Hemisphere, Africa, and the Pacific Islands.[53]

Part of this picture comes from classical Greek accounts treating marvelous things and strange tales—often referred to as *teratologia,* from the word *terata,* as in wondrous signs, marvels, portents, as well as monstrous and deformed creatures. Aristotle examines *terata* while discussing strange births in his study on the generation of animals, offering physical explanations for how prodigious creatures come into being. Notably the early Abbasid translation of Aristotle renders the Greek terms *terata* and *teratōdes* as *ʿajāʾib* and *ʿajīb,* as in *awlād ʿajība,* "marvelous offspring." Following the same track, Ibn Sīnā, in his reading of Aristotle on animal generation, categorizes creatures born with multiple limbs and rare forms as *ʿajāʾib,* evoking the strange and marvelous.[54]

In contrast, the Latin translation of Aristotle's *De generatione animalium* made in Toledo by Michael Scot (d. ca. 1232), scholar-in-residence and court astrologer for Holy Roman Emperor Frederick II, finds recourse in the language of monstrosity. Scot worked, not with the original Greek, but instead with the earlier Arabic translation. With the vocabulary of *ʿajāʾib* before him, Scot heard not only *mirabilia,* "marvels to behold," but also "monsters," as in *filii monstruosi,* "monstrous offspring." A later Latin translation based directly on the Greek also links *teras* with *monstrum,* in an association that served as the rational basis for describing how monsters formed. Monstrosity had long been a potent category in classical Latin, entirely at home in Pliny's natural history filled with prodigies, miracles, and monsters.[55]

In the Arabic vocabularies of wonder and rarity the distinction between the monstrous and the sublime could easily fade away. Qazwīnī's collection ends with a short section dedicated to animals possessing strange forms and shapes, *ḥayawānāt gharībat al-ṣuwar wa-l-ashkāl* (see figure C.3). Like Ibn Sīnā before him, when discussing these creatures Qazwīnī evokes wonder and rarity. For Qazwīnī these include peoples or nations, referred to as *umam,* that God created on the far flanks of the earth and on the islands of the sea; many bear uncanny resemblances to regular humans, but they are also quite distinct.

FIGURE 5.6: Dog-headed creatures imprison a group of shipwrecked sailors, from an Ilkhanid copy of the Juwaynī recension.

Next to Gog and Magog with their shaggy manes, sharp teeth, and apocalyptic appetite are small pygmies who are attacked by cranes—an account made famous by Homer and by Aristotle's study of animals—as well as dog-headed islanders (figure 5.6) and string-legged creatures, who are known for taking the unsuspecting for a ride by mounting and wrapping their legs around their victims. Among winged curiosities and people with ears so long they can wrap themselves up in them like blankets, speech can be heard, but it is often unintelligible to outsiders. Black, white, red, olive-hued, covered in hair, and completely naked are frequent attributes. Some islanders, such as those with long trunks and wings, are said to actually be jinn. It is a notion that builds on the theory, well known to Qazwīnī, that before human history, jinn were cast out to the edges of the earth in a primordial battle with angels. There they were confined to remote islands as a divine punishment for wreaking havoc. Many of the same strange creatures feature earlier in *Wonders and Rarities,* in accounts of island populations. The identification with jinn in their variegated forms lends some of these curious creatures a providential history of cosmic proportions.

But not all such oddities were so far away. Qazwīnī also discusses, in this concluding chapter, composite or hybrid animals, *al-ḥayawānāt al-murakkaba,* which are generated through various combinations, such as the mule, the giraffe, and breeds of canines and horses. Yet here too are men that are half bears, and the *nasnās,* who have regular human forms, apart from possessing only one arm and one leg. Surprisingly they can run with incredible alacrity. Qazwīnī's spectrum of existence lends many of these creatures a conceptual logic. Between minerals and plants are fungi, an intermediary species that is part plant, part mineral. Then there is the date palm, which has attributes of both plants and animals. The same goes for horses with their good manners and elephants with their great intelligence. They are closer on the rung to humans than other beasts. Likewise, prophets stand nearer to angels than normal people, in an amazing diversity that is all part of providential order.

The last category of strange animals consists of seldom-occurring creatures. By way of explanation, Qazwīnī notes that physicians hold these forms to be produced in gestation through a rare humoral mixture, *mizāj gharīb,* whereas astrologers believe them to be the result of astral conjunctions during nativity that produce an uncommon horoscope, *ṭāliʿ gharīb.* Here are giants, conjoined twins, talking birds with human heads, flying foxes, and horses with horns.[56]

A variety of forces—climatic, medical, physiognomic, astrological, and even providential—are offered as explanations for these oddities. But all of these creatures are grouped together in the language of rarity, which lends their strange variations a conceptual unity. In the divide between Arabic marvels and Latin monsters could abide distinct categorical and theological conceptions for fathoming difference within the world.

Yet there was also a general consensus that hierarchical order was the structural basis for creation. In Arabic philosophical and scientific discourses, the word *ṭabīʿa,* as in nature or disposition, was used to render the Greek *physis,* as in the physical forces, in Aristotelian terms, that cause motion, action, or being at rest for any given body. In this sense, nature was conceived, not as the sum total of the world in an abstract form, but instead as the physical dispositions and powers that shaped existence.

The concept of *ṭabīʿa* draws on associations of impressing and imprinting, from the verb *ṭabaʿa,* "to imprint," as in God's imprint on the physical world. Thus came the inherent natures, characters, tempers, dispositions, or qualities, known as *ṭabāʾiʿ,* that govern all aspects of creation in harmony. In philosophical terms, both Plato and Aristotle offered a vision of innate, natural hierarchy that separated humans from beasts, and the savage from the civilized. Following the likes of Galen and Ptolemy, Muslim authorities were also quick to ascribe the differences between humans to variations in climate, innate characteristics, balances of the humors, regimes of diet, and the astral configuration of the heavens. These ideas also maintained an enduring appeal among Christians and Jews of various persuasions.[57]

Needless to say, the range of conceptual and social mechanisms used to separate and evaluate people in classical Islamic thought did not follow the same contours as later racial ideologies that indelibly marked the course of modernity. Nor were the various techniques for differentiation stable or static over time. All the same, broad associations between skin color, dispositions, and hierarchical order also shaped understandings of human diversity. This could proceed at times with a kind of racialized logic that tied traits and characteristics to skin pigmentation. While there was a good deal of modulation, the frameworks of climatic determinism known to Qazwīnī took extreme whiteness and blackness to be aberrations endemic only to regions where normal human habitation was all but impossible.[58]

As for group identity, various categories could offer analogues to ethnicity in a hereditary sense, such as *shaʿb* and *qawm*, both of which were used to translate the Greek *ethnos*. Likewise, the concept of *umma*, a people or community, intersected concretely with notions of genealogy, connected conceptually and etymologically to *umm*, the word for mother, source, or origin. The tension between *nasab* and *kasab*—what one inherits versus what one acquires—was an early means for challenging genealogical pedigrees. There was also a range of other mitigating forces felt throughout Islamic universalism, which spoke to all humankind in the Quran and hadith, and which sought to measure people by the degree of their faith and not the color of their skin. Thus, the *umma* of the faithful could convey the soteriological aspirations of a Muslim community that transcended boundaries, united in ideals of belief, if not in actual governance or political reality.[59]

Yet lessons from prophetic history could also be used to account for the branching genealogies that separated the people of the world and their natural dispositions in decidedly hierarchical terms. Not only did Muslims inherit ancient theories of climatic determinism and humoral disposition, but they also shared with Christians and Jews the biblical story of how the three sons of Noah—Shem, Ham, and Japheth—populated the earth after the Flood.

Tug even gently at these lines of descent, and all mischievous variety of prejudice unfolds.[60]

Just how accurate were these catalogues of strange creatures from remote seas or the distant past? The repeated refrain "God knows best" concludes many of Qazwīnī's tales, with pious awe that celebrates both ambiguity and uncertainty. Yet the question was an ancient one. Several classical authorities, such as Aristotle and Galen, raised serious objections to the existence of legendary animals such as the sphinx or the centaur; these doubts, in turn, circulated through influential Arabic translations that informed further developments in natural philosophy.[61]

Many of the curious creatures in Qazwīnī's natural history were known only through earlier written collections. On knowledge of the unseen, *ʿilm bi-ghāʾib,* Jāḥiẓ offered three grades for establishing what could not verified by direct sight or contact. At the highest level were accounts so widely transmitted and universally accepted that there was no need to question their accuracy. Next were reports related by several different transmitters that could be tested against each other. Finally, there was information passed on from only one or two people, who were equally likely to be lying or telling the truth, in which case some degree of doubt must remain.

When discussing the giant *ʿanqāʾ*, Masʿūdī likewise turned to the same problem of authenticity by drawing on methods advanced by jurists and hadith scholars for accrediting reports from the Prophet and the Companions to obtain various degrees of certainty. Most legal scholars in the major cities, he notes, accept well-known reports transmitted by multiple authorities from one group to the next as the basis of law. The accounts on the *ʿanqāʾ* or the half-bodied *nasnās* do not rise to this level of communal verification, because they were reported only by individuals, not by multiple eyewitness authorities. It is not necessary to accept such reports or to believe their authenticity, but they abide nonetheless in the realm of what is conceivably possible, *fī ḥayyiz al-jāʾiz al-mumkin.* This is the case, Masʿūdī concludes, for narratives told by the likes of ancient Israelites on the past and for reports on the wonders of the seas, both of which populate Qazwīnī's *Wonders and Rarities*.[62]

As for the bird large enough to carry marooned merchants to safety, Qazwīnī refers to it as a strange story, even though he observes that "it is not beyond the providential grace of God and His succor." The account features in his treatment of strange islands and maritime creatures. In Qazwīnī's hands the wonders of the ocean are woven into a broader design of natural order. He structures the cosmos, from the heavens through the atmosphere, down to the Earth, through a series of well-defined chapter headings and subdivisions. Such bibliographical orders proved indispensable for encyclopedic endeavors, witnessed by the likes of ʿUmarī with his multivolume encyclopedia that stretches from botany to zoology, covering along the way such topics as geography, history, literature, and civil administration.

A mere drop in the vast ocean of his collection, ʿUmarī concludes the treatment of the wonders of the sea and his selection of Ibn Rabāḥ's *Authentic Accounts* with the wry observation that had this treatise not been written by one notable for another, he would not have included it. Stressing the point, he ends the chapter with an old Arabic proverb: "As for the ocean, nothing stands in the way of those who relate its wonders and report its oddities." It is a well-worn sentiment. Before ʿUmarī, Qazwīnī had evoked this same proverb when speaking of the wonders of the sea, measured in pearls, coral, and ambergris, along with the ships that swiftly set course searching for riches.[63]

6

ALCHEMICAL BODIES

ʿUMARĪ WAS A CLOSE READER OF QAZWĪNĪ. In the span of just a generation, as ʿUmarī was working volume by volume on his encyclopedia, Qazwīnī's *Wonders and Rarities* had obtained considerable renown, consulted and copied as a work of lasting significance well beyond Iraq and the immediate territory controlled by the Ilkhanids. Working in Syria and Egypt, often while in the service of the Mamluk secretariat, ʿUmarī turned to Qazwīnī foremost as an authority on mineralogy, botany, and zoology. He copied for his encyclopedia large sections from the *Wonders,* the simple title he used to reference the collection. Here Qazwīnī's compendium appears alongside authoritative medical writings by the likes of Abū Bakr al-Rāzī (d. 313/925), Ibn Sīnā, and the botanist and pharmacologist of Malaga, Ibn al-Bayṭār (d. 646/1248), who had settled in Cairo.[1]

Qazwīnī followed the accepted Aristotelian teaching that all sublunar matter was composed of the foundational elements of fire, air, water, and earth in various mixtures and combinations. The transformative power of the elements was also a core principle of alchemy. The masters of the elixir—the *aṣḥāb al-iksīr,* as Qazwīnī refers to the alchemists of his day—sought through trial and experience to purify base materials into precious metals. Behind these teachings lay the idea that the physical world was fundamentally interconnected, from macrocosm to microcosm, high to low. These connections were driven by elemental forces of temperature and moisture and by powers of attraction and repulsion, or sympathy and antipathy, which governed creation.

The vision of balance and cosmic interdependence guided prevailing medical theories, ideas about the human body, notions of generation and corruption, and conceptions of ethical comportment. This balance in the universe, in turn, was tied to a theory of hidden forces and special properties that served to explain the efficacious power of talismans, charms, and amulets in medical and scientific terms. It is this underlying vision of existence that led physicians to compose compilations on *khawāṣṣ*, or unique properties of various materials, as well as collections of natural wonders. Such a confluence of interests goes some way toward explaining why the copyist of the surviving manuscript of *Wonders and Rarities* completed in Wasit at the end of Qazwīnī's life was himself a physician.

Qazwīnī describes how hidden properties could be utilized to sustain and improve health, and to produce transformations in the physical world. The mutability of substances and influences drives the medical and occult literature on *khawāṣṣ*, as it builds on a larger set of associations with the body and the soul. In this way, just as physicians tended to ailments, moral philosophers and religious leaders promoted diverse exercises for purifying the soul and improving ethical comportment. Uniting these embodied practices of self-fashioning was a cosmology of hierarchical order. Natural forces governed human traits and dispositions, from the spectrum of characteristics separating the sexes to the temperaments of entire populations. The alchemical language of transformation also offered a means for progressing along a scale of improvement and refinement. Just as the body could be healed, so could ethical traits be refined. These ideals of transformative power inform Qazwīnī's concern for ethical discipline, therapeutic practice, and special properties. This focus guides the third chapter of *Wonders and Rarities*, comprising the mineral, vegetal, and animal kingdoms.

The high-level medical theory of the day generally followed the teachings of Galen, who saw the physical world through the prism of temperamental dispositions. Health was maintained through the equilibrium of four humors flowing in the body: blood, phlegm, yellow bile, and black bile. These humors were aligned along a gradient of temperature and moisture, where the qualities of hot, cold, moist, and dry were products of fire, air, water, and earth. The elemental basis to humoral medicine built on Aristotelian natural philosophy of physical generation and corruption. Following the medical classifications of sanguine, phlegmatic, choleric, and melancholic, the temperamental dispositions were aligned with various forces and substances, from specific bodily organs to the cosmic movements of the stars and planets. External influences of weather, diet, and location, as well as internal functions of the body, could all shape the humoral balance for any individual. A series of elemental connections and sympathetic relations pulsed through all matter, binding the cosmos together.[2]

The theory of temperamental equilibrium offered a systematic means to account for the diverse effects that substances could have on the body. Qazwīnī observes that all beings subject to generation and decay, referred to as *kāʾināt,* were engendered through combinations of the four elements. Qazwīnī also refers to the four elements, namely fire, air, water, and earth, as the *ummahāt*—the mothers, matrices, or sources for physical matter. In turn, processes of elemental generation linked all beings to what Qazwīnī describes as "an amazing interconnected arrangement and wondrous order," created by God.[3]

At the base lay the mineralogical stratum. According to the teachings of Ibn Sīnā and Fakhr al-Dīn al-Rāzī, minerals were composed of strong and weak bonds. Qazwīnī traces mineral generation through elemental forces of water and the quickening heat of the Sun, which work together, generally in the subterranean layers of the Earth. Depending on how malleable, durable, or soluble these bonds were, they formed metals, stones, or oils. The most malleable were the so-called foundational seven metals, or *filizzāt*—namely, gold, silver, copper, iron, tin, lead, and Chinese iron, each of which was generated through various combinations of mercury and sulfur. Following the seven metals are durable gemstones and rocks, which petrify differently according to the mixture of water and earth and various gradations of heat and levels of moisture. Finally there are oily substances, such as sulfur, mercury, tar, and naphtha, which are formed through the contraction and expansion of moisture blocked within the Earth.[4]

The foundations behind the theory of mineralogical formation were rooted in alchemical teachings that explained how to transform various substances.

Writings advanced in the name of the famed alchemist Jābir ibn Ḥayyān (d. ca. 200/815?) of Kufa taught that the correct combination of sulfur and mercury could produce the philosopher's stone, the crucial elixir used for the transmutation of metals. This theory lies at the heart of the alchemical teachings in the early Arabic cosmographical compendium *The Secret of Creation* ascribed to Apollonius, the peripatetic Pythagorean philosopher. Apollonius asserts that metals were formed through a transformative power generated by combining sulfur and mercury.

This was a notable variation on the role the elements played in Aristotelian physics. Though Ibn Sīnā absorbed much of the alchemical reformulation of Aristotle on generation and corruption, he famously rejected the alchemical theory that humans could harness natural forces to independently change the species of a substance in order to generate gold. In contrast, Fakhr al-Dīn al-Rāzī saw alchemy as a tested means of purifying base substances into precious metals, a view that Qazwīnī also held.[5]

Wonders and Rarities promises within its pages a pharmacy of hidden powers to cure, poison, and enchant. In his pharmacological writings, Ibn Sīnā recorded the unique properties, *khawāṣṣ,* and actions, *aʿmāl,* of simple drugs along with the therapeutic applications of substances when mixed together. Minerals, plants, and animals, both common and rare, were known to possess numerous beneficial uses, referred to generally in Arabic as *manāfiʿ*, that could aid in the restoration and maintenance of health. The carefully procured substances stored in an apothecary's cabinet were selected based on the view that the human body was a microcosm of the world, influenced by elemental forces that flowed through creation.[6]

Appeals to direct experience and empirical observation appear throughout the literature on special properties. Much of what Qazwīnī brings together is drawn from early written compilations. Here the authorities he cites most by name are Ibn Sīnā, Aristotle, and Apollonius. The pharmacological sections from Ibn Sīnā's *Canon of Medicine* prove an indispensable resource for Qazwīnī, who trains his focus on special properties and medicinal uses.

In his discussion of metals, stones, and gems, Qazwīnī also draws extensively on the authority of Aristotle, copying from the *Kitāb al-Aḥjār* (Book of stones), a lapidary that circulated in Aristotle's name in Syriac and then widely through an early Arabic translation. The collection is filled with descriptions of unique properties and amazing talismans fashioned from precious jewels.

The Book of Stones was not without its detractors. In his own lapidary, Bīrūnī doubts that Aristotle, the first teacher, as he was known in Arabic, was truly the author of such a work.

But as with many other occult writings ascribed to Aristotle, the treatise came to form part of the physical and theological teachings associated with Aristotelian philosophy as it was studied in the course of late antiquity. Aristotle was also often identified with the imperial adventures of his star pupil, Alexander the Great. This is a theme running throughout *The Book of Stones*, which makes frequent reference to episodes in the adventures of Alexander the Great that include marvelous accounts of various stones and gems.[7]

Aristotle was also credited with uncovering the teachings of the ancient sage Hermes, master of alchemy, astronomy, and the hidden properties of nature. Aristotle's *Book of Stones* forms part of a larger corpus of Arabic Hermetic writings. The messenger god of the Greek pantheon, Hermes had long been equated with the ancient Egyptian deity Thoth, the clever intermediary, the scribe, and the weigher of dead souls. Honored by the epithet Trismegistus, the "thrice-greatest," in Greek, and *al-muthallath bi-l-ḥikma*, "triplicate in wisdom," in Arabic, Hermes and the writings in his name occupied a significant place in the development of the occult sciences.

An early report circulated by the court astronomer Abū Maʿshar al-Balkhī (d. 272/886) recounts three different figures known as Hermes: the first an antediluvian sage who established the sciences and built the pyramids in Egypt, identified with the Quranic prophet Idrīs; the second a Babylonian who renewed the sciences lost during the Flood; the third a philosopher-physician from Egypt and the author of alchemical writings filled with knowledge of lethal drugs and harmful animals. Qazwīnī alludes to this tradition when he relates that many people identified the first Hermes with the prophet Idrīs, known to the Greeks as Enoch. It is this Hermes, Qazwīnī notes, who through divine revelation foretold the Flood and built the pyramids as storehouses to protect the ancient writings on the sciences from impending destruction.[8]

The story of lost writings uncovered by a wise sage features in several Hermetic collections. *The Secret of Creation* written in the name of Apollonius, the master of talismans, opens with an account of writings hidden by Hermes that teach Apollonius the secrets of the cosmos. The motif has an obvious parallel with the hidden writings of the Persian magus Ostanes discovered in a temple in Egypt, which feature in the *Physika kai mystika* (Natural and secret matters) ascribed to Democritus, a foundational work in the early Greek alchemical canon. The theme is also picked up in the famed *Dhakhīrat al-Iskandar* (The treasury of Alexander), a collection on the talismanic arts and astrological teachings, which develops a similar premise of an ancient book of Hermetic wisdom that Aristotle had unearthed and translated for Alexander. It too includes talismans produced by the great sage Apollonius.[9]

Several of the Arabic writings of Aristotle drawn from Hermes take mysterious titles, such as the *Kitāb al-Ustūwwaṭāṣ*, which also speaks of hidden books and ancient talismans connected to the mansions of the Moon. Here Aristotle learns from Hermes that the microcosm and the macrocosm, *al-ʿālam al-ṣaghīr wa-l-ʿālam al-kabīr*, are interconnected from the highest to the lowest. There are hidden influences, beyond physical perception, that bind existence together. These influences are formed through the *ummahāt*, the elemental matrices. This particular category was central to early Arabic alchemical writings in Hermes's name. It came to shape the later development of natural philosophy, picked up by Qazwīnī, among others. These foundational elements, in turn, generate physical matter and are animated by *rūḥāniyyāt*, spiritual forces that govern the cosmos through the will of God. The theory of a spiritual power that unites existence shapes the reception of Aristotelian metaphysics, promoted by the likes of Alexander of Aphrodisias (fl. 200 CE), one of the chief commentators on Aristotle, whose writings were well known in Arabic.[10]

A prevailing picture of Aristotle inherited from late antiquity was that of a philosopher sage who had uncovered the hidden forces of nature. One of the early Arabic Hermetic treatises, the *Kitāb al-Isṭamākhīs*, relates that Aristotle had prepared a collection of talismans and potions for Alexander as he set out on his conquests of the East. Here Aristotle provides Alexander with teachings on how to unlock the secrets of existence by mastering complete or perfect nature, *taḥkīm al-ṭibāʿ al-tāmm*, by purifying the body and the soul to obtain control over the spiritual powers that move the heavens. Collectively the writings in the voice of Aristotle and ascribed to ancient teachings of Hermes follow the power of the stars over the climates of the world and trace the unique properties of minerals, plants, and animals.[11]

The mysterious titles of the early Arabic Hermetic corpus had long been available in Baghdad's book markets. Ibn al-Nadīm references several in his list of works written by Hermes. Many feature in his final chapter on the art of alchemy. The last of the ten chapters that make up his bibliographical survey is dedicated to the *ṣināʿa*, the "art" or "craft," as alchemy was also known.

Ibn al-Nadīm concludes that while some hold that alchemy began in Egypt, others assert that the science started with the Persians, or the Greeks, or in India, or even as far away as China, while still others held this ancient science to have been revealed directly by God. The Arabic word *al-kīmiyāʾ*, whence the English term *alchemy*, parallels *chēmeia* in Greek, which was used to describe techniques for manufacturing tinctures or dyes. Like other physical sciences, alchemy appealed to natural forces that transcended region, language, or creed. According to the physician and alchemist Abū Bakr al-Rāzī, known as Rhazes to Latin scholastics, no one could claim knowledge of philosophy, let alone take the title philosopher, without first developing a command of alchemy. Abū Bakr al-Rāzī's list of those who mastered the art includes the pre-Socratic philosophers Pythagoras and Democritus, as well as Plato, Aristotle, and Galen.[12]

To learn more about the indecipherable scripts of Egypt and the Hermetic secrets they held, Ibn al-Nadīm turned to a book containing "accounts of the Earth with the marvels of buildings, kingdoms, and various nations found therein," composed by a member of the Āl Thawāba, a family of high-ranking secretaries in the Abbasid administration. What he found was a tantalizing report of excavations in the pyramids and temples. These well-trodden monuments brought caliphs, governors, and treasure hunters seeking to unlock ancient secrets and vast riches; they also attracted geographers and were popular in travelers' tales.

Qazwīnī viewed the pyramids to be founts of hidden knowledge. So did he count them among the ancient architectural wonders of the world, an association that traces back through antiquity. The pyramids feature in Ibn Khurdādhbih's treatment of wondrous buildings; they are covered in an indecipherable script containing "every magic and marvel of medicine and the stars." Ibn al-Nadīm read that the temples of Egypt were designed for practicing alchemy, with special rooms for grinding, pulverizing, dissolving, binding, and distilling. Covering the walls were mysterious inscriptions and writings in Chaldean and Coptic. Beneath the floors were hidden treasuries containing the sciences of the art written on parchment, engraved on stone, and etched on gold and copper sheets.[13]

The alchemical arts were coded in the language of astonishment, a discipline whose secret teachings were hidden from sight but whose results were capable of awe-inspiring wonders. The mineralogical focus on special properties was also closely tied to the practical science of talismans. In his treatment of alchemy, Ibn al-Nadīm addresses Balīnās, or Apollonius, one of the ancient authorities on alchemy, with the title "master of talismans." The full honorific given to Apollonius the sage in *The Secret of Creation* is "master of talismans and wonders."[14]

As with Aristotle's lapidary, with its ties to Hermes, Apollonius is a major source for Qazwīnī. He knew the ancient sage as a wandering builder of talismans activated through the conjunction of heavenly and earthly bodies at specific times. In the course of his geography, Qazwīnī records several of the protective devices that Apollonius built for both the Roman emperor and King Qubādh, the legendary founder of the Kayanid dynasty of Iran who erected cities throughout the land. These talismans could turn saline water potable and ward off wild beasts, venomous snakes, and insects. They could also be used to heat baths, safeguard hidden treasures and cities, increase the yields of crops, and prevent inclement weather. Apollonius was even credited with an ingenious water clock built in Constantinople, an automaton that had different statues to mark each hour.[15]

Accounts of these civic wonders feature throughout early Arabic geographical writings, recorded by the likes of Ibn al-Kalbī and Ibn al-Faqīh, who were familiar with the talismans erected by Apollonius for Greek and Persian cities. Yāqūt grumbled that the charms of Apollonius were nothing but "fables meant to deceive the gullible," but others viewed such monumental talismans as ancient and efficacious wonders. In both his natural history and geography, Qazwīnī references several such talismanic statues, figurines, and devices.

Along with rare minerals, plants, animals, and marvelous buildings, Qazwīnī promises in the opening to his geographical gazetteer to list the "strange talismans" that philosophers made in towns and cities across the lands. Many of these protective charms formed part of public spaces, such as the lion statue guarding Hamadan, the traces left of Ctesiphon, and the talismanic statue—built for the Abbasid caliph al-Manṣūr crowning the green dome of the city palace in Baghdad—of a mounted knight bearing a spear. The latter was designed to warn residents of enemy incursions. Qazwīnī notes that it stood watch over Baghdad until 329/941, when the dome collapsed under the force of a powerful storm.[16]

Talismans built to protect cities feature throughout the *Kitāb al-Ṭalāsim al-akbar* (The major book of talismans), an Arabic adaptation of the Greek *Apotelesmata,* written in Apollonius's voice. Ibn al-Nadīm refers to the treatise as a well-known collection. In it Apollonius speaks of harnessing powers, as he traveled Anatolia, the far west, and through Iran, Egypt, Nubia, and Syria.

In addition to driving away all forms of animals and venomous creatures, there are talismans to protect against jinn, demons, and ghouls; to draw water from wells; to produce amnesia in others; to cast foes from their abodes; to increase the profits of merchants; to keep women chaste; to cause crops to increase or die out; to induce sleep on entire populations; and to uncover anything lost or hidden through a gem-incrusted mirror. There also survives an account of talismanic charms Apollonius built to heat a thermal bath in Edessa, which directly echoes the tradition with which Qazwīnī was familiar.[17]

With stones, metals, and jewels, these recipes frequently call for the use of gold, copper, lead, and iron. Mercury and sulfur also appear. The objects and figurines generally mirror forms to match their use: a scorpion to cast out scorpions, a woman to govern women, a serpent for serpents, and a mouse for mice. More elaborate instructions require molds of metals cast together adorned with gems and engraved with mysterious symbols, referred to in the Arabic translation as *qalafṭīriyyāt,* a calque derived from the Greek *phylaktēria,* protective charm or amulet. However, here in the Arabic translation of Apollonius's *Book of Talismans* this word is used to render the Greek *charaktēres,* a term of art for secret cyphers, indecipherable characters, and cryptic diagrams used when summoning demons and angels, ubiquitous in charms and amulets of antiquity. These mysterious symbols became a mainstay of

Islamic talismanic writing, on display in grimoires by the likes of Qurṭubī, Ṭabasī, Fakhr al-Dīn al-Rāzī, and Sakkākī, among others.[18]

Signet rings with precious stones feature in the *Book of Talismans* alongside large statues and small figurines. Apollonius teaches how to fashion talismans at auspicious times in the course of the week with invocations to sacred angelic forces that governed the planets and each hour of the day. Buried underground, worn on the body, and placed in public squares, talismans function by mixing spirit and matter, through the power of *pneumata,* generally referred to in Arabic as *rūḥāniyyāt*—souls or spirits that move the planets and the sublunar world of generation and corruption. Qazwīnī's account of angels as the hidden source for cosmic forces animating creation echoes these views. The connection between cosmic power and physical matter also intersects with the notion, which Qazwīnī also held, that the angelic *rūḥ,* breath or soul, was the basis for all the individual souls of beasts and humans.[19]

Theories of cosmic correspondence and sympathetic forces drew on beliefs and attitudes that were developed by pagans and advanced and transformed along the way by Christians, Jews, Zoroastrians, and Manicheans, to name a few. Amulets designed for diverse purposes in numerous shapes—some resembling animals and humans, others taking the form of phallic symbols of fertility gods—adorned bodies, guarded homes, and lined streets across antiquity.

Just as Christians and Jews addressed the occult powers animating physical objects in monotheistic terms, Muslims came to envelop talismanic and alchemical forces in the vocabulary of the Quran, the Prophet, and his family. In this process Hermes transforms into the Quranic prophet Idrīs, the alchemical teachings of Zosimos of Egypt (fl. 300 CE) come to predict the advent of Muhammad and Islam, and the *Book of Talismans* in the name of Apollonius concludes in its Arabic adaptation with a recipe added by the copyist that calls for fasting and reciting verses from the Quran when preparing a powerful potion.[20]

By Qazwīnī's day, many of the foundational teachings at the core of the Arabic Hermetic corpus had become absorbed into natural philosophy. In *The Concealed Secret,* his handbook of practical magic, Fakhr al-Dīn al-Rāzī explains that talismans function through sympathetic relationships and forces that govern creation. Following a general Avicennan framework, Rāzī notes that the art of producing talismans requires a "knowledge of the states that mix the celestial active faculties with the passive terrestrial faculties in order to produce either something which contravenes customary phenomena or to prevent that which corresponds to custom from occurring." This definition parallels Rāzī's treatment of talismans in his Persian compendium of the sciences, where he notes that talismans function through natural correspondences binding the cosmos.

Rāzī observes that through experience over time philosophers were able to determine the special traits of the degrees of the zodiac and the influences of the planets over the sublunar realm. He further teaches that the celestial forces, or *rūḥāniyyāt*, governing the planets not only had physical influences but also could be mobilized through various practices such as prayers, suffumigations, and incantations connected to particular timings meant to align with the movement of the stars. Rāzī also turned to Apollonius's *Book of Talismans* in his own discussion of the use of hours for casting talismans, as was he intimately familiar with Hermes as a source for ancient wisdom and occult knowledge. In his schema of the sciences, Rāzī grouped together the study of alchemy, mineralogy, talismans, and agriculture in a consecutive sequence. These sciences collectively follow the premise that there are unique properties spread throughout physical matter waiting to be tested and harnessed.[21]

In addition to Ibn Sīnā's *materia medica* from the *Canon*, one of the most important sources for Qazwīnī's study of physical substances is the *Kitāb al-Khawāṣṣ* (The book of special properties) of Apollonius the Wise. Qazwīnī catalogues under the name of Apollonius an array of unique properties and recipes that grant strange powers: The stone found in the eye of the sea turtle when placed under the tongue provides knowledge of the unseen. Eating the meat of a black cat protects against magic. A child who wears a hyena's heart acquires great intelligence and learns with alacrity. Rubbing the excrement of a black chicken on the gate of a house causes enmity to arise among its inhabitants. Whoever applies raven fat mixed with rose oil on the face receives any request made before a ruler. Tie a person's hair around the neck of a mountain swallow and that person will not sleep until either the swallow dies or the hair is removed.

Qazwīnī also cites Apollonius as an authority for cures, powerful drugs, and protecting agents. The smell of fresh fish removes intoxication. A rooster anointed with oil made from castor seeds ceases to crow. The root of the olive tree can cure scorpion stings. Coconut shavings burned in an oil lamp induce sleep. Ten dirhams of saffron ease labor, as does the root of the coriander plant attached to the thigh of a pregnant woman. Dill placed under a pillow removes panic, chewed up it protects against a hot iron, while a knife soaked in dill and vinegar loses its edge. The meat of the swallow improves eyesight and its droppings can dress boils. Whoever drinks the filth from an elephant's ear will not sleep for a week. When pulverized and eaten, the longest bone in a vulture's left wing incites an insatiable appetite. The same bone from the right wing, however, produces a hatred for food. Eating the gizzard of a chicken prevents bedwetting. The dried heart of a crow when ground and mixed with alcohol intoxicates in one drink, while its gallbladder obtains the same effect with just one sip. A dried spider's web reduces a phlegmatic fever, while viper's meat can alleviate leprosy and improve eyesight, and when mixed with oil and applied on the body, it can stop hair from growing. As with many other substances, viper flesh can also be used as an aphrodisiac.

Apollonius tells of potions to seduce, to strengthen a man's carnal vigor, and to decrease a women's wandering eye. The blood of frogs rubbed on the face induces desire in onlookers, when imbibed it increases sexual potency, but when applied to the gums it makes teeth fall out. The citron leaf crushed and mixed with oil is a powerful aphrodisiac, as are watercress seeds ground with almond oil and sugar. Chicken eggs increase a man's sperm and virility, and the heart of the hoopoe boosts a man's power for coition. Drinking the blood of a black cat attracts women. Wearing the spleen of a crow excites passions. Hyena fat applied to the eyebrows makes the wearer irresistible. Here too are potions for controlling women's bodies and their appetite for sex. The hyena's vagina when dried and fed to a woman without her knowledge removes her desire for other men, as does the blood of a swallow. The dried excrement of a vulture increases a woman's fertility, but a woman who drinks the blood of a hare never conceives.

Qazwīnī's treatment of *materia medica* also lists deadly substances recorded on the authority of Apollonius. Consuming lead produces madness. Oleander leaves and branches when cooked with barley generate a poisonous concoction capable of destroying an entire cavalry. Apollonius tells of amazing experiments. A pig tied firmly on the back of a donkey dies if the donkey urinates. Pulverized glass poured in a flask with water and wine causes the water and wine to separate, an experiment that Qazwīnī reports is easy to replicate.[22]

The line between the outlandish and efficacious is hard to draw. Yet appeals to direct experience and tested trials weight these descriptions with gravitas, where the hidden properties preserved by Apollonius abide comfortably next to Ibn Sīnā's pharmacy. The material also closely parallels the diverse ingredients that populate recipes in handbooks of magic. Many of these beliefs, such as the existence of the evil eye or the rabbit's foot as a charm, were so widespread and common across numerous societies that they defy a single origin. Their ubiquity also made them entirely intelligible, interchangeable, and easily transported, from one place and time to the next, passed along with the imprimatur of a universal scientific system that confirmed their reality.[23]

The medical and philosophical writings on *khawāṣṣ* built on earlier works in Greek on *physika,* which covered an array of topics including stones, plants, animals, agriculture, medicine, as well as crafts and tricks—all focused on *physeis,* powers or virtues residing within physical substances. These collections frequently drew from agronomic writings on *geoponika,* a body of learning concerned with practical instructions for the cultivation, care, and uses of plants and animals.

Writings on agriculture circulated in Greek, Latin, Syriac, Armenian, and Middle Persian before making their way into Arabic. Organized under the Arabic rubric of *filāḥa,* the field of husbandry commonly accompanied the study of meteorology and the movement of celestial bodies, alongside the talismanic arts. As with medical pharmacology, agricultural writings followed the view that sympathetic and antipathetic forces governed the interaction of physical substances. Qazwīnī frequently cites an otherwise anonymous book of agriculture that builds directly on this ancient body of learning, drawn from across Mesopotamia and the Mediterranean. As with astrology and talismanic forces, the discussion of unique properties plays a central part in Qazwīnī's agricultural material.[24]

The Arabic term *khāṣṣa* could be used to render the Greek *idios* and *idiotētes,* the unique physical qualities or properties that influence other forms of matter. The concept plays an important role in the medical writings of Galen, who uses the language of *idiotētes arrētoi,* indescribable properties, to refer to phenomena with effects that are discernible but cannot be explained by the manifest forces of the humors in their various combinations and mixtures. Such properties could be known only through experience. They included not only a variety of substances and conditions, but also amulets and charms. In medieval Latin medical and scholastic writings, the concept was generally rendered as *qualitates occultae,* the unseen or occult properties that represented the immediate agent or efficient cause behind a given physical phenomenon. These forces were beyond direct perception and could be witnessed only indirectly through their effects on substances, such as the manifest power of magnets, the effects of various drugs, or the influences of planetary movements.[25]

The notion of influence is central to Ibn Sīnā's definition of *khawāṣṣ* as properties, both active and passive. These properties either are unique to the physical form of a given substance or arise through the interaction of substances with elements and humors. Ibn Sīnā is quick to affirm that the influences of unique properties are not derived from the psychic power of the soul; instead they are to be explained through physical forces of bodies reacting to each other in sympathy or antipathy. From this follows Ibn Sīnā's explanation, which Qazwīnī repeats, that the elemental magic of *nīranjāt* functions through the influence of unique properties in physical substances.

In his Persian treatment of the topic, Qazwīnī further notes that *nīranjāt* are activated through the mixture of elemental bodies and *ruḥānī,* or spiritual, powers. The art, he explains, was developed by the ancient Chaldeans of Babylon, who held that angels and demons worked within physical bodies; these powers in turn could be harnessed through various combinations of prayers and sacrifices to produce amazing results. This cosmic framework for understanding the material world echoes ancient Zoroastrian ideas of dualistic forces animating creation. As with talismans, Qazwīnī includes recipes for *nīranjāt.* These involve cyphers, incantations, various substances mixed together, incense, and suffumigation, performed at specific astral timings.

In extraordinary influences, *āthār gharība,* Ibn Sīnā sees correspondences pulsing through celestial bodies and terrestrial substances. Taken together such forces can produce effects so amazing that one would almost judge them to be beyond the natural order, *khārija ʿan al-majrā l-ṭabīʿī.* The sympathetic and antipathetic powers of unique properties alone are innumerable. Ibn Sīnā lists as mere examples the force of magnets over iron, stones repelled by vinegar, and the ability of *kahrabāʾ*, or amber—the *ēlektron* of Greek lapidaries—to attract straw when rubbed. For Ibn Sīnā the topic of extraordinary physical influences also connects to what "the philosophers refer to as natural magic [*al-siḥr al-ṭabīʿī*]," where the force of love can bind together lovers, song can move the body, and speech can sway an audience.[26]

Well before Qazwīnī, alchemists and physicians had devoted treatises to indexing the unique properties of various substances. Many of the cures, recipes, and pathologies have ancient pedigrees repeated in Greek alchemical, medical, and agronomic literature. Some substances were rather pedestrian, such as the flowering peony common to Asia and the Mediterranean. When worn on the body, Qazwīnī reports, peony protects against seizures. Here Qazwīnī uses the medical category *ṣarʿ*—a term that was also identified with demon possession, though many physicians took their cue from Galen and Hippocrates, who preferred physiological explanations to account for seizures through humoral imbalances induced by excessive bile and phlegm.[27]

Qazwīnī adds that the peony remedy had been tested and verified, citing the authority of Ibn Sīnā. The famed physician and alchemist Abū Bakr al-Rāzī in his encyclopedia of medicine, also confirms peony's unique power. He evokes the authority of Galen, who recorded in the pharmacological compendium *On Simple Drugs* his personal experience with an amulet made of peony that protected a child against seizures. Galen, the great physician from the Aegean city of Pergamon, had verified that seizures would start up again when the amulet was removed and would desist only when the amulet was placed back around the neck.

For Galen, what distinguished the science of medicine from mere superstition was the empirical basis of knowledge generated through trial and observation. Galen rejects an array of remedies as superstitious magic. However, it is not that the use of amulets is irrational; rather the underlying physical mechanism behind their efficacy remains hidden. Abū Bakr al-Rāzī, like Galen before him, was also quick to distinguish between quacks and real physicians, and he warned of soothsayers and charlatans who masqueraded as doctors duping gullible men and women with incantations and sleight-of-hand deceptions.[28]

Among Abū Bakr al-Rāzī's many writings is a treatise dedicated to cataloguing unique properties. In it he draws extensively from Greek sources on medicine and agriculture, as well as a host of early Arabic authorities. The treatise also includes a section on talismans and concludes with an account of various wonders of the world. Rāzī opens with a defense against skeptics who doubt the strange and uncanny powers found in substances. He implores those with independent reason to suspend judgment until the validity of any given report can be verified.

Rather than reject everything whose causes remain unknown, Abū Bakr al-Rāzī encourages experiment. There is no one in our time, he argues, who can explain why the scammony plant relieves excessive yellow bile, but it would be foolhardy to reject its medicinal power, which is known through experience. Many substances—using the Arabic word *jawhar*, a standard calque on the Greek *ousia*—have amazing, beneficial effects, the causes of which the human intellect fails to grasp. In response to those who claim we should cast out everything we do not fathom, he cautions that much would be lost, for "if even a small fraction of these claims turns out to be authentic, there would still obtain a great and wonderful benefit." It is an argument that Qazwīnī uses to similar ends in his own catalogue, which he acknowledges is modest compared to the innumerable properties waiting to be uncovered.[29]

Qazwīnī includes many of the same wondrous reports recorded in Abū Bakr al-Rāzī's treatise and cites Rāzī's authority on several occasions. For his part, Rāzī quotes Aristotle, Galen, Hermes, Apollonius, and the Byzantine physician Alexander of Tralles (d. ca. 605), among others, and he draws copiously from Greek agronomic collections. Much of Rāzī's material, such as references to the phoenix that rises from the dead and the giant sea creature mistaken for an island, stretches back to antiquity. But he also includes more recent accounts, such as stones found in the land of the Turks that can cause violent storms. Bīrūnī ridicules Rāzī's report of rain stones as nonsense, though Qazwīnī is much more open-minded and recounts a demonstration of their unique power to induce rain held in the court of the Khwārazm-Shāh.

Rāzī also references the belief that magnets lose their ability to attract iron when rubbed with garlic, but that they regain their power when soaked in vinegar, an idea repeated much earlier by the likes of Plutarch and Ptolemy. It turns out that magnets are not actually affected by garlic, onions, or vinegar, but this did not stop the claim from circulating, carrying with it the air of empirical authority. Qazwīnī cites this unique character of magnetic stone as an example of strange properties so numerous that a single life would not be sufficient to count them all, repeating the refrain that there is no reason to abandon all these reports merely because there may be doubt concerning some of them.[30]

So it is with the dazzling emerald, which Abū Bakr al-Rāzī notes can blind vipers. Again, Bīrūnī, highly skeptical, observes that so many storytellers have repeated this claim that it now even features in books on unique properties—a clear jab at the likes of Rāzī, whose vast writings Bīrūnī followed closely and sought to rival. Bīrūnī even performed a series of tests over several months on vipers to ascertain the blinding effect of emeralds, all to no avail. Qazwīnī, by contrast, is content to include the oft-repeated claim, citing Abū Bakr al-Rāzī as his source.[31]

Most leading medical authorities of the day agreed that amulets and talismans were in some fashion efficacious, even if they debated the mechanisms behind their power. No greater authority could be invoked than Galen himself, who taught so much, as Abū Bakr al-Rāzī asserts. For this, Rāzī paraphrases Alexander of Tralles, who made the same argument, quoting Galen directly: "I used to think that incantations were nothing other than the fables of old women, but as I grew older I learned there is some truth to them, for I saw a man who had been stung by a scorpion perform an incantation and was then cured."

The Arabic for incantations, *ruqā,* parallels *epōdai,* and *khurāfāt,* for fables, nicely captures the sense of *mythoi* in the original Greek of Alexander's *Therapeutica,* a source with which Rāzī was intimately familiar. Rāzī continues by citing Alexander's own experience with a copper signet ring used as a powerful remedy, in an example Alexander uses to further argue that incantations and amulets are efficacious. Alexander supports this point with the authority of none other than Galen, "the most divine."[32]

Galen's *On Simple Drugs* addresses a cognate case of a man who claimed to be able to kill a scorpion through an incantation. Qazwīnī cites this anecdote in his treatment of the marvels of the human body and the unique benefits and uses of human anatomy. While witnessing the performance, Galen, ever the empiricist, noticed that as the man recited the incantation he spat over the scorpion; through a further trial with only spitting, Galen then deduced that it was the spit and not the incantation that was lethal. From this observation, Galen concluded that human saliva could harm poisonous creatures. The anecdote fits well with Qazwīnī's abiding interest in the unique properties of substances, as it bolstered strange phenomena with an empirical basis of scientific knowledge. This claim, as with many others rooted in the authority of antiquity, endures through cumulative appeals to direct experience. The unique property of human saliva is affirmed not only by Galen but also by a host of later medical authorities. Qazwīnī's focus here is not on the efficacy of amulets, which he never doubts, but on the diverse wonders of all aspects of existence, from the sublime heavens to the base matter of saliva.[33]

By Qazwīnī's day, Galen had long been celebrated in Arabic as a "leader of physicians and a worker of wonders" who could "manifest marvels." This image follows the reception of Galen in both Greek and Arabic. It resonates with Galen's vast command of unique properties and the underlying forces of medicaments and amulets. It also connects to Galen's argument, well known in Arabic, that the wondrous design of creation—from the legs of a gnat to the great movement of the heavens—was itself the best proof for the wisdom of a divine Creator.[34]

The appeal to wonder features prominently in Galen's study of human anatomy, the *De usu partium corporis humani* (On the uses of the parts of the human body), known through an Arabic translation as *Manāfiʿ al-aʿḍāʾ* (The beneficial functions of the limbs). Like many of Galen's writings, the translation was produced with the aid and supervision of the Nestorian physician and lead translator Ḥunayn ibn Isḥāq (d. 260/873) in the context of the Abbasid translation movement. It was one of the most influential anatomical works to circulate in Arabic, explicitly directed not only to philosophers and physicians but also to the field of theology.

Ḥunayn notes in his record of the Galenic works he translated that the collection celebrates the wisdom of God through a study of human anatomy. In it, Galen evokes wonder and awe in the sight of *thaumata,* marvels, that can be found in the most mundane aspects of the human body and that serve as confirmation of the wisdom, or *sophia,* of the Creator. Ḥunayn renders Galen's *thaumata* into Arabic as *ʿajāʾib* and *sophia* as *ḥikma.* Both the Greek and the Arabic evoke wonders and, in the context of science and philosophy, the language of divine order and natural marvel.[35]

Qazwīnī develops similar sentiments on human anatomy as a microcosm of creation and as a testament to divine wisdom. The argument from design as a proof for an intelligent Creator played an important role in the development of monotheistic views of God. Ghazālī engages with Galen on the matter. Following this well-established path Fakhr al-Dīn al-Rāzī turns to the marvels of the cosmos and the amazing properties of existence, from human anatomy to the movement of the heavens and everything in between, as a teleological argument for a wise Creator.[36]

Several prominent physicians and philosophers developed writings on the marvels of the world that evoke this sensibility. In addition to Abū Bakr al-Rāzī, who mixes in his treatise on unique properties a section on wonders and talismans, there is Ibn al-Jazzār (d. 369/979~), the renowned physician from the city of Qayrawān in North Africa, who penned a short work on *khawāṣṣ,* which drew from Abū Bakr al-Rāzī, Aristotle's lapidary, and a host of other ancient sources, including the great sage Hermes. Ibn al-Jazzār also developed the argument, well known to Qazwīnī, that the unique properties of substances could be known empirically through trial and experience. A key value in the development of the practical sciences—from botany, medicine, and minerology, to alchemy, astrology, and talismans—is the premium placed on empirical knowledge.

The corollary to trial and experience was the widely held belief that the secrets of nature could also be known by revealed knowledge that had been passed down from ancient prophets, in a line that invariably stretched through Hermes back to Adam. Through Latin translations, Ibn al-Jazzār's treatise joined other Arabic collections drawn from ancient lore that played a key role in the development of medieval Christian writings on occult properties. Like many others, Ibn al-Jazzār penned a work on the wonders of the world.

Similarly, Bīrūnī, the sage of the Ghaznavid court, did not shy away from the language of wonder and rarity in his diverse writings on geography, mineralogy, and ancient history. He also began collecting various accounts for a treatise entitled *Kitāb al-ʿAjāʾib al-ṭabīʿiyya wa-l-gharāʾib al-ṣināʿiyya* (Book on natural wonders and prodigious crafts), which, though by all measures left incomplete, addressed incantations, elemental magic, and talismans, among other topics. In his surviving writings, Bīrūnī evinces a good deal of knowledge of, if also skepticism toward, the claims of enchanters and the strange powers of nature, the authenticity of which could be investigated only through direct experience.[37]

Collections on wonders by physicians and natural philosophers were well known to Qazwīnī. Among the list of sources that Qazwīnī cites extensively both in his geography and his natural history is the *Tuḥfat al-gharāʾib* (Gift of rarities). Though Qazwīnī never identifies the author, the compendium survives in several Persian manuscripts, one of which is ascribed to Abū Jaʿfar Muḥammad Ibn Ayyūb al-Ṭabarī (fl. c. 502/1108), a mathematician and astronomer from Iran. Celebrating the wonders of creation and the rarities of existence, the collection draws from the observations of "philosophers, sages, and travelers of the world." Qazwīnī turns to the work frequently in his treatment of strange phenomena found in islands, mountains, rivers, and springs.

In addition to marvels by land and sea, the *Gift of Rarities* opens up a range of practical skills employed by druggists, artisans, and masters of various trades. Here are ingredients for aromatics; recipes drawn from the special properties of minerals, plants, and animals; directions for making inks, pigments, and dyes; cleaning agents for removing stains; ways to uncover the deceptions of thieves; techniques for trapping animals; methods for summoning demons and jinn; instructions on how to enchant others through spells and amulets; tests for virginity and pregnancy; as well as alchemical transmutations, talismans, subterfuges, tricks, and games.[38]

In *Wonders and Rarities*, the catenation of practical techniques and remedies suggests a field of therapeutic possibilities for mobilizing the orders of creation. Catalogues of unique properties focused on the transformative forces in creation that were generated through influences and elemental reactions in physical matter. The study of ethics and moral philosophy, in contrast, sought to harness the soul to influence the body and society. These topics are woven throughout Qazwīnī's natural history, featuring not only in his study of human traits but also in his broader understanding of the capacity of the soul to produce magic and miracle.

Qazwīnī was intimately familiar with ethical exercises and spiritual disciplines that sought the refinement of character. In the push and pull between nature and nurture, the study of ethics offered possibilities for self-transformation. The range of problems and concerns that the study of unique properties could address provides one index for tracking the desires and anxieties of male elites, measured in the numerous recipes for increasing the virility and the prowess of men, while controlling the sexual appetite of women. As a field, moral philosophy was also largely directed to men, and took cosmic hierarchical order as a basic predicate for all existence.[39]

For classical virtue ethics, the metaphysics of hierarchy advanced male dominance and rule as a natural basis for domestic order and political life. The authority of men at home and in governance was supported not only through recourse to the Quran and the sayings of the Prophet, but also through a corpus of medical and philosophical writings. Virtue ethics, referred to as *akhlāq*, the traits or characteristics of individuals, formed a branch of practical philosophy that asserted a natural basis for social hierarchy.

When Qazwīnī speaks of contemplating the wonders of existence as an ethical pursuit, he builds on a long tradition in moral philosophy that saw corporeal and intellectual discipline as a means of spiritual refinement. Ghazālī, Rāzī, and Ṭūsī, among others, all contributed important writings on ethics, which turned to the discipline of the body for purifying the cognitive power of the soul, or *nafs*. As a term of art, the central category of *nafs* also carried with it the sense of mind and self.

In the framework of Aristotelian ethics, the idea of *telos*, as both end and completion, holds a special importance. It is directly related to the central concept of *entelecheia*, the work of bringing into full actuality the ideal self through a process of self-fulfillment. It is a concept generally referred to in Arabic and Persian as *takmīl*—to render perfect, complete, or actualized. These writings privilege above all an ideal male subject whose body is marked for purification and perfection.

The gendered values that animate practical philosophy advance the authority of men as heads of families and as leaders of society. To be sure, elite women participated in various aspects of public life, across time and place, as patrons, religious authorities, and poets, and not merely as mothers, queens, or lovers. Yet a range of metaphysical, legal, and social strictures also drastically curtailed women's access to arenas beyond the walls of the home or the palace.[40]

As with Plato's yoking together of body and state, *sōma* and *polis,* the metaphor of the body politic is a central feature of Islamic moral philosophy. It is a conceit with incredible traction. "The excellent city is like a perfect, healthy body whose limbs all work in concert," explains the philosopher Fārābī. Like Plato, Fārābī viewed the perfect state as governed by an enlightened ruler who could lead and guide all classes of society, from the ignorant masses to the educated elite. Justice, in such a framework, is obtained by balancing heterogeneous and unequal factions in complete harmony, where individuals know their place and act according to and within their own capacities. These ideas were well known to Qazwīnī, who also drew on the ancient metaphor of the human as a microcosm, *ʿālam ṣaghīr,* and as a mirror for the political order of the state, with the mind ruling over the body.

For Qazwīnī, the hierarchy of cognition starts with the rational soul, which reigns as the sovereign, and then proceeds to the faculty of reason, which operates as the chief minister, overseeing all the senses. In turn, the senses function as an army, with the common faculty of perception—*al-ḥiss al-mushtarak* in Arabic and the *sensus communis* in Latin—working as an intelligence officer who gathers information from the senses. Perception serves as an intermediary to the faculty of reason, which selects what reports to accept and what to reject. The tongue all the while serves as a court interpreter, which translates perceptions and thoughts to the external world. Qazwīnī makes a similar parallel for the cosmos, with the Sun as king, the other planets as ministers and courtly officials of the state, and the stars as the regions of the land, all attuned to the astrological influences of the heavens.[41]

Writings on ethics promised individual and social refinement. As Ṭūsī notes in the opening to his widely read manual on the good life, theoretical philosophy culminates in the abstract analysis of the soul as a force in nature, whereas practical philosophy teaches how to actualize the power of the intellect. Starting with *tahdhīb al-akhlāq,* the individual refinement of character or virtues, the path of perfection progresses to the organization of domestic economy, and finally ends in the study of governance. This ultimate field of applied philosophy focuses on *siyāsa*, with the meaning of discipline and punishment, as in the classical Persian expression *siyāsat kardan*—to discipline or punish and thereby to rule.[42]

The study of ethics intersected directly with the vocabulary of dispositions and traits, generally referred to as *akhlāq*, from *khulq*, meaning character trait, related to the verb *khalaqa*, "to create, fashion, or mold," and *makhlūq*, "created," a central category for Qazwīnī's cosmology and vision of nature as the product of a divine Creator. Some of these characteristics were malleable and could be altered through the discipline of the soul. Others, however, were largely predetermined through innate predispositions. Thus, Qazwīnī explains that although climate and temperature influence the traits of people across the world, in a very broad sense individuals can also acquire ethical characteristics and moral behaviors through habituation and practices of refinement.

These ideas emerge from teachings of classical moral philosophy, which developed *hexis*, or disposition, as a key category advanced by the likes of Plato, Aristotle, and Galen. This Greek term was generally rendered in Arabic as *malaka* or *hay'a*, and in Latin as *habitus*. Just as Qazwīnī held that humans have a collective habitus, *al-hay'a al-ijtimāʿiyya*, which compels them to work together, he also saw that individuals possess a range of dispositions in the soul that can push them toward either good or evil actions or traits. Qazwīnī explains a trait, or *khulq*, as a firmly rooted disposition or habitus, *hay'a rāsikha*, of the soul "from which actions proceed with ease, without the need for thinking or reflecting." The habitus by which noble actions flow in accord with reason and law, *ʿaqlan wa-sharʿan*, is called a good trait.

Traits can be essential to an individual or acquired through habituation. Qazwīnī's definition of ethical behavior advances a harmony between reason and law in decidedly Islamic terms as a basis for determining the good. He draws his definition almost word for word from Ghazālī's explanation of habitus in his magnum opus of ethical and spiritual endeavor, *The Revival of the Religious Sciences*, which in turn builds on earlier reflections on Aristotelian moral philosophy by the likes of Farābī, Miskawayh, and Ibn Sīnā.[43]

Dispositions in the soul can harden through habit or be acquired through the refinement of character. By relinquishing attachments to the physical world through devotion and ethical pursuit, individuals can strengthen their own psychic resolve and spiritual force. Qazwīnī argues that the contemplation of the wonders of creation is an ethical method for refining the self in order to attain higher levels of awareness of existence. Through such disciplines of the body and the mind, those among the ranks of prophets and saints can obtain nearly angelic states of gnosis.[44]

Qazwīnī was also familiar with the key teachings of the masters of talismans on how to use the spirit, or *rūḥ*, referred to as *al-ṭibāʿ al-tāmm,* perfect nature, to draw on celestial forces of the stars, harnessing, as it were, one's innate spiritual self. It is an idea with ancient roots and it featured in an array of early Arabic philosophical, occult, and alchemical writings, including the corpus of Hermetic teachings, *The Secret of Creation* in the name of Apollonius, and the handbook of talismans and astral magic, *The Aim of the Sage* ascribed to Maslama al-Qurṭubī. Through the Latin translation of *The Aim* the concept had important afterlives in Christendom as *natura completa.* For Fakhr al-Dīn al-Rāzī, perfect nature serves as a foundational concept both in his work on astral magic and in his metaphysical writings on the power of the rational soul. The category also played a pivotal role in the Illuminationist philosophy of Shihāb al-Dīn al-Suhrawardī. Qazwīnī knew that Rāzī was intimately familiarly with Suhrawardī's metaphysics, and there is much which links the two, beyond just the power of the soul. Qazwīnī's account of perfect nature features in his treatment of the amazing influences that virtuous souls, *al-nufūs al-fāḍila,* can produce.[45]

Qazwīnī explains that the "people of truth," which is to say Ashʿarī theologians, hold that souls differ according to their substances. There are lofty, luminous souls that have knowledge of the realm of spirits, the *ʿālam al-arwāḥ*. From forces emanating through this spiritual realm, these souls can produce amazing phenomena. At the other end of the spectrum are heavy, turbid souls enamored with physical bodies, which have no knowledge of the realm of the spirits.

Moreover, Qazwīnī explains, some philosophers maintain that the rational soul is of one genus that comprises many species of individuals, which differ one from the other only in number. In turn, each individual is like the progeny of a celestial spirit. Qazwīnī further notes that masters of talismans hold that this celestial spirit—which Fakhr al-Dīn Rāzī refers to as an angel—is called "perfect nature." This celestial spirit governs the betterment or repair of individual souls through intimate prayers, revelations, and angelic inspiration in the mind. By this point, perfect nature, a crucial concept drawn from the Arabic teachings that circulated in the name of the prophet Hermes, had been integrated into high-level philosophical discussions on the power of the soul and the cosmos. The idea notably resonates with the ancient notion that each individual is coupled with a divine double, generally referred to in Greek philosophy as a personal *daimōn*—a kind of guardian spirit with access to otherworldly knowledge and power. While *daimones* were often identified in Arabic as jinn, the language of perfect nature shifts the discussion of the divine double into the celestial realm of angelic spirits.[46]

Importantly, Qazwīnī's concise treatment of the topic directly parallels Rāzī's formulation, without mentioning the master by name. How to unleash perfect nature is a central concern in Rāzī's theory of prophecy, as does it feature throughout his handbook of magic, which offers techniques for drawing on the spirits that move the heavens. Rites with supplications and suffumigations are met with practices meant to cleanse the body and purify the soul so as to create noetic connections with celestial powers. Qazwīnī further develops these activities of self-discipline as a basis for occult power in his Persian version of *Wonders and Rarities*, where he has sections on talismans and the elemental magic of *nīranjāt* that are otherwise missing from the surviving Arabic redactions.[47]

For all the promise of transformation—whether through unique properties, talismanic powers, and alchemical arts, or through spiritual disciplines and ethical refinement—the forces that governed the form and shape of the cosmos were in a basic sense inescapable. Qazwīnī took it as a foundational principle that differences in human characteristics could be explained through physical conditions in the world. The movement of the stars influenced gestation and health, and with them the dispositions that governed nativity, in decidedly astrological terms.

An increase in heat in the womb produced males, while its absence resulted in females. This measure of temperature generated a spectrum of traits, from the masculine, *tadhkīr*, to the feminine, *ta'nīth*. This range of characteristics could be expressed in both males and females. If innate or natural heat, *ḥarāra gharīza,* is perfected in the womb at the moment of conception, a male is produced with complete organs and vigorous masculinity. Along this gradation, a decrease in heat produces males who resemble females with feminine traits and natures, females who resemble men with manly traits and natures, and ultimately effeminate females.

The balance of temperature also served as an explanation for the formation of the *khunthā,* or intersex offspring, which Qazwīnī identifies as an extraordinary or strange, rarely occurring condition, *ḥāla gharība baʿīda.* Sexual traits, dispositions, affects, and orientations were all tied to physical forces in nature. The register evoked by the word *gharīb*—strange, rare, peculiar, odd, queer—represented a broader continuum of sex differentiation, supported in the leading medical discourse of the day by the likes of Abū Bakr al-Rāzī and Ibn Sīnā. The male stood for the normal and optimal, against which everything else could be measured, in some degree, as a deviation. These ideas also lent a cosmological basis to the casual misogyny that surfaces in *Wonders and Rarities,* where women are characterized as ontologically and ethically distinct from men—sources of temptation, and objects to be desired and controlled, capable of all forms of cunning and deception.[48]

Similar natural dispositions inform Qazwīnī's notions of human diversity. Across the climes, the gradation of temperature and the force of stars bore a direct influence on the people of the world. The blacks of Africa, Qazwīnī explained, lived too close to the Sun, which through excessive heat burned their skin into charcoal and thinned out their bodies. This was the cause behind their *akhlāq,* traits that made them akin to savages and wild beasts. By the same logic, Qazwīnī observed that the regions in the north, occupied by the Slavs and the Rus, suffered the inverse effect due to their extreme distance from the Sun. Pale and unripe in their whiteness, with lank blond hair, and massive dull bodies, they lumber around with the traits of livestock.

These ideas of environmental disposition, which were also wedded to the gradation of reason, served to naturalize all forms of prejudice and bigotry. The theory of climatic determinism supported by Ptolemy, Galen, and Hippocrates had long been offered as a means for affirming differences of capacities and dispositions regulating groups. Climate explained not only why some were savages, but why others were inherently more skilled and intelligent.

Qazwīnī follows an established principle common throughout early Islamic geographical writings that the people of Persia, living as they do in the center of the climes, enjoyed perfectly sound intellects, superior minds, healthy bodies, masterful dispositions for every craft, the most beautiful forms, the most measured manners, and the most knowledge for managing all matters. The ancient philosophers had developed similar notions of the world, though with Athens at its center.[49]

At the heart of human capacity was the soul, which determined the cognitive ability to fathom complexity. The more one was imbued with the power of the rational soul, the greater one's intelligence. Qazwīnī inherited this theory of cognitive variation from Ibn Sīnā's psychology of the intellect. The rational soul invests prophets and saints with the power to work miracles. It guides all forms of divination, from soothsayers and fortunetellers to those with the knowledge of *firāsa,* the art of physiognomy, which held that exterior physical features could be read to assay a person's inner character or thoughts.[50]

Just as intelligence and alacrity could be explained by the force of the rational soul driving the cognitive capacities of the intellect, idiocy and stupidity were the result of the uneven distribution of reason among men and women, which could also be affected by climatic variation. In turn, the endowment of the rational soul, Qazwīnī affirms, is what gives humans as a group a superior rank over animals, plants, and minerals. As the basis for cognition and perception, the soul could influence other bodies in ways that were distinct from the forces of attraction and repulsion that governed the unique properties spread in unequal parts throughout creation.

As with the writings on the ethical purification of the self to obtain higher levels of gnosis, much of the literature on properties also blurs the line between magic and medicine. Accounts of occult forces feature both in medical compendia and in writings on talismans, amulets, and incantations. This too is a legacy inherited from late antiquity, where Galen and Aristotle came to stand for masters of nature both hidden and visible.

With the Abbasid translation of Greek and Syriac materials into Arabic, there emerged a sophisticated vocabulary for fathoming magic and miracle that built upon earlier philosophical and scientific writings. Notably, the *Enneads* of Plotinus offered a theory of magic as a physical force that could be studied and harnessed. Large portions of the collection feature in an Arabic paraphrase prepared for Kindī and his circle. This work came to be known in Arabic generally as the *Theology of Aristotle*. Many of its ideas, in turn, influenced the likes of Fārābī and Ibn Sīnā.

Although the Arabic paraphrase only contains a selection of the *Enneads*, it preserves a significant discussion of cosmic sympathy and the natural power of magic. Plotinus holds that "true magic"—*hē alēthinē mageia* in Greek and *al-siḥr al-ḥaqq* in the Arabic paraphrase—is the magic of the cosmos, composed of love and strife. It occurs either through sympathy and the concord of similar things, or through opposition and divergence found in plurality and variance. The theory is based on a vision of the sympathetic and antipathetic forces of affection, *philia*, and discord, *neikos*—rendered in Arabic as *maḥabba* and *ghalaba*—which govern the physical world. It is a philosophy of interconnected existence.

Everything works in concord. When one part moves, a series of reactions ensue, as if the entire cosmos were one being. The paraphrase explains that a learned magician seeks to resemble the universe and practice its works to the extent of his capability, making use of affection in one place and strife in another and employing drugs and natural devices, *al-ḥiyal al-ṭabīʿiyya*, which are scattered throughout terrestrial substances, *munbaththa fī l-ashyāʾ al-arḍiyya*. The various techniques of self-purification that the magician or wonderworker must undertake in order to master otherworldly forces were common topics in handbooks of occult learning. They find ready parallels in ethical and spiritual practices that seek to discipline the body as a vessel for the soul.

The treatise further notes, "We say that in terrestrial substances are powers that produce amazing affects [*al-afāʿil al-ʿajība*], and it is from celestial bodies that people acquire these powers, for when these effects are produced they occur through the help of heavenly bodies." Furthermore, incantations, supplications, and tricks are nothing compared to natural substances possessing wonderful powers, *al-ashyāʾ al-ṭabīʿiyya dhawāt al-quwā l-ʿajība*, that can cause marvelous effects on other substances in the cosmos. This expression of magic, Plotinus explains, is like a river, which discriminates against neither good nor evil intentions. It is instead a natural force that binds together the heavens and the Earth.[51]

For Plotinus, *daimones*—the individual guardian spirits ubiquitous to the natural orders of antiquity—play an important role in the cosmos. In the history of Christianity these entities came to be identified increasingly with maleficent forces, hence the English word "demon." Yet their moral status in antiquity was by no means so neatly delineated, as they could function as a kind of divine double and offer means for achieving unseen powers and greater states of knowledge. In the Arabic paraphrase, Plotinus's *daimones* are rendered as jinn in a common equivalence. Thus, a central text for the Arabic picture of Aristotle explains that by natural disposition, *bi-l-ṭabīʿa,* jinn are influenced by speech and answer those who summon them. In this way, they serve a parallel function to stars as sources of hidden powers which can be mastered through prayers and ritual offerings.[52]

At the heart of the forces of sympathy and antipathy as physical explanations for the power of magic is a philosophy of "the One and the many." In Plotinus's view of existence, the One is equivalent to the Good. As everything ultimately emanates from the One, there is no independent evil in the cosmos. At the highest scale from the One flows the intellect, followed then by the psyche or soul. While Plotinus is critical of the claims of judicial astrology, he develops a theory where the stars themselves are ensouled, in a kind of celestial empire of the heavens. This proceeds in a great scale of being that binds all existence, from the heavens down through the Earth. By drawing on the power of celestial bodies, the magician merely harnesses innate forces in nature.

A key concern, as developed by the students of Plotinus, is theurgy, so-called divine work, through which philosophers strive by incantations and ritual acts of self-purification to obtain otherworldly powers and higher states of gnosis. This vision of the cosmos played a role in the development of what has been referred to since the eighteenth century as Neoplatonism, though the term was not used by Plotinus or any of the numerous acolytes who followed after him. The theory of cosmic interconnection emanating from the One, in turn, left an unmistakable imprint on the development of Islamic philosophy and theology.

It also coincided with the circulation of pseudo-epigraphic writings, many associated with Aristotle and Hermes, which promoted magic broadly and occult powers particularly as natural forces that could be observed and mastered. Knowledge of magic represents the highest stage of learning in the epistles of the Brethren of Purity, an encyclopedic collection associated with the metaphysical teachings of Kindī and his students. This view guides the medical and philosophical thinking of Abū Bakr al-Rāzī, and it also shapes Qurṭubī's famed handbook of magic, *The Aim of the Sage*. What unites these writings is a vision of magic and marvel as natural phenomena that can be described in scientific terms.[53]

These developments, needless to say, preceded the advent of Ibn Sīnā's psychology, which advanced the core teachings of Platonic emanationism in important ways. Perhaps more than anyone, Ibn Sīnā gave these ideas their most authoritative form among Sunni religious authorities, as represented particularly by such towering Ashʿarī theologians as Ghazālī and Fakhr al-Dīn al-Rāzī. The philosophical arguments that lent magic and miracle a rational basis in natural philosophy also shaped the Illuminationist philosophy of Suhrawardī, who engaged extensively with Avicennan metaphysics. For his part, Qazwīnī knew the writings of Suhrawardī, whose teachings would gain increasing authority in the coming centuries.[54]

This worldview leads Qazwīnī to include magic as a natural phenomenon within the physical world. It is a vision also entirely consonant with the way Aristotle features in *Wonders and Rarities* as a master of special properties and Hermetic wisdom. Aristotle teaches that the eagle stone, the famed *aetitēs* of antiquity, found in the nest of eagles, can aid in childbirth and if placed under the tongue can help to defeat any adversary in a dispute. Aristotle further describes green malachite, well known to Hermes, as having many amazing properties including the abilities to cure scorpion bites and increase sexual appetite. Then there is the flying stone. Whoever possesses it, Aristotle explains, can subjugate demons and learn their knowledge.[55]

Many of the gems that Qazwīnī encountered in Aristotle's *Book of Stones* reference the adventures of Aristotle's leading pupil, Alexander the Great. What is to be done in the valley of the serpents guarding diamonds in Sarandīb, whose gaze kills anyone who chances to look in their direction? On Aristotle's advice Alexander dispatches them with the trick of iron mirrors. Pieces of meat are thrown down from above, sticking to the diamonds below. Eagles then safely retrieve the gem-encrusted morsels.

Then there are the sex stones that Alexander discovered in Africa. When brought near human or beast they increase appetite for coition. Alexander wisely forbade his men from taking them, lest they be overrun with uncontrollable desire. He had one of the stones broken and found within it a scorpion and its image, the underlying talismanic force behind the stone. But there were other stones whose powers proved indispensable. On conquering India, Alexander found a gem that canceled out magic and guarded against jinn. He ordered his men to wear it to protect them from the deceptions of demons. There was the ocean stone that Alexander found in the encircling sea of darkness, which guarded against drowning. Aristotle learned from Hermes that it could cure debilitating illnesses and paralysis. Another stone encountered in the dark encircling sea would grant wisdom. Whoever wore it while having sex would produce a wise son. It also protected against the evil eye and could be used to cure leprosy.[56]

Alexander was amazed by stones and their occult properties, Qazwīnī reports, quoting Aristotle. In the early Arabic corpus of Hermetic writings, Alexander learns from Aristotle techniques for mastering perfect nature to harness the celestial spirits that moved the stars. As Alexander conquered the world, stone after stone made it into the imperial treasury. Spells, amazing objects, and powerful talismans feature throughout the various adventures told of Alexander's exploits in Greek, Syriac, Armenian, and Ethiopic, not to mention all the major vernaculars of Latin Christendom. The cycle of adventures was well known in Arabic and Persian, where Alexander features as a pious king set on a divine mission to enlighten the inhabitants of the world and to uncover along the way the wonders of existence.

As an ancient witness to strange peoples and rare substances, Alexander features frequently in writings on marvels, and Qazwīnī's *Wonders and Rarities* is no exception. Qazwīnī knew Alexander also by the Quranic honorific Dhū l-Qarnayn, "The Two-Horned," the title of a righteous king whose mission to the spread the message of God across the earth appears in the Quran and in early Islamic tradition. Not only did Alexander traverse the known world, but as *The Book of Stones* written in Aristotle's name attests, he also sailed into the encircling sea of darkness, the great barrier that defined the terrestrial limits of human knowledge. Other than reports of a giant cosmic mountain—an abode of jinn and demons—at the edges of its shores, nothing beyond the encircling ocean was known.

PART III

Distant Shores

It might feel daunting, to travel so far. Yet with everything so closely connected, Qazwīnī made it only a matter of how to navigate the pages. Collectively the chapters, subdivisions, and individual entries gave the object of inquiry—nothing less than the entire cosmos—coherence and unity. There were many ways to read a collection with such clear signposts guiding the way, to pause on one entry or to jump ahead to another. And from a certain vantage, as mere abstraction, the tome itself could signify an entire metonym for creation, a disposition, a sensibility, a set of values and ideals that were readily recognizable and intelligible for generations.

Many of the most dramatic parts of this story lie ahead, following the reception of *Wonders and Rarities* across time and place as it transformed into a venerated classic. From village notables to mighty emperors, the promise of a world within a book would delight and edify in new and unforeseen ways. But there is also much that is lost along the way. In due course the massive edifice of learning and authority that made Qazwīnī's sense of wonder so enduring would be whittled away, piece by piece.

As the limits of knowledge shift, fissures form between the two halves of the globe, between the Earth and the Sun, and in the relationships between magic, miracle, and science. These unequal divisions separate the past from present. They produce their own remainders in the material that cannot be integrated—left out, unassimilated. In the transformations that unfold,

Qazwīnī would come to be strange and curious for reasons that his many readers over the ages would have strained to imagine. But such feelings of alienation were still years in the making. First oceans had to be crossed, lands conquered, and fortunes made, before the round edges of a new world would come into view.

7

LONG DIVIDED

ON THE HORIZON OF THE SETTING SUN lies a distant shore. Encircled by a sea of darkness, a vast island marks the western boundary, beyond Africa to the south and the Franks to the north. At the farthest edge of the map rise the towering summits of Qāf, an unreachable region of high peaks. Here dwell jinn and demons. For their part, these undulating mountains support the entire edifice. Like a grape floating in a goblet or a yolk suspended in the white of a cosmic egg, half of the Earth is exposed to air, the other half is submerged beneath the water of an encircling ocean. Only a quarter of the exposed Earth is suitable for human habitation. The rest either is covered in water or experiences extremities in temperature that make it inhospitable to life. The mountains of Qāf propping up the world rest on the back of a bull, standing on a leviathan. The whole weight of it all is stacked on the shoulders of an angel (figure 7.1).[1]

This particular map of the world features in an Ottoman Turkish translation of Qazwīnī's *Wonders and Rarities* produced by the madrasa professor Musliḥüddīn Sürūrī (d. 969/1562) for the prince Muṣṭafā (d. 960/1553), the ill-fated son of the tenth Ottoman sultan, Süleymān the Magnificent (d. 974/1566). Works on wonders of the world and geographies of distant lands had become particularly fashionable among the Ottoman elite, who by this period commanded one of the most powerful naval fleets in the world. The illustration of an angel supporting a whale and a bull offers a picture of the cosmos that neatly fuses Ptolemy's division of the inhabitable quarter of the Earth into seven climes with the ancient belief that the heavens and the Earth rested on towering mountains and cosmic creatures.[2]

FIGURE 7.1: Diagram of the world surrounded by the mountains of Qāf, supported on cosmic creatures, with the south oriented toward the top, and outline of the New World to the right. Both the Kaʿba and the wall of Gog and Magog can be made out in the map. Turkish translation by Sürūrī dated 960/1553.

The image of divine beings or angelic forces supporting the world finds parallels in countless creation stories. A variety of pillars, including mountains, bulls, whales, angels, snakes, and turtles, populate cosmic theories from antiquity. The story of Atlas with the Earth on his shoulders has affinities to ancient Near Eastern myths as well as Hindu, Buddhist, and Jain cosmologies. The mountainous superstructure of Qāf, which has echoes in Iranian lore, also resonates in the Indian subcontinent with Mount Meru bearing the universe in a picture of the cosmos found across Asia. The early Arabic accounts, which Qazwīnī draws from, build on these heavenly orders, often mixing and matching, recast in the voices of early Muslim authorities.[3]

While there were several combinations, early preachers frequently describe a bull supporting the entire mass, balanced on the back of a giant whale, often identified as Bahamūt, a name derived from *Behemoth*, mentioned in the Bible. Yāqūt addresses the story with a good deal of suspicion, ascribing it to storytellers of questionable reliability. For his part, Qazwīnī relates the account on the authority of Wahb ibn Munabbih, where the Earth is supported on a colossal angel standing on a mammoth ruby on the back of a giant bull, with thousands of radiant eyes, snouts, and tongues, on top of a whale, supported by a sea above the darkness, past which nothing is known of what might extend beyond. These angelic pillars are so immense that all the seas of the Earth could fit within one nostril of the bull like a single grain in a vast desert.[4]

Although much of the image in this manuscript would have been intelligible, Qazwīnī certainly would not have recognized the distant western shore that appears on this particular Ottoman map, produced at some point shortly after Sürūrī's death. Sürūrī abandoned his translation of Qazwīnī's *Wonders and Rarities* after his patron, the prince Muṣṭafā, was executed. Sürūrī had sought, at the prince's request, to produce a more faithful and fluid rendition of Qazwīnī's cosmography, which had already appeared in several earlier Turkish and Persian translations. Yet his translation also expanded on Qazwīnī's text in notable ways, with florid prose patterned by strong currents of mystical piety. Though Sürūrī never finished the project, which was meant to also include Qazwīnī's geography, *Remnants of the Regions,* it found many readers, if measured solely in surviving manuscript copies.

The diagram that features in this particular manuscript differs from other copies of the period for its inclusion of a detailed world map. Like the many other maps of classical Islamic geography, this diagram is oriented with the south at the top and the north at the bottom, a long-standing cartographical practice. At the heart of the map a black rectangle represents the Kaʿba of Mecca, the sacred center of the world, while to the far northeast the wall of Gog and Magog built by Alexander the Great can be made out. On the western shores of the Earth an outline of the New World stretches into the distance. Qazwīnī was aware that the outer ocean encircling the world was filled with innumerable islands and countless creatures, unknown to all humankind. Yet he had no knowledge of a vast land in the west.

This diagram of the cosmos formed part of the early Ottoman reception of European cartographical and historical accounts on the discovery and conquest of the New World, as the Western Hemisphere was frequently called. The map accompanying Sürūrī's translation represented an amalgam of various cosmic orders and cartographical models, and reflected the dynamic and ever-changing trajectories by which Qazwīnī's natural history circulated. Yet the models Qazwīnī knew well continued to circulate, shaped by Ptolemy and classical Arabic geography, with the world surrounded by the unreachable mountains of Qāf (figure 7.2).

FIGURE 7.2: A map of the world surrounded by the mountains of Qāf, from a copy of Qazwīnī's *Wonders*, likely produced in the seventeenth century.

FIGURE 7.3: Map of the world with outlines of both the Far East and the West, including the New World, with the south oriented toward the top, from Qazwīnī's *Wonders*, copied in Ottoman Egypt.

But sensibilities were also changing. Take, for instance, a richly illuminated Arabic copy of *Wonders and Rarities* produced in Ottoman-ruled Egypt in 973/1565 for a high-ranking dignitary. Qazwīnī's schematic map of the world is replaced with one shaped by the new cartography that had occupied the Ottoman elite, with their seaborne adventures in the East and their keen awareness of the discoveries in the West (figure 7.3).

As Qazwīnī's natural history traveled, knowledge of the world also transformed. Over time, after the discovery of the New World, the geographical model that Qazwīnī had inherited would be profoundly challenged. As a boundary marker, the great encircling ocean represented the limits of human knowledge, shrouded in mystery and terror. Starting in the early twentieth century, the question of whether Muslims reached the Western Hemisphere before Columbus occupied an array of reformists, who sought to reclaim for the history of Islamic civilization precedence in all areas of science and learning. Yet the classical authorities of Islamic geography were unanimous in their belief not only that the encircling ocean was impassable but also that no one knew with certainly what lay across its shore. Many modern reformists, from Aḥmad Zakī to Fuat Sezgin, would turn to stories of westward expeditions for evidence that Muslims reached the New World before Europeans. Yet to read the classical Arabic anecdotes of treacherous journeys in the sea of darkness as testaments of early contacts with the Americas is to skip past all the elements of wondrous uncertainty that guide them.

Qazwīnī himself is quite explicit about the limits of what could be known beyond the encircling ocean. Faced with towering waves of the deep, sailors hugged the shoreline and dared not venture into the open seas. In this Qazwīnī echoes a long line of earlier geographers. "If you thought about it," Qazwīnī counsels, "you would recognize that humans are bounded by the seven climes." Only God knows what extends past the horizons.[5]

What exactly lay beyond the edge of the Earth was unknown to the authorities of antiquity. The Western Hemisphere was absent from the classical tradition of Islamic cosmography, which in scientific circles hewed closely to Ptolemy's astronomical and geographical picture of the world. Despite all the influence of the Ptolemaic model, other theories still circulated. For the form and shape of the Earth, Qazwīnī places the opinions of ancient philosophers on geography and astronomy alongside accounts of the cosmos related by Muslim storytellers, theologians, and traditional religious authorities.

Qazwīnī is content to relate the contradictory reports of the Earth and the varying accounts of its shape and location without trying to harmonize them. In the competing visions of the ancients the Earth took many forms: a flat, four-cornered plane, a shield, a drum, and even a half sphere. The accepted wisdom, though, was that the Earth was a globe. Following Ptolemy's example from the *Almagest,* Qazwīnī noted that the sphericity of the Earth could be demonstrated by how a ship at sea spots the summits of lofty peaks before making out the base of mountains on the coast. He also recounts that Ptolemy was able to calculate the circumference of the Earth by observing the different timings for an eclipse as it appeared in two cities whose distance by land had been previously measured. This produced the measurement for a degree of latitude from which, Qazwīnī explains, Ptolemy was able to estimate the Earth's circumference.[6]

While there was general consensus that the Earth was in the center of the cosmos, Qazwīnī was also familiar with the position of the followers of Pythagoras who argued that the Earth was not in the middle of the universe but instead spun in a rotating orbit. The apparent motion of the heavens, they reasoned, was caused by the Earth's orbiting movement and not by the perceived motion of the stars. In the fullness of time, this ancient theory would gain support with the development of heliocentric models to account for the perceived movement of Sun. But the Pythagorean theory of planetary motion found no acceptance in the authoritative teachings of Islamic natural philosophy. On this matter Qazwīnī adhered to the likes of Ibn Sīnā and Fakhr al-Dīn al-Rāzī, who followed Aristotle and Ptolemy, picturing the Earth not only as the center of the cosmos but as entirely stationary.[7]

Another point on which Qazwīnī expresses no doubt is the matter of the encircling ocean, *al-baḥr al-muḥīṭ,* that surrounded the inhabitable quarter of the Earth. Early Arabic and Persian geographers drew their idea of the inhabitable territories of the Earth from the ancient Greek concept of the *oikoumenē,* often translated into Arabic as *al-arḍ al-maʿmūra,* and into Persian as *ābādānī,* terms that carry the sense of inhabitable land and of settlement. This inhabited quarter was surrounded by the encircling ocean, known in Greek as *Ōkeanos,* a term Qazwīnī would find reading Bīrūnī among others. This outer sea marked the terrestrial limit of human knowledge.[8]

The impenetrability of these waters, also referred to as the sea of darkness, or the sea of pitch, is a theme that runs throughout classical Islamic geography. The notion of a western boundary beyond which knowledge ends is a central feature of the cosmic picture of the world dating back to antiquity. The idea is captured concisely in the story of the pillars that Hercules erected flanking the Strait of Gibraltar, as part of his famed labors performed across the world.

These boundary markers appear throughout Greek accounts of the world and were well known in early Arabic geographical writings, where the pillars often transformed into idols or talismans. Masʿūdī, famed encyclopedist of Fatimid Cairo, for instance, describes a lighthouse of brass and stone that the king Hercules built at the border shared between the Mediterranean and the encircling ocean. The edifice was covered with inscriptions and statues that indicated by their hands that "there is no path for all those who enter this sea." Masʿūdī notes, "No habitation is found there, no rational beings reside there, its extent is unmeasured and its end is uncharted, its limit unknown, it is the green, encircling ocean of shadows." On this, Qazwīnī accepted the unknowability of the encircling waters beyond the Mediterranean. He also was acquainted with talismanic statues at the edge of the western shore that warned the foolhardy not to venture too far from land.[9]

Masʿūdī explains that these Herculean statues, which he refers to elsewhere as idols, were famous because they marked the western limit of the known world and were discussed by such authorities as Aristotle, Alexander of Aphrodisias, and Ptolemy.[10] In ancient Greek geographical configurations of the world, the pillars identified the western boundary of terrestrial knowledge, symbolizing an outer barrier of human knowledge. Early Arabic geographical literature often approached the statues, alternatively said to have been built by Alexander the Great, as marvelous talismans used to protect sailors, impede foreign invasion, and define the limit of terrestrial habitation.[11]

For all the challenges, exploration of the Atlantic has ancient roots. There are several Greek and Latin accounts of voyages beyond the pillars drawn from both historical events and seafaring tales. These sources collectively preserve what amounts to a record of ongoing exploration along the shores of the Atlantic, from both the north and the south. This includes encounters with the volcanic Canary Islands, which form part of a group of Atlantic archipelagoes known today as Macaronesia—a modern neologism, drawn from the Greek *makarōn nēsoi*, the "Islands of the Blessed." These archipelagoes cover a vast maritime region from the Azores to Cape Verde, and reflect the multiple locations of the Elysian paradise of heroes, identified with shifting archipelagoes at the far western edge of the known world. Referred to in early Arabic sources as the Eternal Islands, *al-jazāʾir al-khālidāt*, or the Islands of Felicity, *saʿādāt*, and in Latin as the *insulae fortunatae*, the sublime isles off the coast of Africa took their name from the ancient belief in a terrestrial paradise located along the western horizon. The identification of these isolated lands where fabled heroes live on in eternity with a series of actual islands reflects a well-established practice of situating the strange, exotic, and paradisiacal along the outer limits of the known world.[12]

The pillars of Hercules feature throughout medieval Christian configurations of the Earth as admonitory boundaries. Dante (d. 1321) famously condemned the wandering Odysseus to the bowels of hell for his *folle volo,* a mad flight trespassing Hercules's barrier beyond which humans should never cross, "che l'uom più oltre non si metta." In the wake of the Iberian conquest of the Americas, this symbol of cosmic limitation was powerfully reconfigured in the imperial propaganda of the Habsburg dynasty. The invention of a columnar emblem with the motto *plus ultra,* further beyond, used by Charles V (r. 1558)—the Holy Roman Emperor, head of the Habsburg house, and ruler of large swaths of newly subjugated territories in the Western Hemisphere—altered the Herculean limit, the proverbial *ne plus ultra,* to suit the territorial ambitions and proselytizing zeal of an ever-expanding empire.[13]

Limits of human knowledge are a structural feature of early Arabic geographical writing. It is a theme, for instance, treated in the Arabic geography of the Sicilian scholar Abū ʿAbd Allāh al-Idrīsī (d. 560/1165), dedicated to the Norman king Roger II (d. 1154). Idrīsī's description of the Earth's topography progresses by clime, moving west to east and south to north across Ptolemy's seven climatic divisions of the world. As with Ptolemy before him, Idrīsī starts at the southwestern limit of the inhabitable quarter of the Earth, using the Islands of the Blessed as the prime meridian, a geographical tradition of measurement known to Qazwīnī. The marker of the prime meridian off the coast of Africa features throughout Islamic geographical models indebted to the Ptolemaic climatic system for measuring the known world.[14]

However, the notion that the western ocean could theoretically be traversed also circulated in Arabic sources. In his treatment of the sphericity of the Earth in his cosmological study *De caelo* (On the heavens), Aristotle argued that by traveling west beyond the pillars, one would be able to reach India. He premised his idea on the observation that elephants were found in both India and Africa, and thus the two regions were likely not that far apart. This theory made its way into Arabic during the Abbasid translation movement. Early Arabic geographical authorities, however, generally rejected the possibility of traversing the Earth in favor of the Ptolemaic model that set the western limits of the inhabitable world at the Atlantic coast of Africa. They did, however, in the main uphold the view that the southern tip of Africa was theoretically at least navigable, in contrast to Ptolemy, who promoted the idea that Africa could not be circumnavigated.[15]

Accounts of adventures beyond the pillars of Hercules form an established feature in Arabic geographical discourse. Masʿūdī refers to such seafaring adventures, noting that there are marvelous reports, *akhbār ʿajība,* associated with the encircling ocean, made by those who threw caution to the wind and risked utter perdition, *man gharrara wa-khāṭara bi-nafsihi.* While many failed, some lived to relate what they witnessed. Masʿūdī cites an account famous among the people of al-Andalus of a certain Khashkhāsh, whose name means poppy, from Cordoba. Khashkhāsh assembled a group from the city, provisioned boats, and set out into the open sea. The expedition was absent for some time before it finally returned home with a huge booty. While Masʿūdī offers no more information, there is reason to believe that the adventurer in question was the same Khashkhāsh the Sailor who led an Umayyad squadron against Viking invasions of the Iberian coast in 245/859, only to perish with a large number of Muslims during the attack.[16]

Abū ʿUbayd al-Bakrī (d. 487/1094), a famed statesman and geographer from Cordoba, explains, after recording the same story, that only part of the encircling sea is navigable—namely, up to the large island of Britannia to the far north; along the coast of Africa to the south, referred to as al-Sūdān, namely, the Land of the Blacks; and to the Eternal Islands to the west. Here Bakrī relates the commonly held view that nothing else is known of the open ocean west of Gibraltar.[17]

Then there is Idrīsī's report of the mariners from Lisbon. As with other geographical authorities, Idrīsī observes that what lies beyond the dark western sea is uncharted. No one has produced an authentic account of its outermost edge because of the difficulty of the crossing. The encircling ocean is murky, with bottomless depths that can suddenly expose razor-sharp reefs. It is filled with impenetrable crashing waves that tower up like mountains.

Skilled sailors avoid the open ocean altogether, preferring to travel safely, never losing sight of the coast.[18]

Despite the fierce conditions, the western edge also promised riches, a point that Idrīsī develops in his description of Lisbon, which, at the farthest limit of land, is battered by the encircling ocean. During the winter months, waves hurl nuggets of gold ashore, which the locals collect. This is one of the marvels of the world that Idrīsī had witnessed with his own eyes.[19]

The enticement of maritime bounty led eight cousins of Lisbon on a mission to find the edge of the western sea and uncover the marvels therein. Idrīsī describes the cousins as *mugharrirūn,* an adjective that suggests intrepid, if foolish, thrill-seekers. It evokes Masʿūdī's description of sailors willing to risk their lives, *gharra bi-nafsihi,* in the open sea. Having provisioned a boat with food and water to last for months, the cousins embark at the first sighting of the constellation of Taurus, with the arrival of an eastern wind. After traveling eleven days in the open sea, they were overwhelmed by thick waves, darkness, and reefs. Changing course, they sailed due south for another twelve days, reaching an island with a spring of flowing water and wild fig trees, inhabited only by sheep whose meat was bitter and inedible.[20]

Leaving the island of sheep behind, they set out in the same southerly direction for another twelve days, at which point they caught sight of an island with settlements and cultivation. Suddenly they were surrounded by skiffs and then seized by tall, fair-skinned islanders, who had long straight hair on their heads but very little hair on their bodies. The women were incredibly beautiful. But this was of little solace, as the sailors were now captives. After days, an Arabic-speaking interpreter appeared to hear their plight. Brought before the king of the island, the sailors related their ordeal and their quest to discover the wonders of the sea and information about its outer edge.

Hearing this, the king let out a good laugh. Speaking through his interpreter, the monarch explained that his father years before had sent out a group of slaves to discover what lay beyond the ocean's limit; after a month's voyage the expedition was overcome by complete darkness, forced to return, having gained nothing for their labor. With goodwill established, the king decided to free the eight captives from Lisbon. Once a western wind picked up, the cousins were tied, blindfolded, and placed out in a skiff that led them, after three days and nights, to the mainland, where they were left with their hands tied behind their backs. At daybreak, after a night in fetters, they heard some rustling and called out for help. Surrounded by a group of Berbers, they related their tale. "I am sorry [*asafī*] to say," was the reply, "but you are some two months journey from your homeland."[21]

The entire story builds up to a folk etymology. As Idrīsī explains, it is from this phrase, *asafī*—meaning "my regret"—that the Moroccan city of Āsafī, known today as Safi, took its name, for it was here, at the farthest western harbor, that the Lisbon sailors finally made landfall. Replete with notable flourishes, the story continued to inspire wonder for centuries. It features, for instance, in the immensely popular natural history by Ibn al-Wardī (d. 861/1457), who elsewhere copies extensively from Qazwīnī's *Wonders and Rarities*. Many of the tale's details draw on other descriptions associated with far-flung lands and distant peoples. The mention of the fair-skinned natives, *rijāl shuqr*, is often applied to inhabitants of exotic islands, such as Qazwīnī's account of fair-skinned natives from Java, whose heads grow on their chests—nations of strange people with roots in classical lore. Qazwīnī also has stories of sailors imprisoned on islands, as does he tell of remote islands populated by sheep.[22]

The adventure of the foolhardy sailors of Lisbon may well have at its core an actual voyage or series of journeys in the western sea. Yet there is much to suggest that the anecdote is also a pastiche drawn from earlier accounts of maritime adventures. The number of cousins, the days at sea, the vague limit of a western boundary, the island populated by sheep with bitter meat, the fair-skinned natives and beautiful women, the serendipity of an Arabic translator, an earlier failed mission to uncover the western limit, and finally the return to the mainland and with it an origin for the name of a city—these details in their totality shape the adventure in a discrete set of expectations designed to instill pleasure, provoke curiosity, and induce a sense of wonder while beholding the strange limits of knowledge. Above all, the high-sea adventure stresses the strange creatures inhabiting the edges of the world, the futility of searching for the outer boundary of the encircling ocean, and the great dangers of navigation on the open sea.

This is all the more apparent when read in the larger context of islands as prime locations for the horrifying or sublime. Similar themes find expression in the later Latin *isolario* tradition on marvelous islands, which has its roots in classical letters, as do they feature in earlier Arabic and Persian writings on the wonders of the sea. In the descriptive structure of his geography, Idrīsī sets the adventure directly alongside a series of other wondrous islands off the western coast of North Africa and the Iberian Peninsula.[23]

Delightful in their strange isolation, these islands offer terrifying glimpses of savage violence. The island of al-Saʿālī is populated with creatures that appear to be human, apart from their exposed fangs, their eyes like flashing lightning, and their legs like burnt planks. They speak an incomprehensible language and ride creatures of the sea in battle. There is no difference between the female and male in appearance, save for their genitalia. The male have no facial hair; they all wear clothes made of leaves.

Nearby is the island of Ḥasrān, with giant mountains and a population of small brown people with wide faces, huge ears, and beards that reach their knees. Then there is the island of Qalhān, inhabited by creatures similar to humans except that they have the heads of aquatic beasts and plunge deep into the ocean to hunt for food. And finally there is the island of the two sorcerer brothers. Through their witchcraft they waylay ships, kill passengers, and take their valuables. This continued until the Almighty intervened and punished the two by turning them into stone monoliths that still stand on the shore. Once liberated from the sorcerers' tyranny, a group of people came to inhabit the island, which Idrīsī explains is across from the harbor of Āsafī. In fact, it is so close that smoke from the island can be seen from the shore on a clear day.

Idrīsī identifies the island of the two sorcerers off the North African coast as the very same spot that was reached by the reckless cousins of Lisbon, as related in their curious story, *qiṣṣa gharība*. Moving closer to land, Idrīsī briefly mentions an unrealized plan to reach the populated island of the sorcerers during the reign of the Almoravid ruler of North Africa and al-Andalus, ʿAlī ibn Yūsuf ibn Tāshufīn (d. 537/1143).[24]

The Canaries, it would appear, were not so far off. Yet the information on the western islands, even those close to the coast, appears not to have been drawn from any sustained maritime or economic contacts. Rather, as the often-exotic inhabitants attest, Idrīsī's treatment of the western sea was pieced together from wondrous stories that circulated widely across various languages, in the long shadow cast by the reception and reconfiguration of misshapen creatures from classical antiquity. The monstrous peoples along the frontiers, who populate the writings of Herodotus, Pliny, and others, are in many cases drawn from legends that stretch even further afield.[25]

Among Idrīsī's list of western islands is one that was plagued by a dragon, which devoured any person, bull, or donkey that crossed its path. Finally, Alexander, during his seaborne adventures, was able to slaughter the beast, through his cunning. Qazwīnī offers leading theories on dragons, known in Arabic as *tinnīn,* with an account of their physical forms and voracious appetite, along with their capacity to swim the seas and fly through clouds. He too relates a very similar story of Alexander's exploits, but set in the Indian Ocean, in a spot he identifies as Dragon Island (see figure I.1).[26]

By their own interconnected logic, the wonders of islands were often interchangeable, run through with both sorcery and savagery. For Qazwīnī, as with Idrīsī and many others before him, the islands in the west parallel the islands in the east through the reduplicative resemblances that guide climatic determinism. Just as Hercules set his pillars marking the edge of the west, Alexander was famed for setting similar markers in the east. The curious nations found on the margins not only are symbols of untamed difference that cannot be fully assimilated, they also offer mirrors of humanity at the antipodes of the inhabitable world. The strangeness of the edges can be replicated, reconfigured, and redeployed, from the islands of the Indian Ocean and the beasts of the Atlantic to the barbarous tribes of the north and the savages of the south.

Stories of western exploration continued to both delight and warn. The Mamluk administrator ʿUmarī preserves a telling example that also affirms the fundamental impenetrability of the encircling ocean. The account came to ʿUmarī by way of Abū l-Ḥasan ʿAlī ibn Amīr Ḥājib (d. 739/1338~), Mamluk governor of Cairo. Ibn Amīr Ḥājib was acquainted with the fabulously wealthy Muslim ruler of the powerful Mali Empire in West Africa, Mansā Mūsā (d. 738/1337), who took up residence in Egypt during his pilgrimage to Mecca in 724/1324.[27]

The lavish generosity that the Mali king bestowed on the elite of Cairo flooded the markets with so much gold that the precious metal lost much of its value for a considerable time. Ibn Amīr Ḥājib learned more of the gold mines during the course of conversations with the Mali ruler. For his treatment of the kingdom, ʿUmarī weaves together reports from several informants, and examines the customs, manners, and history of the Mali kings, touching on such topics as the recent conversion and improper religious practices of the court, man-eating infidels, magic practiced by witches, the peculiar sexual mores of the Mansā people, and the poisons derived from plants and venomous animals in the region.

These accounts supply further information on an area of the African continent that was largely unknown to the previous literary corpus of geographical writing. For his part, Qazwīnī presents Africa entirely through earlier reports. There were cannibals, lands of gold, as well as the source of the Nile in Ptolemy's Mountain of the Moon. Qazwīnī knew that Ptolemy had attempted to procure more details through informants, who found a desolate land with no inhabitants or animals, scorched due to its proximity to the Sun. In contrast, ʿUmarī offers much more detail of the regions southwest of the Mamluk domains.[28]

The expanse of the Mali kingdom along the shores of the encircling ocean is also a noted theme. The topic comes up in a discussion of succession, when Ibn Amīr Ḥājib asks Mansā Mūsā how he had ascended the throne. Mansā Mūsā replied that kingship was inherited from among a family of rulers. As for the previous king, he was convinced that the edge of the encircling ocean indeed could be reached. So he prepared two hundred boats provisioned with enough gold and supplies to last for several years, commanding the sailors not to return until they reached the end of the sea or ran out of supplies trying.

A long period passed with no news. Finally a single boat emerged on the horizon. The king soon learned that the expedition had met a fierce current that pulled all the boats toward a vast watery abyss. Only one of the two hundred boats was able to escape its force. Nothing was known of the fate of the others. Hearing this, the king determined to set out for himself to discover what truly lay on the other side. With a second expedition prepared, the ruler delegated Mansā Mūsā to accede to the throne should he not return. When nothing was heard from the king who set sail, Mansā Mūsā became the ruler of the Mali kingdom.

Once again there is the vast ocean, a mission to discover what lay beyond the horizon's edge, the terrifying power of the sea, and the ultimate failure of the expedition. All these elements form part of a discrete set of conventions on the limits of the encircling ocean. Such explorations are portrayed as imprudent expressions of curiosity. In Mansā Mūsā's account this is taken to the extreme, with a king so obsessed by the ocean's call that he ultimately abandons everything.

As for the historical circumstances, it is of note that the succession of Mali kings had involved several usurpations of power, as the kingship passed back and forth between rival clans. This was also the case for Mansā Mūsā, who established a new line of succession when he seized the kingdom from the family of Qū, who had themselves taken it from Sākūra, the slave usurper. Regardless of whether or not the adventure actually took place, the story of the king's foolhardy charge into the abysmal sea quite literally made Mansā Mūsā's predecessor disappear.[29]

These explorations in the sea of darkness take on new significance only with the subsequent Iberian missions to the New World. The question of whether earlier seafarers made contact with the Western Hemisphere before the famed mission in 1492 came to occupy both Christians and Muslims and continued to inspire theories of ancient exploration to the New World. Yet the classical Arabic material preserves no record of contacts with western lands, much less knowledge of their existence. Beyond vague accounts of the Eternal Islands, and strange adventures steeped in the lore of wondrous tales, the image of the encircling ocean as an unfathomable expanse of darkness endured. This image, immediately recognizable to Qazwīnī and his readers, circulated long after the Christian conquests across the Atlantic.

The ensuing centuries witnessed an unprecedented increase in activity across the shores of Africa and the Iberian Peninsula, as Christian sailors and merchants began to compete in the region. Knowledge of explorations in the Canaries was known to Muslim elites. In the geographical section of his world history, Ibn Khaldūn makes reference to new forms of navigational charts when discussing the Eternal Islands. Recently, Ibn Khaldūn notes, some boats of the Franks reached and plundered the islands, battling with the native inhabitants and enslaving many of them. This general outline reflects the early stages of the explorations, led first by Genoese merchants in the beginning of the fourteenth century. These expeditions often sought slaves from among the Guanches, the native inhabitants of the Canaries, as they became known in Christian accounts from the period.[30]

Ibn Khaldūn mentions that some of these captives were then sold into slavery along the western coast of Morocco and came into the possession of the Marinid sultan. While in captivity they gained knowledge of Arabic, by which they could communicate something of their previous lives. In devotional terms, this consisted of worshiping the Sun, "as they had no knowledge of religion [*dīn*] nor had they been reached by any missionary activity [*lam tablughhum daʿwatun*]," Ibn Khaldūn explains. Stressing the difficulty of locating these islands, Ibn Khaldūn says they can only be reached by chance, as nautical information about the area is entirely lacking. He then contrasts this with the navigation in the Mediterranean, where the prevailing wind patterns and their directions had been well charted.

To this broad outline Ibn Khaldūn also adds that sailors have charts that represent the Mediterranean and its respective shores in proper order, reflecting closely their actual physical forms, depicting prevailing wind directions and where they lead. Sailors rely on such charts, known as a *kunbāṣ*, for their journeys at sea. As nothing like this exists for the encircling ocean, Ibn Khaldūn notes, ships do not enter it; instead they travel along the coast, as the weather that gathers on the encircling sea and the mists that cling to the surface of the water hamper ships from traversing it. The Eternal Islands are difficult to find, as information on them is scarce.[31]

Ibn Khaldūn's passing note about the Christian exploration of the Canary Islands and his description of cartographic practices point to navigational techniques that circulated throughout the Mediterranean. The term *kunbāṣ*, as the name for the cartographic chart used in the region during the period, is of Romance derivation. It parallels the Spanish *compas*, the Italian *compasso*, and the Latin *compassus*, used to describe the grid diagram of wind patterns on nautical charts.[32]

The *kunbāṣ* is not an allusion to the mariner's magnetic needle, which was referred to as a compass only years later. It points rather to the older sense of a compass as a drawing tool, a pair of dividers used to delineate circles and circumferences. The term suggests the patterned rhumb lines featured in the compass rose, also known as a wind rose, representing prevailing wind patterns configured with respect to the cardinal directions used for navigation.[33]

During the period when Qazwīnī was writing his geography and cosmography, a distinct means for charting and navigating coastal regions was emerging in the Mediterranean. Generally referred to today as portolan charts, these maps reflect the navigational techniques and measurements that had developed in the course of broader exchanges among Christians, Jews, and Muslims throughout the region. In terms of representations, these nautical charts highlight coastal features, often in impressive detail.[34]

In addition to coastal complexity, among the most striking characteristics are the radiant networks of crossing rhumb patterns. Ibn Khaldūn's description aligns with this new nautical cartography of rhumb lines designed specifically around the wind directions. Intersecting at different coastal points, these lines are designed to plot a ship's course. The cartographic innovation of the nautical chart drew on techniques of measurement and navigation that developed as Muslim potentates, traders, and scholars dominated the military, commercial, and intellectual traffic around much of the Mediterranean. During this period Christian merchants, mercenaries, and pilgrims were also taking to the sea with increased frequency.[35]

The sectional maps accompanying Idrīsī's geography offer important background to these cartographic innovations, as they relate to the Mediterranean and its position in the larger world. Notably, Idrīsī's maps form a network of lines drawn from the Ptolemaic model of seven parallel longitudinal climes, from south to north. Ptolemy's geography was largely unknown in Latin Christendom during the period. Idrīsī's addition of ten latitudinal dividers progressing west to east, across each of the seven climes, created what appears to have been a novel graticule for a total of seventy maps comprising the inhabitable world.[36]

In addition to his sectional maps, Idrīsī also produced for the Norman monarch in Sicily, a map of the world on a giant disc made of pure silver, noting the oceans, major rivers, inhabited regions, wastelands, and distances from place to place. Idrīsī dedicated his geography to Roger II, whom he praises as the best ruler the Roman Empire had ever seen, an Imam of his people, and a philosopher king. Idrīsī made a concerted effort to obtain accurate cartographical measurements and descriptions. His maps and the accompanying geographical information of place names and features had shaped further developments in cartography in Italy and beyond. For all these exchanges and points of connection, though, geographical writings in Latin Christendom often ignored Arabic materials. Nonetheless, there were many obvious points of synergy, reflected in a shared technical koine of methods, vocabulary, and information.[37]

An even earlier reference to these new nautical techniques features in ʿUmarī's account of the Mediterranean. ʿUmarī refers to the grids accompanying portolan charts as *qunbāṣ*, with a slightly different spelling of the same term known later to Ibn Khaldūn. Furthermore, he provides a good deal of information on the use of directional lines for navigation based on the patterns of prevailing winds. He even supplies a diagram of interlocking patterns for primary and secondary wind directions used alongside maps. His observations offer one of the earliest accounts for explaining the function of navigational grids on nautical charts. ʿUmarī's description is derived from discussion with Abū Muḥammad al-Anṣārī, a ship's captain originally from Cordoba. Here the Latinate etymology of the Arabic reflects contacts forged in exchanges of nautical argot in such contact zones as the shores of Iberia and the coasts of Sicily and Malta.[38]

More than anything, the Romance origin of the *qunbāṣ* points to the porous boundaries of Arabic maritime vocabulary, which was drawn from multiple sources at various points. The same proves true for many nautical terms in medieval Latin, Italian, Spanish, and Portuguese that derive from Arabic. With numerous areas of confluence across the Mediterranean, through commercial networks, military alliances, and pilgrimage routes, the mariner's chart represented a shared technology for navigation and cartographic representation.[39]

Perhaps most striking in Ibn Khaldūn's observations on the wind rose as a basis for navigation is his insistence that the Eternal Islands remained unknowable and inaccessible, as they lay beyond the map's edge in the impenetrable fog of the encircling sea. His lack of information on the region is also reflected in his own world map, which draws heavily on Idrīsī's model. These cartographical materials are much more sophisticated than the rough maps produced in the earliest manuscripts of Qazwīnī's writings. The classical maps, nonetheless, were delimited by the dominant theory of the encircling ocean. ʿUmarī and Ibn Khaldūn, like Qazwīnī and Idrīsī before them, only had a vague notion of the location of these western islands. This contrasts with the ongoing Iberian colonization and settlement of the Canary Islands during this period. Neither the Nasrid dynasty in Granada nor the Marinids of North Africa participated in this western expansion into the Atlantic.

There were, however, efforts to reach these islands. After moving to Egypt in 784/1382, Ibn Khaldūn went on to relate further details about the Eternal Islands to his disciple in Cairo, Taqī l-Dīn al-Maqrīzī (d. 845/1442), the prolific Mamluk historian and geographer. Ibn Khaldūn recounted to his disciple that around the year 740/1339, the Marinid sultan Abū l-Ḥasan (d. 752/1352) held an audience with a group of Genoese sailors in the port city of Ceuta on the Moroccan side of the Strait of Gibraltar.

Traveling in war galleys, the Genoese went out with the goal of discovering what lay in the ocean and sailing around the inhabited land. They made landfall on the Eternal Islands, where they found a people who went around largely naked. The Genoese occupied the island by force and took several natives captive, two of whom they sold to the Marinid sultan. Ibn Khaldūn also observes that no mention or notice of the mission of Islam had reached the native inhabitants, who had no weapons or clothes, fought only with stones, and prostrated to the Sun.

When the following Marinid sultan, Abū ʿInān Fāris (d. 759/1358), ascended the throne, he was determined to reach the islands. He charged his captain of the navy with the task. A war galley provisioned with supplies and men set off from the city of Azemmour on the western shores of Morocco and remained at sea for some two months. The mission, however, failed to bring back any news of the islands. The young Ibn Khaldūn, who in 757/1355 had joined the entourage of Sultan Abū ʿInān in Fez, notes that he was present at the court when the captain delivered his report.

The sailors had traveled on the open ocean until mists began to gather. Fearing certain destruction, as their ship was about to be torn apart, they returned without finding the islands they sought. The captain remained at court for some days relating the wonders witnessed at sea. On one of these occasions he mentioned having seen a green bird, at which point the sultan protested. Birds need fresh water and land, so the islands could not have been far off. Startled, the captain could not proffer a reply quickly enough. The sultan immediately ordered him stripped and lashed some five hundred times for having fallen short of his mission.[40]

During the period, interest in gaining access to remote lands features in the far-flung journeys of the North African merchant Ibn Baṭṭūṭa (d. ca. 779/1377), whose famed travelogue recounting his adventures in the East and through the Indian Ocean was commissioned by the same Marinid sultan in Fez who had failed to reach the Eternal Islands. The Arabic title of Ibn Baṭṭūṭa's journey, *Tuḥfat al-nuẓẓār fī gharāʾib al-amṣār wa-ʿajāʾib al-asfār* (Gift to onlookers into rarities of cities and the wonders of journeys), gives a sense of the wide audience that the language of the marvelous could command.[41]

Despite these global contacts, the inability to compete with western expansion into the Atlantic had notable consequences. Foremost, by conquering the Canaries, Spain and Portugal gained a foothold for further economic and political penetration along the African coast, which enabled Christian merchants and mercenaries to circumvent the rival Muslim dynasties of the region. It also led to a growing navigational sophistication, as Christians not only reached the islands repeatedly but also surveyed them with increasing precision.

In 1339 the Italian cartographer Angelino Dulcert of Mallorca produced a nautical chart with a fairly accurate placement of the Canary Islands of Fuerteventura and Lanzarote. On Dulcert's map, the latter takes its name and its Genoese shield of arms from Lancelotto Malocello (fl. 1312), the Genoese navigator who had landed there some three years earlier and claimed the island, as it were, for the Italian city-state. The nautical chart of 1367 by the Venetian cartographers Domenico and Francesco Pizigani reveals even greater detail and reflects sustained contact with the islands off the African coast and the cartographic knowledge accumulated in the interceding years. A similar pattern follows with the royal Catalan atlas of 1375, with further identifications of islands in the western archipelago.[42]

In addition to information from nautical exploration, these maps draw from the larger historical imagination of the Atlantic as a wondrous sea of marvels, with references to the disappearing Island of Brasil and the Islands of Saint Brendan, famous from the adventures of the Irish missionary. In this

FIGURE 7.4: Arabic portolan chart of the Mediterranean with details of the Canary Islands and the western coast of Africa lined with ships at the bottom, on parchment, early sixteenth century.

vein the Pizigani nautical chart also features a statue on an island beyond the Strait of Gibraltar. At the western edge of the map a statue bears a caption warning sailors that they cannot go farther into the vile seas. The reference to the impenetrability of the encircling ocean and to the statue of Hercules has direct parallels with the accounts preserved by the likes of Masʿūdī, Idrīsī, and Qazwīnī. Such points of contact indicate a wide geographic body of knowledge that circulated across the Mediterranean in Arabic and various Romance languages.[43]

Yet during this period of heightened Christian explorations beyond the straits, the lack of geographical information in Arabic on these islands is telling. Qazwīnī knew the Eternal Islands for their abundant fruit and fertile soil and as the limit of knowledge, an image that endured for some time to come. Even with increased cartographic precision, major lacunae remained. A portolan chart produced in Tunis, dated 816/1413~, by Aḥmad ibn Sulaymān al-Ṭanjī, for instance, beautifully covers the Mediterranean, the Black Sea, and stretches north to the British Isles, and south along the western coast of North Africa. Absent, however, is any reference to the Canary Islands, which at this point were very present in rival Christian cartographies. The Tunisian map, one of the few Arabic portolan charts of this period to survive, ends at the southwestern corner of the ocean, leaving out the very archipelagoes that were the launching ground for subsequent Iberian conquests across the Atlantic.

Within a century, cartographers and navigators working in the employ of the Ottoman court came to address these oversights and many more. The Canaries would appear in detailed portolan charts that became fashionable among Ottoman cartographers (figure 7.4), who saw, like their Christian counterparts in the west, mapmaking as an imperial endeavor. While territorial ambitions had long been a feature of Islamic geography, these maps in the Ottoman court were quite distinct. In the years to come, new modes for representing the world would emerge that were entirely foreign to the models of the cosmos known to Qazwīnī. Beyond the encircling ocean and past the seven climes that had long bounded all humanity, a series of ruptures would change everything.[44]

8

ACROSS THE GLOBE

WORLDS UPON WORLDS, populated with creatures that only God could know. For Qazwīnī, as there was no telling what one might find beyond the encircling ocean and past the emerald-green mountains of Qāf, the possibility of multiple worlds could not be ruled out. It is an idea with firm roots in both classical philosophy and Islamic theology. As no one could be certain, the debate continued whether there were other worlds populated with untold creatures. Confronting the magnitude of existence, "human intellects are weak and the human sciences are insignificant," Fakhr al-Dīn Rāzī concluded while defending the possibility of multiple worlds. The truth, he explains, is what God mentions in divine scripture: "You have only been given a small portion of knowledge."[1]

Such humility conditioned a good degree of flexibility when confronted with new information. An attitude of probabilistic reasoning gave the language of wonder an enduring appeal. The edges of human knowledge were filled with boundless potential. The Quran, after all, spoke of savages bottled up behind a giant wall at the end of the world by Dhū l-Qarnayn, the Possessor of Two Horns, long identified by Muslim authorities as Alexander the Great. So it was not such a wild idea that there would be other strange possibilities at the margins of existence.

As with many questions on the limits of human knowledge, Qazwīnī turned to the adventures of Alexander for insight into what was to be found in the encircling sea. While there were many other stories about journeys in the sea of darkness, Alexander's encounter with a people from a vast land far beyond the horizon circulated both before and after Qazwīnī memorialized it in *Wonders and Rarities*. The story was also frequently illustrated (see figure 9.1). Its significance, however, dramatically changed as knowledge of the New World made its way to various Muslim courts.

Alexander was an ideal model for tales of exploration. A master of perfect nature and a collector of occult knowledge, the world conqueror long stood as an emblem for universal sovereignty. Collections of wonders, geographical compendiums, and manuals of astrology often addressed imperial aspirations. As the Ottoman Empire expanded in territory, the appetite for knowledge of distant lands shared by sultans, courtiers, and high-ranking scholars also grew. Books and maps documenting an age of ever-new discoveries would come to circulate in the capital city of Istanbul and beyond, some brought in from the printing presses of Christians west of the Bosporus, others composed at home. Geographical knowledge expanded from Africa through the Indian Ocean and the far reaches of China to the shores of the New World. Yet through it all the classical models of learning, which Qazwīnī gathered into a concise form, continued to appeal, though often read in new ways and for distinct ends.

Preceding these transformations, the body of writings on natural history and geography that *Wonders and Rarities* represents became increasingly connected to the circulation of Persian learning. The rise of Persian as a prestigious language of scholarship was tied to learning and patronage that connected courtiers and scholars from Bengal to the Balkans. These networks brought together religious authorities, diplomats, and functionaries, as well as poets and litterateurs who shared common sources for knowledge of the world and its history. The development of Ottoman Turkish, in many ways, followed the precedent of Persian as a language of prestige. All the while Qazwīnī continued to attract new readers with promises of strange powers and distant shores.

The question of certain knowledge when confronted with the strange and remote is a theme that is immediately legible in the phrase *ʿajāʾib wa-gharāʾib*. Qazwīnī's generous attitude to the realm of the possible gave license to include both the ordinary and the outlandish. Balancing the authority of science with levity and awe, Qazwīnī mixes formal taxonomy with the pleasure of storytelling and the feelings of uncertainty and perplexity in the face of divine transcendence.

The division of the cosmos into constituent parts could make the heterogeneity more manageable. Where classification of all existence can span tomes of philosophical inquiry, Qazwīnī's digest quickly reduces the world into a concise form. The classificatory model defines the spine of the work, reflected in the division of creation into hierarchical orders. This taxonomic system is immediately legible in Qazwīnī's table of contents as well as in his definition of created beings, *makhlūqāt,* found in the title of the collection.

Creation is everything other than God, Qazwīnī explains in true scholastic fashion. It is either sufficient in itself or dependent on something else. If it is self-sufficient, it is either located in space or it is not. If it is located in space, it is a body. If it is not, it is a spiritual substance that is connected to bodies, in which case it is the soul. If it is not connected to bodies, it is free from desire and anger, in which case it is an angel; if not, it is a jinn.

As for that which is not self-sufficient, these are either spiritual accidents in the philosophical sense of nonessential attributes, such as knowledge and capacity, or they are physical accidents, identified with the ten standard differentiae of the Aristotelian categories of logic that can be predicated of a proposition: substance, quantity, quality, location, time, association, action, influence, possession, and position. Included in all physical accidents that cannot be predicated in such terms are emotions, perceptions, and gradations of cognition. This concise schema reflects divisions of existence that would have been memorized and recited in various contexts of education.[2]

The confidence that the system exudes is betrayed only by the persistent incapacity to say with certainty the scope, extent, or reality of any given phenomenon that might be beyond direct perception or experience. In this, Qazwīnī weighs scriptural knowledge and traditional lore with the philosophical authority of natural science, and he is quick to recognize the limits of human capacity before an all-knowing God. The hierarchical classification of the sciences and the continued debate over what branches of learning should take precedence in any unified theory of knowledge was an enduring feature of these tensions.

Following the parameters of natural philosophy, the study of cosmography became a standard feature of scholastic theology. It is a tradition that Qazwīnī inherits and that in turn shaped the immediate legibility of *Wonders and Rarities* as grounded in both philosophical authority and religious piety. In the succeeding centuries, major compendiums of theology consistently engaged with natural philosophy, drawing on astronomy and geography in discussions shaped by the scholastic absorption and promotion of scientific endeavors.

There were those such as ʿAbd al-Raḥmān al-Ījī (d. 756/1355), a Shāfiʿī jurist and Ashʿarī theologian of Iran, who were circumspect about the potential of what could be known of the cosmic order merely from empirical observation. In an influential theological summa, Ījī provides a fairly sophisticated overview of astronomy and Earth science. Discussions of rainbows and the formation of mountains accompany theories about the shape and location of the Earth and the movement of the heavens. Yet Ījī also insists on the ultimate conjectural character of the theories and models that guided the scientific study of the cosmos.[3]

Others, such as Quṭb al-Dīn al-Shīrāzī (d. 710/1311), express a good deal more confidence. An adept of Suhrawardī's philosophy of cosmic illumination, Shīrāzī studied under Ṭūsī in Maragha and was in his own right an important scholar of astronomy, optics, and mathematics. He wrote extensively across the religious sciences, on the Quran, jurisprudence, and theology. He also taught in various madrasas and served as a judge. As with Ṭūsī before him, Shīrāzī developed a sustained critique of the Ptolemaic theory of planetary motion. Likewise, his technical astronomical works were copied and read in madrasas. Furthermore, he produced an encyclopedia in Persian in which he sought to develop a unified field of knowledge. While he drew extensively from Ibn Sīnā's classifications of the sciences, his own formulation reflects a further attempt to situate religious learning within the frameworks of natural philosophy.[4]

Some forms of knowledge, Shīrāzī explains, are universal and unchanging regardless of time, place, or religion. This is the case, for instance, with astronomy, mathematics, and ethics. Other branches, however, are conditioned by time and place, such as religious law. These two forms of knowledge Shīrāzī identifies with the philosophical and religious sciences, respectively. Philosophy, *ḥikma,* seeks both to understand everything as it is in the world, *dānistan-i hama-yi chīzhāst chūnānkih hast,* and to conduct affairs according to how they ought to be, all to the extent that it is humanly possible. Shīrāzī also explains this effort "to see the world as it is" using the technical Avicennan expression *fī nafs al-amr,* in the truth of the matter, at the heart of the very phenomenon itself. In the language of logic, Shīrāzī defines *ʿilm,* meaning knowledge or science, as conception of the realities of existence and assent to the principles and necessities of existence such as they really are, to the extent of human capacity.[5]

Thus, with regard to the nature of the world, the fields of philosophy and religion have overlapping objects of concern situated in the world itself. Both branches address knowledge of God, the reality of prophecy, and the state of life after death. Yet unlike philosophy, the religious sciences include scriptural and rational forms of knowledge. The division between *naql* and *ʿaql,* tradition and reason, for Shīrāzī, is thus not between religion and philosophy. Rather, religious sciences are themselves composed of both rational and traditional sources of learning. Notably, the principles or foundations of religious knowledge, the *uṣūl-i ʿilm-i dīnī,* have reason at their core. While these principles may be confirmed or supported by traditional or scriptural knowledge, they need not be, for they are in and of themselves derived from rational observation and contemplation. The branches, or *furūʿ,* of religious sciences, in contrast, are drawn from revealed knowledge rooted above all in the authority of the Quran and the sayings of the Prophet.

Shīrāzī offers four divisions for the principles of religious science. Starting with the unity and attributes of God, these principles are followed by the study of divine actions and the subtleties of creation, and conclude with the nature and wisdom of prophecy and scripture. All of these forms of knowledge can be obtained through rational means, a conclusion that is born from a thorough engagement with Avicennan philosophy. Importantly, at the heart of these principles of religious science lies the demonstrative power of observing the world as it is.

By pondering and scrutinizing the secret subtleties of creation, Shīrāzī contends, knowledge obtains the fullness of potential, increasing in wisdom, toward a greater state of perfection. "The more one contemplates and the more informed one is of the wonders, rarities, and marvels behind the actions of all created things," he counsels, "the more one's knowledge reaches perfection and wisdom increases, by observing the sky, Earth, the celestial throne and pedestal, the realms of the heavens, the bodies of the fixed stars, the moving planets, the subtleties of the three derivatives of matter—namely, minerals, plants, and animals."

Take a look at how leaves grow on trees: "whether large or small, veins run through them all, beginning to end, from each individual branch the leaves spread out alternating right to left, flowing through the core of the tree out of sight, from high to low, divided branch by branch." It all proceeds with *ḥikma*, divine design or wisdom, giving nutrition throughout the tree, radiating across branch and leaf, divided with the exact amount of benefit and measure necessary, *ba-qadr-i maṣlaḥat va-andāza-yi ḥājat*. And if in one single leaf there is such wisdom and design, imagine the rest of creation, "from the heaven and Earth and everything in between, from minerals, plants, animals, all with such amazing variety of divine purposes [*chih ḥikmathā-yi badīʿ-i gūnāgūn*]." Thus follows the multiple meanings of *ḥikma* as philosophy and wisdom, as well as divine design and hidden intention.[6]

The image of the leaf is not chosen arbitrarily. It reflects in its own right the very structure inherent in Shīrāzī's classification of knowledge. The hierarchical framework referred to as *shajara*, or "tree" in Arabic, draws on the Porphyrian schema for logical taxonomy. It serves as an abiding feature for the classification of the sciences, both before and after Shīrāzī employed it as the structural backbone for his encyclopedia. The elevation of contemplating the wonders of existence as a constituent foundation of religious science, however, is one of the many reasons works such as Qazwīnī's *Wonders and Rarities* came to enjoy such prominence, reflecting both pious and scientific forms of contemplation and speculation, organized along the lines of the philosophical classification of existence.

Shīrāzī's framework advances religious knowledge with its fusion of scripture, reason, and accounts of the hereafter, as the most excellent form of learning. It takes as its object a rigorous study of the world in a form shaped by the philosophical tradition of the natural sciences. The practical end to this body of knowledge is the proper way of being in the world. As with many before him, Shīrāzī's cosmos of the celestial movements, elements, and physical forces also features talismans, amulets, alchemy, judicial astrology, and the natural power of the soul to produce miracles.[7]

Numerous collections of natural history circulated well before Qazwīnī turned to wonder as a means to unify existence, though none perhaps with the same attention to scholastic learning or systematic classification. Several of these were written in Persian, a language which had become a prestigious vehicle for historical, literary, religious, and scientific pursuits. The *Gift of Rarities*, the Persian compendium of practical wonders and natural paradoxes that circulated in name of the astronomer Ibn Ayyūb al-Ṭabarī, leaves a lasting imprint on both Qazwīnī's natural history and his geography.

Likewise, Qazwīnī was familiar with the *Nuzhatnāma-yi ʿalāʾī* (The supreme book of delight) by Shahmardān Ibn Abī l-Khayr (fl. 490/1096), a Persian natural history composed for Abū Kālījār Garshāsp (d. 536/1141), the regional Kakuyid ruler of Yazd in central Iran. The title "Supreme" in the collection was taken from Garshāsp's honorific, *ʿAlāʾ al-Dawla,* "Supreme Lord of the State." It is also a conscientious nod to the *Dānishnāma-yi ʿalāʾī* (The supreme book of philosophy), a Persian manual of metaphysics that Ibn Sīnā wrote a generation earlier while in Isfahan for a progenitor of the local feudal lords, who also took the same honorific, "Supreme." Shahmardān worked as an administrative official and wrote widely on alchemy and astral knowledge. His natural history also exhibits an abiding interest in the occult sciences. Its sections on talismans, charms, and the unique properties of various substances draw on such authorities as the "the Sufi philosopher" and renowned alchemist, Jābir ibn Ḥayyān.[8]

While Qazwīnī references the *Gift of Wonders* and the *Supreme Book of Delight,* he makes no mention of the natural history of wonders by Muḥammad ibn Maḥmūd al-Ṭūsī, dedicated to the last Saljuq ruler of Persia, Ṭughril ibn Arslan (d. 590/1194), which was finished at some point after 562/1166. Notably, Ṭūsī entitles his collection the *ʿAjāʾib al-makhlūqāt va-gharāʾib al-mawjūdāt,* the same title Qazwīnī would use some hundred years later for his own cosmography. The compendiums of Shahmardān and Ṭūsī each circulated through illuminated manuscripts. Ṭūsī's book of wonder was further translated into Ottoman Turkish.

The title *Wonders and Rarities* was itself not unique to Ṭūsī either, who elsewhere refers to his work as the *ʿAjāʾib-nāma* (Book of wonders) and the *Jām-i gītī-numā* (The world-revealing goblet), a reference to the magical goblet of the legendary Iranian king Jamshīd in which the whole world could be seen. In light of the points of intersection between the two works, Qazwīnī appears to have also been familiar with Ṭūsī's natural history. And although the collections are quite distinct in arrangement, organization, and tone, they both start with the heavens and progress downward to cover the minutia found on Earth. Ṭūsī is more concerned with weaving in anecdotes and morals, befitting a mirror of princes, but he also exhibits a similar view of wonder at the natural world as a basis for knowledge. Here too are talismans, charms, and strange creatures from faraway lands, as well as countless stories drawn from the adventures of Alexander the Great.[9]

Qazwīnī was also intimately versed in Persian classics. Allusions to masters of Persian verse flow through his own Persian redaction of *Wonders and Rarities*. Qazwīnī also references in his Arabic geography Firdawsī's epic the *Shāh-nāma* (Book of kings), the mystical poetry of ʿUnṣurī (d. ca. 431/1039), the classical love story of Vīs and Rāmīn by Jurjānī (fl. 441/1050), and the sweeping spiritual epics of high gnosis by Niẓāmī Ganjavī. Qazwīnī's vision of world history was deeply shaped by ancient Persian lore. Both Farhād's unrequited love for Shīrīn and Afrīdūn's victory against the evil tyrant Bīwarāsb—two stories immediately recognizable in the pantheon of Persian letters—feature in *Wonders and Rarities*. Qazwīnī also recognized that Alexander the Great, the student of Aristotle who traveled the world to master the science of unique properties, was actually an ancient Iranian king, the secret half-brother of Darius. This widely held claim served to make Alexander the world conqueror, cursed destroyer of ancient Iranian libraries, into a paragon of Persian kingship.[10]

The cultivation of Persian letters also meant the promotion of particular forms of etiquette and bodies of knowledge. These fields of learning and comportment called on ancient history, sophisticated poetry, and philosophical inquiry of high gnosis. Persian became a *lingua franca* in the courts of powerful rulers and among the secretariat class of scribes, functionaries, and bureaucrats. Yet the court was not the only audience for early Persian writing. Numerous Persian commentaries and line-by-line translations of the Quran were produced within madrasas in eastern Iran and Central Asia. Scholars beyond the court also turned to Persian collections of poetry, philosophy, and metaphysics. As a written language, early New Persian emerged in the shadow of Arabic, only some three centuries after the Arab conquests of the Sasanian Empire and the lands past the Oxus River. When Muslim scholars and poets ultimately took to writing in Persian, they did so using the Arabic script, first in Khurasan and Central Asia.

While Arabic remained an esteemed language of scripture and scientific investigation, Persian became ever more an authoritative vehicle for writing. This included Persian translations of Qazwīnī's *Wonders and Rarities*. Qazwīnī may well have produced the earliest Persian redaction himself, though the surviving manuscript evidence leaves much uncertain. It is clear, however, that this version circulated first within a courtly context, dedicated to an otherwise unidentified patron, ʿIzz al-Dīn Shāhpūr ibn ʿUthmān, whose honorifics, *ṣāḥib al-ʿālam,* "master of the world," *sayyid al-akābir,* "lord of the mighty," and *mālik al-ṣudūr,* "commander of chiefs," suggest an administrator or man of arms.[11]

The earliest Persian redaction contains an entire chapter on human life that is otherwise absent from the surviving Arabic recensions of the compilation. The extra material is in keeping with general interests in practical knowledge that shaped earlier Persian compendiums of natural wonders, and it appears to have been integral to the original structure and design of the work. The chapter opens with the diverse customs, practices, and beliefs among the peoples of the world. Differences of habit and disposition are shaped by climatic determinism, though there are also efforts to trace all humanity back to the progeny of Noah. Accounts of idol worship, divination, and magic are mixed with observations on the diverse models of kingship, various crafts and sciences, as well as a range of sexual mores.[12]

One of the central arguments developed here is that human settlement thrives through the collective mastery of crafts so that every need can be satisfied through mutual cooperation, a theme Qazwīnī also explores in his geographical gazetteer. From this follows an overview of the professions necessary for social order and the marvels associated with them: agriculture, animal husbandry, hunting, weaving, architecture, metallurgy, carpentry, and commerce. These lead to further degrees of refinement: arithmetic, writing, poetics, music, medicine, cosmetic beautification, the removal of physical defects for both men and women, judicial astrology, the use of the astrolabe, magic squares produced by numerical patterns, talismans, sympathetic magic of *nīranjāt,* subterfuges, and illusions. Collectively these chapters speak to practical arts, ethical disciplines of the self, and occult practices of purifying the body and soul in order to wield extraordinary powers.[13]

On the whole, this body of material and the sources cited are entirely consonant with Qazwīnī's Arabic natural history. Echoes of the same arguments, statements, and stories can be also heard in his descriptive geography. The focus on hidden properties and occult knowledge conform with Qazwīnī's view of the world. While this chapter appears, by all measure, to be original to the collection, it remains to be seen exactly why it did not survive in any of the Arabic recensions. Climatic determinism was not particularly controversial.

The discussion of incantations may have invited some opprobrium. But given the prominence of talismans and occult properties throughout Qazwīnī's Arabic geography and natural history, this does not seem particularly likely. Qazwīnī's frank treatment here of an array of sexual practices may also not have suited all tastes.

Whatever the case may be, the recipes for spells, charms, and talismans, interspersed with amorous accounts of love potions and seductions, provided a frequent inspiration for painters and readers alike. In this way the earliest Persian reception of *Wonders and Rarities* linked the work even more explicitly to occult learning and tales of sexual encounters. The section on talismans frequently features elaborate paintings that drew on iconography common to handbooks of astral magic (figure 8.1), just as the erotic paintings formed part of a broader repertoire on the carnal arts.[14]

FIGURE 8.1: Talisman of Jupiter mounted on an eagle clutching a phallus, Persian translation of Qazwīnī's *Wonders* copied 974/1566.

Like their Arabic counterparts, many of the earliest Persian manuscripts were lavishly illuminated. In the field of painted books of natural history, Qazwīnī also shared much with Shahmardān and Ṭūsī, whose Persian natural histories circulated in presentation copies that included splendid illuminations and fine calligraphy. They all took on topics of occult learning, while hidden in their pages were often illustrations of erotic encounters.

Qazwīnī built on earlier precedent. Similarly, later writers of natural history, such as the Ilkhanid statesman Ḥamd Allāh al-Mustawfī (d. ca. 744/1344), turned to Qazwīnī as a model. Mustawfī's Persian collection, the *Nuzhat al-qulūb* (Delight of the minds), drew extensively from Qazwīnī, covering astronomy, Earth science, geography, botany, mineralogy, and zoology. The model of wonders of various regions also forms a defined conceit, from Iran to the far reaches of the world. Throughout Qazwīnī proves indispensable. Mustawfī's work too circulated in countless manuscript copies, many adorned with illuminations common to both Arabic and Persian manuscripts of natural history. Across the broad stretch of Persian letters, Mustawfī's collection traveled through courts and libraries, often set alongside Qazwīnī's compendium.[15]

The desire for encyclopedic totality had long captivated the interests of potentates and their courtiers. So it was that throughout the centuries Qazwīnī's name would be feted by courts and mighty patrons, in imperial and provincial workshops busy with calligraphers, painters, and bookbinders. Through gifts, conquests, and new commissions, *Wonders and Rarities* continued to move. Illuminated copies produced under the Timurids, the dynasty founded by Tīmūr (d. 807/1405)—the mighty Turkic warlord who had united an empire across Iran and Central Asia cast in the mold of the great Mongol conquerors—not only made their way into royal Ottoman archives and institutions of learning (figure 8.2), but also served as models for courtly productions as far afield as the Deccan in south India.[16]

Compendiums of natural wonders grandly evoked the celestial and terrestrial, all the while they could easily fit within the pages of a single tome. Yet there were tastes for even greater ambitions with multivolume collections on a vast scale, which cumulatively sought to cover every detail across time and

FIGURE 8.2: Detail of a bifolium frontispiece removed from a Persian translation, copied in the Timurid court of Shiraz in 824/1421, with the title of Qazwīnī's *Wonders* in the central medallion framed by two angels, an *ʿanqā*ʾ grasping an elephant, and a *karkadan*, generally identified with the rhinoceros. By the beginning of the sixteenth century, the once complete manuscript formed part of the royal Ottoman treasury.

place. The monumental histories of all creation, every utterance ascribed to the Prophet, sprawling dictionaries of strange lemmas and rare expressions, the argots of peoples long past, medicinal recipes and ingredients both quotidian and remote, summae that spoke for all knowledge and science—these volumes occupied public libraries and royal collections.[17]

Qazwīnī's collection, in contrast, was eminently mobile. Copied, translated, expanded, and redacted, it transformed as it traveled, read by mystics and courtiers, princes and preachers. Some updated the work with new oddities, such as the conjoined twins born in Egypt in 768/1367~ who attracted the attention of the Mamluk elite in Cairo. Many copies were quite humble, shorn of illuminations, often produced in contexts of devotion, such as a manuscript copied by an imam at a congregational mosque in a southern outpost on the Nile in 790/1388, or the exemplar finished in 986/1578 in the sacred precinct of al-Aqṣā Mosque in Jerusalem by a pious scribe who notes at the end of the manuscript that he had copied in his life countless volumes on Quran exegesis, hadith, jurisprudence, and lexicography. This mobility also ensured that Qazwīnī joined a pantheon of works penned by Muslims that crossed sectarian divides, read and circulated by Christians, Jews, Zoroastrians, and Hindus.[18]

Qazwīnī's influence can be felt not only in Persian and Turkish manuscripts but also in later Arabic collections of natural history, such as the immensely popular *Kitāb al-Ḥayawān al-kubrā* (The large book on animals) by Kamāl al-Dīn al-Damīrī (d. 808/1405), a professor at the Azhar madrasa of Cairo. Drawing on poetry, parables, compendiums of history, science, talismans, medicine, and law, Damīrī turned to Qazwīnī as an authority on a variety of topics. Volumes of both works were frequently shelved side by side in madrasa libraries and royal archives.[19]

Sometimes other materials were added to the same manuscript, such as the *Akhbār al-zamān* (The reports of time), an Arabic compendium of wonders that long circulated under Masʿūdī's name and made for an obvious companion to Qazwīnī's natural history. Expansions to the text itself were not uncommon. A copyist in Cairo, for instance, evidently was not content with Qazwīnī's treatment of Egypt. To remedy the matter, he wove seamlessly into the body of the text extensive passages on the pharaonic past taken from the likes of Ibrāhīm ibn Waṣīf Shāh (fl. 350/961?) and Taqī l-Dīn al-Maqrīzī.[20]

A similar process was at work in an abridgment in Anatolian Turkish ascribed to the Sufi preacher Yāzıcıoğlu Aḥmed Bīcān (fl. 870/1465). Written in 857/1453 after the Ottoman conquest of Constantinople, the introduction tells that Bīcān was inspired to undertake the work through a vision of his deceased spiritual master, the saint Ḥācı Bayrām Velī (d. 833/1430), founder of the Bayrāmiyye order in Anatolia. Bīcān reports to have uncovered an ancient work on the wonders of the world composed during the time of Alexander the Great, which was translated from Hebrew into Arabic during the days of Imām Shāfiʿī.

No mention of Qazwīnī is made here. But on close inspection, the collection reveals an intimate familiarity with earlier Arabic collections of natural wonders, Qazwīnī included, with extensive paraphrases and notable expansions woven throughout. A concern for ancient mysteries and apocalyptic destruction often leads the work in new directions. Aḥmed Bīcān made a career for himself as a preacher warning of the end of time. He expressed these concerns vividly in the *Dürr-i meknūn* (The hidden pearls), a widely read account of the universe and the impending apocalypse. His *Müntehā* (Epilogue), a Turkish commentary on Ibn al-ʿArabī's mystical treatise the *Fuṣūṣ al-ḥikam* (Bezels of wisdom), celebrates the recent conquest of Constantinople as a portent of the final days.[21]

The wonders of the cosmos could easily accompany reflections on the ultimate end of the world. Earlier eschatological themes drawn from natural marvels feature throughout the popular Arabic natural history by Ibn al-Wardī, entitled *Kharīdat al-ʿajāʾib* (The pure pearl of wonders). The work opens with a dedication to the warlord Shāhīn al-Muʾayyadī, who was at the time serving as the Mamluk deputy over the hilltop citadel of Aleppo.

The city was an important staging ground in the course of the ongoing military conflicts between the Mamluks of Syria and Egypt and the Ottomans of Anatolia. A sense of impending conflict can be felt throughout the collection. Written in 822/1419, Ibn al-Wardī presented his patron with a map of the world and a list of his sources. These include Ptolemy's geography, the book of wonders by Masʿūdī, and the astronomical memoir of Naṣīr al-Dīn Ṭūsī. In this list, Ibn al-Wardī makes no mention of Qazwīnī, though he does cite him as an authority later in the collection. But as with the broader circulation and repurposing of earlier materials across the canons of classical letters, Ibn al-Wardī draws extensively from Qazwīnī, often without ever acknowledging it.

For all the interconnections, the work's focus and form differ from Qazwīnī in significant ways. *The Pure Pearl of Wonders* opens with the two diagrams, one of the entire world surrounded by the mountains of Qāf and the other of a *qibla* map marking the direction of prayer for every region. Yet unlike the *Wonders and Rarities*, Ibn al-Wardī's collection did not circulate as an illuminated manuscript, which may well have suited the pious tastes of those who were less interested in figural representations.

Even more notable, Ibn al-Wardī concludes his catalogue of existence with a detailed treatment of how it would all come to end. Quoting from scripture and tradition, Ibn al-Wardī foretells that Muslims would finally seize Constantinople. Shortly after, the Sun would rise in the west and set in the east, the hordes of Gog and Magog would descend across the Earth, Mecca would be lost, the world would be destroyed, the cosmic trumpet would ring out, as the final judgment would separate the saved from the damned. Lest the message of pious humility before impending destruction be lost on his readers, Ibn al-Wardī concludes with a list of synonyms for the Resurrection. This is followed in many manuscripts with a poem by the Egyptian preacher and Sufi ʿAbd al-ʿAzīz al-Dīrīnī (d. ca. 697/1297) warning of the punishments facing sinners and the rewards awaiting the faithful, come the final hour.[22]

It is a sentiment largely absent from Qazwīnī's concerns, but it appealed to many, as measured in the countless copies of Ibn al-Wardī's compendium for humble readers and splendid courts. His work also found a steady audience through Ottoman translations. It was not that far-fetched, after all, given the old predictions that the world would come to an end soon after the conquest of Constantinople. In Ibn al-Wardī's life, the Ottomans of Anatolia had been making steady progress against the last Byzantine stronghold across the Bosporus. It was only a matter of time before the imperial city fell.[23]

The Ottomans captured Constantinople, now Istanbul, in 857/1453. Shortly after, they established for themselves a new palace complex adjacent to the ancient Hagia Sophia basilica, with terraced gardens overlooking the Golden Horn. Over the course of their conquests, the sultans of Anatolia had amassed a treasury of books and relics.

In 908/1503, the Ottoman sultan Bāyezīd II (d. 918/1512) commissioned his royal librarian Hayreddīn ʿAṭūfī (d. 948/1541) to make an inventory of all the manuscripts held in the imperial collection. The titles for some five thousand works reflect wide-ranging topics across numerous branches of learning that guided the interests of the court. Here were books of philosophy, medicine, chronicles, biographies, encyclopedias, and multivolume commentaries on the Quran and jurisprudence, alongside tomes of occult learning, as well as geographies and collections of wonders and rarities.

The number of works on natural history and geography in the inventory is relatively modest when compared to the shelves dedicated to other volumes. Titles on the wonders of the creation and the maps of the climes, however, are listed next to works that appealed to the practical sensibilities of courtly education and the political arts. These include biographies and histories, techniques of war, matters of rule, governance, and discipline, as well as horsemanship, veterinary science, and falconry.

Several works marked as historical are dedicated to the adventures of Alexander the Great and the cycle of marvelous tales from *The Thousand and One Nights*. The organization of titles suggests ways natural history and geography could intersect with writings meant to edify and entertain. ʿAṭūfī's classification also indicates that the lines between history and epic could often blur, where fabulous feats of heroes were celebrated in ever-increasing displays of marvels.[24]

The language of wonder also features in the treatment of the occult sciences. ʿAṭūfī brings together dream interpretation, physiognomy, alchemy, mineralogy, geomancy, omens, elemental magic, talismans, enchantment, amulets, apocalyptic divination, books of amazing crafts and mechanical devices, sorcery, and amazing phenomena, *al-umūr al-ʿajība,* concluding with works on astral divination. While natural histories and geographies were catalogued separately from the occult sciences in the royal library, they drew on cognate sensibilities of awe and astonishment long associated with occult learning. Illustrations of astral figures, jinn, and strange creatures from faraway lands were not unique to compendiums of wonders. They also populate illuminated works used for augury and bibliomancy, known as *fāl-nāma,* which became popular within the Ottoman palace and were used to read omens and predict the future.[25]

Among the natural histories, the most common title in the inventory is *ʿajaʾib al-makhlūqāt wa-gharāʾib al-mawjūdāt,* the wonders of things created and rarities of matters existent. ʿAṭūfī identifies three exemplars explicitly with Qazwīnī. In addition to Arabic copies, there are Persian and Turkish translations. The Ottomans had long cultivated an attachment to these compendiums. A generation earlier, Meḥmed I (d. 824/1421) was presented with an illustrated translation into Anatolian Turkish of Ṭūsī's *Wonders and Rarities,* which was directed to the courtly education of princes through a good use of morality tales mixed with a healthy interest in talismans and incantations.[26]

ʿAṭūfī emphasizes that several of these compilations were written *min qibali l-tawārīkh*—from the perspective of historical events—lending more logic to their classification within a broader section of works on history and statecraft. The identification with history was far-reaching. On the title page of the Wasit manuscript of Qazwīnī's *Wonders and Rarities* a later hand added the same phrase, "from the perspective of historical events," on the frontispiece page beneath the main title (see figure 4.2). As a classificatory marker, the identification highlights the historical value of studying the physical world, its relationship to cosmic time, and the study of the Earth and cosmos as a means for tracking historical processes of creation.

Several titles in ʿAṭūfī's entry are also dedicated to strange stories drawn from across the sea and India. Others have clear eschatological content, such as a book on wonders that focuses on events to come at the end time and the conditions for the final hour according to the Sunni creed. Many of the geographies cover the known world, whereas others focus only on the climes, kingdoms, or regions of Islam. Ptolemy's geography caps off the list, with three copies in Arabic translation.[27]

Bāyezīd II's father, the conqueror of Constantinople, Meḥmed II (d. 886/1481), famously consulted the maps of Ptolemy's *Geographia* and commissioned an Arabic translation. Ptolemy's geography had originally made its way into Arabic early, on the orders of the Abbasid caliph al-Maʾmūn; but this translation appears not have circulated for some time. Commissioned in Istanbul, the new translation opens with lavish praise of Meḥmed II, the "supreme king and mighty sultan, depth of the sciences and sea of philosophy, mine of justice and equanimity, proof of Islam," who is also "the destroyer of all the enemies of God almighty, crusher of all the foundations of their baseless religion." Like the many dynasties before them, the Ottoman elite turned to geography and cosmography to articulate their own imperial ambitions. With control of the great capital of the Eastern Roman Empire, Meḥmed II could claim an imperial dominion like no other Muslim potentate before him.

The translation begins with Ptolemy's definition of geography, his pursuit of accurate knowledge of the world, and his critique of merchants, who spin long yarns with no basis so that their journeys sound amazing. For all his skepticism, Ptolemy also recognizes strange animals to be found in remote lands and the vast expanse of geographical information that could consume more than a lifetime of research. Here, too, the prime meridian starts with the Islands of the Blessed past the Strait of Hercules. Also featured here is Ptolemy's claim that the Nile finds its origin in the Mountain of the Moon—an identification found throughout the course of early Islamic geographical writing and well known to Qazwīnī. The translation and maps in the copy that was once held in Bāyezīd II's collection, before it made its way as a pious endowment for the nearby Aya Sofya madrasa, also include material not found in Ptolemy's original. The map of the Arabian Peninsula features small depictions of the Kaʿba of Mecca and the tomb of the Prophet in Medina, and the geography discusses the savage tribes of Gog and Magog to the far north—details entirely foreign to Ptolemy.[28]

The courtly context for the study of wondrous lands and distant nations is consistent with the classification of the sciences taken up by ʿIṣām al-Dīn Ṭaşköprīzāde (d. 968/1561), a high-ranking madrasa professor and judge in the service of the Ottoman state. Ṭaşköprīzāde includes a section on *musāmarat al-mulūk*, a field of learning focused on evening entertainments for kings. These contain "stories, anecdotes, sermons, morals, parables, the strange phenomena of the climes and the wonders of the regions, and other such matters which kings prefer." The entry forms part of a discussion on Arabic language and literature and suggests a general affinity with ʿAṭūfī's treatment of natural histories in his earlier inventory for the imperial library.

But then again, wonders were notoriously hard to pin down or to figure out where exactly they belonged. Ṭaşköprīzāde returns to the topic in his treatment of *ʿilm al-hayʾa*, cosmography, which falls within his discussion of the theoretical natural sciences. The science of properties unique to the different regions, the *khawāṣṣ al-aqālīm*, forms a subsection of cosmography, which studies both the heavens and the Earth. It is a branch of learning that examines the beneficial, harmful, and extraordinary phenomena spread across the world.

Qazwīnī's *Wonders and Rarities* features in this subsection. "But there is a better book on this topic," Ṭaşköprīzāde notes. "I couldn't recall its name, then I asked a friend of mine, who reminded me that it was *The Pure Pearl of Wonders* by Ibn al-Wardī." In the context of pious education, this sentiment may well account for the numerous manuscript copies in Arabic and Turkish of Ibn al-Wardī's cosmography, with its focus on the end days. A close reader of

Ghazālī's metaphysics and discipline of the soul, Ṭaşköprīzāde returns later in his classification to the topic of bewilderment, referencing Ghazālī's treatise on the wonders of God's design, while also reflecting on astonishment at God's creation as a spiritual exercise for obtaining divine gnosis. In Ṭaşköprīzāde's classifications of the sciences, wonders of existence could inspire both kings and the pious, just as they also took measure of geographical diversity across the world.[29]

The conceptual register of wonder and rarity had also long been connected to the occult sciences. There was the judge Ibn Manyās (fl. 841/1438) who wrote an Arabic treatise on the wonders of spells and talismans, as well as a compendium in three parts on jurisprudence, arithmetic, and the wonders of minerals, plants, and animals, written in Anatolian Turkish for Murād II (d. 855/1451), entitled *Aʿcebü l-ʿüccāb* (The most wondrous of all wonders). The judge's collection also weaves in discussions of talismans, magical squares, and various practices of divination with accounts of demons and jinn and cyphers for harnessing hidden forces. Ṭaşköprīzāde knew Ibn Manyās to have had a distinguished career as a madrasa professor and to have obtained renown for gathering material on talismans, charms, and spells not found anywhere else.[30]

During this period, astrologers skilled in geomancy and the arts of divination also had a readymade framework for describing the power they could harness through the vocabulary of wonders. The Egyptian jurist and court astrologer Ibn Zunbul al-Maḥallī (fl. 981/1573) made a career for himself in Cairo and Istanbul foreseeing all matters hidden. He a wrote a work on geomancy, with the same title as Qazwīnī's *Wonders and Rarities,* revealing the secrets of the art with its detailed calculations and a sustained attention to climatic variation and the movements of the heavens. In addition to this compendium of divination, Ibn Zunbul also developed an interest in the marvels of creation with his own natural history, illuminated with wondrous phenomena from across the world, following quite consciously a mold that Qazwīnī had established centuries earlier.[31]

For all these innovations and adaptations, the classical collections on natural marvels continued to travel, copied and read for various ends in Arabic, Persian, and Turkish. But the world was quickly changing. While Iberian naval expeditions set off across Africa and the Western Hemisphere, the Ottomans were forging an expansive empire. Their territories came to stretch across the Levant, the Hijaz, and North Africa, and deep into the Balkans. With war galleys and squadrons, Ottoman forces spread out from their ancestral stronghold in central Anatolia across the seas. A series of conquests followed: Constantinople, the Black Sea, and the Aegean. This culminated in 923/1517 with the destruction of the entire Mamluk state of Egypt, Syria, and Arabia.

The fall of the Mamluks also put control of the sacred cities of Mecca and Medina into the hands of the Ottoman sultans, who proclaimed themselves Sunni guardians of Islam. All the while, they continued to expand their naval supremacy across the Mediterranean, through the Red Sea, into the Indian Ocean, and along the eastern coast of Africa, where they vied with the Portuguese and the Dutch for control of the spice trade. While they did not cross the encircling sea into the New World, Ottoman notables were well aware of its existence.[32]

In Latin the conceptual neologism of the New World—the *mundus novus*, the *novus orbis*—provided a distinct way of thinking about human history, geography, and providential design. Temporally, the old world, the *vetus mundus*, came to stand for the limitations of knowledge; in contrast, spatial access to the Americas opened the globe in its round expanse. Yet for all its profound implications, the ruptures occasioned by the discoveries and subsequent conquests were built on earlier patterns of exploration and mercantile exploitation.[33]

The maritime expansion of the Iberian kingdoms followed the historical movement of the *reconquista*, which saw Christian forces take possession of an ever-expanding territory long ruled by Muslims across the Iberian Peninsula. As with the Crusades to the Holy Land, this process was well known to Qazwīnī, who documents in his geography the many towns and cities of al-Andalus, the Arabic term for the peninsula, now in the hands of Christians. For the native population of the Canaries, contacts with Christian explorers led to massacres, forced enslavement, and sustained persecution. This settlement, in turn, laid the bureaucratic, mercantile, and religious foundations for the further colonial administration and subjugation of the indigenous populations of the Americas.[34]

In addition to the flow of exotic goods, the Atlantic islands served as way stations for trafficking slaves from the African coast and beyond, a role that became even more prominent in the following centuries. This pattern of colonization and enslavement, in turn, formed the basis for further Iberian penetration across the Atlantic. For each of his four voyages, Columbus first passed through the Canaries on his way to the Americas. Subsequent conquistadors followed similar itineraries, stopping at various islands in the archipelagoes in the pitched movement back and forth across the Atlantic in the wake of the first contacts.[35]

The technologies of navigation that gave life to New World cartography helped to animate this frenzied exchange of bodies and commodities. Maps came to spatially represent not only overseas possessions but new horizons for conquest and trade. The radiating navigational system of the wind rose plotted coastal features, often punctuated with flags that visually marked territorial claims. As with the earlier portolan charts of the Mediterranean, many maps of newly discovered lands were populated with savage nations, exotic flora and fauna, and untold riches waiting to be had.[36]

The earliest Iberian chroniclers of the conquests of the Americas portrayed their bloody struggles using a sacred language drawn from the battles of the *reconquista* between Christians and Moors. Their accounts fill the idolatrous land of the New World not so much with priests or temples, but with *alfaquis* and *mesquitas*—Spanish words for Muslim religious authorities and for their mosques, from the Arabic *faqīh* and *masjid,* respectively.[37]

The Spanish monarchy would enact legislation forcibly excluding all religious outsiders, particularly Jews and Muslims and their brethren who had converted to Christianity, from traveling to the New World. Several motivations guided these royal statutes, including a desire to curtail the introduction of heterodoxy among the native populations, who were subject to the missionary zeal of the Catholic Church. Similarly, the competing mercantile and imperial forces that pitted Ottoman naval power against the Spanish and the Portuguese also shaped territorial demarcation in explicitly religious terms.[38]

The juridical justification for territorial possession underwriting the conquests was rooted in ecclesiastical jurisprudence developed during the course of the Crusades, based on the global authority of the Catholic Church and its mission to care for the souls of both Christians and infidels alike. The legal rationale for the seizure of territory from savage nations, *barbaricae nationes,* immediately proffered by the Vatican bull of 1493 in direct response to the voyage of Columbus, granted Ferdinand and Isabel, the Catholic monarchs of Aragon and Castile, legal claims to all islands and lands discovered or to be discovered, *omnes insulas et terras firmas inventas et inveniendas.*

For Ferdinand and Isabel, the year 1492 marked not only the first voyage to the New World but also their conquest of Granada and the expulsion of Jews from the land through the bureaucratic arm of the newly founded Holy Office of the Inquisition to root out heresy. In the legal language of discovery, the Catholic monarchs were granted papal authority to take possession of all the gold, spices, and other precious items they encountered, *aurum, aromata, et aliae quamplurimae res pretiosae . . . reperiuntur.* This was based on their legal responsibility, as representatives of the Church, to convert the native inhabitants to the Catholic faith. They were well qualified for this charge, the papal decree explained, as evinced by their recovery of Granada from the tyranny of the Saracens.[39]

The conquistadors and the ecclesiastical officials and administrators who followed found readymade analogues for describing what they viewed as the savagery of the natives in the well-worn representation of the sect of Mohamet as bestial and irrational.[40] To be sure, Muslims came to share a part in the early transatlantic migration, although almost entirely as slaves, taken captive in Iberia or the West African coast, or as converts, often of dubious Catholic faith, who had managed to evade the regulations prohibiting their travel to the New World.[41]

Like all such imaginative and legislative boundaries, there were other forms of passage. In the following years the Ottoman elite kept up-to-date with the latest developments in cartography and geography. The first, and by far the most stunning, cartographic testament to survive of this movement is preserved in a fragment of a map completed in 919/1513 by Aḥmed Muḥyīddīn Pīrī (d. 960/1553), who is known today simply as Piri Reis, from his title *re'īs* as an admiral in the navy (figure 8.3).

The remaining section of what was once a world map lines the Atlantic Ocean and its islands with carracks and caravels sailing across shores populated by strange creatures and filled with detailed inscriptions. The eastern portion of the fragment features part of the Iberian Peninsula, the northern

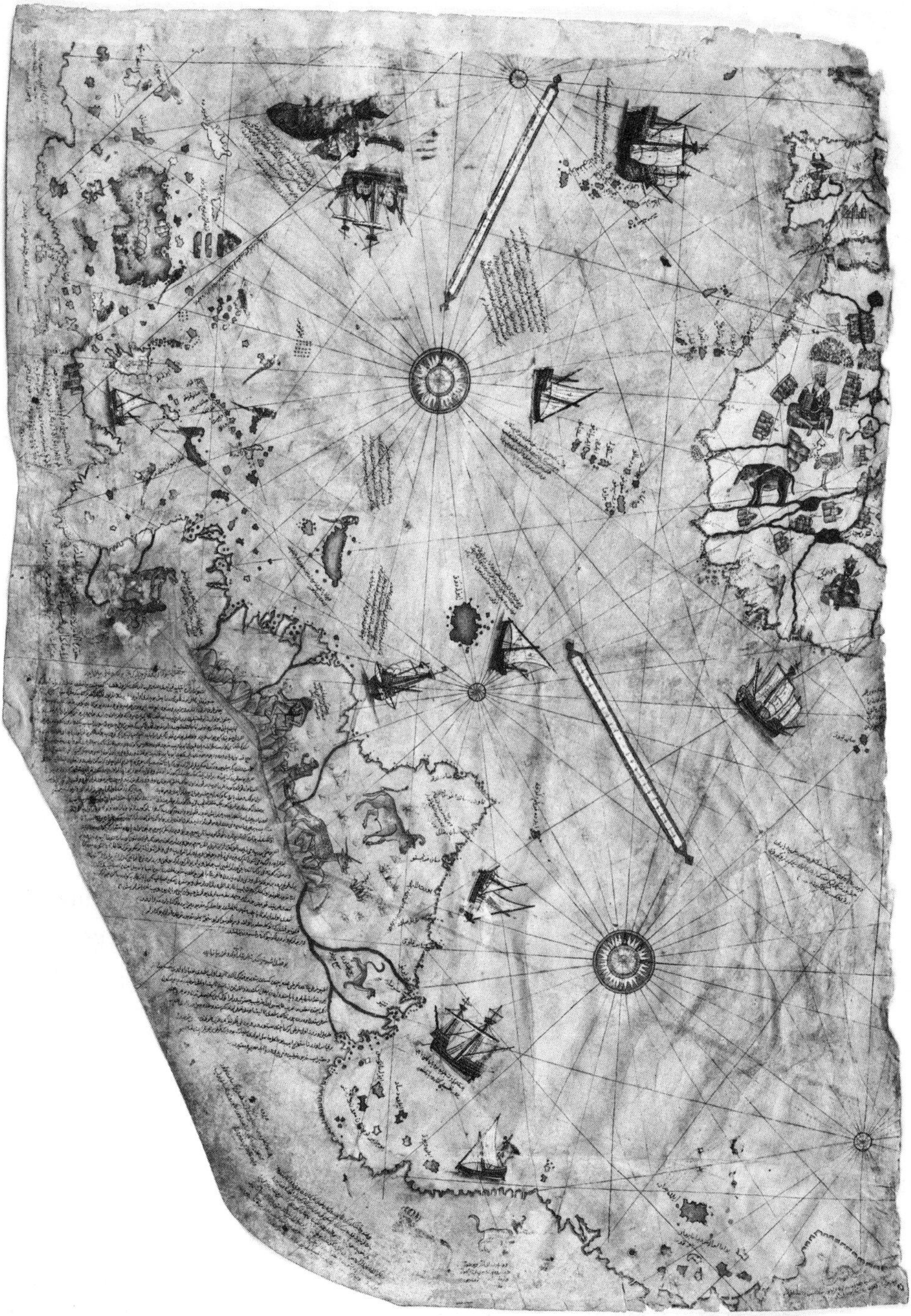

FIGURE 8.3: Surviving western portion of the Piri Reis world map, on parchment, dated 919/1513, with strange creatures in the New World, including a dog-headed figure and a man whose head is on his chest.

coast of France, and a slice of the African coast. Yet it is the vellum chart's western shore that has drawn the most attention. It preserves a clear outline of the American continent and the Caribbean Islands in a shape that, while clearly discernible, reflects the fluid state of Iberian cartographic knowledge of the New World at the beginning of the sixteenth century.

In his book on navigation, the *Kitāb-ı Baḥriyye* (Book of maritime knowledge), completed in 932/1525, Piri Reis mentions that he gifted a map to Sultan Selīm Khān (d. 926/1520) in Cairo. The sultan's presence in Egypt coincided with his victory over the Mamluks, which in turn opened the way for the Ottoman penetration of the Red Sea, and with it direct conflict with the growing naval power of the Portuguese in the Indian Ocean and the Spice Islands. According to Piri Reis, his gift to the sovereign offered an unrivaled level of detail, for he drew on earlier maps of the oceans of India and China, the likes of which had never been seen in Ottoman lands. The passage stresses the eastern seas, which were the theater of ongoing navel conflict for maritime supremacy, and an area of geographic interest for Ottoman ambitions.[42]

The royal map rediscovered centuries later in the Topkapı Palace may well have been a fragment from the map Piri Reis presented in Cairo, though it would have been presented some three years after its dated colophon. There is reason to believe that the far western section of the map, and with it the American continent, survived the vicissitudes of time precisely because it was the Eastern Hemisphere that was of greatest interest and use to the Ottomans during this period, given their ongoing efforts to disrupt the rapid Portuguese mercantile and military expansion into the Indian Ocean, Persian Gulf, and Red Sea. In terms of political implications, these maritime rivalries drew the Aceh Sultanate of Sumatra under the protection of the Sublime Porte in Istanbul.[43]

The map itself bears testament to the countless acts of translation that made maritime expansion possible. Local navigators and go-betweens would prove instrumental to the history of Spanish and Portuguese exploration. Interpreters who could speak Arabic were at times highly prized. These included Jewish *conversos*, Muslim slaves, and Christians who, through experience in North Africa or in the East, had picked up some Arabic, often as merchants or captives. For Columbus, who viewed the western passage as a means of ultimately regaining Jerusalem under the banner of Christendom, this famously meant bringing interpreters who could communicate in Arabic with the Great Khan, whom he was sure he would meet on the other side of the ocean. Interpreters ready to use Arabic and a smattering of other languages were also on-board Portuguese ships journeying across the shores of Africa and the Indian Ocean.[44]

Both in his map and in his subsequent *Book of Maritime Knowledge*, Piri Reis credits Columbus with the discovery of these new lands, which he refers to as "Antilya," the mythical island of Iberian lore, known as Antilles, which Columbus thought he would encounter on his voyage. Columbus failed to fully grasp the actual size, geographic location, or significance of the lands he reached, and he remained convinced that he had found the western passage to Asia.[45]

Modern theories promoted by Fuat Sezgin and others have turned to Piri Reis's map as proof that Muslims reached the Americas before Columbus. The place names and coastal features of the Western Hemisphere that Piri Reis uses, however, are all the product of earlier Iberian explorations. Piri Reis himself says as much. Several of the inscriptions that accompany the carracks and caravels sailing the Atlantic explicitly refer to Genoese and Portuguese expeditions that have clear correlations to historical events.[46]

In one inscription of the mainland of the Western Hemisphere, Piri Reis notes that he drafted his entire map after consulting some twenty maps, including four recent Portuguese maps of Sind, Hind, and China, and one map on the western region drawn by Columbus.[47] Another inscription gives a fairly accurate digest of Columbus's voyages, noting that Columbus, a Genoese infidel, was the first to find these regions. Again, this same inscription, the longest of all, notes that the islands and coasts on the Western Hemisphere derive from a map that Columbus made. This account draws directly from a Spanish slave, who was said to have sailed three times with Columbus across the western sea.[48]

The note relates that the slave was in the possession of Gāzī Kemāl (d. 916/1510), Piri Reis's uncle, an admiral in the Ottoman navy. The informant may have been taken prisoner in the Ottoman naval siege of Valencia, led by Gāzī Kemāl in 906/1501, when several Spanish vessels were captured. In addition to slaves and exotic booty from remote lands, this may well have included cartographic materials on the New World.[49]

Beyond the captive's tale, further sources on the Iberian transatlantic voyages informed Piri Reis's cartographic projection for the entire region. The detailed representation of South America, including the broad turn of the Brazilian coast, along with the well-informed descriptions of Portuguese expeditions across the Atlantic, particularly the voyage of Álvares Cabral, which reached the coast of Brazil in 1500, all point to materials well after Columbus. Through it all, Piri Reis demonstrates an impressive command of the historical developments shaping the discovery and colonization of the New World, in cartographic and political terms.[50]

The inscription next to the depiction of sailors kindling a fire on the back of a whale in the middle of the Atlantic is emblematic of these lines of exchange. Here the story of Ṣan Vulrāndān, who in ancient times was said to have sailed the seven seas, is featured. The inscription notes that the crew lit a fire on what they thought was dry land. But what they believed was an island turned out to be a giant fish that plunged back into the ocean as soon as its back began to burn (figure 8.4). A very similar story appears in countless Arabic and Persian accounts of the marvels of the sea, including Qazwīnī's *Wonders and Rarities*.

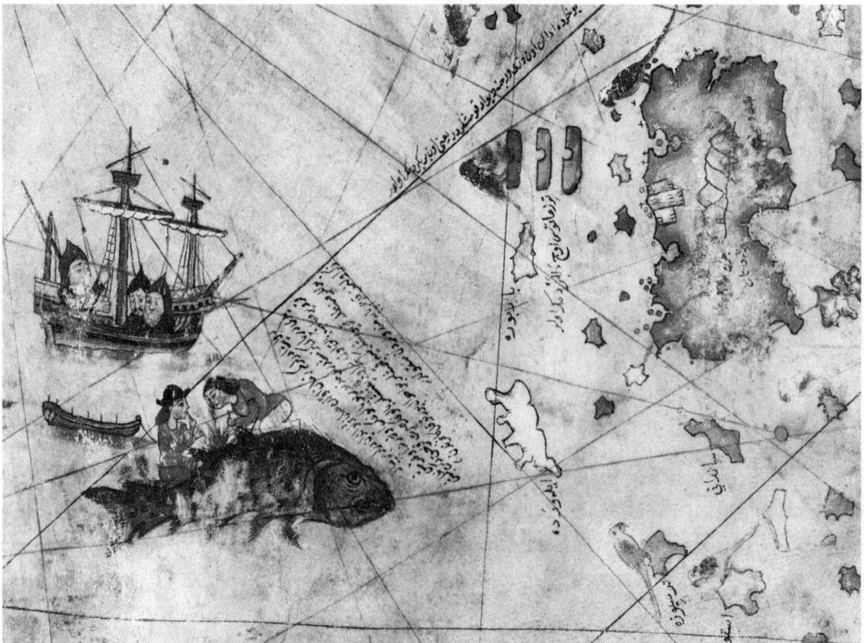

FIGURE 8.4: Detail of sailors setting fire on a sea creature mistaken for an island, the map of Piri Reis.

However, Piri Reis clearly drew this tale from Christian sources. Noting that the sailors fled to their ship, the map states that this event was not related by the Portuguese infidels, but instead features in older *mappae mundi*, referred to in a Turkish calque as *yāpāmūndolar*. The reference to the adventurer of the seven seas, while a bit garbled, points to the anecdote as it had become attached to the maritime feats of the Irish Saint Brendan, which also tells the tale of a fire built on the back of giant sea beast mistaken for an island to celebrate an Easter meal.[51]

Exotic flora and fauna, strange natives and distant rulers, castles and fortified cities, mountains and tributaries line the map. The reference to *mappae mundi* further links these figural forms to the cartographic representations developed in Latin Christendom throughout the previous centuries. While the visual idiom closely aligns with the figural forms of miniature painting in Arabic, Persian, and Turkish book culture, the cartographic configuration of wondrous diversity finds its closest analogues in the late medieval European corpus of world maps and Mediterranean portolan charts.

The competing interests of the Spanish and Portuguese are referenced with an allusion to the famed Treaty of Tordesillas of 1494, which marked a line of control dividing the territory discovered or to be discovered between the two competing seafaring empires. The Portuguese mission to find a trade route around Africa to India and the Spice Islands is also mentioned. Gold, pearls, exotic spices, plants, metals, and strange creatures appear in terms that directly echo the first accounts of the New World. Similarly, Piri Reis grasps the linguistic difficulty of trading with savage natives who go about naked and have no religion or *mezheb*. He observes that the Spanish and Portuguese were forced to communicate with the locals by *işāret*, gesturing and pointing. Famously, for the first contacts the interpreters who had been brought along were not of much help. Practices of cannibalism, which occupied many of the early Iberian accounts, are also noted.[52]

An inscription relates that Columbus learned from a book that at the end of the western sea lay islands, shores, many kinds of precious metals, and a mountain of jewels. Later, in his *Book of Maritime Knowledge*, Piri Reis once more returns to this mysterious source, explaining in a rather curious anecdote that Columbus learned of these new lands from a book documenting everything that Alexander the Great saw during his journeys at sea.[53]

According to Piri Reis, this book containing all the knowledge of maritime navigation, *deryā ʿilmini*, was originally composed in Egypt, presumably in

Alexandria. The reference has echoes of Aḥmed Bīcān's preface to his reworking of Qazwīnī's cosmography, where he discusses a book of the world's wonders composed in Egypt during the reign of Alexander. This work, Piri Reis explains, fell into the hands of Christians, referred to as Franks, who had fled the Arab conquests of Egypt in the seventh century. Thus, in Piri Reis's telling, Christians gained knowledge of the New World by studying Alexander's global exploits.

Piri Reis concluded that at some point the work was translated into the language of the Franks. This book of hidden secrets fits into the long-standing connection between Alexander and secret knowledge of the world, common throughout occult writings, such as the *Sirr al-asrār* (The secret of all secrets) and the *Dhakhīra* (The treasury), which were attached to Alexander's name. The reference also echoes the early and often conflicting accounts of the inspiration for Columbus's western voyages. Most notable is the insistence on Alexander the Great, the paradigm of the philosopher king, as the ultimate source of geographic knowledge for the entire world and, with it, the Western Hemisphere.[54]

In the account of his third voyage of 1498, Columbus observes that the possibility of a western passage to the Indies had been known since antiquity. As proof, he paraphrases extensively from the *Imago mundi* by the French astrologer Cardinal Pierre d'Ailly (d. 1420). This Latin cosmography appears to have also been the basis for Columbus's underestimation of the circumference of the Earth and his belief in the proximity of Europe to Asia, moving west across the Atlantic. The distance separating these two regions, Columbus argues, is not that great, for Aristotle observes that one could easily pass from Spain to the Indies by moving across the western ocean.

Columbus repeats d'Ailly's argument that Aristotle could have known many secrets of the world, including its true size, from his role as tutor to Alexander the Great. The argument draws on the notion that Alexander circled the entire world and thus was able to provide Aristotle with concrete knowledge of the true shape of the Earth.[55] The importance of d'Ailly's cosmography for Columbus was well known to the Dominican priest and historian of the West Indies, Fray Bartolomé de las Casas (d. 1566). After seeing Columbus's annotated copy of the *Imago mundi,* las Casas wrote that of all the materials that shaped Columbus's theory of a western passage, the work by Pierre d'Ailly was by far the most influential.[56] A copy of d'Ailly's cosmography survives from Columbus's personal library with his own extensive annotations along the margins, including notes on the passage in question.[57]

Piri Reis's comments about the importance of Alexander for discovering the New World echo a larger debate in European letters on whether classical authorities had any awareness of the Western Hemisphere. The story of the translation of Alexandrian nautical knowledge also reflects a repurposing of the argument over precedence. The account pictures rival Christians pilfering the exploits of the god-fearing Alexander, a pious ruler from the pantheon of ancient believers, who in his mission to conquer the world prefigured the coming of Muhammad.[58]

This account can be contrasted with Piri Reis's earlier world map, where he notes that among the cartographic materials he consulted were charts composed during the time of Alexander, which covered the inhabited quarter of the world. The term "inhabited quarter" used when describing these ancient maps comes from classical models of the Earth and thus would likely exclude the Western Hemisphere. Nonetheless, the notion that Alexander, referred to often simply by his Quranic epithet Dhū l-Qarnayn, had knowledge of these distant lands became a notable argument in the early Muslim treatment of the New World. Above all, the claim served to connect Muslims to an ancient and global history of exploration.[59]

The theory that Alexander crossed into the Western Hemisphere makes its way into an Ottoman history of the New World, entitled *Ḥadīs-i nev,* meaning the new or fresh account, report, or revelation, also known as *Tārīh-i yeñi dünyā* (The history of the New World), commissioned in 991/1583 for Sultan Murād III (d. 1003/1595). The work was composed under the supervision of the courtier Meḥmed ibn Ḥasan Suʿūdī (d. 999/1591), who styled himself as a follower of Masʿūdī, the great geographer and encyclopedist.[60]

The collection introduced the history of the Americas and their Iberian discovery and conquest to an Ottoman readership in and beyond the court. As with Piri Reis's generous use of sources, the *New Report* drew extensively from Christian historical materials, reflecting the centrality of Istanbul as an entrepôt for merchants, sailors, scholars, and diplomats brought together from diverse communities and distant regions to trade and engage in commercial and political endeavors. Surviving in multiple illuminated manuscripts, the collection was also one of the first Ottoman books to appear in print, published in 1142/1730, with accompanying woodcut illustrations and maps, by İbrāhīm Müteferriqa (d. 1160/1747), a Hungarian convert to Islam who set up the first imperially chartered Ottoman printing press in Istanbul.[61]

The work consists of three sections. The first two serve as a critical introduction to geography, with summaries and revaluations of classical accounts of the world, taken from earlier Arabic, Persian, and Turkish materials. The final and largest section, however, has drawn the most attention. Here the New World is unveiled with an Ottoman translation of early accounts by Pietro Martire d'Anghiera (d. 1526), Gonzalo Fernández de Oviedo y Valdés (d. 1557), Francisco López de Gómara (d. ca. 1559), and Agustín de Zárate (fl. 1560), all of which were accessed, in turn, through secondary Italian translations.

This separate portion of the work focusing on the Iberian conquests was later translated into Persian and circulated in several illuminated manuscripts in Iran and India during the late eighteenth and early nineteenth centuries. The title of the Persian translation, *Tārīkh-i yangī dunyā* (History of the New World), preserves the Ottoman nomenclature. The Persian phrase *yangī dunyā* was one of the most common means to refer to the New World during this period, pointing to the Ottoman provenance of a good deal of geographical knowledge about the Americas as it spread among Persian readers.[62]

In his critical introduction, Suʿūdī situates these new discoveries in the frameworks of established geography. Here he turns to Alexander the Great's knowledge of the New World. This opening section reflects an extended effort to recalibrate classical geographical knowledge. In particular, Suʿūdī highlights the difficulty of synthesizing the discovery of the Western Hemisphere with the long-standing Ptolemaic system of climes that had dominated the study of geography for centuries. Suʿūdī attempts to strike a balance between accepting established accounts at face value and questioning others in light of information brought from the New World. He engages extensively with classical authorities, particularly Ptolemy, Masʿūdī, Bakrī, and Naṣīr al-Dīn Ṭūsī, as well both Qazwīnī and Ibn al-Wardī.

Such is the case with the high-seas adventures of Alexander Dhū l-Qarnayn, which are taken as a precursor to the later discovery of the New World. Suʿūdī relates that during a voyage through the Caspian Sea, Dhū l-Qarnayn encountered a ship from a distant land whose crew spoke a language utterly incomprehensible to him, like nothing he had ever heard. After a successful plan to train a translator to discover the origins of this mysterious vessel, Dhū l-Qarnayn learns that the crew were from a far-off region ruled by a king greater than any other on earth. Suʿūdī accepts the authenticity of this story, which he quotes from Ibn al-Wardī. But he rejects the Caspian Sea as the site for this encounter, arguing that the true location of this adventure was the Atlantic, the *baḥr-i okyānūs,* and that the ship appeared from the New World, *dünyā-yı cedīd.*[63]

A nearly identical account features in Qazwīnī's *Wonders and Rarities* as an example of a marvelous story, *ḥikāya ʿajība,* concerning the encircling sea. Dhū l-Qarnayn sought to learn what lay at the edge of the encircling ocean, so he sent out a ship provisioned for an entire year, during which time his men encountered nothing other than the surface of the sea. Just as they were running out of provisions, the crew spotted a vessel commanded by a people whose language was totally unintelligible. They exchanged one of their own men for one of the women from the mysterious ship. A wedding ensues and the child of the union learns to speak both languages. Dhū l-Qarnayn learns that the ship came from a kingdom much larger than his own, with a more powerful king who also, in an uncanny parallel, sent out a mission to uncover what lay beyond the other side of the sea.

The anecdote has notable parallels with other stories of exploration in the sea of darkness, with provisions made and strange encounters. The solution of intermarriage also echoes anecdotes of other unions meant to learn how to communicate with strangers. A few folios later Qazwīnī gives the story of how a king once married a woman to a recently discovered merman, referred to as *insān al-māʾ,* or *shaykh al-baḥr,* the Old Man of the Sea. The son of the union was, in turn, able to communicate on his father's behalf. Like the tale of the Old Man of the Sea, the story of Alexander and the encircling ocean took on a life of its own.[64]

The adventure connects Alexander's pursuit of global knowledge with his ability to discover solutions to strange problems. In Ibn al-Wardī's telling, the story appears alongside a discussion of the Caspian. It is likely for this reason Suʿūdī believed the location for the adventure had been wrongly identified. In contrast, Qazwīnī's version—which Suʿūdī does not reference—occurs in the encircling ocean and thus it gives the theory that Alexander made contact with people from the New World even greater mooring. It is an idea that would gain incredible traction over time. Yet other than this allusion to Alexander's contact with the Western Hemisphere, Suʿūdī, as with Piri Reis before him, stresses the precedence of Columbus, and absorbs the master narratives of discovery that were fully operative in the early Iberian historical writings on the New World.[65]

An appetite for exotica guides the aesthetic principles of Suʿūdī's collection, with its figural and textual display of marvelous cities, people, and flora and fauna in a register that draws directly from the authority of natural history and descriptive geography. The pleasure that comes from entertaining the strange and wondrous, as well as the violent and abhorrent, offers several points of overlap with the primary Iberian materials, which in many ways were motivated by a desire to possess, consume, and tame the consummate difference of these hitherto unknown regions.

On occasion Suʿūdī questions earlier authorities. Yet his criticism is often slight compared to his broad acceptance of what he takes to be established knowledge of the world and its wonders. It is not that Alexander's strange encounter did not take place, rather it just did not occur in the Caspian. In this way the New World offers Suʿūdī and his readers a space at the farthest edge of the map where strange tales abide and wondrous creatures thrive unabated.[66]

The first illustration in the *New Report* foregrounds the idiom of wonder common in earlier Arabic and Persian works on geography and natural history. Here are pictured the arboreal creatures from the Wāqwāq islands, where women grow from trees, drawn from the marvels of the Indian Ocean. The exotic women of Wāqwāq are a stock feature in Arabic, Persian, and Turkish geographical accounts, and appear notably in Qazwīnī's *Wonders and Rarities*. Both the island and its trees are often said to receive their names through an onomatopoetic association with the sound *wāqwāq* that is heard when the dangling female fruits fall to the ground after having ripened. Sexual pleasure is ready for the taking, though the women only last a couple of days before rotting.[67]

Like other mighty emperors, Murād maintained a healthy interest in both remote lands and occult learning. Not only did Suʿūdī present a history and geography of the New World, but he also went on to dedicate to the sultan a work of astral magic and divination, the *Meṭāliʿü s-saʿādet wa-yenābīʿü s-siyādet* (Ascensions of felicity and sources of sovereignty). Surviving are two lavishly illuminated copies that were given as gifts to Murād's daughters. Suʿūdī's compendium draws from earlier materials adapted into Ottoman Turkish. In the Islamic calendar the dawn of a new millennium approached and with it a renewed attention to cosmic cycles of time. Many of the motifs and stories echo Qazwīnī's natural history, as does the appeal to wonder and rarity. The cosmic teachings of Ibn al-ʿArabī make an appearance, along with accounts of geomancy and other forms of sortilege. The cycle of paintings and text are designed to augur fortunes: the Wāqwāq tree, the Old Man of the Sea, Gog and Magog, Alexander in the land of darkness, the valley of serpents and diamonds, a full cycle of planetary images, astral degrees, and lunar mansions, all wrapped up with powerful demons and jinn.[68]

Suʿūdī and his readers saw the New World as filled with marvels. The illustrations accompanying the presentation volumes pick up this sentiment through strange vistas of cityscapes and studies of armadillos, prickly pears, birds of paradise, and bison. Here as well is a merman in the western seas that would have been entirely at home in Qazwīnī's natural history (figure 8.5). In the Turkish translation and its later Persian adaptation, wonders permeate the language of discovery.[69]

The phrase "wonders and rarities" also serves as a classificatory model for treating curious accounts and strange phenomena described in ways that often do not find ready analogues in the original Iberian source material. Although wonder was itself a motivating force in the Christian accounts of the conquests, the sensibilities are quite distinct. Lifting the veil from the marvelous possessions of others heightened the position of the Ottomans as outsiders to the process of conquest and colonization. This space of distance, in turn, infused the New World with a mixture of curiosity and envy absent from the original Christian sources.[70]

In this way the translation entirely weeds out the Christian justifications for domination. And yet the promotion of astonishment in the *New Report* appears most concerned with cultivating an explicitly imperial desire for further territorial possession. The responses of pleasure and appetite closely follow the *mirabilia* of untold riches—in human flesh and precious metals—promised

FIGURE 8.5: A half-fish, half-man off of the Island of Cubagua, from Suʿūdī's *New Report*, copied in the early seventeenth century.

by the conquistadors. Here the desire for possession, however, is placed directly at the foot of the sultan, who is repeatedly incited to take up arms against the Christian infidels and carry the victorious banner of Islam to the New World.

The *New Report* advances an attitude of shocked curiosity at the wanton savagery of the natives, often adopting the perspective of the Christian chronicles. The natives perform ritual offerings to the devil, fashion hideous idols in his form, and on occasion even have sexual intercourse with demons, from which issue offspring with horns. But the real moral outrage is directed to the Christians themselves, as purveyors of discord and disbelief. The Ottoman translation notes that in the conquests, thousands fell to the sword and many more were enslaved. A prayer is added that, by God's aid, one day this bounteous land would be filled with Islamic rites, *şeʿāyer-i islāmiyye,* and joined with the other Ottoman domains, *memālik-i ʿosmāniyye.* This aspiration is set in opposition to the Catholic church's mission to convert the natives; it reflects imperial ambitions of expansion in religious terms that mirror the proselytizing zeal of the conquistadors.[71]

In *The Book of Maritime Knowledge,* Piri Reis makes similar statements regarding the natives in his treatment of the wonders of the western sea. These are lands festering with savages, cannibals, and a group of people whose eyes are one full span apart. A specimen from these strange creatures features in his earlier world map (see figure 8.2), in a pictorial form that directly resembles the headless monsters of classical antiquity, known in Greek as *akephaloi* and in Latin as *blemmyae,* whose faces grow on their chests.[72]

Piri Reis situates these creatures spatially next to the famed nation of dog-headed men. Both populate Qazwīnī's *Wonders and Rarities* and commonly feature in early Arabic and Persian geographical literature and cosmological lore. For Piri Reis, though, the strangeness of the wild natives, who live like animals and follow no religious ceremonies, *āʾyīn-i dīn,* is outmatched by the nefarious barbarity of the Christian infidels, who claim the superiority of their own religion over those who possess none, *bī-dīnler,* all in an effort to lead these savages to infernal perdition.[73]

The stress placed on the bestial and primitive state of the natives is notable; as Muslim theologians of various stripes historically upheld the belief that animals have souls, and are, by natural, instinctive disposition, or *fiṭra,* believers in God.[74] Official Catholic theology took a very different route on the question of the bestial souls of animals and their capacity for faith. The Catholic church's recognition that Indians had souls and were indeed human did not come until *Sublimis Deus* (The sublime god), the papal bull of 1537. The decree established the legal obligation to convert the native inhabitants of the New World to Christianity, for as merely dumb brutes they would be incapable of receiving the Catholic faith.[75]

The differing theological attitudes separating normative articulations of Christianity and Islam during the period are even more intriguing when the question of evil is raised. At least in the authoritative language of Islamic theology, the devil's role in eschatological terms was significantly diminished in the face of the ultimate transcendence of an all-powerful God. A good number of orthodox Muslim religious authorities consistently challenged the perpetuity of hell itself. Such attitudes toward the limited autonomy of evil can be juxtaposed with the diabolical powers that early Christian missionaries frequently imagined were at work in the New World. These fears would come to guide witch trials at home and abroad. In contrast, Islamic theological visions of humanity did not proceed from the notion of original sin or in the belief that the material world was a chief basis for demonic temptation. Rather, diversity of existence in all its hierarchical order provided proof of God's majestic design.[76]

All the same, like their Iberian rivals the Ottoman elite also viewed the conversion of the local population as a matter of moral necessity. But savages with no religion—as the native inhabitants of the New World were generally viewed—were quite different from the Christians and Jews in Ottoman lands who enjoyed a fair measure of autonomy and protection. With the conquest of the Byzantine and Mamluk dominions, a large number of non-Muslims were absorbed into Ottoman territories.

Merchants and religious minorities from Europe also found protection in the port cities of the new empire. When facing edicts of expulsion from Spain, many Sephardic Jews took refuge in Ottoman lands. The ideal of a sovereign who ruled with magnanimity and justice over diverse religious communities had long been a feature of Islamic political and juridical thought, even before the Mongols took it on as an imperial policy. Similar ideals can be heard in justifications for global hegemony advanced by ancient emperors of Rome. Aspirations to world peace also found expression in European letters in the novel notion of being "cosmopolitan"—a word conjured up by the French humanist Guillaume Postel (d. 1581) in the face of rising Ottoman power.[77]

Just as the twin imperial legacies drawn from Rome and Mecca shaped Ottoman languages of universalism and dominion, they also informed Suʿūdī's appetite for conquest across the Atlantic. The primary audience for all of these aspirations was no less important. Rulers and ministers were the principal recipients of Piri Reis's materials and Suʿūdī's collection on the discovery of the New World. The language of conquest fits into wider commercial and military conflicts staged on a global scale.

To the east, the Indian Ocean and the Malay Archipelago had taken on a new significance in the preceding decades as a theater for Ottoman naval efforts to check Portuguese incursions into the Red Sea and along the eastern coast of Africa to vie for trade from the Spice Islands. The Ottoman military support of the Sultanate of Aceh in Sumatra during this period reflects the ideological and commercial ties binding the imperial center of Istanbul with eastern peripheries in the Indian Ocean.[78]

In addition to religious motivations, there were material considerations—namely, the massive booty to be had. With their mighty navy, the Ottomans maintained a near monopoly over commerce crossing the Mediterranean. They were also keenly aware of the economic implications of trade in the Atlantic. Suʿūdī's collection displays the aura of majestic abundance through the wealth, measured in spices, gems, and bodies, flowing out of the New World.

The account of the conquest of Mexico led by Hernán Cortés (d. 1547) directly raises the issue of gaining access to these riches. It concludes with a long inventory of the treasure seized upon the collapse of the Aztec Empire, measured in gold, silver, rare jewels, and native artifacts. The chronicler de Gómara records a petition sent along with the Aztec fortune on behalf of Cortés that more bishops, clerics, and priests be brought to New Spain, to oversee the conversion of the Indians. The letter ends with the request that recent converts, *tornadizos*—that is, New Christians—be banned from making the journey.[79]

The exclusion of Muslims from the enterprise of proselytization in the New World may have motivated the decision to drop the passage altogether from the Ottoman collection. For the translator, the immense wealth of the Aztec spoils is used as a rallying call to arms in decidedly imperial terms. Here, rather than citing Christian ecclesiastical authorities, the translation summons the missionary might of Islam, and implores the sultan to expand the holy war westward across the Atlantic, to take possession of all the treasures seized by the infidels, and to fill that bounteous land with *envār-ı şeʿāyer-i islāmiyye,* the light of Islamic religious rites.[80]

Notwithstanding these full-throated petitions, Murād did not chase the infidels to the western end of the world. Despite a crushing naval loss at the Battle of Lepanto off the shores of Greece in 979/1571 to a confederacy of Catholic states led by the Habsburg Empire, the Ottomans were able to rebuild their fleet. By the time Murād came to power, Ottoman dominance in the Mediterranean was largely restored. As a means of placing further economic pressure on his Catholic adversaries, Murād continued earlier Ottoman diplomatic efforts with Protestants and pursued commercial relations with

Elizabeth I (d. 1603), the Protestant Queen of England. During this period the Sublime Porte also first granted English merchants safe conduct and trading rights. This would form the basis for the charted Levant Company, which gave the English a lasting foothold in the East.[81]

For all the temptations of the New World, Murād did not direct his attention to the Atlantic, nor he did take up Suʿūdī's suggestion that the Ottomans build a canal between the Red Sea and the Mediterranean to strengthen their hold on the Indian Ocean and to quicken trade. Rather, the Ottomans engaged in what became a protracted and costly war against the Safavids, the rival Shia dynasty in neighboring Iran. The territorial gains won against fellow Muslims—generally branded as heretics for their adherence to Shia rituals and beliefs—were ultimately short lived. And with the definitive defeat of the admiral Mīr ʿAlī Beg at Mombasa in 997/1589, Ottoman naval power also failed to check the Portuguese dominance in the Indian Ocean and along the eastern coast of Africa. Ottoman aspirations for a seaborne empire in the East also soon faded. Yet the frenzied movement of bodies and ideas across seas and over land would only increase.[82]

9

ON THE EDGE

THE SOUND OF DRUMS KEEPS THE PACE, as an acrobat, contorted into knots, balances with perfect poise on a tightrope high above. Nearby a trainer sends a goat up onto a rickety platform, as a crowd below gathers, gasping and waving their arms. Around the corner two wrestlers locked in combat push and pull each other to the ground, one wielding a club, the other a stone. To the side, the commotion quiets down under the shade of a tree. At its base is a painting of Jesus surrounded by devotees—one embraces his feet, others exchange words of reverence. Not far off a mighty lord sits on a throne calmly taking it all in, one leg crossed over the other, with a pillow behind his back, under a royal canopy, surrounded by servants. With the fine precision of a master hand, these scenes come alive in the delicate work of *khatambandī,* the art of marquetry, each expression and gesture pieced together with paper-thin slices of ivory, inlaid on the shell of a tiny hazelnut.

The world-seeing eyes of connoisseurs had never come across anything like it, the precision, the artistry, the expanse of emotions and movements captured *mise en abyme* on something so small to behold. At least this is what a marginal note records with ekphrastic delight in a tiny hand, at times barely legible, on the side of a folio hidden deep in a royal copy of Qazwīnī's *Wonders and Rarities.* The manuscript had formed part of the private collection of Muḥammad ʿĀdil Shāh (d. 1066/1656), vassal ruler of the kingdom of Bijapur in the Deccan plateau of southwestern India. The margins of the presentation copy are lined with treatises, recipes, and anecdotes that were added to the manuscript by a later reader at some point near the beginning of the eighteenth century.

Qazwīnī's *Wonders and Rarities* traveled east across the same routes of prestige and learning that took Muslim elites through India. Along the way, it transformed into a distinctly Indian repository for history and cosmography. The languages of wonder and awe, for which Qazwīnī had become a concise emblem, took on new meanings in a land long associated with the strange and remote.

With claims to comprehensive knowledge, couched in the measured terms of curiosity and admiration, Qazwīnī could address imperial audiences and private aspirations. Decked in Indian attire, his natural history continued to open up a world of authoritative learning, offering a vehicle for absorbing new information, values, and sensibilities. This capaciousness was well suited to a land where notables of countless faiths and confessions intermingled, united by ideals of sovereignty, courtly disciplines of etiquette, and the refinement of Persian letters. Along the edges also came Christian merchants and missionaries from among the *farang*, bearing maps and paintings, some to seek fortunes, others to save souls. Across these contact zones, they would encounter unimaginable wealth, strange customs, and all varieties of enchantment.

As for the remarkable hazelnut referenced in the margins of the royal Bijapur copy of *Wonders and Rarities*, an artisan had gifted it to the powerful Mughal emperor Jahāngīr (d. 1037/1627), who was so astonished by the tour de force that he showered the craftsman with rupees. In his daily memoirs, Jahāngīr describes the nimble artistry of ivory work on the hazelnut as the utmost of rarities. Jahāngīr had inherited the tradition of imperial autobiographical writing from his paternal grandfather, Bābur (d. 937/1530), the founder of the Mughal dynasty in India.

Through Bābur, the royal family traced their lineage directly to Tīmūr and the dynasty founded in his name. The Mughals of India kept up the patronage of learning and art that had flourished under the Timurids, as did they exercise their own power through the language of universal sovereignty. Styling themselves as divine emperors ruling equitably over all faiths, the house of Tīmūr controlled from India an empire that stretched past the mountains of Afghanistan and up to the peaks of Kashmir, through the Indus valley along the Gangetic plain, spilling into Bengal, down across the western Ghats, and reaching into the Deccan plateau.[1]

Jahāngīr was both a collector and a hunter. Manuscripts, metalwork, globes, fine paintings, rare jewels—all passed through his hands, as did weighty matchlock muskets, which he prized for bringing down wild beasts and rare creatures. Traders sought fortunes in the court of the mighty emperor—whose name meant "world conqueror"—bearing gifts from distant lands. As with his father, Akbar (d. 1014/1605), many came to court to present themselves before Jahāngīr's splendid visage: Jesuit priests, Dutch and Portuguese merchants, Brahmin astrologers, painters and poets from Iran, scholars and philosophers both Sunni and Shia, Yogis, Sufis, dignitaries, jugglers, and acrobats. Songs in Persian and Hindi were recited over libations and feasts, before performances of skilled dancers.[2]

Then there was the embassy of Sir Thomas Roe (d. 1644), the first English ambassador to the Mughal court. While Jahāngīr poignantly neglects to mention the English mission in his own memoirs, Roe left a detailed account of his frustrated attempts to secure trading rights for the newly founded English trading factory in Surat. By the time he landed in India on the shores of Gujarat, Roe had already led an expedition for Prince Henry of Wales (d. 1612) through the West Indies and up the Amazon in search for riches and El Dorado, the fabled city that had captured the imagination of the early conquistadors. Securing the graces of the Great Mughal would prove no less elusive.

Christian missionaries and merchants often found the court to be exotic and confusing, a mirrored labyrinth of foreign etiquette, manners, and skills. These could be at once familiar, in the shared knowledge of Aristotle, Ptolemy, Galen, Avicenna, and Alexander the Great, and yet also utterly unrecognizable in the grand tolerance toward multiple faiths and creeds, the mastery of Persian letters and Indic cosmography, and the appeals to Turkic values and genealogies.

Though Roe would frequently speak of the wonders of India, the *farang* were themselves objects of bemusement and curiosity. Led through a pleasure palace en route to a banquet, a visitor might pause to take in murals of French kings and Christian princes paying obeisance. Roe could one day present a painting to Jahāngīr in the style of the European masters, only to see it the next day perfectly replicated by the emperor's artisans. The royal atelier had taken on a taste for the naturalism found in the prints and paintings peddled by Christian merchants who frequently boasted of their command over the mimetic arts. But Jahāngīr's artists also could comfortably reproduce the nativity of Jesus or the tender countenance of Mary.[3]

Bearing gifts of rare wine, fine paintings, and musical instruments, Roe presented himself as a courtier whose business was the pleasure of the court, hoping to secure trading interests by means of gifts of gem-encrusted goblets, potent liquor, and dainty feathered hats for the harem. But with no real mastery of Persian, the English embassy, which came equipped with a jester, priest, and painter, evoked the language of pleasure and wonder through objects or interpreters. With competition from the Dutch and Portuguese, as well as local middlemen, the market for exotica was already quite crowded.

Although Roe fails to make an appearance in Jahāngīr's autobiography, many other rare occurrences feature, such as a bearded woman, or the strange products and creatures brought to him from the Portuguese trading post of Goa. There was the delicious pineapple that the Mughals had begun to cultivate. Jahāngīr knew it as *annanās,* from the Portuguese *ananás,* a loanword imported from the Amazon. Then there was that motley colored bird. Jahāngīr ordered his painters to produce a likeness of one of these marvelous animals to include in his memoir—a turkey from the New World. Jahāngīr described the amazing creature evoking the language of wonder and rarity. The portrait was rendered with a striking attention to form. In the paintings he commissioned, Jahāngīr cultivated, if not strict naturalism, then a deep appreciation for fidelity and symbolic value. In this he shared with his father a fascination for capturing the veritable likeness of the world.[4]

The image of India and its ocean as populated by strange creatures, mysterious powers, rare gems, and spices was not unique to the far-off Franks, who had soaked up ancient writings about the marvels of the East and the adventures of Alexander the Great. It is an idea repeated throughout Arabic and Persian letters. Like Roe, who held rather exotic notions about Indians, Qazwīnī knew India to be a land of marvels and magicians. But this image found new relevance as Muslims in India—from native families and immigrant communities—encountered rishis, Yogis, and Brahmins skilled in astral divination, and confronted images and stories of ghosts, exorcisms, potent mantras, demonic *rākṣas,* and deities with multiple arms and heads.[5]

Cast in the web of the sky god Indra, the art of sorcery and deception, referred to in Sanskrit as *indrajāla,* literally "the net of Indra," had an ancient pedigree in Indian metaphysics. But just as wandering sadhus worked wonders, Muslim dervishes and saints could perform miracles. Scholars in madrasas and dargahs had ready explanations in natural terms for these worlds of magic and miracle in the metaphysical teachings of Ibn Sīnā and in the immensely influential writings of Ibn al-ʿArabī, the great philosopher and mystic of Murcia.

The heirs of Ibn al-ʿArabī took his argument that "existence is one unitary reality [*inna l-wujūda ḥaqīqatun wāḥidatun*]" and developed it into a full-fledged theory focused on the unity of being, referred to in Arabic as *waḥdat al-wujūd.* The expression turns the philosophical doctrine of the necessary existence of God—encapsulated in the phrase *wājib al-wujūd*—into a gnostic position on the ultimate reality of all existence. This theory, which featured in the development of Islamic metaphysics in India and beyond, was also wedded to the ideal of a perfect human, *al-insān al-kāmil,* as the culmination of creation and as an archetype for prophets and saints. As a philosophical position, the theory of unitary being collapsed the perceived distance separating God and creation into an emanationist form of monist reality. The metaphysics of unity also offered a further basis for fathoming magic and miracle as expressions of the divine made manifest in the material world, harnessed through spiritual stages of gnosis.[6]

In the lands of Hindustan, Qazwīnī's *Wonders and Rarities* was embraced with open arms. While his natural history of wonders circulated in Arabic, Persian, and Turkish, the sheer number of manuscripts in India is astounding. Copies are found in Lahore, Srinagar, Delhi, Aligarh, Agra, Rampur, Lucknow, Allahabad, Deoband, Patna, Calcutta, Hyderabad, Bijapur, Mumbai, Jodhpur, Jaipur—the list goes on. Most were made locally. Although Arabic manuscripts appear, the vast majority circulated in Persian. This proliferation is itself a testament to larger transformations in book culture that followed an expanding circle of readers of Arabic, Persian, and Turkish manuscripts across a vast stretch of land—from the far reaches of Africa to the distant shores of Asia.[7]

Some copies of Qazwīnī's natural history in India were modest, produced by rushed hands without illustrations; others were opulent works of art for powerful patrons who could pay a premium to have painters illuminate and gild each folio. Many of the paintings in the Persian translations exhibit local styles and vernacular expressions, whereas the Arabic manuscripts tend to hew more closely to earlier models that had long circulated outside of India. Fine presentation copies were also imported from Iran and Central Asia, such as an illuminated manuscript once kept in the royal library of the Timurid emperor Shāhrukh (d. 850/1447) that bears the personal seal of Amānat Khān Shīrāzī (d. ca. 1054/1644), émigré to the Mughal court, whose calligraphy adorns the Taj Mahal.[8]

Over the centuries, as tastes and knowledge ebbed and flowed, and the number of readers grew, Qazwīnī's *Wonders and Rarities* continued to travel. Along the way, it was copied and imitated, transformed, and adapted, illuminated for royal treasuries and put up for sale in the quick movement of markets, at times bending to new styles, at others cleaving close to earlier forms. It traveled the same routes taken by physicians, pilgrims, dignitaries, and merchants by land and sea, connecting India—a region renowned for unfathomable riches—to an expansive network of cities, ports, and villages.

In this movement of bodies and ideas, the distant realms of the *farang* had for many remained largely out of sight, a strange and insignificant land of semi-barbarians at the edge of the known world. In due course this would profoundly change. But through it all, the worlds of the strange and uncanny evoked in books of natural wonders, with the emotive and conceptual frameworks that made them intelligible, continued to find ready audiences and generous patrons.[9]

With the power to behold the entire Earth, wonder was the coin of powerful emperors and mighty lords, as it was for mystics and preachers who could fashion careers in appeals to the uncustomary. In 1022/1613, two years before Roe set out to negotiate a treatise for the shareholders of the newly minted

East India Company, the governor of Khurasan in the Afghan city of Herat commissioned a sumptuous manuscript of poetry, illuminated with gold and delicate paintings. The narrative poem was an adaptation of Qazwīnī's *Wonders and Rarities* woven with Sufi precepts of piety in the form of a *masnavī*, the preeminent genre of Persian verse for narrative, allegorical, and mystical poetry. The patron of the collection was a Turkman governor of Herat in the service of the Safavids, the powerful Shia dynasty centered in Iran, which had risen to power in the wake of Timurid fragmentation.

The painters of Herat, who enjoyed the patronage of Tīmūr and his descendants, had long been renowned for their technical artistry—frequently imitated but seldom surpassed. They had taken on the canonical poems of Persian letters—Firdawsī's *Book of Kings*, Niẓāmī's *Quintet*, Ḥāfiẓ's collection of verse—opening up spellbinding vistas with legend, allegory, and unrivaled sophistication. Several artists from Herat made their way into the Mughal atelier, celebrated for their ability to conjure enchanted landscapes populated with undulating peaks, wispy clouds, and plaintive trees.

In the distance along the borders of the paintings are perched birds and mountain ibexes, as multicolored leaves evoke natural fascination, inviting beyond the storied palaces, citadels, and gentle rivers that foreground the main action. The buildings are decorated with geometrical designs, tapestries adorned with images of delicate animals, and patterned carpets. The colors and strokes come together promising, on one folio, pleasure gardens, fine music, food, and sweet wine, and on the next, the brutal savagery of demons and warriors. Subtly set within are instructions on how to react before it all, found in the expressions of perplexity and astonishment on the faces of onlookers set at the edge of the action. The gesture of their hands, often referred to in Persian as the *angusht-i taʿajjub*, the finger of astonishment, points at once to the mouth in awe and to the very matter that is itself the wonder to behold.

The *masnavī* adaptation of Qazwīnī deposited in the royal archive of Herat was originally penned by the itinerant Sufi poet Fakhr al-Dīn Āzarī (d. 866/1472), who had gained the favor of emperor Shāhrukh, son of Tīmūr. Āzarī's travels took him through Iran to Mecca and then to India, where he served as poet laureate to the court of Aḥmad Shāh Bahmanī (d. 838/1435) in the Deccan, before returning to his hometown of Isfarā'īn in Khurasan.[10]

Composed as a poetic response, the adaptation loosely follows *Wonders and Rarities*, using it as an inspiration rather than as a basis for a close paraphrase.

The poem moves from the Earth up to the heavens, inverting Qazwīnī's order. The crisp couplets also weave together a didactic argument for pious reverence in the face of divine sublimity, drawing on the classics of Persian verse, all quite absent from the original work. Along the way can be heard references to the *Manṭiq al-ṭayr* (The conference of the birds), a narrative allegory of divine union by Farīd al-Dīn ʿAṭṭār (d. ca. 617/1220), and to the famed *masnavī* of Jalāl al-Dīn Rūmī (d. 672/1273), who was known throughout Persian letters by the simple honorific *mawlānā*, "our master."

But through it all, echoes of Qazwīnī can be heard, such as the definition of wonder as a perplexity that arises from human incapacity and of rarity as a seldom-occurring phenomenon. The poem frequently turns to magic and talismans in terms that give a good sense of how Qazwīnī was read over time. These themes are put into the language of piety, focused on the majesty of God's divine wonders, from the mountains of Qāf to the movement of the heavens. Describing sorcery and alchemy as nothing without divine power, merely the "play of children," Āzarī addresses God directly:

> The alchemy of magic and spells
> is a highway robber on the path of liberation
> Your being is the talisman sealing the treasury of existence
> Without You there is no alchemy.

Along the Sufi path, awe and perplexity offer a means for obtaining a deeper state of gnosis. "Those rarities are all pointers and those wonders are all allusions," Āzarī explains, for they are lessons that lead to higher truth in the marrow of divine purity and eternal speech. As the paintings in the royal manuscript of Herat demonstrate, the path of knowledge is filled with fine wine, music, and entertainment, the tree of Wāqwāq, string-legged creatures, the wall against Gog and Magog, animals copulating, beasts in battle, and Alexander's mission in the encircling ocean (figure 9.1).[11]

Āzarī's poem also circulated widely, across Iran into Ottoman lands, and throughout India, and was often lavishly illustrated. There continued an enduring appetite for natural histories. Throughout the Timurid period, illuminated copies of Qazwīnī's *Wonders and Rarities* in Persian were produced in commercial workshops of Iran, particularly in Shiraz and Tabriz, as well as in major cities of Central Asia. These exhibit a great variety and adaptation,

FIGURE 9.1: Ship sent by Alexander into the encircling ocean encounters another boat, with two spectators looking on from a balcony set into the scene from Āzarī's poetic adaptation of Qazwīnī's *Wonders*, copied 1022/1613 in Herat.

appealing to the tastes of local notables and patrons, many well outside the inner circles of the ruling elite. Some took a greater interest in anatomy and the stars, others gravitated toward demons, jinn, talismans, and sex magic.[12]

By the time the illuminated copy of Āzarī's poem was presented to the royal archive of the Safavid governor in Herat, Qazwīnī's *Wonders and Rarities* had been a staple in the city. At this point Herat was in the hands of Shāh ʿAbbās (d. 1038/1629), the powerful ruler of the Safavid dynasty who was vying with Jahāngīr for control over Afghanistan. Even though Shāh ʿAbbās was himself a native of Herat, he set up the capital of his dynasty in Isfahan, the heartland of Persia.

After staving off Ottoman encroachment to the west, Shāh ʿAbbās turned his attention to Mughal control over the Afghan city of Kandahar. While Jahāngīr makes no mention of the English mission, he spends a good deal of time on the diplomatic exchanges with his "brother" of Iran, whom he viewed as a lesser rival. Among the rare gifts received from the shah of Iran was a painting of Tīmūr, Jahāngīr's forefather, done in a style that looked as though it had been executed by none other than Bihzād (d. ca. 942/1535), the master of Herat.[13]

In his own diplomatic mission to the Safavid capital in Isfahan, Jahāngīr had sent a court painter to render the likeness of his rival. One of the symbolically charged paintings, which Jahāngīr commissioned for an album before the Safavids ultimately captured Kandahar, takes on cosmic proportions. Here Shāh ʿAbbās humbly seeks the great emperor's embrace. The two stand on top of a globe, a lion under Jahāngīr's feet and a radiant sun of divine glory behind him, with his brother above a meek sheep, seeking succor and protection (figure 9.2a).

The map of the globe renders coastlines and mountain peaks with delicate detail. Toponyms of numerous cities and regions spread across the contours of the map in a minuscule hand, which required a patient and focused eye to read. Most of the attention is given to India: Agra, Delhi, Surat, Bijapur, Karnataka, Golconda, Ceylon, moving west past Lahore, Multan, Tabriz,

FIGURE 9.2A AND 9.2B: Jahāngīr embraces Shāh ʿAbbās, standing on the globe. The Wall of China can be made out as well as the continent of North America in the far northeast. Painted circa 1618.

Baghdad, the honored city of Mecca, Cairo, the Ḥabash and Zanjibar beneath the Nile, the Berbers along the north coast of Africa, and jungle dwellers, *jangaliyān,* in the south toward the equator. To the north are Portugal, France, Venice, the Majār of Hungary. Moving eastward across the high steppe are the Qipchāq and Qalmāk confederations, and then finally the kingdom of China, separated by a giant wall from the northern Khitai (figure 9.2b).[14]

Among the many gifts that Roe presented to Jahāngīr was a finely gilded atlas by the Dutch cartographer Gerardus Mercator (d. 1594). Roe mentions that he had purchased the latest version to take with him to India. This would have been the enlarged edition published in Amsterdam in 1613 by Jodocus Hondius (d. 1612), who expanded upon the plates of Mercator's original cartographical projections. The map of China by Hondius has a feature that had become common to European cartography of the period—a depiction of the Wall of China (figure 9.3). Of the countless

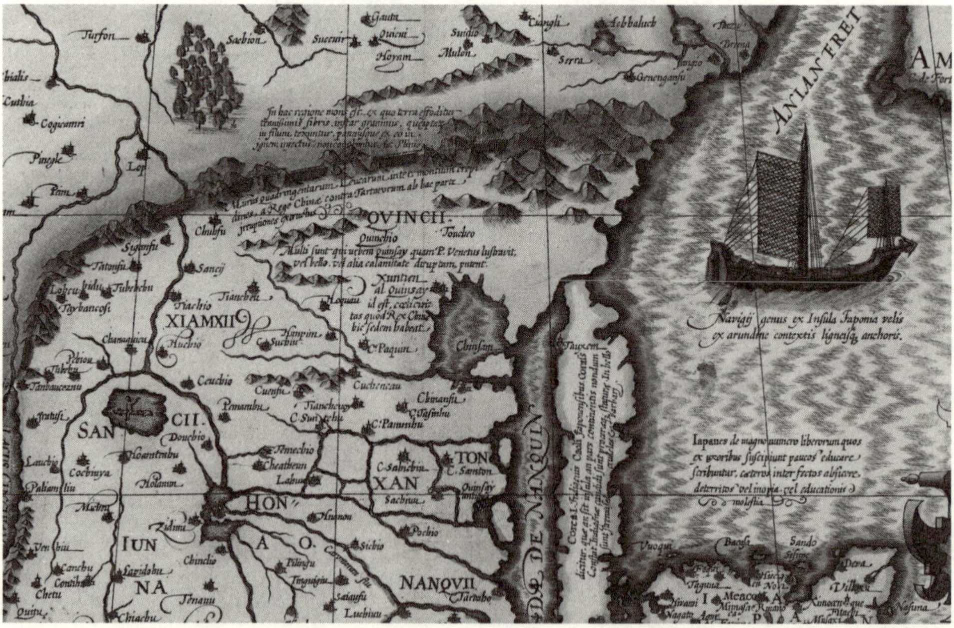

FIGURE 9.3: Detail of the wall separating China from the Tartars, in a sectional map of China from the atlas of Gerard Mercator and Jodocus Hondius, printed in 1613. The upper right-hand corner of the full map also features a sliver of North America.

images and place names that appear in the Dutch atlas, this landmark evidently caught the attention of the painters in Jahāngīr's court. As with the globe beneath the emperor's feet, which has a giant landmass residing along the northeastern horizon, Hondius depicts a vast land in the upper northeast corner of the map, which he identifies as belonging to America.[15]

Roe commented that none of the religious scholars, whom he refers to as *mulaies,* in the court could understand the atlas. Jahāngīr had it returned to him. With their new grand vistas obtained through the maritime expeditions across the world, Christian missionaries and merchants hoped to both awe and diminish the potentates of the East whom they encountered. Yet the Mughal painters had clearly mastered the message of representing and possessing the entire globe, which became a trope throughout the paintings of the period. Earlier maps of the world brought in from Jesuit priests had already made their way as gifts in Akbar's court.[16]

In addition to the symbol of the globe in albums, the court had an appetite for mural paintings. Already in Jahāngīr's day Mughal artists produced a map of the two hemispheres of the world that could cover an entire wall. Along with strange aquatic creatures, here too a giant barrier at the edge of the globe can be made out. As with other acts of repurposing the mimetic capital imported from the distant West, these new ways of reading the world were coded in the dominant modes of classical Islamic thought. The massive gates at the edge of the globe could at once could be the Wall of China, as it was depicted in map books drawn from the early Jesuit missions to the East, as it could be the barrier against Gog and Magog, long known to Muslim cartographers.[17]

Jahāngīr also commissioned allegorical paintings of himself standing on top of the globe, facing down foes both real and imagined. The topographies are loosely inspired by the new maps of the world, while also drawing on ancient cosmic orders. Among the phantasmagoric lands populated by a menagerie of tame beasts, in one such painting the loose outline of the Eastern Hemisphere can be discerned. But here as well, though, are pillars of the Earth supported on the shoulders of a giant ox resting on the back of a leviathan. As with its adaptation in the Ottoman court to suit new cartographic information, this vision of the globe was both pliable and immediately intelligible to the Mughal elite as a means of conceiving the world in cosmic proportions.[18]

For Jahāngīr and those who attended his assemblies, the new worlds that the *farang* spoke of to the far west would not have been particularly novel. During the reign of Jahāngīr's father a general idea had already formed of explorations at the edge of the globe. As with their missions to China and Japan, the Jesuits at Akbar's court also brought along maps and atlases to win over heathens to the true Christian faith. Knowledge of geography might not have secured many converts, but it could certainly be easily adapted and repurposed.

Akbar's court chronicler, Abū l-Fazl ʿAllāmī (d. 1010/1602), offers a telling account in the *Āʾīn-i Akbarī* (Institutes of Akbar), a gazetteer of the Mughal Empire that included an entire geography of the world, commissioned as part of the royal chronicle of Akbar's reign. After describing the inhabited world as it was known to him, Abū l-Fazl turns to a story from Alexander's adventures as a preface to his discussion of the New World. It echoes closely the account of Alexander's journeys through the encircling ocean that Qazwīnī popularized.

Having conquered the northern quarter of the globe, Alexander sought information about the other quarters and seas. So he supplied a ship for a six-month journey. Eventually, Abū l-Fazl explains, Alexander's party came upon a boat commanded by a people with an unintelligible language. A fight ensued, and Alexander, victorious, had his men marry some of the female captives. As the children of these marriages came to speak both languages, Alexander was able to learn that the people he encountered on the high seas had come from a distant kingdom beyond the edges of the known world.[19]

Abū l-Fazl then makes the following observation: "Today, authoritative scholars hold that like the north, the southern part of the globe is also inhabited. Recently, the *farang* seized an extremely vast island, widely populated, in the direction of the south, which they have termed a New World.'" It happened that a ship lost at sea was driven there. The inhabitants of that land saw a rider come into view. "Taking the man and the horse to be one large creature, they were overcome with trepidation and for this reason their kingdom was quickly captured." The reference echoes a story circulated in the earliest Iberian conquest narratives that the natives at first contact mistook conquistadors mounted on horseback for deities.[20]

Abū l-Fazl's information derives from Jesuits and other Portuguese traders at Akbar's court. This helps explain the identification of the New World with the south, given the Portuguese focus on the southern half of the Western Hemisphere. It also stands in direct opposition to the accepted theory that no human habitation could exist in the southern regions of the Earth. Just as Suʿūdī did in the Ottoman court, Abū l-Fazl turned to the adventures of Alexander as a means to account for ancient contacts with this strange new land. Both Abū l-Fazl and Suʿūdī wrote for royal audiences during the same generation. A vast distance separated them, stretching from north India to the shores of the Bosporus. Yet they shared stories of Alexander's exploits, drawn through sweeping poetic romances, books of natural wonders, and histories of the world.

The anecdote of contact along the rim of the world serves to confirm that the great conqueror, Alexander—long a paradigm of Islamic learning and kingship—truly came to know of the inhabitants of the entire globe. It builds on the long-held belief, known to Qazwīnī and countless others, that Alexander, of ancient Persian stock, pupil of Aristotle, traveled through the lands of darkness and reached the setting Sun in the west, while conquering Rome, China, the Turks, and India, acquiring along the way knowledge of the unique hidden properties spread throughout creation.[21]

Like the Ottoman sultans, Akbar also drew on Alexander as a model of imperial rule. Muslim emperors and princes had long fashioned themselves as a second Alexander or Solomon, who through cosmic, divinely granted power were able to unlock the secrets of nature and rule equitably with wisdom and justice over vast territories. Yet unlike the Ottomans, Akbar could claim that he followed a long line of mighty men who, like Alexander, had made their way to India. It is a list that Abū l-Fazl draws back to Adam, the progenitor of all humankind, who landed on the Indian island of Sarandīb after having been expelled from the garden of Eden.

In contrast to the faraway sultans of Istanbul, Akbar could directly converse with Brahmins just as Alexander was famously known to have done. The story of Alexander's dialogues with Indian sages forms a centerpiece of his adventures in the East, as told in the earliest writings on his exploits in Arabic and Persian letters. The dialogues trace back to antiquity with Greek accounts of Alexander's exchanges with the *gymnosophistai,* the naked wise men of India. Akbar knew the story well, as it was repeated both in the classical sources that circulated in his court and in the chronicles produced specifically for him.[22]

Another idea that Akbar would have heard was the counsel that Aristotle gave to Alexander. Just emperors and perfect guardians were responsible for harmoniously ministering to the affairs of everyone—*umūr-i maṣāliḥ-i jumhūr.* To rule justly, emperors had to obtain a reasonable degree of certainty about any given matter before passing judgment, for they held the station of divine determination and decree.

The Arabic word *maṣāliḥ* has the sense of affairs, matters, or personal interest as the basis of general well-being and social good, connected to the notion of *iṣlāḥ,* rectification, amelioration, repair, and *ṣāliḥ,* the good, upright, and sound. Just as *maṣāliḥ* conveys these meanings in Persian, like many other Arabic words it takes on further valances. Connected with the sense of balance and harmony, the word also comes to convey beneficial compounds, agents, or ingredients in drugs that could restore health, and in this sense it is the basis for the modern English word *masala* derived from Hindustani, as in the spices that give food its perfect taste. This notion of *maṣāliḥ,* and with it the ideals of universal justice and knowledge, famously animate the Mughal ideology of magnanimity, civility, and tolerance toward other religions. This is captured in the expression *ṣulḥ-i kull,* which conveyed the sense of universal harmony and equilibrium. Akbar promoted this policy in the face of the staggering diversity of beliefs, practices, languages, and customs within his dominions.[23]

Like many royals before him, Akbar cultivated a taste for curiosity in matters worldly and divine. According to the courtier ʿAbd al-Qādir Badāʾūnī (d. ca. 1024/1615), royal assemblies were held where learned scholars and authorities of various faiths would join in conversation, discussing and investigating day and night the secrets of the sciences, the fine points of wisdom, with wonders reported and rarities related that could fill volumes. Strange gifts were highly sought after, such as a musical organ imported from the lands of the *farang*—one of the very marvels of creation—whose melodic workings captivated the court.[24]

For Akbar a key facet to universal rule was the cultivation of encyclopedic knowledge. This included the translation of ancient Sanskrit texts. Among the most ambitious projects that Akbar commissioned were the multivolume

Persian translations of the *Mahābhārata* and the *Rāmāyana*. These works were collaborative efforts that employed the skills of Brahmin pundits alongside high-ranking Persian litterateurs. As with all acts of translation, the sustained engagement with Sanskrit texts required significant interpretive leaps and creative acts of adaptation.

The vocabulary of wonder, astonishment, perplexity, and awe greet the feats of deities and demons that line the pantheon of Sanskrit letters, told in Persian with attention to talismans, sorcerers, and magical oaths. The kaleidoscopic picture of the cosmos in the *Razm-nāma* (The book of combat), the Persian title given to the sprawling epic of the *Mahābhārata,* is rendered intelligible through coordinates of equivalence. Beyond seas of milk, honey, and oil lies the familiar boundary of the *daryā-yi muḥīṭ,* the encircling ocean, an equivalence that could reasonably be inferred from the Sanskrit *sāgara.* Though *sapta varṣāṇi,* the seven regions of the Earth, unfold across a foreign geography of strange islands and far-off regions, the division is entirely legible, rendered as the seven climes taught in classical Islamic geography.

Just as Akbar knew that the wise saint Khiḍr joined Alexander on a quest for the waters of immortality hidden in the high peaks of Qāf, the Sanskrit epic speaks of these same mystical mountains. There is Kailash, the abode of the deities, and a pool of nectar, *amṛta,* and further beyond is Meru, the pillar for the entire cosmos. The Persian translation often blends into commentary. It describes these mountains as home to countless *ʿajāʾib o-gharāʾib,* wonders and rarities, and adroitly identifies the entire cosmic structure with the mountains of Qāf, home to the waters of life.[25]

Across these pages are countless counsels to kings instructing universal ethics through penance and purification, with the promise of divine knowledge and power. The core teachings of the Indian sages follow the path of monotheism, *tawḥīd,* and gnosis, *maʿrifat,* and demonstrate that spiritual liberation, *najāt,* is itself an expression of divine sovereignty. The techniques of liberation are frequently conveyed through tales of the strange and marvelous. Many of the stories feature creatures that would have been recognizable to the readers of Qazwīnī's cosmography.

Here is Garuda in the form of a giant eagle that can fly off with both an elephant and a giant tortoise in its talons. Later on are strange populations of savages with ears large enough to serve as blankets, a whole people who all hop about with only one leg, and a kingdom populated only by women. Similar accounts make their way to Qazwīnī, but from earlier Arabic sources and not directly from the *Mahābhārata,* reflecting an ancient tapestry of strange people and creatures. The value of astonishment and wonder as an aesthetic basis for cognition also finds important expressions in the history of Sanskrit letters. In the Mughal context this all could fit perfectly into a repertoire that turned to an appreciation of the strange and uncanny as a means for cultivating ethical perfection, holding the mirror of divine sovereignty up to all creation in its delightful heterogeneity.[26]

The presentation copies of the Persian translations of Sanskrit teachings were illustrated with lavish paintings that called upon those who leafed through them to behold in awe the unlocked secrets of ancient Indian knowledge. Akbar's painters also turned to Bābur's autobiography, translated into Persian from the original Chaghatai. They showcased battles and adventures, as well as the natural beauty of India, with fine illustrations of the flora and fauna found in Hindustan. Qazwīnī's book of natural wonders offered an obvious inspiration.

The same atelier that painted the sage Vasiṣṭha before the deity Rāma wrapped in serpents also produced a large collection of astrological images used for divination that drew from the *Hidden Secret,* Rāzī's manual of astral magic. Populated with demons, strange creatures, and images of the zodiac, this collection of images of astral degrees was used for predicting the future. Above all, it highlights that the visual realm of painting could serve as a means, not only for appreciating the various signs of creation, but also for learning how to harness them.[27]

Far and wide, Muslim courts continued to cultivate various forms of astral knowledge. Framed as an inquiry into the wonders of God's creations, Rāzī's book of astral divination and talismanic magic was quickly translated into Persian for the ruler of the Delhi Sultanate, the "right hand of the caliph," Abū l-Muẓaffar Iltutmish (d. 633/1236). Centuries later, after the Maragha observatory had all but disappeared, Ulugh Beg (d. 853/1449), grandson of Tīmūr, funded a massive observatory complex in Samarqand, which picked up the work of Naṣīr al-Dīn Ṭūsī and his cohort.

The astronomical observations produced in Samarqand for Ulugh Beg circulated in India and served as the basis for further calculations. The Ottoman sultan Murād III, an aficionado of talismans, astrology, and New World lore, also founded an imperial observatory in Istanbul, which despite its size was rather short-lived. But in India the body of astronomical calculations, which stretched back to the work of Maragha and Samarqand, continued to be refined well into the eighteenth century with a series of observatories established throughout north India for the Mughal emperor Muḥammad Shāh (d. 1161/1748) by the Rajput vassal ruler Maharaja Jai Singh (d. 1743).[28]

Hindu astrologers and the art of *jyotiṣa,* the preeminent field of Indic astral science, were well known to Muslim rulers and their courtiers in Hindustan. Shortly before Akbar commissioned translations of Sanskrit texts and Persian works of astral learning, the regional ruler of Bijapur, ʿAlī ʿĀdil Shāh (d. 987/1579), had occupied himself with similar pursuits. These interests are on full display in the *Nujūm al-ʿulūm* (The stars of the sciences), a sumptuous, illuminated encyclopedia produced in his court.

Although the manuscript record of the collection is fragmentary, a royal copy dated 978/1570 preserves large portions of the Persian compendium, which focuses on magic and the astral sciences as well as the arts of war and animal husbandry. The work draws from a body of knowledge deeply connected to Indian cosmography, portions of which appear to have been written by ʿAlī ʿĀdil Shāh himself. Like his grandfather and great-grandfather, ʿAlī ʿĀdil Shāh was also a Shia and had the names of the twelve Imams publicly recited in the Friday sermons. But interest in cosmic knowledge and the occult sciences transcended sectarian professions or religious commitments.[29]

In addition to a range of Indic material, *The Stars of the Sciences* also showcases an entire treatise on jinn magic by the son of the famed scholar and statesman Sirāj al-Dīn al-Sakkākī. Writings associated with Sakkākī circulated in India in both Persian and Arabic, as did Ṭabasī's book of spells, Qurṭubī's *Aim of the Sage,* and the *Treasury* of Alexander. In the Bijapur manuscript, this section on subjugating astral and planetary spirits, demons, jinn, and angels illuminates in lavish detail the art of casting a *mandal,* the sorcerer's circle (figure 9.4). The cycle of paintings depicts an enchanter, referred to as the master of the summons or assembly, at work. In the illustrations the master conjures spirits while seated within a *mandal,* reciting sacred formulas, having prepared various recipes to be thrown into a sacred fire, with angels and demons standing outside the magical boundaries of the diagram waiting to be commanded for any purpose.

These illustrations closely follow the directions provided in the text itself. The spells are often noted as proven to be efficacious. In addition to drawing

FIGURE 9.4: Casting a *mandal* to subjugate Saturn, as represented by a multi-armed deity mounted on an elephant, *Nujūm al-ʿulūm,* copied 978/1570, Bijapur, India.

the *mandal* in the form of concentric circles or rectangular patterns, incantations to bind spirits usually involve the recitation of various sacred phrases known for their blessings, taken from the Quran, the names of God, and other pious supplications, the invocation of strange demonic and angelic names, the sacrifice of various animals, the suffumigation of incense, medicaments, and intoxicants, the use and placement of specific tools, and the writing of ciphers and seals, generally to be conducted at specific astral timings.[30]

Many of the terms and concepts in Islamic magic betray ancient roots with exotic origins. The term *mandal,* which Ṭabasī, Rāzī, and Sakkākī use as a term of art for the sorcerer's circle, is from the Sanskrit *maṇḍala,* meaning a disk, circle, halo, orbit, and, by extension, a region, both cosmic and terrestrial, connected also to the zodiac and used as a general label for cosmic diagrams prominent in various Buddhist, Jain, and Hindu ritual practices for summoning deities and astral powers.[31]

The art of casting a *mandal* had long been calibrated in the language of Islamic piety as an efficacious technique for harnessing spirits, well before Muslim religious authorities crossed the Indian subcontinent and reintroduced it in a form entirely shaped by the language of Solomonic magic and Quranic theurgy. When Muslim authorities in South Asia began to draw extensively on esoteric dimensions of Yoga and Tantra, both before and after the production of such texts as *The Stars of the Sciences,* they already possessed a vocabulary of astral knowledge and occult power that had been drawn from Indian learning centuries earlier. The Bijapur compendium for ʿAlī ʿĀdil Shāh seamlessly juxtaposes recipes for subjugating spirits by Sakkākī's son with astral diagrams and charts taken from Indic cosmography, highlighting a view of astral knowledge as a universal field of learning that, like practical philosophy and the physical sciences, transcended sectarian commitments.[32]

The array of topics promised in the introduction to the fragmentary compendium is breathtaking: astral science and the structure of the cosmos, talismans, elemental magic, Persian and Hindi spells, alchemy, sorcery, invocations to God and angels for summoning spirits, mechanical devices, agriculture, medicine, cures, poisons, music, the art of Yogis, the statecraft of kings and ministers, the deceptions of spies, the wonders of the regions, physiognomy and the various types of people in the world with their religions and customs, love and poetry, the powers of precious stones and minerals, Quranic exegesis, hadith, arithmetic, grammar, the astrolabe, the chancellery arts, storytelling, the vocabularies of Arabic, Persian, Turkish, Hindi, Sanskrit, *farangī,* and the local Kannada and Telugu, calligraphy, catalogues of swords, daggers, maces, and muskets, the care of horses and of elephants. As the list goes on, so do appeals to wonder and rarity. In addition to references to Hermes, there is Apollonius and an entire treatise in his name on the talismanic arts, with a cycle of images detailing how to fashion charms and spells.[33]

"I was a hidden treasure and wanted to be known, so I created creation." The collection opens its treatment of the heavens with this widely quoted Arabic expression spoken in the voice of God and woven into countless writings on philosophy and mysticism. The seven heavens are populated with angels of various shapes and sizes, the throne of God, and the divine pen, which encodes all being in cosmic letters, a notion of language and writing alive in the occult sciences of the day. The heavens are surrounded by an encircling orb, the sphere of all spheres, known as *aṭlas,* which sets it all in motion.[34]

As with the rich iconography that illuminates its pages, the compendium presents a picture of the world that is drawn from an array of sources. Large sections build on Sanskrit astrological and divinatory writings. Detailed instructions are offered to determine the influence of the heavens on any particular act or object. Specific devices are used, in the form of diagrams and charts, referred to *chakr,* from the Sanskrit *cakram.* These diagrams are drawn from techniques at the heart of Tantric philosophy. This material intersects directly with an influential collection of the astral sciences of Tantra by the Jain scholar Narapati (fl. 1177) from Gujarat, which in turn circulated in multiple Indian languages.[35]

Yet a close reading of the collection also reveals Qazwīnī's unmistakable mark. There is the pronounced emotive appeal to wonder and astonishment as a part of a larger process for obtaining ever greater stages of knowledge. There are the images of divine beings and earthly creatures with immediate analogues in the cycle of paintings that made illuminated books of natural wonders so famous. Then there are entire passages on the structure of the cosmos that are drawn word for word from Qazwīnī's *Wonders and Rarities.* Not only does the visual register of angels, planets, and the cycle of northern and southern constellations directly overlap, but *The Stars of the Sciences* incorporates large passages copied from Qazwīnī, and cites *Wonders and Rarities* as a primary source. In this broad array of material, Qazwīnī's collection sits comfortably alongside Indic and Islamic bodies of learning, from the visual cycle of astral degrees used for divination, the *mandals* for incantations, and the talismanic charms of Apollonius, to the cosmic creatures, demons, and deities from the universe of Tantric science.[36]

By this point Qazwīnī's cosmography had circulated in multiple copies across the Deccan, in both Arabic and Persian. Yet the Persian text used for ʿAlī ʿĀdil Shāh's ambitious compendium was unique to the court of Bijapur. The compilers of the encyclopedia needed to look no farther than to the translation that had been commissioned in 954/1547 by his father, Sultan Ibrāhīm ʿĀdil Shāh (d. 965/1558).

Ibrāhīm had publicly embraced Sunni rituals and jurisprudence, unlike his son, father, or grandfather, the founder of the dynasty. This was a significant departure, for the court had important diplomatic alliances with the Safavids, the major Shia dynasty in Iran. So did Ibrāhīm have to contend with many high-ranking Shia notables from emigrant families who were embedded in the administration and the military.

But there were many factions in Bijapur to address. These included a local Sunni aristocracy with strong ties to influential pietistic movements in the region. In the Deccan the Chishtī mystical order had grown particularly powerful with their brand of universalism drawn from the teachings of the local saint Sayyid Muḥammad Ḥusaynī (d. 825/1422), known by the honorific Banda-navāz Gīsū-darāz, or "the generous one with long forelocks," a preeminent interpreter of Ibn al-ʿArabī in Persian. Populating the court in Bijapur were also Abyssinian mercenaries and slaves, Persian and Turkmen officials who had migrated from Iran and Central Asia, Brahmin functionaries, and a strong force of Maratha cavalry.[37]

Even as the ʿĀdil Shāhī dynasty expanded and consolidated power over rival Muslim potentates in the Deccan plateau, they faced the threat of insurrections by local Hindu Rajas, internal factions, and the increasing power of the Mughals to the north. Then there was the additional menace of the Portuguese on the coast, who had first reached Calicut in 1498 with the naval expedition of Vasco da Gama (d. 1524) to uncover a route to the Indies by circumnavigating Africa. For many of these voyages of conquest and plunder, native informants and local pilots proved indispensable. In the following years the Portuguese Estado da Índia centered in Goa had established a series of fortifications and outposts along the western coast that formed part of a vast maritime empire.[38]

As with other Muslim dynasties, the sultans of Bijapur sponsored lavish illuminated manuscripts and paintings, as well as grand public works proclaiming their power and magnificence, with aqueducts, congregational mosques, shrines and tombs, citadels and battlements, splendid palaces and dazzling gardens. In the same year that the new translation of Qazwīnī's *Wonders and Rarities* was presented to the "feet of the imperial throne," an ornate inscription on a royal water tank in the Panhala fort of Maharashtra was completed in the name of Ibrāhīm ʿĀdil Shāh, who boasts sovereignty over the world. The inscription is adorned with a quatrain by the celebrated Persian sage ʿUmar Khayyām (d. 517/1123), from far-off Nishapur, which raises a goblet to life's pleasures:

> In the heart a tree of grief mustn't sustain,
> A book of joy should be ever recited,
> wine to be drunk, the heart's desire sated,
> How so little in the world will remain.[39]

Qazwīnī's *Wonders and Rarities* would certainly have been a joyous book to behold in a garden of earthly pleasures. But not a hint of temporal transience, suggested so frequently in Khayyām's poetry, greets the sultan in the opening dedication to the Persian translation commissioned in his name. Rather Ibrāhīm ʿĀdil Shāh is feted with the orotund prolixity that was then quite fashionable in Persian letters.

After a florid benediction praising God and His messenger, the translation turns to the royal patron with an ornate tribute that mixes Persian and Arabic in rhymed prose:

> The mighty emperor, refuge of the world, celestial orb of the court of the wise sultanate and caliphate, the shadow of God on the Earth, the protectorate of God among His creation, the caliph of God's world, the just, excellent, perfect, powerful, noble sultan, served by the lords of the kingdom, the glorious and commanding, the sword and the pen, reviver of the prophetic customs, guardian of the mighty law of the Chosen Prophet, the straight balance of order and cohesion, clasp of all matters encompassed and fathomed, he who raises the standards of justice and equity, and sets down the laws with rectitude and integrity, through whom God lifts up the banners of the Islamic empire as a grace to all of His creation, and as blessing upon them, of the highest station, who strips from the admixture of time's passage all matter of strife and corruption with the strike of his sword and the piercing edge of his pen, possessor of sacred soul and all form of human excellence, wise master of affairs of the sultanate and caliphate through the secrets of the literal and metaphoric, known as the Friend of God, Abū l-Muẓaffar Ibrāhīm ʿĀdil Shāh, son of the Just King, grandson of the Just King the Gallant Fighter, may God support him through His mighty succor in matters of the state, the execution of justice, and the preservation of the kingdom all through his mighty defense against error, sedition, and corruption.[40]

The emphasis on justice and equity plays on the title of the dynasty, the ʿĀdil Shāhīs, the "Just Kings," which Ibrāhīm had inherited from his father. The universal claims of world dominion for a regional dynasty give some insight into the self-conception of a court confronted with multiple adversaries at home and farther afield with appeals to divine kingship shrouded in the authority of piety. The verbosity of the entire translation speaks to the status of Persian as a language of sophistication and erudition, both in the court and beyond.

In India, courtly patronage of Persian literature stretches back to the days of the Delhi Sultanate, followed by the expansive Bahmanid kingdom of the south, out of which several competing successor states arose, including the ʿĀdil Shāhīs. Persian could take a functionary, an aspiring scholar, or an enterprising merchant from Bengal to beyond the Bosporus. Along with other classics of Arabic and Persian letters, *Wonders and Rarities* featured in a tapestry of cities and regions whose cross-stitched patterns of learning and prestige came to define what it meant to be cultured.

In this economy of grandiloquence, Qazwīnī too is raised up to the highest echelon of sagacity, celebrated in the introduction to the ʿĀdil Shāhī translation with an Arabic encomium that stretches to the stars: "Our master, possessor of complete blessing, the most perfect sage, the learned mufti of the sects, quintessence of scholars, model for philosophers, authority among the hadith transmitters, imam for Quran interpreters, the most judicious of all jurists and savants, inheritor of the sciences of the prophets and apostles, the most erudite of the authoritative scholars, proof of God to creation." The translation itself is feted as a book of great eloquence, *balāghat-niṣāb,* translated from the language of the Arabs into the language of the Persians, to be deposited into the imperial treasury.

But such royal commissions were not meant to be kept locked away, read only by a coterie of the highest elite. The very purpose of these literary endeavors was to exalt the name of the powerful patrons far and wide, to have their memories repeated and raised high through the fame of others. And thus the Bijapur translation of Qazwīnī spread across the Deccan and beyond, copied for both the modest and the mighty.[41]

The ʿĀdil Shāhī rulers came to promote themselves as descendants from the Ottoman royal line. For his part Ibrāhīm ʿĀdil Shāh presented himself as a champion of Sunni orthodoxy. His grandson, who also took the regnal title Ibrāhīm ʿĀdil Shāh II (d. 1035/1626), ran his court as a Sunni monarch, and advanced an ideal of a Sunni caliphate very much in keeping with his grandfather's precedence. He continued the tradition of lavishly supporting music, poetry, and the visual arts, and he took an abiding interest in Indic philosophy and history. Under his rule the local vernacular of Deccani Urdu grew in prominence as a literary language. The court historian, Muḥammad Qāsim Astarābādī (fl. 1018/1609), known by the pen name *Firishta,* the angel, celebrated Ibrāhīm ʿĀdil Shāh II as a model ruler and a just king in terms that echo the florid encomia in vogue among other courts during the period.[42]

Ibrāhīm II extended his control over large swaths of south India. But like his forefathers, he had to contend with the Mughals to the north, who came to view the kings of the Deccan as petty vassals. The *farangī* traders in their costal enclaves also remained a nuisance. Firishta took notice of the Portuguese attacks against pilgrim ships setting sail from India. The attacks had hemmed in Akbar, who refused to pay the foreign marauders for their protection.

The notorious barbarity of the Portuguese, with their rape and pillage and their attempts at mass conversion cast as holy crusade, allowed both the Dutch and the English to present their own mercantile interests as wedded to a policy that kept religion out of the affairs of trade. By then, though, Firishta notes, the Christians had not only established forts circumscribing the coast of India—at Mangalore, Mylapore, Nagapatnam, and Bengal—but they had also made their way to Ceylon and Sumatra, and even as far as China.[43]

As for Firishta, both Aceh and Malacca formed part of the geographical horizons for Akbar's court historian, Abū l-Fazl. The Mughals were no longer confined to the landlocked dynasty they had inherited from Bābur. Under Akbar, they had gained control of Gujarat and its ports. Bengal would follow. With Jahāngīr and his successors, Shāhjahān and ʿĀlamgīr, the Deccan would be pulled further into the Mughal orbit. Not only did greater contacts ensue with *farangī* merchants, mercenaries, and missionaries, but also there came more information on the far eastern islands. There is Ṭāhir Muḥammad Sabzwārī (fl. 1015/1606), a courtier in the Mughal court, who presented a world history that contained reports on the Malay Archipelago, as well as anecdotes of the dynasties of the *farang.* As with other remote regions, the pleasure and perplexity of wonder and rarity continued to shape how distant lands and people were fathomed and represented, all with a firm grasp that these places were indeed quite real.[44]

Writing in 1142/1730 Muḥammad Shafīᶜ, known as Wārid, a chronicler in the reign of the Mughal ruler Muḥammad Shāh, also balanced the Christians of the West with the islands of the East. He dedicated a section of his history to regions of the world not adequately covered by previous writers. From the rare spices of Java and their spectacular parrots, he moves to Sumatra and the Sultanate of Aceh, where in the jungles still dwelled savage cannibals. In the archipelagoes and beautiful coasts can be found camphor, powerful drugs, rare fruits, dazzling jewelry, white elephants, and sorcerers. Though the waters are overrun with muskets and warships of the *farang,* in the ports of Burma and Sumatra cargo comes and goes from all over the world.

In the regions of the *farang* are accounts of magical caves and strange technical artistry. Wārid contrasts how Muslims were prohibited from setting foot in Castile with the way Portuguese merchants were treated in India. Ingratiating themselves with Ismāᶜīl ᶜĀdil Shāh through amazing presents and wondrous gifts of fine craftsmanship, they secured a foothold on the coast. In the span of merely a few years they seized some seventy-four islands that originally belonged to Hindustan.

Even for remote Indian islands Qazwīnī's *Wonders and Rarities* still proved valuable. On the island of Barṭāyīl were savages of short stature. The sound of drums in the distance could be heard. Commerce could take place, but only through carefully choreographed movements of silent trade, where merchants exchanged goods without ever coming into contact with the natives. It is an account drawn from Qazwīnī's natural history that ultimately stretches back to the earliest sources of Arabic geography. Wārid again turns to Qazwīnī for Palace Island, warning readers not to disembark there, for the island was populated with dog-headed creatures that lure in unsuspecting sailors who never make it out alive (see figure 5.6).[45]

Readers of *Wonders and Rarities* came to anticipate such strange stories about the edges of distant lands. What the world had to offer could bend to the preconceptions of what one expected to find. By this point Qazwīnī's compendium was not only a staple in India for imagining strange lands and faraway places. It had also become a means for fashioning a particularly Indian vision of the world and its history.

An illustrated copy of *Wonders and Rarities* made its way to the treasury of Muḥammad ʿĀdil Shāh, the great-grandson of Ibrāhīm ʿĀdil Shāh, who had originally commissioned the translation. Years later a scribe filled its generous margins with treatises and anecdotes (see figure 5.2). Sometime after that, in 1751, another reader thought it would be helpful to provide a detailed table of contents outlining all the subjects covered in the copious marginal notes.

An epistle on medicaments opens the series of marginalia, picking up where Qazwīnī cites the Quranic verse "look at what is in heaven and on Earth," offering cures listed for various ailments. To the Quranic verse "the decree of God, the powerful, the wise" in Qazwīnī's definition of creation, there is added a short *fāl-nāma*, a tract on Quranic divination, in the name of the eminent north Indian Sufi saint Shāh Sharaf al-Dīn ibn Yaḥyā Manerī (d. 782/1381). When turning to Qazwīnī's introductory division of all existence, the margins are filled with the introduction to the *Mujmal al-ḥikma* (A précis of philosophy), a widely read Persian adaptation of the famed epistles of the Brethren of Purity, which celebrates the two worlds of the heavens and the Earth, and all the wonders contained therein. Qazwīnī's taxonomy of existence is quickly followed by the classification of the sciences modeled on Ibn Sīnā's famous treatment of the topic. Alongside Qazwīnī's discussion of the planets unfolds a digest of sacred words Indians use to subjugate astral forces. Beside dog-headed men and other strange islanders is a Persian treatise commenting on the mystical philosophy of Ibn al-ʿArabī and the elevated station of the *insān-i kāmil*, the perfect human, in the cosmic orders of creation.[46]

In the margins, ancient history abounds with appeals to philosophy, science, and medicine. Socrates offers counsel. Alexander learns the accounts of the philosophers about the true shape of the world with the teachings of Plato, Porphyry, Hermes, Apollonius, and Aristotle. There are tales of the prophets. Next to Qazwīnī's section on climatic geography, a marginal note tells of the famous sons of Noah. They scattered across the lands and gave birth to all the different peoples of the world—a biblical idea that endured over time and was often used for rather insidious ends to rank the various people of the world. The names of the seven sleepers in the Quranic story of those who slumbered in a cave for hundreds of years are given, while ʿAlī, the

Leader of the Faithful, offers divinatory maxims and instructions on the magical use of the alphabet.

Unfolded along the margins is an entire digest of world history starting with Adam, the first prophet, and culminating with Muhammad, the last. Then come the rightly-guided caliphs and the kings of ancient Persia, the names of the Umayyad and Abbasid caliphs, followed by the dynasties of the Samanids, Ghaznavids, Saljuqs, the Khwārazm-Shāh and all the descendants of Genghis Khan, who ended caliphal rule in Iraq. A detour leads through the Rajas in north India, followed by the sultans of Delhi. This culminates with the reigns of the Mughal emperors from Bābur through the death of Bahādur Shāh (d. 1124/1712), the successor of ʿĀlamgīr (d. 1118/1707).[47]

The series of reigns and dates helps to identify the sequence of notes and epistles in the margins of the royal Bijapur copy with the beginning of the eighteenth century. The focus on north India and the notable absence of Muslim dynasties in the south suggests that the compilation of marginalia was not made in the Deccan, but had been added well after the passing of Muḥammad ʿĀdil Shāh, who had lived on as a vassal of the Mughals.

Mughal history is celebrated throughout. A series of wonders from Jahāngīr's memoirs is recorded, and his grandson, the ill-fated prince Dārā Shikoh (d. 1069/1659), features. In the section on the stations of the Moon, there appears a long quotation from Dārā Shikoh's *Majmaʿ al-baḥrayn* (The confluence of the oceans), a widely read comparative study that identifies conceptual parallels between Islamic mystical philosophy and an eclectic array of Indic philosophical concepts, from Vedanta and the metaphysics of nondualism to discussions of Vaishnava devotion. Dārā was renowned for his sustained engagements with Indic thought, following in the steps of his great-grandfather Akbar.

Along the margins features a quotation on cosmic order: "According to the mystics of the people of India," Dārā Shikoh explains, "*brahmāṇḍa* is the word for the macrocosm." It is the cosmic egg of Brahma. Though Qazwīnī addressed the heavenly placement of the Earth situated like a yolk floating in an egg, the series of equivalences laid out in the marginal notes would have been quite unfamiliar to him.

In the selection drawn from the *Confluence of the Oceans,* Dārā Shikoh follows a vision of the cosmos as composed of the divine being, known in

Sanskrit as *mahāpuruṣa,* the great person or body. The section in question is Dārā's treatment of *mukti,* or liberation. A series of equivalences follow. The devils and demons are at the feet of the Great Body. The encircling ocean, identified with *samudra,* is situated at the navel. The sacred rivers of the Ganga, Jamuna, and Sarasvati flow through the mighty veins. Angels spread out across the hands and upper limbs. The Veda, which is truth and integrity, are recited on the lips. All light originates in the Sun of the cosmic eyes, while the noble face gazes upon the entirety of creation. Within the abode are the forms of every individual person, while the perfect human, the *insān-i kāmil* of the gnostic sages, is a dwelling of seclusion for the cosmic being. All this unfolds seamlessly across the margins of the manuscript.[48]

Qazwīnī's book of all creation synthesizes an array of disparate material and makes a perfect home for further currents in Islamic thought that also draw upon diverse sources of knowledge. Later in the margins excerpts appear from Dārā's dialogues with the Hindu mystic Bābā Lāl, as well as questions and answers between Krishna and Arjun from the *Bhagavad Gita,* redacted in Persian by Ṣūfī Sharīf. A Chishtī scholar who followed the teaching that all existence is one, Ṣūfī Sharīf served as tutor to Shāh Jahān before he became emperor.

Place names of Indian cities and villages line the landscape of its pages, while stories in Urdu are woven into the Persian notes. On display is a vision of the world particular to India. It was a view nurtured over the course of centuries, as Muslims cultivated a unique understanding and appreciation of geography, language, and history throughout the subcontinent. Sustaining this vision were contacts and exchanges with Hindus from an array of backgrounds. In these acts of equivalence also emerged the idea of a universal religious truth, which helped to cultivate a vibrant language of ecumenical tolerance. This view of religious diversity repeatedly struck the *farang,* who sought out the riches of India, to be not only strange but even abhorrent.[49]

Yet throughout the full panoply of material collected on the edges of the volume, the most consistent theme is the power of occult learning to harness hidden forces in the world. This comes alongside references to classical mystical piety. Mawlānā Rūmī makes an appearance; so does Niẓāmī, and an epistle on Naqshbandī Sufi practice. These flow seamlessly into treatises on geomancy, the unique power of numbers, and diagrams of magical squares. Here are instructions on how to craft talismans for numerous purposes, manners of performing divination and harnessing occult powers, with prayers to planets and secret alphabets, cyphers, magical seals, and the alchemical arts. Spells to defeat armies and to ward off the plague follow recipes for prolonging life, aphrodisiacs and amulets to increase sexual potency, and the erotic arts of women. On the edge of one folio is a note that the faculty of astonishment is unique to humans, a potent summation for the entire collection.[50]

Many of these themes of occult learning and sex magic come out in the original Arabic. The earliest Persian redaction developed these topics in even greater depth, with an entire section on the arts and sciences, which took on amulets, the talismanic arts, and an array of erotic practices and stories. Even in the Bijapur translation, which lacks this section, naked angels abound delighting and tantalizing. Unveiling nude bodies in sundry positions became a stock feature for many illuminated manuscripts of natural history. Often these focused on women's bodies, though naked men make a steady appearance. The Indians who share their women is a section from the earliest Persian recension, which when illuminated goes from the erotic to the graphic. The discussion of the wiles of women also provides similar inspiration. But one page with lovers greets another with murderous demons. The fluid boundaries between the strange and concupiscent fascinate and enthrall (see figure I.2).

Illuminated copies peppered with erotica circulated widely among Persian readers, back and forth between Iran, Central Asia, and India. One examplar of the Persian translation with the chapter on the arts finished in the year 1041/1632 in the Safavid capital of Isfahan made its way into the Mughal court during the same period the marginal notes were added to the royal Bijapur manuscript at the beginning of the eighteenth century. The emotive register of beholding seductive bodies and unseen forms follows the pleasure of entertaining the whole world unveiled. The promise of exotic titillation was woven throughout the text of *Wonders and Rarities* from the very beginning, even if earlier illuminated copies of the work were more restrained.[51]

Given the popularity of Qazwīnī's natural history, it is not surprising that the cycle of marginalia added to the Bijapur copy also circulated widely. Several later manuscripts of the translation originally commissioned by Ibrāhīm ʿĀdil Shāh are filled with the very same notes—Jahāngīr's hazelnut and all.

FIGURE 9.5: Painting of Qazwīnī in Indian style at the colophon of a Persian translation with a forged seal in the name of Qazwīnī.

One of these copies concludes with a painting. A solitary, gray-bearded man, wearing a green turban, associated with noble linages, puts the finishing touches on a book, seated next to a bolster. The bottom of the folio is marked by a seal impressed in black ink (figure 9.5). Seals were a common feature of manuscripts as markers of ownership. This one reads "the seal written by ʿImād al-Dīn Zakariyyāʾ ibn Muḥammad ibn Maḥmūd al-Kammūnī." The honorific "Cumin seller" identifies this as the hand of Qazwīnī himself, whose full name appears in this very form in the opening to *Wonders and Rarities*.[52]

A year in the seal can also be made out, 953/1547. The name immediately conveys that this was an authorial copy and that the image is none other than Qazwīnī himself. But the date cannot be right. It comes some three hundred years after Qazwīnī first began to redact his book of wonders that would travel so widely. It does, however, precede by a few months the year mentioned in the introduction to this Persian translation, which states when the royal ʿĀdil Shāhī commission was completed. It is on the basis of this date that some even came to think that Qazwīnī was a later Indian Muslim who wrote in the court of Ibrāhīm ʿĀdil Shāh.[53]

This is not the seal of the author, and this manuscript was not completed in the middle of the sixteenth century. It is a sparsely illuminated copy transcribed much later, in a careful pen, most likely in north India—although this is just a guess. Perhaps through a clever ruse, the copyist sought to increase its value, to pass it off as an original authorial copy, with a painting of what Qazwīnī would have looked like in the elegant attire of a noble Muslim sage from India. As with all things strange, marvelous, and rare, written so many years removed, the value of such handiwork lies not so much in what is true or real, but in what we are willing to entertain or are ready to behold.

CODA

Acts of Enchantment

SO MUCH HAD TO BE DONE. The braided garlands, the coconuts with betel leaves and nutmeg, the precious stones and diamond-encrusted jewelry of gold and silver—it all had to be carefully taken out from the crates and arranged. The costly pavilion, with its intricate carpet and patterned canopy, needed to be pitched, and the divan covered in green velvet spotted with gold hauled up from the ground floor. It had to be just right for the prince to recline in perfect repose, surrounded by supplicating courtiers and slaves holding large fans. In the corner went the inlaid chess set, next to the tiger skin and the finely tooled boxes of ivory and ebony, by the brass hookah with the shining amphora and that sumptuous oil lamp.

In the boxes were spices, grains, raw fibers, and supple fabrics. Some were filled with incense, tinctures, perfumes, resins, and vegetable dyes, others with bolts of silk, cotton, and cashmere. The peacock feathers went to the side, next to the elegant display cases and the collection of weapons—gauntlets, chain mail, helmets inlaid with turquoise, lances of ivory, matchlocks, delicately engraved battle axes, silver quivers of arrows fletched in colorful arrays. The bazaar still needed tending, where the mannequins of villagers were posed in their native attires next to figurines of elephants, snakes, and monkeys in the broad foliage of talipot fronds, beside the models of mosques, temples, and a climbing pole, on the miniature grounds of a festival.

Many hands moved the curios along, from the artisans who had been commissioned for the exhibition by a phalanx of committees in India to the go-betweens who shipped out the cargo from distant ports and the laborers who lugged the chests filled with raw and manufactured goods, including the life-sized dolls of princes and paupers and the accessories that would bring the exotic tableau to life. Up they went, over the stairs of the still-unfinished Palais de l'industrie—a sprawling edifice erected especially for the occasion, overlooking the Seine to one side and the Champs-Élysées to the other.[1]

The officials running the gallery assigned to the Honorable East India Company wanted to make an impression at the Universal Exhibition of Paris in 1855. It was a global fair of arts and industry for all the world. In the great game of nations, the event was also a staging ground for the empires of Europe to showcase their technological prowess and their control over large swaths of the world in the form colonies, protectorates, dependencies, and dominions.

The galleries reserved for the East India Company were larger than some spaces assigned for entire countries. In charge was Dr. John Forbes Royle (d. 1858), a botanist in the employ of the Company who had made a career studying Indian crops for their medicinal and economic value. Four years earlier Royle had organized the Company's arcade at the Great Exhibition of London in the Crystal Palace. He knew exactly what spectators would hope to see when encountering the Orient, a land of both wonders and raw materials that could be used in countless applications.[2]

If compared only to the square footage that had been allotted in the Crystal Palace of Hyde Park, the world exhibition of 1855 was a considerably more modest affair for the East India Company. But their displays in Paris were a smashing success, both the ethnographic gallery in the main palace and the large exhibition for natural products of India and the Indian archipelago in the northeastern end of the annex along the river. The jury awarded imperial medals to Dr. Royle for all his efforts curating and explaining the collection and to the Company with its high reputation for power and commercial wealth.

When all was ready, the spectators reacted just as the organizers had hoped: with awe and astonishment, transported to a distant world of enchantment, to the land of *The Thousand and One Nights,* beholding every conceivable variety of arabesque and fantastic carving, admiring the beauty and originality of the details so vividly struck in the mind's eye. Accompanying these adulations, the local press and guides also saw a moral behind these imported luxuries in the tried lessons that travelers and scholars of the Orient had to teach.

Visitors to the Company's gallery could not help but deplore, one Parisian pamphlet noted, the state of misery into which these peoples have fallen. Abject in their luxuries, the rich princes take on an idiotic pride born of ill-conceived pretensions to superiority, while the rest go around almost naked, painfully earning a handful of laughs. There is the degradation of women, the contempt for the lives of children, and the disdain for liberty in favor of political despotism, aristocratic fetishism, and religious superstition.[3]

For the audiences in Hyde Park and along the Seine, the script for how to stage, describe, and view the Orient was immediately recognizable. By then all the emotive gestures had been well rehearsed. Wonder at the luxuries and curiosities seized from conquered lands could be reduced into immediately recognizable symbols: the giant diamond known as the *Koh-i noor,* the mountain of light, taken from Lahore; or the automaton organ of a tiger devouring an officer of the East India Company—a trophy won in the brutal defeat of Tīpū Sulṭān (d. 1213/1799) of Mysore. Such curated objects offered metonyms for territorial control and possession.

The spectacle of Eastern wealth and rarity was also meant to elicit approbation at the backward and primitive conditions of the vanquished. The cost of freedom, the argument went, required a good measure of colonial rule and schooling in the lessons of European superiority. The baroque language used to describe the exotic marvels of the East was accompanied by new categories and modes for mastering the world. "Civilization" was still a fresh neologism, chemistry and astronomy were making great efforts to distinguish themselves from alchemy and astrology, and the modern dividing line between religion and science was only now truly beginning to take hold.

A close observer, one trained in the languages of the Orient, may have picked out in the gallery of the East India Company in Paris—from the clutter of conquest and the tributes sent in by subjugated Hindu rajas and Muslim princes—a singular copy of Qazwīnī's *Wonders and Rarities.* The sumptuously illuminated manuscript had been especially commissioned for the occasion by the Punjab Committee of Lahore, run by Charles Raikes (d. 1885), the commissioner and superintendent for the district. With a wink, Raikes notes in the manuscript that the collection "is looked upon by the Mahamedan 'Savants' of India as a work of great authority."

Someone with a keen eye for Persian who took time to leaf through the folios could read along the margins the amazing story of Jahāngīr's hazelnut

FIGURE C.1: Angels in the shape of a lion and an eagle, with marginal commentary from an epistle on the metaphysics of the cosmos from a copy commissioned by Charles Raikes in Lahore in 1854 for the East India Company's display in the World Exhibition in Paris, 1855.

or Dārā Shikoh's discussion of the cosmic egg of creation. But there was so much more to do in Paris than linger with the native literature, and Qazwīnī's appearance was merely as a prop for staging the lavish Orient, nestled alongside specimens of Indian calligraphy and other indigenous crafts (figure C.1).[4]

By then leading Orientalists would have recognized Qazwīnī's book of curiosities. In European printing houses, *Wonders and Rarities* would come to represent all that was strange in the Orient, still stuck in a world before Columbus and Copernicus. Nonetheless, Qazwīnī was also mined as a valuable source of ancient lore. While stationed in India, the naturalist Captain Newbold had prepared a précis of the natural history, with an eye toward the utility of the collection for knowledge of the East, published in 1844. Soon after, the German Orientalist Ferdinand Wüstenfeld (d. 1899) finished a workable, if deeply unsatisfactory, *editio princeps* of Qazwīnī's Arabic cosmography and geography.

During the eighteenth and nineteenth centuries, Qazwīnī's collection, like other writings on wonders, continued to circulate in Arabic, Persian, Turkish, and then Urdu, traveling alongside compendiums of talismans, arts of geomancy, and collections of spells. Just as Orientals were for Europeans objects of curiosity, the language of wonder provided a means for depicting the Christian *farang* as strange people who possessed powers of fascination and their distant land of Europe as a prime abode of marvels. Throughout it all, Muslim elites followed broader transformations in learning and science and developed novel ways to account for the many changes afoot. One of the chief responses in reformist circles would be to cull out vast swaths of Islamic learning as retrograde or superstitious. Yet the enduring relevance of *Wonders and Rarities* also underscores the diverse means by which magic and miracle would continue to be rationalized through the authority of natural science.

Well before the rise of modern Orientalism, Latin translations of Arabic works of science and philosophy played a significant role in the formalization of medieval scholasticism; these translations touched on numerous fields, from astronomy to zoology, as well as broad areas of medical and occult learning. The study of Arabic, Hebrew, and Syriac took on renewed attention in the course of the Renaissance and Reformation, as theologians moved beyond the Latin vulgate to decipher the original languages of the Bible. In Catholic circles there were also increasing contacts with Maronite Christians from the Levant who contributed to the development of Arabic studies. Likewise, natural philosophers, particularly in university cities such as Padua, Bologna, Paris, Oxford, and Leiden, had long sustained an interest in Arabic letters—particularly in the fields of philosophy, medicine, astrology, and alchemy—as keys to unlocking the ancient wisdom of the past.[5]

This spirit of inquiry led the French Protestant, biblical antiquarian, and scholar of Oriental languages Samuel Bochart (d. 1667) directly to Qazwīnī. While Bochart was rather hostile to Muhammadans, he also found value in their writings. Qazwīnī's *Wonders and Rarities* was a pivotal source for Bochart's

Hierozoicon (Sacred zoology), a sprawling philological study of animals mentioned in the Bible. Bochart worked on Qazwīnī's natural history from manuscripts, copying large portions by hand as he refined his Latin translations of several passages.

In addition to Qazwīnī, Bochart also drew on Ibn Sīnā, along with several other Arabic authorities, and a full array of Greek, Latin, Hebrew, and Syriac materials. It was an age when ideas could circulate in print, as new information on the natural world came from far-flung commercial ventures and imperial missions. One consequence of this terrestrial knowledge was an effort to reassess classical accounts of natural history. Arabic, Persian, and Turkish sources also offered a means to better fathom the world in general and the East in particular.

As Bochart saw it, there was a lot of carving out that had to be done between fact and fiction. Qazwīnī features throughout Bochart's discussion of animals considered to be real, *animalia vera considerantur,* such as the hyena, the camel, the crocodile, and the giraffe—a creature that for some time had puzzled biblical scholars in Latin Christendom. Qazwīnī also proves indispensable for his treatment of the unicorn, which Bochart identifies as a rhinoceros, known to Qazwīnī as the *karkadan.*

But in addition to what could be verified, Bochart was also concerned with determining which animals were merely fictive creations. Qazwīnī repeatedly appears in Bochart's final treatment of dubious or fabulous creatures, *de dubiis vel fabulosis animalibus.* Bochart links classical accounts of strange animals found in the likes of Pliny with later Arabic sources. Thus, Qazwīnī's *ʿanqāʾ* parallels the griffin, while for Bochart the *nasnās*—often pictured in Arabic as monopods—echo Pliny's *monocerote,* single-horned monsters.

Although Bochart drew copiously from Qazwīnī, he considered the work, which he titled *Miracula rerum creatarum* (Marvels of created things), to be a *liber valde fabulosus,* a rather fabulous book, in the sense of containing mythical or fictive accounts of dubious pedigree. The description echoes Bochart's claim that Arabs have a strong proclivity toward fables, *ut sunt in fabulas valde proclives.* It was a sentiment with polemical roots that came to play an outsized role in the ways scholars in Latin Christendom would come to understand the East.[6]

With the publication of the *Hierozoicon,* Alkazuinius—Bochart's name for Qazwīnī—features in the subsequent development of humanist letters as an authority quoted in encyclopedias, natural histories, and biblical exegesis. Soon after, Qazwīnī's *Wonders and Rarities* made a passing appearance in the *Bibliothèque orientale* by Barthélemy d'Herbelot (d. 1695), a scholar of Arabic who taught in Paris. According to d'Herbelot, the *Agaib almakhloukát* was a natural history that treated in the first part the heavens, stars, and meteors, and in the second the earth, the seas, metals, plants, and animals.[7]

Here as elsewhere, d'Herbelot took his information from the Arabic bibliographical survey by the Ottoman polymath Kātib Çelebī (d. 1067/1657), also known by the honorific Ḥājjī Khalīfa. But Ḥājjī Khalīfa was no simple antiquarian, though that might be the impression gained from leafing through d'Herbelot's *Oriental Library*. For Ḥājjī Khalīfa also developed an abiding interest in geography, crystalized in his massive *Cihānnümā* (Guide to the world), which he worked on over the course of his career.

The Ottoman statesmen read the latest atlases and geographical compendia imported from European merchants. The redaction of Mercator's *Atlas minor* by Hondius lays the foundation, which Ḥājjī Khalīfa had translated into Turkish. Suʿūdī's *New Report* on the New World also proves indispensable. Classical Arabic and Persian authorities on geography and cosmography feature, including Qazwīnī. Lining the horizon are Japan, China, archipelagoes across the Indian Ocean, Africa, and the Americas.[8]

But d'Herbelot was not interested in how Orientals were part of a wider world or how they continued to engage with the learning of the day. He offered a hermetically sealed land of Muhammadan letters untouched by any awareness of broader developments across the globe. For succeeding generations of scholars, d'Herbelot's *Oriental Library* came to define the horizons of the East. The heavy tome made its way into all the major libraries of Europe. As a universal dictionary it sought to produce an entire world, promising to contain everything touching on the knowledge of the peoples of the Orient with their histories, traditions, true or false, religions, sects, politics, government, laws, customs, morals, wars, revolutions, empires, sciences, arts, theology, mythology, magic, physics, ethics, medicine, mathematics, natural history, chronology, geography, astronomical observations, grammar, and rhetoric. The purview offered nothing less than the Orient itself within a single volume of over a thousand printed pages.[9]

D'Herbelot's dictionary was published and edited posthumously by fellow Orientalist Antoine Galland (d. 1715), who is most famous for introducing into European letters a French translation of *The Thousand and One Nights*. Galland took liberties expanding and obfuscating his Arabic source material. Yet through his translation, the *Nights* and its strange tales of genies, sorcery, and sexual delights transformed into the most preeminent source for conceiving of the Orient—a vast land that stretched from North Africa to the Far East.[10]

This imagined geography of exotic difference made the East into everything the West was not. Islam was tyrannical and oppressive; its adherents superstitious, decadent, childish, ignorant, feminine, deviant, and deceptive. With all the magical twists and turns, Orientalists came to see in the *Nights* "the common soul of Islam"—to be read as a vast trove of Oriental mores, which unveiled the "religious tenets, superstitious opinions, customs, laws, and domestic habits of the followers of Mahummud."[11]

All the while, manuscript copies of *Wonders and Rarities* kept reaching major universities and private collections—London, Oxford, Cambridge, Paris, Stockholm, Leiden, Berlin, Leipzig, Gotha, Madrid, Rome, Vienna, St. Petersburg, the list goes on. As with other sources of classical Arabic and Persian, Qazwīnī featured in the earliest teaching anthologies. Some merely repeated what Bochart had excerpted. Others added new material. This was the case with the extensive quotations of Qazwīnī in the *Chrestomathie arabe* edited by Antoine Isaac Silvestre de Sacy (d. 1838), one of the pioneer Orientalists in Paris. The collection included an extensive selection of Arabic materials with accompanying French translations. Although Qazwīnī's book of marvels was filled with "puerile and fabulous descriptions," it was the "best treatise on natural history that the Orientals possess." Several passages, an editorial note observes, find direct connections with *The Thousand and One Nights*. The parallels concern the adventures of Sindbad the sailor, which turn out to be anything but straightforward. Nonetheless, the magical world of the *Nights* offered ballast for anchoring Qazwīnī in the fantastic and macabre world of Oriental adventures.[12]

Along the way, there was also general interest in Qazwīnī's natural history as a repository for ancient learning. The German astronomer Christian Ludwig Ideler (d. 1846) edited and translated Qazwīnī's treatment of the fixed stars in 1809. Through Qazwīnī, Ideler could identify the many Arabic star names and their meanings, which had long shaped the firmament above Europe. Acrab, Algol, Betelgeuse, Rasalgethi, Sadr, Vega, and countless other luminaries drew their medieval Latin names from earlier Arabic star catalogues. Qazwīnī helped unravel their history.[13]

"The Pliny of the Orient" was one common epithet bestowed on Qazwīnī, in an effort to make intelligible the value of *Wonders and Rarities* as it related in some general sense to the development of European letters. Bochart had already set the groundwork for this particular reception. Pliny's multivolume natural history was in many senses much more encyclopedic, but his interest in monsters and curiosities, couched in the language of science, found in Qazwīnī a ready analogue. The common refrain that Qazwīnī's natural history was a childish collection of fables continued, even while others found a good deal of utility in it. This was the case in both Europe and the colonial outposts, where attitudes toward Oriental knowledge could modulate considerably.[14]

One such colonial official posted in the Orient was Dr. Patrick Russell (d. 1805). Qazwīnī makes a passing appearance, via Bochart, in Russell's *The Natural History of Aleppo*. A Scottish physician and naturalist, Russell was stationed in Aleppo in the employ of the Levant Company before he made his way to India as a servant of the East India Company. Despite the regional scope, with a focus on an Ottoman provincial capital and English mercantile interests, Russell offered a grand study of ethnography, botany, zoology, and medicine, styling the work as a natural history for the entire Orient.

In addition to plants and animals, Russell spends a good deal of time assessing Muslims and their knowledge. Readers are invited to witness the inferior intellectual condition of Eastern learning and simplicity: mathematics is seldom understood; astronomy is neglected; natural history and the experimental physics have made no progress; history is not much read; Muslims show no curiosity for other countries; their physicians are quacks; of chemistry little is known, save for pretensions of alchemical transmutation. Sorcery, however, is widely practiced, along with the attendant arts of astrology, geomancy, augury, and sortilege; belief in the power of charms, talismans, and amulets reigns supreme.[15]

Russell's account of the stagnant state of Eastern learning was not novel. It reflected what by this point was a common attack on the backward condition of Orientals. His line of comparison was also not particularly fair. The Ottoman elite had for some time kept abreast of advances in geographical and medical knowledge, as did learned scholars and travelers from across Iran and India. The pioneering chemical and medical teachings of Paracelsus (d. 1541) circulated in Arabic, Turkish, and Persian compendiums well before Russell's critique, and the New World had long made its appearance in various forms across numerous geographical texts penned by Muslims.[16]

As for astronomy, inquisitive minds had already begun talking about a heliocentric universe. As with many areas of inquiry, Islamic scientific and philosophical writings exerted an enduring, if often unrecognized, influence on the science of the stars. It turned out that one needed to look no further than to Naṣīr al-Dīn Ṭūsī for a challenge to Ptolemy's theory of planetary movement. Ṭūsī's mathematical models played an important role in later astronomical developments, shaping in direct ways the theories advanced by Copernicus to account for how the planets rotated around the Sun.[17]

The theory of heliocentrism, though, was not limited to European developments. Qazwīnī was familiar with the idea through the transmission of ancient Greek theories of the cosmos, as were countless other authorities, including Ḥājjī Khalīfa. Moreover, the heliocentric model of the universe advanced in Europe was known in Ottoman scientific circles by the end of the seventeenth century.[18]

Over the coming centuries the vast Ottoman Empire bled territory to various European powers, but leading scholars in Istanbul and beyond kept up-to-date on the latest technological and military advances made in the capital cities to the west of the Bosporus. All the same, Russell's self-serving assessment circulated widely in European letters and was frequently copied by readers who had come to see Qazwīnī's universe of marvelous tales as one more footnote to the moribund state of learning among Muslims, which, as the argument went, had not progressed since the Mongol destruction of Baghdad.

In measures of comparison, the imperial and colonial languages of superiority were designed to mask more than reveal. It remains to be seen just how disenchanted European learning of the day really was, concerned as it was with theologies that sought out Noah's Flood as a dividing line for world history and that couched philosophical humanism in teleological arguments of divine design. In many circles, the Bible remained a powerful tool for probing the depth of the heavens made visible through the telescope and for fathoming the minute bodies revealed by the microscope.[19]

Like many other Orientalists and physicians, Russell was initiated into the universal brotherhood of the Freemasons. While in Aleppo he frequented the Freemason lodge established there by officials of the Levant Company. Natural philosophers, colonial administrators, and Orientalists crossed paths with Freemasons, who helped to shape Dutch, British, and French imperialism. With forts, churches, and mission schools came masonic lodges.

As a secret society, Freemasonry has roots in the long history of Christian esotericism. Pronounced strains of universal brotherhood and religious tolerance were central to its wide appeal during an age that witnessed the gradual rise of liberalism and egalitarianism as positive values. But also attractive were hidden rituals of initiation, Solomonic mythology, deep strains of Hermeticism, and a deism based in a transcendent vision of God as the Great Architect of the Universe. Throughout the period, Freemasons found in the Orient an endless fount of secret learning that promised the key to the Adamic origin of languages, religions, and humanity itself.[20]

In its idealized form, the liberal masonic value of universal equality and fraternity was ecumenical in spirit, as Masons drew inspiration from various religions of the world, looking to the antiquities of the East for perennial wisdom. The reality proved more difficult, as the debate in the nineteenth century over admitting Muslims, Hindus, Sikhs, and Parsees into the masonic brotherhood attests. Native elites, including prominent Muslims, did make their way into lodges, although not in large numbers. What they would find was a secret society wrapped in a language of emanationism and a doctrine of the eternal soul. The point deserves noting, as the intellectual chasm was not nearly as great as Russell would suggest or the oppositional language justifying the colonial enterprise would demand.[21]

For all the antagonism, there were also scholars and officials who held romanticized views of the East as the ultimate source of universal spirituality. Many of these commonalities were the product of the *arabum studia,* through the introduction, absorption, and naturalization of Arabic philosophical and scientific learning during the course of medieval Latin scholasticism. These conceptual areas of confluence made the metaphysical and ethical system undergirding Islamic philosophy easily intelligible to those whose business was the Orient.

But whose business it exactly was depended on context. Countless go-betweens, dragomen, munshis, scholars, and mercenaries made the whole enterprise possible. In the hands of Orientalists, Qazwīnī's universe was useful for fixing a timeless, never-changing East. Such a view required looking past all the people at home and abroad who were much more complex than simple caricatures could allow. As manuscripts of Qazwīnī's *Wonders and Rarities* were shipped off in crates back with other spolia, so too came aboard natives, some willingly, many more, though, in chains, from across the Indian Ocean and along the coasts of Africa. Among the many who journeyed the seas were, here and there, elite Muslims who wished to see for themselves the strange lands of the *farang*.

In 1775, some twenty years before Russell's scathing attacks, a munshi in the service of the East India Company, Mīr Muḥammad Ḥusayn Iṣfahānī (d. 1205/1791) set sail from Bengal to Europe. The language he used to describe what he saw there was that of wonders and rarities, a standard register for describing distant lands. But so much had changed—from the Battle of Plassey in 1757, which gave the Company control over almost all of the territory that had once been in the hands of Mughals, to the new knowledge used to master the world.

Embarking on an East Indiaman laden with cargo, Muḥammad Ḥusayn made his way around the Cape of Good Hope, through violent storms, to Lisbon, and then to London. There he spent a year studying English, while touring museums, libraries, and shipyards. Like many other notables who crossed exotic shores, Muḥammad Ḥusayn wrote about what he saw.[22]

Muḥammad Ḥusayn's writings offer an early example of a genre that would gain considerable attention in the coming years. The number of tomes steadily grew, dedicated to documenting the curiosities found in the distant countries of Europe. Some came from travelers of modest means, others from those who wielded great power. Most, but not all, were written by men.[23]

There were changes afoot, Muḥammad Ḥusayn knew, in geography, with the discovery of the Americas and countless islands, and in astronomy, with new research on the movement of the planets and the stars. He recognized that these fresh accounts of the world made the teachings of ancient Greek philosophy obsolete. The sciences of the real, the *ʿulūm-i ḥaqīqiyya,* were his goal, which at this point were closely associated with the mathematical study of the cosmos, for these "did not change from one faith or religion to the next." During his travels Muḥammad Ḥusayn followed the latest investigations of European scholars, particularly in the fields of astronomy, medicine, and anatomy.[24]

In addition to his report on sciences and curious customs, Muḥammad Ḥusayn is noteworthy for comments made by ʿAbd al-Laṭīf Shushtarī (d. 1220/1805), another traveler of the day. Originally from Iran, Shushtarī had sought patronage in India. He too recorded an account of his journeys across Persia and South Asia, which included observations on the telescope, Copernican astronomy, Newtonian gravity, the Spanish conquest of the Americas, the conversion to Christianity and forced conscription of Pacific islanders into European armies, parliamentary government, and the exploitative trading policies of the East India Company—all framed by the emotive register of wonders and rarities. Among the many pieces of information that Shushtarī picked up in his travels was the rumor that the ancestors of Columbus were originally from the Arabian Peninsula and that in addition to the languages of the *farang,* the famed navigator knew Arabic. Theories of such connections and intimacies, both real and imagined, would over time become increasingly appealing to Muslim reformists who sought to account for what Europe owed Islam.[25]

As for Muḥammad Ḥusayn, Shushtarī knew him to be a master of the *funūn-i ʿilmī,* the scientific disciplines, particularly the *ʿaqliyyāt,* the rational sciences. Yet Shushtarī states, rather derisively, that Muḥammad Ḥusayn did not uphold any *mazhab* or *millat,* creed or profession. "Sometimes he counted himself a Shia, or a philosopher, or a theologian, but he did not consider himself to be of any sect, and was famous for his corrupt beliefs." The prudent were wary of him, Shushtarī concludes.[26]

Was it crossing India, Europe, or some combination therein that provided the coordinates for such a state of confessional nonconformity? Or was this simply slander born of some personal animus? Shushtarī does not offer much help in his passing observations. There are numerous examples of libertines and antinomianism in the course of Islamic history. But in many respects this was a different age for a certain class of intellectuals who mixed in and out of courts, jostling among Hindus, Christians, Jews, and Muslims of various stripes.

Along with the allure of a new science was the promise of patronage and advancement. Muḥammad Ḥusayn hoped that traveling abroad would benefit his career with the British. But the mercantile and imperial mechanisms that drew the world into greater commerce did so frequently through coercive means. The often-mundane exercises of physical and economic violence proceeded with an ideology of the new and modern, spoken in the measured language of progress, enlightenment, and liberty.[27]

Within a hundred years after Muḥammad Ḥusayn set sail, the *farang* had gained almost total control over the entire subcontinent. Along with the Dutch and the French, the British also extended their domains throughout Asia, Africa, and the Pacific. Despite successive waves of reform, the Ottoman Empire faced external threats and internal unrest, while the newly minted Qajar dynasty of Persia shed ever more territory to imperial Russian forces. Through it all, the Spanish and Portuguese did their best to hold on to territories in the Americas, Africa, and Asia.

The cataclysmic events that brought about these changes are generally not in dispute. But what drove it all, other than the cold pursuit of power? Was it technological superiority that gave a decisive advantage? What about social values, innate dispositions, or even divine providence?

When answering these questions, exhibits with gunpowder, the magnetic compass, and the printing press proved to be early metonyms for ingenuity, superiority, and the advance of mastery. Eventually the steam engine, the telegraph, and standardized time joined the list. Such technological determinism often invoked a unique spirit of discovery and liberty. Theories of race and civilizational hierarchy came to gild the entire endeavor with an aura of benevolent self-sacrifice. On scrutiny, all these answers have proven painfully wanting.[28]

The cannon, the musket, the magnetic compass, the horse—these are explanations that Muḥammad Ḥusayn offers for the conquest of the Americas. But he pauses at the riches flowing out of the New World, which gave rise to further plunder. There was nothing inevitable about the initial subjugation of the Western Hemisphere as it played out, but the wealth of knowledge and power that issued from it demanded new ways of seeing the world.[29]

There are many adjectives that describe Muḥammad Ḥusayn—philosopher, theologian, freethinker, scientist, poet, translator, informant, courtier. While in the employ of the East India Company in Calcutta, before he set sail, Muḥammad Ḥusayn helped to translate into Persian a highly influential Arabic legal compendium. The translation, sponsored by Warren Hastings (d. 1818), the first governor general of India, would come to play an outsized role in the colonial construction of Islamic jurisprudence, styled in the Indian subcontinent as Anglo-Muhammadan law.[30] Muḥammad Ḥusayn also prepared a Persian grammar dedicated to officers in the Company so they could further advance in the language of culture and court administration, as did he procure Persian manuscripts for his British patrons.[31]

On his return from the land of the *farang,* Muḥammad Ḥusayn's British patrons commended him to the nawab of Murshidabad, a vassal of the East India Company. There the local court showed little interest in his proposals to initiate a translation project of European works of science into Persian. Everyone was preoccupied, according to the contemporary Ghulām Ḥusayn Khān Ṭabāṭabā'ī (fl. 1230/1815), with the growing controversies surrounding the "brave Governor" Warren Hastings and the accusations leveled against him of bribery and profiteering.[32]

But Muḥammad Ḥusayn continued his service for Hastings and was ultimately appointed agent of the Company in the court of the Niẓām of Hyderabad, a vital ally of the British. There he regularly sent letters, presented decrees, and relayed gifts—jewels, candelabras, and clocks passed before the court—all while supplying details on the movements of Tīpū Sulṭān, the Marathas, and the French, information that greatly benefited the military aspirations of his employers.

Yet Muḥammad Ḥusayn also complained of not receiving from the Company his salary in proper course or in the correct amount, which left him unable to pay his creditors. Although Hastings held the "highest opinion" of his integrity, there were rumblings from the Court of Directors of the Company in London about placing even a "very intelligent Native" in such an important position.[33] In 1791 Muḥammad Ḥusayn, by then quite ill, died in

Benares en route to his home in Lucknow, after having just secured the arrears in payment owed to him for years of service. Ṭabāṭabāʾī lavishes praise on both Hastings and Muḥammad Ḥusayn while lamenting the general lack of interest in the beneficial sciences that Muḥammad Ḥusayn was ready to translate.[34] Yet what could Muḥammad Ḥusayn have had to offer that would have changed anything? Copernicus, the solar system, Newton, and the limitless expanse of the universe seen through the telescope? "Our world compared to the countless worlds above," he exclaimed, "is merely one grain of sand on the shore of an ocean." It is a sentiment Qazwīnī expressed repeatedly before the infinitude of creation, though with no telescope to guide his awe.[35]

Did it really matter, though, whether the Earth orbited the Sun, when the classical models of spherical trigonometry for centuries had done a fine job calculating the movement of the heavens? The Sun still rose in the east and set in the west. To be sure, in certain circles there was great interest in the newfangled sciences of the *farang*. There was also in other corners a good deal of skepticism. But all this would require further techniques for unwrapping the self-aggrandizing rhetoric of European supremacy in which it was all bound. New modes of reading and reclaiming would be necessary, with new vocabularies and new categories of knowledge.[36]

In the succeeding generations, that hope for translation came to pass on a scale hitherto unimaginable, both through a steady regime of colonial education and by successive reform movements that sought to reclaim and repurpose the master's tools for their own ends. There were primers and catechisms in colonial classrooms, with their demonstration globes and world maps. Some readers updated old natural histories with marginal notes correcting or emendating with information about Europe, the Americas, and the solar system.[37] Other treatises in geography, astronomy, chemistry, economics, history, and literature were sponsored by new societies for reform, such as the Scientific Society founded in 1864 by Sir Sayyid Aḥmad Khān (d. 1315/1898) to promote the translation of European learning into Urdu.

The scientific society was established in the wake of the unsuccessful revolt of 1857 against Company rule, which saw Hindus, Muslims, and Sikhs take up arms in an unsuccessful attempt to break free of foreign rule. Many sought to diagnose the causes of the uprising, including Aḥmad Khān himself. Charles Raikes, who had sent the lavish copy of Qazwīnī's *Wonders and Rarities* from Lahore to the world fair, also weighed in on the matter. The universal exhibition in Paris was merely three years before the slaughter of countless English settlers in villages and cities across north India and then the unrelenting retribution by Company regimens, filled with native sepoys, against local populations. The uprising resulted in the formal dissolution of the Mughal dynasty and the transference of Company rule in India directly to the British Crown.

A common theme among colonial officials was that the natives failed to appreciate all that European learning had to offer. There were many proposals to remedy this. In the fall of 1863, dailies in England, Scotland, and India ran announcements advertising a competition for the best essay comparing "the influence of Greek learning on the Arabs, under the Abbaside Caliphs of Bagdad and the Ommyade Caliphs of Cordova, with the subsequent influence of Arabian learning on the reviving European mind after the Dark Ages."

The regulations stipulated that the comparison should "infer the probable influence which the mature intellect of Europe should exercise in its turn, now that it is once more brought in contact with the Muhammadan mind in India." The essays were to be written in Hindustani, either in the "Persian or Roman character" and accompanied with an English translation. The author of the best essay, to be chosen by committee, would receive what was then the handsome sum of 500 rupees.[38] Talk of the competition picked up in Muslim reformist circles. Aḥmad Khān praised the initiative.[39]

The prize committee, consisting of both British officials and Muslims in their employ, received two submissions. One came from Edward Rehatsek (d. 1891), a Hungarian Orientalist who had settled in India in 1847. In addition to his English original, Rehatsek supplied a Persian translation, as he was not sufficiently competent to produce an Urdu version, contrary to the stipulations of the prize. Otherwise, Rehatsek hewed closely to the premise of the competition, following the rise and fall of science and reason across "Muhammadan civilization." The category "civilization" was still a relatively novel term, but its comparative use had already been set in motion. Rehatsek concluded that Muslims would benefit tremendously from British rule by obtaining "a good secular education." The word "secular" was another term that would capture the sensibilities of a modern age.[40]

The other submission, by Sayyid Karāmat-ʿAlī (d. 1290/1873), took a rather different tack. A recognized leader among the Shia of Bengal, Karāmat-ʿAlī was a prominent member of the newly formed Mahomedan Literary Society of Calcutta and of Aḥmad Khān's Scientific Society. Like many within the Muslim elite after the events of the failed uprising of 1857, Karāmat-ʿAlī also called for the general acceptance of British rule.[41]

Karāmat-ʿAlī submitted his essay, entitled *Risāla-yi bayān-i makhaz-i ʿulūm* (An epistle elucidating the origin of the sciences), first in Urdu, which he followed up with an English translation made with the assistance of two of his leading pupils. The essay exhibits a thorough engagement with current trends in science and presents an admirable grasp of the historical reception of classical Greek learning in Arabic. Yet, instead of following the paternalistic argument of influence and decay, Karāmat-ʿAlī insists that the advances of modern science were all ultimately foretold in the Quran and the sayings of the Prophet and the Imams, and thus were merely confirmations of the prophetic truth of Islam.

Karāmat-ʿAlī dispenses with Ptolemaic geography and cosmography as not only invalid but disproven by Quranic authority. From the heliocentric structure of the universe to comets, sunspots, and the rings of Saturn, he sees the prophetic authority of Islam as entirely in concord with modern scientific learning. Much of the essay unfolds as an exegetical exercise, citing verse after verse to prove that "the Quran is filled with mathematical and natural science [*Qur'ān meṇ riyāzī o ṭabīʿī ka ʿilm bhara huwa hai*]," marveling that "all the scientific learning of European scholars can be found in the Quran [*ḥukamā'-i farang ke ḥikmatein sab is qur'ān se miltī haiṇ*]."[42]

Karāmat-ʿAlī is quick to highlight the question of Islamic precedence even in such speculative arenas as the question of extraterrestrial life. This comes out in his engagement with the question of multiple worlds, a topic of growing interest with the flood of astronomical advances and new means of viewing the cosmos and our place in it. The debate over whether there were other inhabited worlds took on new urgency with optical improvements in telescopes.[43]

As for the plurality of worlds, Karāmat-ʿAlī argues that the heavens and the stars are inhabited, *asmān-o-sitāre maʿmūr-o-ābād haiṇ*, and that they contain everything found on Earth and more. He does this by drawing on the authority of the Quran and the corpus of sayings ascribed to the Prophet and Imams. Further, he claims that while the old Ptolemaic model holds that the stars and planets were uninhabited, the majority of European scholars now follow Muslims on this matter, *ḥukamāʾ-i farang ghālib mānand-i musalmān*. The plurality of worlds has many points of confluence with earlier Islamic cosmography, well beyond Qazwīnī's treatment of the topic. Many Muslim authorities had long embraced the possibility of multiple worlds in clear contradistinction to the Aristotelian insistence on the singularity of the universe.[44]

Ultimately, Karāmat-ʿAlī concludes that the origins of science are to be found in prophetic knowledge and in the divine guidance of religious learning. This leads to a rejection of the divide between religion and science as a categorical solecism, itself a well-worn argument designed to balance revelation and reason. To the question of benefits Muslims might obtain from European learning, Karāmat-ʿAlī coyly retorts that modern science may grant Muslims a greater appreciation of the truth of their own religion.

Needless to say, these arguments for the precedence and the continued relevance of prophetic authority were not entirely well received. The Orientalist and colonial administrator William Muir (d. 1905), who sat on the prize committee, objected that so little did this treatise "touch on the topic of the advantages to be gained by the Muhammadan mind in India from European learning" that "it can hardly be said to have quite fulfilled the conditions of the essay."[45]

In contrast, the head professor of the Calcutta madrasa, Mawlvī Muḥammad Vajīh, who served as one of the two native Muslims on the committee, retorted that portions of Rehatsek's essay were themselves "contrary to the tenets of Islam, irrelevant to the question, and simply not true [*khalāʾif-i ʿaqāʾid-i islām ast va-khārij az suʾāl va-ghayr ḥaqīqī*]." Ultimately, as Rehatsek failed to produce an Urdu translation, the committee divided the prize money between the two submissions equally.

Karāmat-ʿAlī's treatise represents an early trend that would come to shape Islamic reform, concerned with demonstrating the concord between the Quran and the discoveries of modern science. Often referred to as *iʿjāz ʿilmī*, scientific inimitability, this field of writing is filled with treatises and multi-volume commentaries in numerous languages and is designed to prove that the Quran is fully congruent with science and that, furthermore, its very miraculous nature is demonstrated through amazing allusions to modern scientific knowledge. As a broad formation, these defensive sensibilities formed in response to the rise of European hegemony and are quite distinct from the view promoted by Qazwīnī, following the likes of Fakhr al-Dīn al-Rāzī and others, who saw numerous examples in which the Quran was in complete harmony with the truths of natural philosophy.[46]

Throughout the transformations of the nineteenth century, Qazwīnī continued to find new audiences. The vocabulary of the marvelous and strange offered a ready means for describing the exotic practices and customs of the *farang*. Increasingly, printed lithographs began to circulate, handwritten onto metal plates and mass-produced through printing presses. In 1864, the year Aḥmad Khān established his Scientific Society and Karāmat-ʿAlī penned his essay on the prophetic origin of science, a lithograph copy of the *Kitāb-i ʿAjīb va-gharīb* (Book of wonder and rarity), also known as the *Jahān-numā* (The world revealer), was published in the north Indian city of Meerut.

This Urdu cosmography was originally written in the court of Lucknow for Nawab Muḥammad ʿAlī Shāh (d. 1258/1842), ruler of Awadh, which was at this point a protectorate of the Company. The collection draws extensively from Qazwīnī's *Wonders and Rarities*. The picture offered of the world is shaped by a classical repertoire, with its seven climes, the mountains of Qāf, and the story of Adam landing in Ceylon. But there are distant regions and amazing talismans, as well as Alexander's adventures in the sea of darkness, and the lands beyond, which the *farang* now control.[47]

The collection ends with an Urdu translation of a magic show for the emperor Jahāngīr. The lavish set piece originally written in Persian is drawn from the spurious memoirs that circulated in Jahāngīr's name soon after his death. Widely read, the earliest versions open with an account of the amazing feats of Bengali acrobats. In Mughal writings of the time, Bengalis were often pictured as an exotic people from a distant, if recently conquered, land; the same also held true for the mountainous lands of Assam to the northeast of Bengal known as Kāmarūp, famed for sorcerers and magicians. Notably in the Urdu book of wonders for the nawab of Awadh, the Bengali troop expands to include *farangī* performers in their midst.[48]

The show is staged as an ever-increasing spectacle of juggling, contortions, sleight-of-hand deceptions, sparks of fire, dismemberment, parades of endless fruit and flesh, and a chain suspended in the air on which animals ascend only to disappear. The account concludes that Jahāngīr had never seen the likes of it. These marvelous performances could not be explained away merely as *chashm-bandī*, simple optical illusions, which the uneducated masses could replicate. They must have been produced, the emperor reasoned, through *ʿilm-i sīmiyāʾī*, the science of natural magic, which the Illuminationist philosophers of Iran had discussed. There are those in Europe in the regions of the *farang*, the passage concludes, who excel in this art; through the purification of the soul and sanctification of the heart, they can produce all manner of wonder and rarity.[49]

The image of Europeans as exotic wonderworkers was its own legerdemain. The foreigners came bearing fine-tooled clocks with hidden mechanic levers, looking glasses for spying distant objects, and movable type for printing presses. All of it was imported, then adapted, imitated, and repurposed. The emotive states demanded by something marvelous to behold could lend the entire enterprise of exchange, gift giving, and commerce an air of pleasure and entertainment. Yet often the circulation of these goods and commodities was underwritten by rank exploitation and unchecked profiteering.

The representation of Europeans as strange curiosities offered its own form of control over what would become increasingly uneven conditions for exchange. This is the image that features in nineteenth-century illuminated copies of the *Shujāʿ-i Ḥaydarī*, a Persian book on the wonders of the world written in the name of a certain Muḥammad Ḥaydar. The collection purports to have been commissioned during the reign of Jahāngīr for prince Shujāʿ (d. 1071/1661), son of Shāhjahān, and deposited in the royal library. Yet anachronisms fill its pages, and uncouth phrasing and clunky syntax betray a work written far from an imperial audience. But beyond its rough edges is a world immediately recognizable from writings on islands and strange people. India is a land of wonders, and Alexander is a trustworthy guide.

Accounts of the *farang*, known for their amazing crafts and the great court of the pope, are set alongside savages, magicians, sanyasis, idols, talismans, strange creatures, and delicious fruit. People are frequently described by the color of their skin—red, white, brown, black, olive hued. In France, which like Kashmir blooms with flowers, can be found fairy-countenanced concubines that the *farang* buy and sell. The women wear no veils and can enter any person's house they wish. The descriptions are a pastiche of ideas evidently drawn from no direct knowledge or experience.

A manuscript produced in the first part of the century in north India illustrates European men and women posed with their strange, native attire in immediately recognizable forms (figure C.2). They feature as objects of curiosity alongside rare animals, bare-chested idolaters, holy men, brutes from exotic shores, and naked couples in sundry positions and inclinations. The taxonomical character of the paintings—representing the inhabitants of the world—is meant to both contain and titillate. The illustrations serve in many ways an analogous function to the classifications that came to occupy works of modern European ethnography and political theory, which were, nonetheless, couched in a very different language of science, race, and nation.[50]

The *Shujāʿ-i Ḥaydarī* also made it to print in a lithograph edition published in a north Indian printing house in 1865. By then Qazwīnī's *Wonders and Rarities* had also appeared in a printed edition nearly two decades earlier, adorned with splendid illustrations in the Qajar style by the publisher and artist Mīrzā ʿAlīqulī Khuʾī (fl. 1271/1855). Khuʾī ran one of the most prolific lithograph publishing houses in Tehran. As with earlier manuscript copies, Qazwīnī's *Wonders and Rarities* enjoyed a circulation in lithographic editions that also contained explicitly erotic content nestled in the pages alongside demons, talismans, and magic spells.[51]

In 1866 the industrious Hindu printer Munshi Nawal Kishore (d. 1895) published a lavishly illustrated lithograph of Qazwīnī's natural history in Lucknow, finished with finely colored paintings (figure C.3). Shortly after, an Urdu translation appeared in print, celebrated as a *kitāb-i lā javāb*, an incredible book. By the frequent print runs and editions, *Wonders and Rarities* was indeed an amazing item. It is a point made in a later colloquial, or *rozmarra*, Urdu translation also published by Nawal Kishore. The publisher's notice celebrates the collection as a peerless and unmatchable work of history

FIGURE C.2: A couple from Portugal pictured with a lion and a civet cat, from a north Indian manuscript of the *Shujāʿ-i Ḥaydarī*, copied in the early nineteenth century.

filled with eyewitness accounts on the sciences and the arts—a book continually in demand, with multiple editions that the public is always quick to snatch up.[52]

The printing press had long been heralded as a revolutionary technology, expanding access to the written page as never before. Colonial officials and

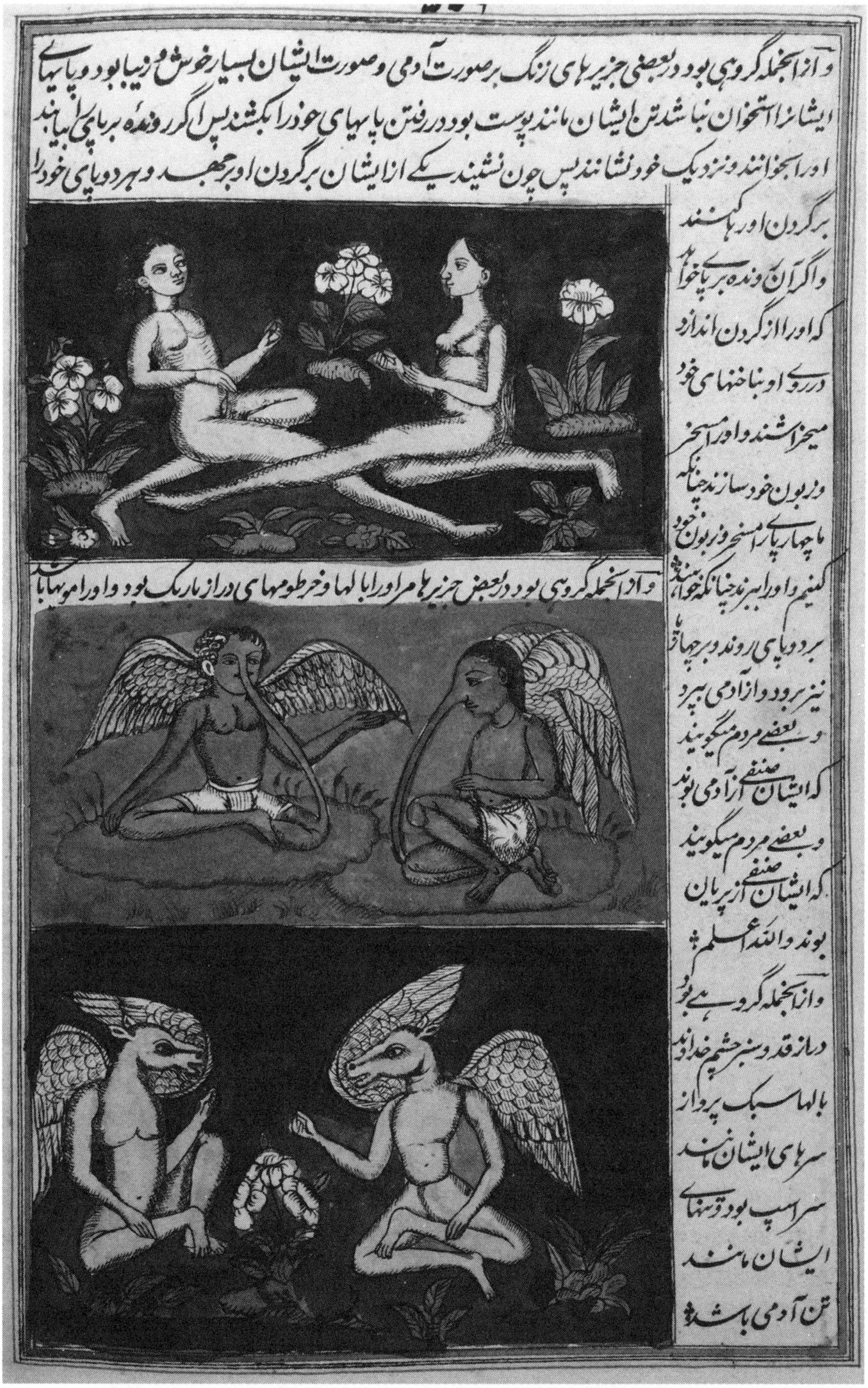

FIGURE C.3: Creatures with strange forms, in a hand-colored Persian lithograph edition of Qazwīnī's *Wonders*, published by Nawal Kishore in 1866.

missionaries described the press as an agent for social improvement, held alongside the magnetic compass and gunpowder as reasons for European progress. It was an idea advanced by countless reformists, such as Aḥmad Khān, who saw printing as a means for uplifting Indians out of the bondage of ignorance. In Aligarh he ran a steam engine printing press, first rendering Urdu with clunky movable type and then with elegant lithograph plates.[53]

The printing press also served as a yardstick for measuring the backwardness of Orientals, who were known to prefer, on the whole, handwritten manuscripts. According to this logic, the late acquisition and adaptation of print was itself a cause of intellectual stagnation. Such reasoning viewed technology not only as shaping modes of production and consumption, but also as forming the very character of society itself, rendering the same results predictably, with uniform consistency.

These notions, however widespread, failed to address the symbolic value associated with calligraphy in the lives of scribes and their readers, as well as the economic structures that bound scribes to patrons. Manuscript culture persisted long after the wide adaptation of print technology in the nineteenth century. When the printing press finally took root, after short-lived earlier attempts in the previous century, it was lithography that emerged as a privileged model for printing Arabic, Turkish, Persian, Urdu, and Malay. This was not for the lack of movable type, but precisely because the lithograph could replicate the visual idiom and authority of calligraphy in all its dexterous elegance.

The unique codex certainly could not match in sheer quantity the number of books manufactured in print that flooded markets and created along the way new classes of readers and new ways of reading. Yet the triumphalism following the rise of print technology was often used as a cudgel to provincialize the very textualized knowledge systems that the importation of print was designed to displace. While Aḥmad Khān hoped to use the printing press to introduce the latest learning from Europe into Urdu, other presses were content to follow the demands of the market. Codes of law, which had practical utility for settling disputes, proved to have immense value with long print runs. So did works of medicine, and with them all variety of writings on sorcery, geomancy, divination, dream interpretation, and talismans.[54]

Then there was the general interest in the natural sciences. Prints of Urdu catechisms of geography and natural history, sold as primers for schools, had long been filled with lessons on Copernicus and Columbus and the casual bigotry of European learning. The terrestrial teachings of missionary and colonial education could not be clearer. When asked *ahl-i jāpān kā dīn kaisā hai,* "What is the religion of Japanese?," students would recite, *but parast haiṇ*, "They are idol worshipers." What is the government of the Persians if not "despotic." Who are the Arabs, whose learning was once so esteemed, if not "exceedingly ignorant." How now are the inhabitants of India subject to the emperor of England? All in unison: *sārī khilqat ārām aur khūbī se rehatī haī,* "everyone lives securely and comfortably." Much better, it turns out, than under the previous government.[55]

But along with all the schooling in the European sciences and matters of measuring the circumference of the Earth and dividing the hemispheres into halves and quarters, the classical models of the universe continued to appeal. Following the adoption of the printing press, illustrated copies of Qazwīnī's *Wonders and Rarities* in Persian, Urdu, and Turkish began to appear in numerous lithograph editions.

Nawal Kishore offered his readers a catalogue of all the books he published in Lucknow and Kanpur. There are separate entries for stories, histories, medicine, law, grammar, poetry, accounting, mathematics, publications on Indian religions, and writings in English and Sanskrit. His catalogue provides a good sense of how books on natural wonders were seen, classified under the heading *majmūʿāt-i ʿilm va-ghayru kutub-i nāyāb-i zamāne,* compendiums of science and other rare books of the age.

Here Qazwīnī's collection features prominently in both Urdu and Persian. So do other titles on natural history and geography that followed Qazwīnī's lead, such as the *Maʿlūmāt al-āfāq* (The sciences of the horizons) by Amīn al-Dīn Khān al-Harawī (fl. 1123/1711), an illustrated book of wonders produced in Mughal India. Nawal Kishore also published this collection in both Persian and Urdu, with illustrations that unveil the sacred landscapes of Mecca and Medina and other holy shrines, alongside demons, a maiden torn from the belly of a fish, the distant mountains of Qāf, elephants carried off by giant birds, monopods, dog-headed men, women growing on trees, enchanted islands and mountains, massive serpents, seafaring adventurers—all drawn across the seven climes and the newly discovered lands beyond.[56]

Alongside these entries in Nawal Kishore's catalogue of imprints features the Urdu travelogue of Yūsuf Khān Kambalposh (fl. 1254/1838), entitled

ʿAjāʾibāt-i farang (Wonders of Europeans), which follows a tried model of viewing distant lands through the emotive cadence of wonder and awe, taking in museums, royal gardens, circus performances, hot air balloons, all while beholding the curious habits of the fairy-faced women walking the streets of London and Paris.[57]

But next to geography, natural history, and travelogues Nawal Kishore also listed collections of talismans in the name of Alexander and Solomon. Again, the association with Qazwīnī is not so far off. Nor is it that strange to find another title classified here: the *Ṭilism-i farang* (The European charm), an Urdu translation of an influential English treatise on the science of mesmerism, which taught techniques on how to hypnotize and anesthetize through the power of animal magnetism. The theory had incredible traction both in Europe and across the colonies of the East. The science of mesmerism and the broader movement of spiritualism with which it was connected were quite far from Qazwīnī. But here his teachings on the power of the soul remained in many ways entirely relevant.[58]

Like other publishers of the day, Nawal Kishore had a healthy print run of epic adventures and marvelous stories, which he classified separately as *qiṣṣe-jāt,* narratives or tales. His Urdu editions of *The Thousand and One Nights, Layla and Majnūn, Khusraw and Shīrīn,* and Firdawsī's *Book of Kings* are representative, as are the amazing adventures of Amīr Ḥamza, the uncle of the Prophet. Several of the titles bear the name *dāstān,* a category increasingly identified in the period with the English genre of the romance, as in the chivalric romances of medieval lore.

As a literary form, the Urdu *dāstān* was immensely popular in India, crafted by professional storytellers, or *dāstāngos,* who could adroitly weave nights of entertainment through entangled tales of unfathomable feats and suspenseful intrigue. Their written collections filled volumes with jinn, demons, fairies, enchanted castles, strange creatures and plants, mighty sorcerers, and powerful talismans. While there may have been a clear line for Nawal Kishore between natural histories of wonder and stories of epic adventures, the worlds they evoked were profoundly connected.[59]

From classical letters to modern compilations, the trusted scientific writings on talismans and prodigies lent the marvelous landscapes conjured by storytellers a conceptual mooring in a cosmos governed by occult forces and hidden sympathies. The work of disenchantment—carving out clearly, once and for all, fiction from fact—would seek to radically upend all this. Throughout the period, the sensibilities that expressed amazement at talismans or powerful jinn—not as fictions of the imagination, but as testaments to the marvelous workings of divine design—came to face increasing scrutiny.

In the course of early Islamic reform movements, the banishment of superstition stands out as one of the many solutions for healing society. The guiding argument of reformist writings produced after the uprising of 1857 held that, to overcome the realities of European domination, the ignorant decay of the past had to be cut off. For Orientalists as well as Muslim reformists, decline became a mode of seeing the world and with it the movement of history.[60]

The identification of practices and beliefs with superstitious ignorance offered one of the main strategies across various reform movements. In this sense Aḥmad Khān's strident stance against superstition is exemplary, with his repeated rejection of the supernatural, which he rendered with the modern Urdu calque *mā fawq-i fiṭrat*. Above all, Aḥmad Khān advocated a thorough reexamination of miracles and supernatural beings such as angels, jinn, and demons.[61]

Among his many efforts, he famously argued that, by a simple lexical confusion, the jinn of pre-Islamic Arabia were, in reality, just savage and wild people, *waḥshī o-jangalī insān*. He likened them to various aboriginals studied by European ethnographers. There are echoes here of the idea, well known to Qazwīnī, that the jinn, in all their sundry shapes and sizes, inhabited far-flung islands. Yet Aḥmad Khān's equation was born of a different age. The novelty of the explanation lies in his linking jinn to primitive barbarity, reshaping Quranic cosmography to fit colonial discourses of civilizational superiority.[62]

Like many reformers of the period, Aḥmad Khān's censure also included magic. He presents a skeptical portrayal of magic as mere trickery and deception, while he is quick to note that the historical persecution of witches was one of the many follies of Europe, *ḥamāqat-i yūrop*. This line of thought is entirely consistent with contemporary attitudes of British colonial officials and intellectuals, who held that the persecutions of hapless folk magic across Europe and the Americas were chapters from a darker age of ignorance. For Aḥmad Khān, the lack of any sustained effort to persecute witchcraft in Islamic history was a testament to the enlightened sensibilities of Muslim authorities. Nonetheless, after a full anatomy of the arts of sorcery, magic is rejected as having no basis in reality and as entirely distinct from the pure religion of Islam, *ṭhēṭh mazhab-i islām*.[63]

Yet it is not often remembered that Aḥmad Khān also maintained that the human psyche had the power to influence the physical world. Drawing on Avicennan metaphysics alongside the latest scientific theories of animal magnetism, Aḥmad Khān argued that in the human soul was a kind of electric or magnetic force or faculty that could exercise a power over others. The turn to electricity and magnetism as explanations for occult power dovetails with Aḥmad Khān's belief in mesmerism.[64]

In European scientific and religious circles of the period, mesmerism offered a way to naturalize occult and spiritual powers. Aḥmad Khān notably identifies the science of mesmerism with *ʿilm al-sīmiyāʾ*, a term that Orientalists of the day frequently rendered as "natural magic" and often associated with spiritualism. He also draws on Ḥājjī Khalīfa, who separated natural magic from mere sorcery, viewing it as a means for harnessing occult powers to influence physical phenomena.[65]

Also important for Aḥmad Khān was the old Arabic compendium on the natural sciences by the Brethren of Purity, which had a notable influence in Urdu letters. The epistles describe the capacity of the rational soul to exercise a power over the animal soul through subtle spiritual influences, *āthār laṭīfa rūḥāniyya*. The physical mechanism behind these forces is the universal soul that emanates throughout existence. This vision of creation posits the capacity of the soul to influence other bodies.[66]

These natural forces, Aḥmad Khān observes, were identified by the likes of Ibn Sīnā and the Illuminationist philosopher Suhrawardī long before the Austrian physician Franz Mesmer (d. 1815) lent his name to the phenomenon. This line of argument leads to the conclusion that the occult powers of mesmerism offer a natural basis for prophetic miracles. It is a vision entirely in keeping with Qazwīnī's own understanding of magic and miracle rooted in Avicennan metaphysics.[67]

By this point, the architecture for such a confluence of associations was well established. The study of mesmerism was a significant chapter in the course of scientific experiments in India, where the colonial space served quite concretely as a laboratory for experimentation. In 1845 the Scottish physician James Esdaile (d. 1859), while in Bengal as a resident surgeon for the Company, began experimenting on natives by using hypnotic techniques of mesmerism as a basis for anesthesia, with the goal of conducting painless surgeries.

India was viewed as a prime location for such experiments. Running through the literature on animal magnetism was the notion that Orientals, along with women and the feeble-minded, were more "susceptible of the Mesmeric influence." Esdaile wrote vivid accounts of hypnotizing Indians with mesmeric procedures that were, he claimed, entirely empirical and scientific and unrelated to magic. He sharply contrasts his ability to harness the occult forces of nature with the superstitious rites of "moollahs and fuqueers," whose rituals may prove at times efficacious through the "Mesmeric influence," even though these witless magicians fail to grasp the true cause of their powers.[68]

The natives, nonetheless, were quick adepts. Mesmerism had an important afterlife in the subcontinent, even if only measured in the volumes of Urdu, Hindi, and Bengali publications dedicated to the topic. While Aḥmad Khān was organizing translations for his Scientific Society, Nawal Kishore published in Lucknow an Urdu translation of the lengthy study by William Gregory (d. 1858) on animal magnetism. Gregory was a prominent Scottish advocate of mesmerism and a professor of chemistry at the University of Edinburgh.

The translation took the title *Ṭilism-i farang*. For Nawal Kishore this "European talisman" could sit comfortably alongside Qazwīnī and his other publications on natural history. The title page also adds: *yaʿnī siḥr-i ḥalāl*, "that is to say licit magic"—a category that has significant resonance in classical theological and legal discourses. The *Ṭilism-i farang* introduced an array of practices and procedures brought together in the field of mesmerism, with its theory of animal magnetism as the natural basis for clairvoyance, prophecy, and hypnosis.[69]

Furthermore, Gregory's manual promised a scientific explanation of the occult forces that primitive societies otherwise identified falsely as the witchcraft of infernal spirits. Highlighting the scientific reception of the work, the *Ṭilism-i farang* was abridged shortly after its original publication, in a process explicitly likened to the summary by Ibn al-Nafīs (d. 687/1288) of Ibn Sīnā's famed medical compendium, the *Canon*. The abridgment was followed by several other works in Urdu that turned to mesmeric techniques as scientific explanations for a panoply of occult phenomena.[70]

British colonial officials frequently noted that titles on mesmerism were in great demand among the local population. The topic prominently features in the popular Urdu digest *ʿAql o-shuʿūr* (Intellect and sagacity) by Mawlvī Sayyid Niẓām al-Dīn, published in 1873, also by Nawal Kishore. The title page advertises diagrams and illustrations and affirms in English that it is suitable "for Indian Girls, Boys, Ladies and Gentlemen." The lithograph engravings open up a landscape of jinn, angels, demons, and talismans, as well as maps of the world, new and old. The Ptolemaic universe follows a discussion of the solar system. Topics include the steamship, the telegraph, electricity, the camera, the locomotive, the thermometer, sign language, as well as diverse forms of calligraphy, typography, and drawing.[71]

The famous scene of European magicians at Jahāngīr's court also appears, just as a sentiment of wonder and awe shaped by classical writings of natural history runs throughout. A section on the marvels of the world follows Alexander's adventures through the land of darkness. Reflections on wonders and rarities open up modern technological and scientific fields of learning. Talismans and geomancy are entirely at home alongside references to Galileo and Copernicus. The entry on mesmerism observes that the discipline was founded in Europe. It appears after sections on the wonders of faraway regions and a discussion of harnessing the occult forces of jinn. In this sense, the title of the Urdu translation on the mesmeric arts, the *Ṭilism-i farang*, is rather fitting. The science of mesmerism is the foreign charm that authorizes Islamic occult learning through the authority of modern European science.[72]

Both Aḥmad Khān and William Gregory enjoyed the patronage of George Douglas (d. 1900), Duke of Argyll, who seriously entertained the reality of mesmeric powers. Needless to say, throughout the nineteenth century numerous European doctors and scientists criticized mesmerism as a form of quackery. Yet this skepticism was also met by a cavalcade of colonial officials, doctors, and travelers, at home and abroad, who believed the mesmeric power of animal magnetism offered a scientific explanation for the powerful charms and incantations encountered in the Orient.[73]

Efforts to universalize occult forces within the physical fabric of nature famously animated the Theosophical Society, founded in 1875, composed of Indians and Europeans alike. Many also took an interest in perennial philosophy and the quest for universal religion, often sought in the eternal, never-changing wisdom of the East. The various currents of spiritualism, furthermore, were frequently vehicles for radical politics of anticolonialism and antimaterialism.[74]

As in Europe, during the late nineteenth and early twentieth century, accounts of conjuring the dead, clairvoyance, mesmerism, hypnosis, and telepathy appeared in Arabic, Turkish, Persian, and Urdu newspapers and journals, along with eyewitness accounts of séances. Skeptics and satirists were quick to ridicule it all as mere trickery. Numerous others turned to the spirit world, discussed in translations and digests of treatises on spiritualism, as a confirmation of Islamic metaphysics.[75] The empirical claims of nineteenth-century European spiritualists had afterlives among Islamic reformists, from such distinct figures as Farīd Wajdī (d. 1954), Egyptian modernist and chief editor of the Azhar university journal, to the revolutionary leader of Iran, Ayatollah Khomeini (d. 1989). Both sought to mobilize the hidden forces of miracles and spirits of classical Islamic cosmography through the currency of modern scientific authority.[76]

As for Aḥmad Khān, his use of animal magnetism reflects efforts to harmonize religion with nature. This all resonates with his famous sobriquet "the naturalist," or *naicharī*, a novel term that some of his Muslim opponents sought to wield against him, though he wore it as a badge of honor. When Aḥmad Khān turned to the topic of mesmerism, he did so in an avowedly Islamic framework. The theory of the subtle workings of the soul as an explanation for occult phenomena had long been advanced by Muslim authorities.[77]

The claim was entirely particularistic, as he sought not only to give priority to classical Islamic metaphysics, but also to Islamicize modern science. Aḥmad Khān's conceptual vocabulary was profoundly informed by earlier philosophical attempts to naturalize miracles and prophecy. Ibn Sīnā, Suhrawardī, and Fakhr al-Dīn al-Rāzī argued that miracles fit into the very fabric of nature through the capacity of the soul to influence other bodies, as developed specifically in the philosophical framework of emanationism. It is an explanation that Qazwīnī helped to spread across the globe.

Despite the many points where Qazwīnī's natural history could still prove useful, for reformists the model of the world that such works represented had to be abandoned. It is a topic taken up by the nawab Sayyid Muḥsin al-Mulk (d. 1907), a close associate of Aḥmad Khān, who was also a member of the larger Aligarh reform movement and had long served in the retinue of the Niẓām of Hyderabad. The nawab delivered a series of lectures in Urdu at the end of 1890 while accompanied by Aḥmad Khān in Allahabad. An edited version was published under the title *Muslamānoṇ kī taraqqī aur unke tanazzul ke asbāb* (The progress of Muslims and the causes of their decline).

Above all, Muḥsin al-Mulk identifies superstition as a main impediment to societal advancement. In Urdu superstition was frequently rendered as *waswās,* as a form of delusion, as in *awhām aur wasāwis,* to convey something akin to "delusions and superstitions," often linked to a belief in the power of demons and jinn. Muḥsin al-Mulk recites well-rehearsed explanations for European progress: religious reformation, the printing press, the discovery of the magnetic compass, the use of vernaculars in the form of national languages for disseminating knowledge.

The pairing of the printing press and the Reformation was an established conceit. Yet Muḥsin al-Mulk also raises doubts. The steam engine press could print treatises on modern geography and geology, as well as age-old books of talismanic spells. He laments that rather than using it as a tool for progress, Muslims have turned to the press to print books that are "childish in the extreme," so much so that "it may well be asked what possible good can be done to education by republishing a book in which it is gravely stated that the Earth is supported on the horns of a cow, and that earthquakes are caused by the cow shifting the weight from one horn to the other." Under attack here is the enduring appeal that classical accounts of cosmography continued to wield.[78]

The widespread printing of Qazwīnī's *Wonders and Rarities* is a testament to the lasting diffusion of this particular picture of the world, continuing on through lithograph editions. Printing presses could as easily sell well-worn books of natural wonders with enduring appeal as could they disseminate the latest innovations of modern learning. Indeed, the numerous prints on talismans and exorcism, in a host of languages from Bengali to Ottoman Turkish, highlight the very different readerships that consumed the printed page.

An array of reformists reproached the ancient vision of the Earth as being supported by a series of celestial beings. They followed missionaries and Orientalists with their critique. The notion ridiculed by Muḥsin al-Mulk—that earthquakes were caused by the raging of a cosmic bull—was frequently singled out. The Egyptian satirist and newspaper editor Muḥammad al-Muwayliḥī (d. 1930), at the end of the nineteenth century, lambasted the learned shaykhs of al-Azhar, historically the premier institution of learning in Cairo, for upholding this same view of the cosmos as supported on a bull teetering on a giant whale on the shoulders of an angel.[79]

Muwayliḥī traveled in the reformist circles of Muḥammad ʿAbduh (d. 1905) and his student Rashīd Riḍā (d. 1935), who clothed the occult world of jinn in the vocabulary of scientific empiricism, reconstituted as microbes—an equivalence shaped by the importation and adaptation of the microscope. These processes of exclusion and adaptation would remain salient among modern reformists.[80]

A potent example can be found in the recommendations issued in 1909 by the renowned Egyptian philologist and statesman Aḥmad Zakī Pāshā, who was charged with reforming the library collections of the Ottoman sultans in the imperial capital of Istanbul. He spent months conducting archival research in the Topkapı Palace, which he lyrically referred to as "the enchanting treasury, sealed by talisman." The phrase is of note if only for the exotic image of the past.[81]

In his report for reforming the Ottoman collections, Zakī Pāshā turned to the numerous holdings on magic, incantations, talismans, magical squares, horoscopes, and other such topics. While Zakī Pāshā promoted open access to the bibliographical collections, here he advised that no borrower should be allowed to see these particular materials, save with the special permission of the head librarian. Zakī Pāshā's efforts to keep the occult currents of history hidden from view fits into a broader movement that sought to cultivate and celebrate fields of learning that could be rationalized and utilized within the pressing demands of modernity.[82]

In the year of his publication on library reform, Zakī Pāshā began teaching in the newly founded Egyptian University of Cairo, with courses on the novel topic of *al-ḥaḍāra al-islāmiyya*, Islamic civilization. Here he delivered influential lectures with the revivalist panache of a wide-read bibliophile on the intellectual history of Arabic and Islamic civilization, two categories that he repeatedly yokes together. His lectures gesture to the natural causes of civilizational decline and degeneration as well as advancement and progress, with reflections on writing and memory as generational purveyors of science and learning.

Zakī Pāshā was also one of the first to promote the theory that Muslims discovered the Americas well before Columbus. The argument meant reading past all the uncertainty, pious awe, and unabashed curiosity that once made the wide purview of Qazwīnī's natural history so enjoyable. In his progression through Arabic letters, Zakī Pāshā was quick to distinguish unambiguously alchemy from chemistry and astrology from astronomy—fields that, he notes, nonetheless had incredible influence on European science. The only value in these discredited sciences lies in what they bequeathed to modernity; everything else about them, it would seem, could be forgotten.[83]

In the modern period, the conceptual isolation of such branches of knowledge reveals a good deal about the fault lines structuring reform. Closing off the archive, as it were, was a foundational chapter in the constitution of particular forms of Islamic modernity, marking certain strains as irrational and destined for erasure, while authorizing other bodies of knowledge as scientific and rational. Much of this has meant delimiting and legitimizing particular categories of learning through the now dominant currency of Western scientific authority.

All acts of curation necessitate procedures for exclusion. Excluded here are diverse fields of knowledge now coded as retrograde, many of which—despite all their ambiguity—once occupied quite prominent places. So thorough have these ruptures been that significant portions of history remain largely unintelligible to the diverse frameworks of modernity. Along these pathways of reform and rejection, Qazwīnī's natural history would not be forgotten, only transformed, celebrated in museums and manuscript collections as a repository for lavish paintings and a curious emblem of medieval lore.

FIGURE C.4: Modern map of the Eastern Hemisphere pasted over the original diagram of the world, preceding the account of Alexander's mission in the encircling ocean, from an illuminated Persian manuscript of Qazwīnī's *Wonders* produced in the Deccan.

In the archives of the New Delhi National Museum there is a finely illuminated manuscript of *Wonders and Rarities.* It is an exemplar of the Persian translation originally produced in the ʿĀdil Shāh court. The final page has a date of sale of 977/1569~, but this is surely a later addition, perhaps to increase its value, as the original date has been smudged out and written over by a later hand. Regardless, the copy is centuries old, and the style of illustrations would suggest that it was originally produced in the Deccan.[84]

The tome passed through several hands before making it to the museum. Somewhere along the way, sometime during the height of the British Raj, a reader—evidently unsatisfied with the original schematic diagram of the world—pasted in its place a modern Urdu map of the Eastern Hemisphere from a lithograph print. It is an image that would be immediately recognizable to modern eyes (figure C.4).

The gesture is perhaps not so odd. Readers had been updating old learning from dusty tomes for centuries. Yet from such a vantage, Qazwīnī's cosmos would appear to reflect a distant and illegible vision from the past. To peer beneath the modern map, to see it all as it once was, would involve great powers of the imagination and sustained forces of sympathy. To conjure up such a world, one might even say, could only be done through some strange act of enchantment.

NOTE ON SOURCES AND METHOD

SIGLA AND ABBREVIATIONS

NOTES

ACKNOWLEDGMENTS

ILLUSTRATION CREDITS

INDEX

Note on Sources and Method

If this were a *dāstān*, an epic romance, or a novel woven through with sorcery, it would need no notes. In our modern languages of the fantastic and the fictional, Qazwīnī would certainly be an appropriate inspiration for such endeavors. The Pakistani writer Musharraf Ali Farooqi has demonstrated as much in *The Merman and the Book of Power: A Qissa* (New Delhi, 2019), told as a contemporary *dāstān*, where Qazwīnī and his natural history feature as the main protagonists. Farooqi is keenly aware of the classical tradition, while also diverting from it to meet the expectations of what talismans and magical rings are meant to evoke in the pages of a novel.

Such fictional writing, as we now call it, need not be tethered to the authority of other voices or be confined to the strictures of fragmentary archives. But this is not a book of fiction. The stories told here have taken a different journey, one made possible through sustained attention to the labor of others, much of which has been preserved—sometimes for centuries—in the form of documents and objects. What survives from these distant ages often does so only through happenstance. But the uncertainties of what has been lost create the possibilities to imagine how it all once was.

The historical realities and sensibilities evoked here—often through vignettes, anecdotes, and detailed descriptions—are drawn from disparate materials that reflect a vast pattern of human endeavor. For those wishing to follow this story further, I have made an attempt to document a path through these records in the notes ahead. Here readers will also find references to contemporary scholarship that can help illuminate the way.

Inspired by the marginal commentary of old, the notes proceed with an eye toward laying out sources as well as further materials and readings for others to pursue. Many are bibliographical diversions on a given topic, to offer greater depth or context to the main body of the text. Others index the lexical detritus that spills over in the course of translation.

This book forms part of a wider effort to situate the study of Islam and the East in a global context, to reimagine the past, and to push the boundaries of what it has meant to be human in the world. In telling these stories of nature, wonderment, and empire, I have sought to question inherited categories and unstated assumptions and to rethink what philology and critique may teach. Such work requires new ways of addressing theory and practice as interconnected domains.

For me this has meant privileging clarity and sympathy as ethical sensibilities for cultivating knowledge. With these considerations in mind, I have also reserved for the notes technical discussions of bibliography and textual history. For the sake of simplicity, abbreviations have been used for resources and materials that feature with some frequency. This includes sigla for several manuscript and printed copies of Qazwīnī's *Wonders and Rarities.*

Many of the archival materials at the heart of this project are now easily accessed as digital copies on the Internet. Past generations could never have so readily drawn on the wealth of sources now available. The Gallica website of the Bibliothèque nationale de France has digitized many of the Arabic, Persian, and Turkish manuscripts that feature here. The digitized collections from the state and university libraries in Munich, Berlin, Gotha, Leipzig, and Vienna are also invaluable resources, as is the Yazmalar website for manuscripts from all over Turkey, save Istanbul. The Parliamentary Library of Tehran has made publicly available a trove of digitized Arabic and Persian manuscripts. The Qatar Digital Library likewise hosts important manuscripts from the British Library.

Particularly useful for the field of early Islamic studies in Arabic and Persian is the Noor Digital Library, sponsored by the Computer Research Center of Islamic Sciences in Iran. Likewise, the resources hosted on the Hathi Trust Digital Library and the many materials scanned from the National Digital Library of India initiative open up an otherwise hard to retrieve universe of rare nineteenth-century lithographs.

As the notes make clear, my book is by no means the first to turn to Qazwīnī's *Wonders and Rarities*. Nor, undoubtedly, will it be the last. Qazwīnī has long attracted scholarly attention. When Ferdinand Wüstenfeld published an Arabic edition in 1848, he did not have available to him the earliest dated manuscript of the work, produced in Wasit, which was completed in Qazwīnī's own lifetime. He also drew on much later copies that were defective in a host of obvious ways.

Shortly after Wüstenfeld, there were several significant publications in German, including a substantial, though partial, German translation by Hermann Ethé (Leipzig, 1868). Julius Ruska published a translation and study of the mineralogical section (Kirchhain, 1896) and an extended article on the various recensions of the text. Franz Taeschner wrote a doctoral dissertation on the chapter of the soul, with a translation and notes (Tübingen, 1912). Eilhard Wiedemann authored articles on natural history, many of which were republished by Fuat Sezgin in two volumes of reprints on Qazwīnī (Frankfurt am Main, 1994). Syrinx von Hees published a study of Qazwīnī's life and cosmography (Wiesbaden, 2002) and several articles in German and English that attend to various elements of his writings. Vladimir P. Demidčik wrote widely on Qazwīnī in Russian, including a monograph published posthumously (Moscow, 2004).

Alma Giese translated most of the natural history into German (Stuttgart, 1986). There is an Italian translation by Francesca Bellino (Milan, 2008). During the course of publication, I also learned that Wheeler Thackston is currently preparing an Arabic edition and English translation of both Qazwīnī's natural history and geography. Such work is long overdue.

The numerous illuminated manuscript copies of *Wonders and Rarities* have captivated the interest of art historians. This includes doctoral dissertations by Hans-Caspar Graf von Bothmer on the Wasit manuscript (Munich, 1971), Julie Anne Oeming Badiee (Ann Arbor, MI, 1991) on the so-called Sarre manuscript, and Vivek Gupta (London, 2020) on the Indian reception. Throughout her career, Karin Rührdanz has published several articles in English and German on various illuminated manuscripts and recensions of the natural history. Rührdanz's recent scholarship drew my attention to the Fatih Mosque copy of the early Persian redaction preserved in the Süleymaniye Library of Istanbul. Persis Berlekamp authored a groundbreaking study on the paintings in the Wasit manuscript in a monograph that also examines a range of other manuscripts, including Ṭūsī's natural history (New Haven, CT, 2011). Stefano Carboni produced an invaluable catalogue of the Ilkhanid manuscript of Qazwīnī's natural history housed in the British Library (Edinburgh, 2015).

Like all endeavors of patient research and reflection, my thinking on this topic has been informed by an array of materials and interlocutors. Inevitably, I have lingered longer with some than with others, contemplating here and there endless wonders to behold, while letting so much else slip by. A single book of the whole world, it turns out, can only contain so much.

Sigla and Abbreviations

Copies of Qazwīnī, *ʿAjāʾib al-makhlūqāt*

B Brit., Or. 14140, Arabic redaction dedicated to ʿAlāʾ al-Dīn al-Juwaynī, copied circa 1300.

D Brit., Or. 1621, Persian translation commissioned in 954/1547 for Ibrāhīm ʿĀdil Shāh, copied before 1066/1656.

F Süleymaniye, Fatih 4174, Persian redaction dedicated to Shāhpūr ibn ʿUthmān, copied 699/1300.

M Munich, Bayerische Staatsbibliothek, MS Cod. arab. 464, Arabic, copied 678/1280 in Wasit.

P Persian redaction dedicated to Shāhpūr ibn ʿUthmān, ed. Yūsuf Bayg Bābāpūr and Masʿūd Ghulāmiyya. Qom: Majmaʿ-i Dhakhāʾir-i Islāmī, 2012.

V Leipzig, Universitätsbibliothek, MS Vollers 736, Arabic, copied 790/1388 in Manfalūṭ, Egypt.

W *Die wunder der schöpfung*, vol. 1 of *Zakarija ben Muhammed ben Mahmud el-Cazwini's Kosmographie*, ed. Ferdinand Wüstenfeld, 2 vols. Göttingen: Dieterichschen Buchhandlung, 1848–1849.

Y Süleymaniye, Yeñi Cami 813, Arabic, copied 722/1322.

Abbreviations

Afʿāl Ibn Sīnā, *al-Shifāʾ, al-Ṭabīʿiyyāt:* 4. *al-Afʿāl wa-l-infiʿālāt*, ed. Maḥmūd Qāsim. Cairo: Dār al-Kitāb al-ʿArabī li-l-Ṭibāʿa wa-l-Nashr, 1969.

Alqāb	Ibn al-Fuwaṭī, *Majmaʿ al-ādāb fī muʿjam al-alqāb*, ed. Muḥammad Kāẓim, 6 vols. Tehran: Muʾassasat al-Ṭibāʿa wa-l-Nashr, 1995.
APW	Dimitri Gutas, *Avicenna and the Aristotelian Tradition: Introduction to Reading Avicenna's Philosophical Works*, 2nd ed. Leiden: Brill, 2014.
Āthār	Zakariyyāʾ ibn Muḥammad al-Qazwīnī, *Āthār al-bilād fī l-akhbār al-ʿibād*. Beirut: Dār Ṣādir, 1960.
Berl.	Archives of the Staatsbibliothek, Berlin.
Biḥār	Muḥammad Bāqir al-Majlisī, *Biḥār al-anwār, al-jāmiʿa li-durar akhbār al-aʾimma al-aṭhār*, 2nd ed., 110 vols. Beirut: Muʾassasat al-Wafāʾ, 1983.
BnF	Archives of the Bibliothèque nationale de France, Paris.
Bodl.	Archives of the Bodleian Library, Oxford University.
Brit.	Archives of the British Library, London.
Buldān	Yāqūt al-Rūmī, *Muʿjam al-buldān*, 7 vols. Beirut: Dār Ṣādir, 1955–1957.
CAL	V. P. Demidčik, *Mir čudes v arabskoj literature XIII–XIV vv.: Zakarija al-Kazvini i žanr mirabilij*. Moscow: Vostočnaja literatura, 2004.
CBL	Archives of the Chester Beatty Library, Dublin.
Cosmos	Persis Berlekamp, *Wonder, Image, and Cosmos in Medieval Islam*. New Haven, CT: Yale University Press, 2011.
CPC	*Calendar of Persian Correspondence*, 11 vols. Calcutta: Imperial Record Department; New Delhi: National Archives of India, 1911–1969.
DIA	*Türkiye Diyanet Vakfı İslâm Ansiklopedisi*, 44 vols. Istanbul: Türkiye Diyanet Vakfı, 1988–2012.
Dozy	Reinhart Dozy, *Supplément aux dictionnaires arabes*, 2 vols. Leiden: E.J. Brill, 1881.
Duwal	Barhebraeus = Ibn ʿIbrī, *Tārīkh mukhtaṣar al-duwal*, ed. Anṭūn Ṣāliḥānī. Beirut: al-Maṭbaʿa al-Kāthūlīkiyya, 1890.
EI2	*The Encyclopaedia of Islam: New Edition*, ed. H. A. R. Gibb et al., 13 vols. Leiden: Brill and Luzac, 1954–2009.
EI3	*The Encyclopaedia of Islam, THREE*, ed. Gudrun Krämer et al., Brill Online, 2007–.

EIr	*Encyclopaedia Iranica*, ed. Ehsan Yarshater. London: Routledge and Kegan Paul, 1982–.
Enz.	Syrinx von Hees, *Enzyklopädie als Spiegel des Weltbildes: Qazwīnīs Wunder der Schöpfung*. Wiesbaden: Harrassowitz, 2002.
FPP	Frank Griffel, *The Formation of Post-Classical Philosophy in Islam*. Oxford: Oxford University Press, 2021.
Funūn	*Kitāb Gharāʾib al-funūn wa-mulaḥ al-ʿuyūn* = *An Eleventh-Century Egyptian Guide to the Universe: The Book of Curiosities*, ed. and trans. Yossef Rapoport and Emilie Savage-Smith. Leiden: Brill, 2014.
FW	*Fort William–India House Correspondence and Other Contemporary Papers Relating Thereto*, 21 vols. Delhi: National Archives of India, 1957–1985.
GAS	Fuat Sezgin, *Geschichte des arabischen Schrifttums*, 17 vols. Leiden: E.J. Brill, 1967–1984, vols. 1–9; Frankfurt am Main: Institut für Geschichte der Arabisch-Islamischen Wissenschaften, 2000–2015, vols. 10–17. Vols. 10–13 translated as *Mathematical Geography and Cartography in Islam and Their Continuation in the Occident*, trans. Renate Sarma and Sreeramula Rajeswara Sarma, 4 vols. Frankfurt am Main: Institut für Geschichte der Arabisch-Islamischen Wissenschaften, 2000–2011.
Ghāyat	Abū l-Qāsim Maslama al-Majrīṭī al-Qurṭubī (ascribed), *Ghāyat al-ḥakīm* = *Das Ziel des Weisen*, ed. Hellmut Ritter. Berlin: Teubner, 1933.
GPT	Frank Griffel, *Al-Ghazālī's Philosophical Theology*. Oxford: Oxford University Press, 2009.
Ḥawādith	*al-Ḥawādith al-jāmiʿa wa-l-tajārib al-nāfiʿa fī l-miʾa al-sābiʿa*, ed. Mahdī Najm. Beirut: Dār al-Kutub al-ʿIlmiyya, 2003.
Ḥay.	Abū l-ʿUthmān al-Jāḥiẓ, *Kitāb al-Ḥayawān*, ed. ʿAbd al-Salām Muḥammad Hārūn, 7 vols. Cairo: Maktabat Muṣṭafā l-Bābī l-Ḥalabī, 1938–1945.
HOC	*The History of Cartography*, ed. John B. Harley and David Woodward, 6 vols. Chicago: University of Chicago Press, 1987–2007.
Iḥyāʾ	Abū Ḥāmid al-Ghazālī, *Iḥyāʾ ʿulūm al-dīn*, 16 parts in 4 vols. Cairo: Lajnat Nashr al-Thaqāfa al-Islāmiyya, 1356–1357/1937–1938.

Ikhwān Ikhwān al-Ṣafāʾ, *Rasāʾil*, ed. Buṭrus al-Bustānī, 4 vols., Beirut: Dār Ṣādir, 1957.

Ilāhiyyāt Ibn Sīnā, *al-Shifāʾ, al-Ilāhiyyāt*, ed. Saʿīd Zāyid et al., 2 vols. Cairo: al-Hayʾa al-ʿĀmma li-Shuʾūn al-Maṭābiʿ al-Amīriyya, 1960.

Ilkh. Stefano Carboni, *The Wonders of Creation and the Singularities of Painting: A Study of the Ilkhanid London Qazvīnī*. Edinburgh: Edinburgh University Press, 2015.

Irshād Yāqūt al-Rūmī, *Muʿjam al-udabāʾ = Irshād al-arīb ilā maʿrifat al-adīb*, ed. Iḥsān ʿAbbās, 7 vols. Beirut: Dār al-Gharb al-Islāmī, 1993.

Ishārāt Ibn Sīnā, *al-Ishārāt wa-l-tanbīhāt*, ed. Mujtabā Zāriʿī, Qom: Bustān-i Kitāb-i Qum, 1381sh/2002.

JAOS *Journal of the American Oriental Society.*

JRAS *Journal of the Royal Asiatic Society of Great Britain and Ireland.*

Kashf Ḥājjī Khalīfa, *Kashf al-ẓunūn ʿan asāmī l-kutub wa-l-funūn*, ed. Muḥammad Sharaf al-Dīn Yāltqāyā and Rifʿat Bīlga al-Kilīsī, 2 vols. Istanbul: Wikālat al-Maʿārif al-Jalīla, 1941–1943.

Khalīqa ps. Apollonius of Tyana, *Sirr al-khalīqa wa-ṣanʿat al-ṭabīʿa*, ed. Ursula Weisser. Aleppo: Maʿhad al-Turāth al-ʿIlmī al-ʿArabī, 1979.

KMSI Archives of the Kitābkhāna-yi Majlis-i Shūrā-yi Islāmī, Tehran.

Maʿādin Ibn Sīnā, *al-Shifāʾ, al-Ṭabīʿiyyāt: 5. al-Maʿādin wa-l-āthār al-ʿulwiyya*, ed. ʿAbd al-Ḥalīm Muntaṣir, Saʿīd Zāyid, and ʿAbd Allāh Ismāʿīl. Cairo: al-Hayʾa al-ʿĀmma li-Shuʾūn al-Maṭābiʿ al-Amīriyya, 1965.

Maʿālim Najm al-Dīn al-Kātibī, *Asʾilat Najm al-Dīn al-Kātibī ʿan al-Maʿālim li-Fakhr al-Dīn al-Rāzī maʿ Taʿālīq ʿIzz al-Dawla Ibn Kammūna*, ed. Sabine Schmidtke and Reza Pourjavady. Tehran: Muʾassasa-yi Pizhūhishī-yi Ḥikmat wa-Falsafa-yi Īrān, 2007.

Mabāḥith Fakhr al-Dīn al-Rāzī, *al-Mabāḥith al-mashriqiyya fī ʿilm al-ilāhiyyāt wa-l-ṭabīʿiyyāt*, ed. Muḥammad al-Muʿtaṣim al-Baghdādī. Beirut: Dār al-Kitāb al-ʿArabī, 1990.

Mafātīḥ Fakhr al-Dīn al-Rāzī, *Mafātīḥ al-ghayb*, 32 vols. Cairo: al-Maṭbaʿa al-Bahiyya al-Miṣriyya, 1933–1962.

Maktūm Fakhr al-Dīn al-Rāzī, *al-Sirr al-maktūm*. Berl., Petermann I 207.

Manāfiʿ	Galen, *Manāfiʿ al-aʿḍāʾ*, BnF Arabe 2853; Greek text edited as *De usu partium*, ed. Georg Helmreich, 2 vols. Leipzig: Teubner, 1907–1909; English translation of Greek, *Galen on the Usefulness of the Parts of the Body*, trans. Margaret Tallmadge May, 2 vols. Ithaca, NY: Cornell University Press, 1968.
Masālik	Ibn Khurdādhbih, *al-Masālik wa-l-mamālik*, ed. Michael Jan de Goeje. Leiden: E.J. Brill, 1889.
Maṭālib	Fakhr al-Dīn al-Rāzī, *al-Maṭālib al-ʿāliya min al-ʿilm al-ilāhī*, ed. Aḥmād Ḥijāzī al-Saqqā, 9 vols. Beirut: Dār al-Kitāb al-ʿArabī, 1987.
Mongols	Peter Jackson, *The Mongols and the Islamic World: From Conquest to Conversion*. New Haven, CT: Yale University Press, 2017.
Muḥaṣṣal	Fakhr al-Dīn al-Rāzī, *Muḥaṣṣal afkār al-mutaqaddimīn wa-l-mutaʾakhkhirīn min al-ʿulamāʾ wa-l-ḥukamāʾ*, ed. Ṭāhā ʿAbd al-Raʾūf Saʿd. Cairo: Maktabat al-Kulliyyāt al-Azhariyya, 1978.
Mulakhkhaṣ	Fakhr al-Dīn al-Rāzī, *al-Mulakhkhaṣ fī l-ḥikma*, Berl., Or. oct 623.
Murūj	Abū l-Ḥasan al-Masʿūdī, *Murūj al-dhahab wa-maʿādin al-jawhar*, ed. Charles Pellat, 7 vols. Beirut: al-Jāmiʿa al-Lubnāniyya, 1965–1979.
Musnad	Aḥmad ibn Ḥanbal, *Musnad*, ed. Shuʿayb al-Arnāʾūṭ, 52 vols. Beirut: Muʾassasat al-Risāla, 1993–2001.
Muʿtabar	Abū l-Barakāt al-Baghdādī, *Kitāb al-Muʿtabar fī l-ḥikma*, 3 vols. Hyderabad: Jamʿiyyat Dāʾirat al-Maʿārif al-ʿUthmāniyya, 1357–1358/1938–1939.
Nadīm	Ibn al-Nadīm, *al-Fihrist*, ed. Ayman Fuʾād Sayyid, 4 vols. London: al-Furqan Islamic Heritage Foundation, 2009.
NG	Manfred Ullmann, *Die Natur- und Geheimwissenschaften im Islam*. Leiden: E.J. Brill, 1972.
NLM	Archives of the National Library of Medicine, Maryland.
Nuzhat	Abū ʿAbd Allāh al-Idrīsī, *Nuzhat al-mushtāq fī ikhtirāq al-āfāq*, ed. E. Cerulli et al., 9 fasc. Naples and Rome: Istituto Universitario Orientale, 1970–1984; republished in 2 vols. Cairo: Maktabat al-Thaqāfa al-Dīniyya, 1994.
Observ.	Aydın Sayılı, *The Observatory in Islam and Its Place in the General History of the Observatory*. Ankara: Türk Tarih Kurumu Basımevi, 1960.

PPT	Heidrun Eichner, "The Post-Avicennian Philosophical Tradition and Islamic Orthodoxy: Philosophical and Theological Summae in Context," unpublished Habilitationsschrift. Martin-Luther-Universität, Halle-Wittenberg, 2009.
Q.	Citations to the Quran, by chapter and verse.
Sittīnī	Fakhr al-Dīn al-Rāzī, *Jāmiʿ al-ʿulūm: Sittīnī*, ed. Sayyid ʿAlī Āl-i Dāwūd. Tehran: Thurayyā, 1382sh/2003.
Storey	Charles Ambrose Storey, *Persian Literature: A Bio-Bibliographical Survey*, 2 vols. London: Luzac and Co., 1927–1977.
Süleymaniye	Archives of the Süleymaniye Yazma Eser Kütüphanesi, Istanbul.
Tadhkira	Naṣīr al-Dīn al-Ṭūsī, *al-Tadhkira fī ʿilm al-hayʾa*, published as *Naṣīr al-Dīn al-Ṭūsī's Memoir on Astronomy*, ed. and trans. F. Jamil Ragep, 2 vols. Berlin: Springer, 1993.
TOK	*Treasures of Knowledge: An Inventory of the Ottoman Palace Library (1502/3–1503/4)*, ed. Gülru Necipoğlu, Cemal Kafadar, and Cornell Fleischer, 2 vols. (Leiden: Brill, 2019).
TPMK	Archives of the Topkapı Sarayı Müzesi, Istanbul.
ʿUmarī	Ibn Faḍl Allāh al-ʿUmarī, *Masālik al-abṣār fī mamālik al-amṣār*, ed. Mahdī l-Najm, 27 vols. Beirut: Dār al-Kutub al-ʿIlmiyya, 2010.
ʿUyūn	Ibn Abī Uṣaybiʿa, *ʿUyūn al-anbāʾ fī ṭabaqāt al-aṭibbāʾ*, ed. August Müller, 2 vols. Cairo: al-Maṭbaʿa al-Wahbiyya, 1882–1884.
Wāfī	Ṣalāḥ al-Dīn al-Ṣafadī, *al-Wāfī bi-l-wafayāt*, ed. Hellmut Ritter et al., 32 vols. Various locations and publishers, 1962–2013.
WTI	Fuat Sezgin, *Wissenschaft und Technik im Islam*, 5 vols. Frankfurt am Main: Institut für Geschichte der Arabisch-Islamischen Wissenschaften an der Johann Goethe-Universität, 2003; translated as *Science and Technology in Islam*, trans. Renate Sarma and Sreeramula Rajeswara Sarma, 5 vols. Frankfurt am Main: Institut für Geschichte der Arabisch-Islamischen Wissenschaften an der Johann Goethe-Universität, 2010.
ZGAIW	*Zeitschrift für Geschichte der arabisch-islamischen Wissenschaften.*

Notes

Introduction

1. Thomas John Newbold, "Notice of the *Ajaib-al-Mukhlukat*," *Journal of the Asiatic Society of Bengal* 13, no. 2 (1844): 632–666, at 634. For Chang and Eng Bunker (d. 1874), see Yunte Huang, *Inseparable: The Original Siamese Twins and Their Rendezvous with American History* (New York: Liveright, 2018).

2. For reflections on ambiguity as an epistemic value and moral disposition, see Shahab Ahmed, *What is Islam? The Importance of Being Islamic* (Princeton: Princeton University Press, 2016); Thomas Bauer, *A Culture of Ambiguity: An Alternative History of Islam*, trans. Hinrich Biesterfeldt and Tricia Tunstall (New York: Columbia University Press, 2021); *FPP*, 471–478.

3. Ibn Sīnā, *al-Shifāʾ, al-Manṭiq: 8. al-Khaṭāba*, ed. Muḥammad Salīm Sālim (Cairo: al-Maṭbaʿa al-Amīriyya, 1954), 103–104; cited in Lara Harb, *Arabic Poetics: Aesthetic Experience in Classical Arabic Literature* (Cambridge: Cambridge University Press, 2020), 94.

4. In European medieval and premodern intellectual history, the topic of wonder has received considerable attention. See, for instance, Stephen Greenblatt, "Resonance and Wonder," *Bulletin of the American Academy of Arts and Sciences* 43, no. 4 (1990): 11–34; John Onians, "'I Wonder . . .': A Short History of Amazement," in *Sight and Insight: Essays on Art and Culture in Honour of E. H. Gombrich at 85*, ed. John Onians (London: Phaidon Press, 1994), 11–33; Caroline Walker Bynum, "Wonder," *American Historical Review* 102, no. 1 (1997): 1–26; Lorraine Daston and Katharine Park, *Wonders and the Order of Nature, 1150–1750* (Cambridge, MA: Zone Books, 1998), 9–20. For points of intersection between Islamic philosophy and Latinate discourses, see Michelle Karnes, "Marvels in the Medieval Imagination," *Speculum* 90, no. 2 (2015): 327–365.

5. On intersections with the Quran, see Mohammed Arkoun, "Peut-on parler de merveilleux dans le Coran?," in *L'étrange et le merveilleux dans l'Islam médiéval: Actes du colloque tenu au Collège de France à Paris, en mars 1974*, ed. Mohammed Arkoun et al. (Paris: Éditions J.A., 1978), 1–25; Waḥīd Saʿfī, *al-ʿAjīb wa-l-gharīb fī kutub tafsīr al-Qurʾān* (Damascus: al-Awāʾil, 2006), 33–51. See also Nasser Rabbat, "*ʿAjīb* and *Gharīb:* Artistic Perception in Medieval Arabic Sources," *Medieval History Journal* 9, no. 1 (2006): 99–113. For reflections on awe and wonder as emotions shared with other primates, see Jane Goodall, "Primate Spirituality," in *Encyclopedia*

of Religion and Nature, ed. Bron Taylor et al., 2 vols. (London: Continuum, 2005), 2:1303–1306; Cynthia Willett, *Interspecies Ethics* (New York: Columbia University Press, 2014), 100–130.

6. For wonder as modern historical inquiry, see Bynum, "Wonder," 23–26; and more broadly, Marnie Hughes-Warrington, *History as Wonder: Beginning with Historiography* (New York: Routledge, 2019).

7. In the context of modern philosophy, see Mary-Jane Rubenstein, *Strange Wonder: The Closure of Metaphysics and the Opening of Awe* (New York: Columbia University Press, 2008). For a philosophical exploration, see Sophia Vasalou, *Wonder: A Grammar* (Albany: SUNY Press, 2015).

8. As related by the jurist Muḥammad ibn Idrīs al-Shāfiʿī (d. 204/820) in M fols. 6b, 222b (W 12, 451). Compare Abū Nuʿaym al-Iṣfahānī (d. 430/1038), *Ḥilyat al-awliyāʾ wa-ṭabaqāt al-aṣfiyāʾ,* 10 vols. (Cairo: Maktabat al-Khānjī, 1932–1938), 9:127–128.

9. See Qazwīnī's place in the exhibition catalogues by Marthe Bernus-Taylor and Cécile Jail, eds., *L'étrange et le merveilleux en terres d'Islam* (Paris: Réunion des musées nationau, 2001); and Helga Rebhan, ed., *Die Wunder der Schöpfung: Handschriften der Bayerischen Staatsbibliothek aus dem islamischen Kulturkreis* (Wiesbaden: Harrassowitz, 2010).

10. For Qazwīnī's cosmography as "un objet de curiosité," see the note dated 1834 on the manuscript acquired by the head surgeon of the French army during the conquest of Algeria, Qazwīnī, *ʿAjāʾib,* Bordeaux, France, Bibliothèque de Bordeaux, MS 1130, fol. 2a.

11. See Thomas John Newbold, "Visit to the Bitter Lakes, Isthmus of Suez," *JRAS* 8 (1846): 355–360; Newbold, *Political and Statistical Account of the British Settlements in the Straits of Malacca,* 2 vols. (London: J. Murray, 1839), 2:191–194, 202–204, 351–361, 367–368. Compare Farouk Yahya, *Magic and Divination in Malay Illustrated Manuscripts* (Leiden: Brill, 2015), 14, 50–51, 112, 152–153. See also Horace Hayman Wilson, "Proceedings of the Twenty-Eighth Anniversary Meeting of the Society," *JRAS* 13 (1852): i–xvi, at ii–v; "Foreign and Colonial News: India," *Illustrated London News,* August 3, 1850, 99.

12. The Ring of Gyges from Plato features in the early Persian redaction of Qazwīnī's *ʿAjāʾib* in the section on the talismanic arts, citing Plato's *Kitāb-i Siyāsāt,* F 221 (P 545); compare *Ikhwān,* 4:287. Newbold's description indicates he had this Persian version of the text. Newbold favorably compares Qazwīnī's arguments to the defense of natural theology in the *Bridgewater Treatises* on religion and science, "Notice," 634. See Jonathan Topham, "Beyond the 'Common Context': The Production and Reading of the *Bridgewater Treatises,*" *Isis* 89, no. 2 (1998): 233–262.

13. See the *Arabian Nights* in Tzvetan Todorov, *The Fantastic: A Structural Approach to a Literary Genre,* trans. Richard Howard (Cleveland: Case Western Reserve University Press, 1973), 24–57. Compare Bynum, "Wonder," 2n6; Daston and Parks, *Wonders,* 329–363.

14. Thomas Warton, "Of the Origin of Romantic Fiction in Europe," in *The History of English Poetry,* 3 vols. (London: J. Dodsley, 1774–1781), vol. 1, unpaginated, at 14, 78.

15. For a rejection of Warton's thesis, see John Dunlop (d. 1842), *The History of Fiction,* 3 vols. (London: Longman, 1814), 1:137–140. See also Frederick Nolan (d. 1864), "Origin of the Marvellous or Poetical Machinery of Old England," excerpted in *The Annual Register* 1809 (printed 1811), 840–843.

16. See Marina Warner, *Stranger Magic: Charmed States and the Arabian Nights* (Cambridge, MA: Harvard University Press, 2012), 28–29, 289–322.

17. See Talal Asad, *Formations of the Secular, Christianity, Islam, Modernity* (Stanford: Stanford University Press, 2003), 12–16, 25n9, 170. See also Aziz al-Azmeh, *Islams and Modernities* (London: Verso, 1993), 11–12, 24–26.

18. For cabinets of curiosity, see Daston and Park, *Wonders*, 255–296.

19. For the category of the *ʿajāʾib-ghar*, see Shaila Bhatti, *Translating Museums: A Counterhistory of South Asian Museology* (Walnut Creek, CA: Left Coast Press, 2012), 3–82.

20. For Urdu horoscopes of the nawab, see Salar Jung Museum, MSS. no. Ramal 10–19, dated 1919–1929.

21. Description based on personal notes of the Salar Jung Museum's collections and manuscript reading room (February 2015). On the prayer, see al-Majlisī (d. 1110/1698), *Biḥār*, 20:73.

22. See Yakup Bektas and Roger Sherman, "A Bold New Enterprise: The Istanbul Museum of the History of Science and Technology in Islam," *Technology and Culture* 54, no. 3 (2013): 619–639, at 623. Compare *WTI*, 1:85–179; trans. 1:79–165.

23. See, for instance, Muḥammad ʿAbd Allāh ʿInān, "Amrīkā l-junūbiyya: Hal qāma l-ʿarab bi-iktishāfihā riḥla ʿarabiyya fī baḥr al-ẓulumāt?," *al-Hilāl* 34, no. 9 (1354/1935): 1079–1082; ʿAbd al-ʿAzīz al-Islāmbūlī, "al-ʿAql al-ʿarabī wa-l-thaqāfa al-islāmiyya humā l-mashʿal alladhī baddada l-ẓulām fī urūba wa-fataḥa l-ṭarīq ilā amrīkā," *Minbar al-Islām* 18, no. 1 (1380/1960): 49–55; M. D. W. Jeffreys, "Arabs Discover America before Columbus," *Muslim Digest* 3, no. 11 (1953): 67–74; Mohammed Hamidullah, "Les musulmans en Amerique d'avant Christophe Colomb," *France-Islam: Organe de l'Amicale des musulmans en Europe* 13–14 (1968): 7–14, translated and expanded into Turkish by İhsan Süreyya Sırma, "Kristof Kolomb'dan Önce Müslümanların Amerika'yı Keşfi," *İlk İslam Devleti* (Istanbul: Beyan, 1992), 91–106. See also Hasan Tahsin Fendoğlu, *Modernleşme Bağlamında Osmanlı-Amerika İlişkileri (1786–1929)* (Istanbul: Beyan, 2002), 149–155. In Urdu, see Sayyid Sulaymān Nadvī, "ʿArab o-Amrīka," *Maʿārif* (Dār al-Muṣannifīn, Shiblī Academy, Aʿẓamghar) 43, no. 3 (March 1939): 165–175, who draws on the publications of Leo Wiener. See also Barry Fell, *Saga America* (New York: Times Books, 1980), translated into Arabic as *Iktishāf Amrīkā qabl Kūlumbus*, trans. ʿAbd al-Qādir Muṣṭafā Muḥayshī Fuʾād Kaʿbāzī (Tripoli, Libya: Dār al-Kutub, 1988); Heinke Sudhoff, *Sorry Kolumbus: Seefahrer der Antike entdecken Amerika* (Bergisch Gladbach, Germany: Lübbe, 1990), translated into Arabic as *Maʿdhiratan Kūlūmbūs: Lasta awwal man iktashafa Amrīkā*, trans. Ḥusayn ʿImrān (Riyadh: Maktabat al-ʿUbaykān, 2001).

24. See, for instance, Recep Tayyip Erdoğan, "Türkiye'nin Tarih Boyunca Taşıdığı Misyon, Barışı Hâkim Kılmak ve Zülme Karşı Durmaktır," www.tccb.gov.tr (November 15, 2014); Recep Erdoğan, "Bu Nesil Bilgi ve Hikmeti Beraber Yüklenmelidir; O Zaman Gelecek Daha Farklı Olacaktır," www.tccb.gov.tr (November 18, 2014); "Fuat Sezgin: Amerika'yı Müslümanların keşfettiğini kitabımda yazdım," *Hürriyet*, www.hurriyet.com.tr (November 18, 2014).

25. Description of the Istanbul Museum of History of Science and the gift shop of the Topkapı Sarayı based on personal observation (May 2015). See Fuat Sezgin, *Piri Reis and the Pre-Columbian Discovery of the American Continent by Muslim Seafarers* (Istanbul: Boyut, 2013); published in Turkish as *Amerika kıtasının müslüman denizciler tarafından Kolomb öncesi keşfi ve Pîrî Reis* (Istanbul: Boyut, 2013), a digest of Sezgin's arguments for the Muslim discovery of America in *GAS*, 13:119–165 (trans. 4:113–157).

26. See Bruce Mazlish, *Civilization and Its Contents* (Stanford: Stanford University Press, 2004), 20–48.

27. For further context, see Ahmed El Shamsy, *Rediscovering the Islamic Classics: How Editors and Print Culture Transformed an Intellectual Tradition* (Princeton, NJ: Princeton University Press, 2020), 123–146.

28. See Anwar al-Jindī, *Aḥmad Zakī l-mulaqqab bi-shaykh al-ʿurūba: Ḥayātuhu, ārāʾuhu, āthāruhu* (Cairo: al-Muʾassasa al-Miṣriyya al-ʿĀmma, 1964), 128–130, 288–289. See also Aḥmad Zakī Pāshā (Ahmad Zeki Pacha), "Une seconde tentative des musulmans pour découvrir l'Amérique," *Bulletin de l'Institut d'Égypt* 2 (1920): 57–59.

29. See Karuna Mantena, *Alibis of Empire: Henry Maine and the Ends of Liberal Imperialism* (Princeton, NJ: Princeton University Press, 2010); Lisa Lowe, *The Intimacies of Four Continents* (Durham, NC: Duke University Press, 2015); Joseph Massad, *Islam in Liberalism* (Chicago: University of Chicago Press, 2015); *Islam after Liberalism*, ed. Faisal Devji and Zaheer Kazmi (Oxford: Oxford University Press, 2017).

30. See Gil Anidjar, "Secularism," *Critical Inquiry* 33, no. 1 (2006): 52–77, at 66; cited in Robert Yelle, *The Language of Disenchantment, Protestant Literalism and Colonial Discourse in British India* (Oxford: Oxford University Press, 2013), 3–4.

31. See Nelson Goodman, *Ways of Worldmaking* (Indianapolis: Hackett, 1978); 92–97; compare Israel Scheffler, "The Wonderful Worlds of Goodman," *Synthese* 45, no. 2 (1980): 201–209; Ayesha Ramachandran, *The Worldmakers: Global Imagining in Early Modern Europe* (Chicago: University of Chicago Press, 2015), 6–10. For more on imaginative geography, see Edward Said, *Orientalism* (New York: Vintage, 1978), 54–55.

32. For enchantment and modernity, see Alex Owen, *The Place of Enchantment: British Occultism and the Culture of the Modern* (Chicago: Chicago University Press, 2004); David Martin, *Curious Visions of Modernity: Enchantment, Magic, and the Sacred* (Cambridge, MA: MIT Press, 2011); Jason Ā. Josephson-Storm, *The Myth of Disenchantment: Magic, Modernity, and the Birth of the Human Sciences* (Chicago: University of Chicago Press, 2017). See also Christopher Partridge, *The Reenchantment of the West: Alternative Spiritualities, Sacralization, Popular Culture, and Occulture*, 2 vols. (London: T and T Clark International, 2004–2005); Jeffrey Kripal, *Authors of the Impossible: The Paranormal and the Sacred* (Chicago: University of Chicago Press, 2010).

1: Stranger Lands

1. M fol. 7a. The master is referred to as Imam al-Anṣārī. Qazwīnī relates the same story in *Āthār*, 406–407. Here, Ghazālī's medium is the occultist and religious authority Abū l-Faḍl Muḥammad al-Ṭabasī (d. 482/1089). Imam al-Anṣārī in Qazwīnī's *ʿAjāʾib* might be Imam Abū l-Qāsim al-Anṣārī, whom Ṭabasī refers to as the "great master" (*al-ustādh al-akbar*), or as "my master" throughout his collection of spells and incantations, *al-Shāmil fī l-baḥr al-kāmil*, Berl., Or. fol. 52, at 6; drawing on Ṭabasī's *Shāmil*, Anṣārī also features in *Maktūm*, fols. 167b, 168a, 171b.

2. Q. 72:2. See Abū Jaʿfar al-Ṭabarī (d. 310/923), *Jāmiʿ al-bayān fī taʾwīl al-Qurʾān*, ed. ʿAbd Allāh ibn ʿAbd al-Muḥsin al-Turkī, 26 vols. (Cairo: Hajar, 2001), 23:310–320.

3. See Aḥmad ibn Ḥanbal (d. 241/855), *Musnad*, 6:325–326n2, §3784. See Franz Rosenthal, "The Stranger in Medieval Islam," *Arabica* 44, no. 1 (1997): 35–75.

4. M fol. 1b; Y fol. 1b; compare B fol. 1b; V fol. 2b. For "wa-khayru jalīsin . . . ," see Abū l-Ṭayyib al-Mutanabbī (d. 354/955), in the commentary attributed to ʿAbd Allāh ibn al-Ḥusayn al-ʿUkbarī (d. 616/1219), *al-Tibyān fī sharḥ al-Dīwān*, ed. Kamāl Ṭālib (Beirut: Dār al-Kutub

al-ʿIlmiyya, 1997), 203. Compare Shams al-Dīn Muḥammad al-Ṭūsī (fl. 562/1166), *Kitāb-i ʿAjāʾib-nāma*, ed. Manūchihr Sutūda (Tehran: Nashr-i Kitāb, 1966), 12.

5. In addition to references in the *Āthār*, this sketch of Qazwīnī's life draws from the following: *Alqāb*, 1:313–314, §444, 2:66–67, §1050, 4:261, §3806, 5:372, §5290; the history attributed to Ibn al-Fuwaṭī, *Ḥawādith*, 212, 299; Shams al-Dīn Muḥammad al-Dhahabī (d. 748/1348), *Tārīkh al-islām wa-wafayāt al-mashāhīr wa-l-aʿlām*, ed. ʿUmar ʿAbd al-Salām Tadmurī, 53 vols. (Beirut: Dār al-Kitāb al-ʿArabī, 1987), 51:101, §85; Ibn Taghrībirdī (d. 874/1470), *al-Manhal al-ṣāfī wa-l-mustawfī baʿd al-wāfī*, ed. Muḥammad Muḥammad Amīn, 13 vols. (Cairo: al-Hayʾa al-Miṣriyya al-ʿĀmma li-l-Kitāb, 1984–2009), 5:265, §1042; Ḥājjī Khalīfa; *Kashf*, 1:9, 2:1127–1128. See also *Enz.*, 18–90; *Cosmos*, 10–15, 46–50; *CAL*, 45–51.

6. See *Āthār*, 8 (*al-bilād al-islāmiyya*), 265 (reference to Egypt), 435 (congregational mosque of Qazvin), 502–503 (*al-mamlaka al-islāmiyya* in al-Andalus surrounded by Franks versus *al-mamlaka al-naṣrāniyya* in the Levant surrounded by Muslims), 461 (Mosul), 498 (Franks in the *bilād al-islām*), 509 (Bukhara), 530 (Byzantium), 538 (Ṭarāz), 620 (Bulghār); for references to the *bilād allāh*, see 40 (Sabaʾ), 53 (Ṣīn), 57 (Ghāna), 66 (Yemen), 131 (Yamāma), 199 (Raqqāda), 202 (Sodom), 232 (al-Ghūṭa), 247 (Kirman), 375 (Samarra), 396 (Shādhiyākh), 451 (near Nahāvand), 473 (Nishapur), 482 (Herat), 520 (Gorgan), 538 (Shāsh), 544 (Sogdia).

7. See *Āthār*, 183 (market, Aleppo), 199–200, 545, 568 (waterwheels), 287 (watermills, Abhar), 481 (windmills, Herat), 387 (calligraphic artistry, astrolabes, and globes in the madrasa of Sāwa, as well as hospitals and hospices), 522 (silks, Ganja).

8. *Āthār*, 7–8. On *sōma* and *polis*, see Plato, *Republic*, 372e, 399e, 556e. See Abū Naṣr al-Fārābī, *Mabādiʾ ārāʾ ahl al-madīna al-fāḍila = Al-Farabi on the Perfect State*, ed. and trans. Richard Walzer (Oxford: Clarendon Press, 1985), 230, §4; 434–435 (commentary). See also Miriam Galston, *Politics and Excellence: The Political Philosophy of Alfarabi* (Princeton, NJ: Princeton University Press, 1990), 146–179. For climatic determinism as a basis of political order, see *Ilāhiyyāt*, 2:451–455; Ibn Sīnā (ascribed), *Kitāb al-Siyāsa*, ed. ʿAlī Muḥammad Isbar (Jableh, Syria: Bidāyāt, 2007), 61–63. See also *Iḥyāʾ*, 12:2272, 2274–2276 (128, 130–132); *Mafātīḥ*, 26:199–200, Q. 38:26–29.

9. Ibn al-Athīr (d. 630/1233), *al-Kāmil fī-l-tārīkh*, ed. Muḥammad Yūsuf Daqqāq et al., 11 vols. (Beirut: Dār al-Kutub al-ʿIlmiyya, 1987), 10:409. Compare *Enz.*, 43; *CAL*, 46.

10. See *Mongols*, 71–93.

11. See *Āthār*, 57, 148, 149, 352; Qazwīnī, *Āthār al-bilād fī akhbār al-ʿibād*, ed. Ferdinand Wüstenfeld (Göttingen: Dieterichschen Buchhandlung, 1848), 63, 66; cited in *Enz.*, 62–63.

12. For regions destroyed under the Khwārazm-Shāh, see *Āthār*, 236 (Farghāna), 538 (Shāsh), 558 (Transoxiana); for references to the conquests, see 236 (Farghāna), 395 (Shādhiyākh), 410 (Ṭarūz), 465–466 (Nasā), 482 (Herat), 492 (Āmid), 513 (Baylaqān), 527–528 (Khuwayy), 293 (Ustūnāwand), 564 (Mūghān), 609–610 (Bāshghirt).

13. *Āthār*, 12–13; M fol. 81a (W 148); compare Abū Rayḥān al-Bīrūnī, *Taḥdīd nihāyāt al-amākin li-taṣḥīḥ masāfāt al-masākin*, ed. Pavel Georgievich Bulgakov (Cairo: Maʿhad al-Makhṭūṭāt al-ʿArabiyya, 1962), 134–135; translated as *The Determination of the Coordinates for the Correction of Distances between Cities*, trans. Jamil Ali (Beirut: American University of Beirut, 1967), 101–102.

14. See Claude Gilliot, "Yāḳūt al-Rūmī," *EI2*, 11:264–266; biography by Iḥsān ʿAbbās in *Irshād*, 7:2877–2940. For Yāqūt's travels and Berl., Or. oct. 3377, a manuscript Yāqūt copied, see

Rudolf Sellheim, "Neue Materialien zur Biographie des Yāqūt," in *Schriften und Bilder: drei orientalistische Untersuchungen*, ed. Klaus Ludwig Janert, Rudolf Sellheim, and Hans Striedl (Wiesbaden: Franz Steiner, 1967), 87–118, plates 7–30 and map.

15. See Wadād al-Qāḍī, "Biographical Dictionaries: Inner Structure and Cultural Significance," in *The Book in the Islamic World: The Written Word and Communication in the Middle East*, ed. George Atiyeh (Albany: SUNY Press, 1995), 93–122; al-Qāḍī, "Biographical Dictionaries as the Scholars' Alternative History of the Muslim Community," in *Organizing Knowledge: Encyclopaedic Activities in the Pre-Eighteenth Century Muslim World*, ed. Gerhard Endress (Leiden: Brill, 2006), 23–75; Ruth Roded, *Women in Islamic Biographical Collections* (Boulder: Lynne Reinner, 1994).

16. On text re-use, see Sarah Savant and Matthew Thomas Miller, "'Tell Me Something I Don't Know!': The Place and Politics of Digital Methods in the (Islamicate) Humanities," *International Journal of Middle East Studies* 50, no. 1 (2018): 135–139. See also the KITAB initiative, a digital humanities project led by Savant: kitab-project.org/text-reuse-methods.

17. On Bīrūnī, see *Buldān*, 1:18; compare *Āthār*, 12; Bīrūnī, *Taḥdīd*, 106–107. For passing reference to Yāqūt's compendium, see *Āthār*, 61, 563. Yāqūt appears, likely as a later scribal addition, in W 157, 190, but is missing in M fols. 84a, 94a; Y 85b; B fols. 56b–57a, 64a; V fols. 92a, 105a; F 82, 94; D fols. 183a, 206b–207a. On Qazwīnī and Yāqūt, see Syrinx von Hees, "Neues zum Verhältnis von Qazwīnīs *Āṯār al-bilād* zu Yāqūts *Muʿǧam al-buldān:* Zwei geographische Texte des 13. Jahrhunderts im Vergleich," in *Norm und Abweichung: Akten des 27. Deutschen Orientalistentages (Bonn—28. September bis 2. Oktober 1998)*, ed. Stefan Wild and Hartmut Schild (Würzburg: Ergon, 2001), 425–435.

18. See *Āthār*, 463, 536. A marginal note in Qazwīnī, *ʿAjāʾib al-buldān* = *Āthār*, BnF, Arabe 2238, fol. 1b, observes that Qazwīnī was the disciple of Abharī, a contemporary of ʿAmīdī and Kashshī. For more on Abharī and Qazwīnī, see *Enz.*, 57–60; *Cosmos*, 47–49.

19. On ʿAmīdī, see *Āthār*, 377, 536; Ibn Khallikān (d. 681/1282), *Wafayāt al-aʿyān wa-anbāʾ abnāʾ al-zamān*, ed. Iḥsān ʿAbbās, 8 vols. (Beirut: Dār Ṣādir, 1968–1977), 4:257–259, §603; Larry Miller, "Islamic Disputation Theory: A Study of the Development of Dialectic in Islam from the Tenth through Fourteenth Centuries" (PhD diss., Princeton University, 1984), 143, 148–162; Brannon Wheeler, "al-ʿAmīdī, Rukn al-Dīn," *EI3*.

20. See Hisashi Obuchi, "In the Wake of Faḫr al-Dīn al-Rāzī: A Critical Edition of Zayn al-Dīn al-Kaššī's Introduction to *Ḥadāʾiq al-ḥaqāʾiq*," *Annals of Japan Association for Middle East Studies* 35, no. 1 (2019): 185–207.

21. See Ṣamad Muḥammad, "Athīr al-Dīn Abharī," *Dāʾirat al-Maʿārif-i Buzurg-i Islāmī*, ed. Kāẓim Mūsawī Bujnūrdī (Tehran: Markaz Dāʾirat al-Maʿārif al-Islāmiyya, 1988–), 6:586–590. For Abharī as a student of Fakhr al-Dīn al-Rāzī, see Barhebraeus (d. 685/1286), *Duwal*, 445. See Abharī's transmission of Ibn Sīnā's *Ishārāt* from Quṭb al-Dīn Ibrāhīm al-Miṣrī, via Fakhr al-Dīn al-Rāzī in Ṣalāḥ al-Dīn al-Ṣafadī (d. 764/1363), *Wāfī*, 2:142–143, §497. On Quṭb al-Dīn al-Miṣrī, see *ʿUyūn*, 2:30. See also Gerhard Endress, "Reading Avicenna in the *Madrasa:* Intellectual Genealogies and Chains of Transmission of Philosophy and the Sciences in the Islamic East," in *Arabic Theology, Arabic Philosophy: From the Many to the One*, ed. James E. Montgomery (Leuven: Peeters, 2006), 371–422, at 406–407. For Abharī's influence in scholastic philosophy and his engagement with Rāzī's teachings, see *PPT*, 98–126.

22. See *Āthār*, 377–378.

23. See Kashshī, *Ḥadīqat al-ḥaqāʾiq*, Süleymaniye, Köprülü 864, fols. 3b, 7a, edited in Obuchi, "Wake," 192, 196, §§5, 12.

24. See *APW*, 6–7, 105.

25. See Robert Wisnovsky, "Avicenna's Islamic Reception," in *Interpreting Avicenna: Critical Essays*, ed. Peter Adamson (Cambridge: Cambridge University Press, 2013), 190–213; Wisnovsky, "On the Emergence of Maragha Avicennism," *Oriens* 48 (2018): 263–331. For the boundaries of *falsafa* and *kalām*, see *FPP*, 77–159.

26. Galen develops the argument throughout *De usu partium*. For its culmination in Arabic, see Galen, *Manāfiʿ al-aʿḍāʾ*, BnF, Arabe 2853, fols. 299b–300a (*ḥikmat al-khāliq*, *asrār al-ṭabīʿa*, etc.).

27. For natural law in Stoic philosophical circles, and its reception, contestation, and reformulation among Jews and early Christians, see Christine Hayes, *What's Divine about Divine Law?* (Princeton, NJ: Princeton University Press, 2015). For legal hermeneutics, see Aron Zysow, *The Economy of Certainty: An Introduction to the Typology of Islamic Legal Theory* (Atlanta: Lockwood Press, 2013); David Vishanoff, *The Formation of Islamic Hermeneutics: How Sunni Legal Theorists Imagined a Revealed Law* (New Haven, CT: American Oriental Society, 2011).

28. See Tony Street, "Arabic Logic," in *Handbook of the History of Logic*, vol. 1: *Greek, Indian and Arabic Logic*, ed. Dov Gabbay and John Woods (Amsterdam: Elsevier, 2004), 523–596. For further context, see Khaled El-Rouayheb, *Relational Syllogisms and the History of Arabic Logic, 900–1900* (Leiden: Brill, 2010). For mathematics, see *GAS*, vol. 5; Sonja Brentjes, "Historiographie der Mathematik im islamischen Mittelalter," *Archives internationales d'histoire des sciences* 42 (1992): 27–63.

29. The foundational study on the madrasa is George Makdisi, *The Rise of Colleges: Institutions of Learning in Islam and the West* (Edinburgh: Edinburgh University Press, 1981), 77–80, 137, 281–282. The thesis that foreign sciences and philosophy were successfully excluded from madrasa education has not stood the test of time. See Abdelhamid Sabra, "The Appropriation and Subsequent Naturalization of Greek Science in Medieval Islam: A Preliminary Statement," *History of Science* 25 (1987): 223–243; Sabra, "Situating Arabic Science: Locality versus Essence," *Isis* 87, no. 4 (1996): 654–670; Endress, "Reading." See, more broadly, Sonja Brentjes, *"Orthodoxy," Ancient Sciences, Power, and the Madrasa ("College") in Ayyubid and Early Mamluk Damascus* (Berlin: Max-Planck-Institut für Wissenschaftsgeschichte, 1997); Brentjes, *Teaching and Learning the Sciences in Islamicate Societies (800–1700)* (Turnhout: Brepols, 2018), 75–111.

2: Measures of Authority

1. On the *muʿīd*, see George Makdisi, *The Rise of Colleges: Institutions of Learning in Islam and the West* (Edinburgh: Edinburgh University Press, 1981), 193–195.

2. On Kamāl al-Dīn, see *Āthār*, 463; *ʿUyūn*, 1:307–308; Ibn Khallikān, *Wafayāt al-aʿyān wa-anbāʾ abnāʾ al-zamān*, ed. Iḥsān ʿAbbās, 8 vols. (Beirut: Dār Ṣādir, 1968–1977), 5:311–318, §747 (Jews and Christians reading Torah and Gospel, 316); Tāj al-Dīn al-Subkī, *Ṭabaqāt al-shāfiʿiyya al-kubrā*, ed. Maḥmūd Muḥammad al-Ṭanāḥī and ʿAbd al-Fattāḥ Muḥammad al-Ḥilw, 10 vols. (Cairo: ʿĪsā al-Bābī al-Ḥalabī, 1964–1976), 8:378–386, §1278.

3. See *Duwal*, 477. See Charles Burnett, "Master Theodore, Frederick II's Philosopher," in *Federico II e le Nuove Culture: Atti del XXXI Convegno storico internazionale, Todi, 9–12 ottobre*

1994 (Spoleto: Centro italiano di studi sull'alto Medioevo, 1995), 225–285; Burnett, "Antioch as a Link between Arabic and Latin Culture in the Twelfth and Thirteenth Centuries," in *Occident et Proche-Orient: Contacts scientifiques au temps des croisades*, ed. Isabelle Draelants, Anne Tihon, and Baudouin van den Abeele (Turnhout: Brepols, 2000), 1–78.

4. *Āthār*, 463. See *Enz.*, 59–60; *WTI*, 1:147, trans. 1:135; Dag Nikolaus Hasse, "Mosul and Frederick II Hohenstaufen: Notes on Aṯīraddīn al-Abharī and Sirāǧaddīn al-Urmawī," in Draelants et al., *Occident et Proche-Orient*, 145–163.

5. See Gülru Necipoğlu, "Ornamental Geometries: A Persian Compendium at the Intersection of the Visual Arts and Mathematical Sciences," in *The Arts of Ornamental Geometry: A Persian Compendium on Similar and Interlocking Figures*, ed. Gülru Necipoğlu (Leiden: Brill, 2017), 11–78, at 32, 68nn75–76; Jan P. Hogendijk, "Greek and Arabic Constructions of the Regular Heptagon," *Archive for History of Exact Sciences* 30, no. 3/4 (1984): 197–330, at 214, 224–226, 239–241, 281–282.

6. See Athīr al-Dīn al-Abharī, *Kitāb Īsāghūjī fī ʿilm al-manṭiq*, with commentary of Ḥusām al-Dīn al-Kātī (d. 760/1359), ed. Saʿīd ʿAbd al-Laṭīf Fūda (Beirut: Dār al-Dhakhāʾir, 2015), 30–39, 101–102. See also Khaled El-Rouayheb, *The Development of Arabic Logic (1200–1800)* (Basel: Schwabe, 2019), 47–53. For *ḥads* in Avicennan epistemology as "guessing correctly" or "divining" the middle term of a syllogism, rather than merely intuition, see *APW*, xii–xiii.

7. For Abharī, see Gerhard Endress, "Reading Avicenna in the *Madrasa*," in *Arabic Theology, Arabic Philosophy: From the Many to the One*, ed. James E. Montgomery (Leuven: Peeters, 2006), 397, 407, 411–412. See the table of contents in Athīr al-Dīn al-Abharī, *Hidāyat al-ḥikma* (Karachi: Maktabat al-Bushrā, 2011), 123–125. For Abharī's influence in scholastic philosophy, see *PPT*, 98–126.

8. For the epithet, see Abharī, *Muntahā l-afkār fī ibānat al-asrār*, KMSI 2752, 351 (paginated).

9. See Abharī, *Hidāya*, 113–122; Abharī, *Zubdat al-asrār*, Istanbul, Millet Kütüphanesi, Feyzullah Efendi 1210, fols. 159b–167a; Abharī ends the *Hidāya* noting that the next lesson can be continued with a reading of the *Zubdat*, highlighting that these writings represented different stages of learning.

10. Abharī, *Zubdat al-asrār*, fol. 167b; Abharī, *Talkhīṣ al-ḥaqāʾiq* and *al-Maṭāliʿ* in Süleymaniye, Köprülü 1618, fols. 67b–68a, 101b; Abharī, *Mabdaʾ wa-maʿād*, BnF, Supplément persan 139, fols. 65b–66b. This Persian miscellany was copied in 688/1289 with treatises by Ibn Sīnā, Shihāb al-Dīn al-Suhrawardī, and Fakhr al-Dīn al-Rāzī, among others. For further on *ḥads*, see *APW*, 179–201, esp. 182–184. For Ibn Sīnā's miraculous power of *ḥads*, see Bahmanyār ibn al-Marzubān (d. 458/1066), *al-Taḥṣīl*, ed. Murtaḍā Muṭahharī (Tehran: Dānishgāh-i Tihrān, 1375sh/1996), 816–817; compare *APW*, 200n66.

11. Abharī, *Zubdat*, fols. 106a (causation), 145a–146b (void); Abharī, *Hidāya*, 111–112 (void). See Peter Adamson, "Fakhr al-Dīn al-Rāzī on Void," in *Islamic Philosophy from the 12th to the 14th Century*, ed. Abdelkader Al Ghouz (Göttingen: Bonn University Press, 2018), 307–324; *FPP*, 312, 485–486, 562n32. See also Paul Hullmeine, "Al-Bīrūnī and Avicenna on the Existence of Void and the Plurality of Worlds," *Oriens* 47, no. 1/2 (2019): 114–144.

12. For epicenters, see Abharī, *al-Mukhtaṣar fī ʿilm al-hayʾa*, BnF, Arabe 2515, fols. 25a–30b. For Ptolemy's summary of planetary movement translated as *Kitāb al-Manshūrāt*, see Bernard Goldstein, "The Arabic Version of Ptolemy's *Planetary Hypotheses*," *Transactions of the American Philosophical Society* 57, no. 4 (1967): 3–55. See also Abdelhamid Sabra, "Configuring

the Universe: Aporetic, Problem Solving, and Kinematic Modeling as Themes of Arabic Astronomy," *Perspectives on Science* 6, no. 3 (1998): 288–330; George Saliba, *Islamic Science and the Making of the European Renaissance* (Cambridge, MA: MIT Press, 2007), 74–129. For more on *ʿilm al-hayʾa*, see Robert Morrison, "Cosmology and Cosmic Order in Islamic Astronomy," *Early Science and Medicine* 24, no. 4 (2019): 340–366. Compare the talismanic representation of Mercury in B with *Maktūm*, fol. 49a; see also *Ilkh.*, 56–66.

13. For a study of early Islamic astrolabes and other measurements for timekeeping, see David King, *In Synchrony with the Heavens: Studies in Astronomical Timekeeping and Instrumentation in Medieval Islamic Civilization*, 2 vols. (Leiden: Brill, 2004–2005).

14. See Taro Mimura, "Too Many Arabic Treatises on the Operation of the Astrolabe in the Medieval Islamic World: Athīr Al-Dīn Al-Abharī's *Treatise on Knowing the Astrolabe* and His Editorial Method," *Medieval Encounters* 23 (2017): 365–403.

15. See Abharī, *al-Zīj al-shāmil*, BnF, Arabe 2528, fols. 2b–10a; compare KMSI 6422, fols. 1a–2b (partial introduction), and KMSI 6445, fols. 7a–15a; CBL Ar. 4076. See further, *Kashf*, 2:968–969; David King and Julio Samsó, "Astronomical Handbooks and Tables from the Islamic World (750–1900): An Interim Report," *Suhayl* 2 (2001): 9–105, at 44, §3.3.4.

16. For the fall of Isfahan, see Ibn Abī l-Ḥadīd (d. ca. 655/1257), *Sharḥ Nahj al-balāgha*, ed. Muḥammad Ibrāhīm, 10 vols. (Baghdad: Dār al-Kitāb al-ʿArabī, 2007), 4:348; discussed in John Woods, "A Note on the Mongol Capture of Iṣfahān," *JNES* 36, no. 1 (1977): 49–51; *Mongols*, 82–83.

17. For Shams al-Dīn al-Iṣfahānī, see Subkī, *Ṭabaqāt*, 8:100–103, §1095.

18. See Abharī, *Kashf al-ḥaqāʾiq*, Süleymaniye, Carullah Efendi 1436, fol. 135a. The colophon is transcribed in Ramazan Şeşen, *Mukhtārāt min al-makhṭūṭāt al-ʿarabiyya al-nādira fī maktabat Turkiyā* (Istanbul: İstanbul Araştırma ve Eğitim Vakfı, 1997), 266–272, at 270. See also Hüseyin Sarıoğlu and ʿAbd al-Majīd Naṣīr, "Abharī, Athīr al-Dīn," in *Mawsūʿat aʿlām al-ʿulamāʾ wa-l-udabāʾ al-ʿarab wa-l-muslimīn*, 22 vols. (Beirut: Dār al-Jīl, 2004–2013), 1:187–202, at 197. On the Shāfiʿī madrasa complex, see Stephennie Mulder, "The Mausoleum of Imam al-Shafiʿi," *Muqarnas* 23 (2006): 15–46.

19. On Tāj al-Dīn al-Urmawī, see *Āthār*, 494–495; *Ḥawādith*, 223; *Duwal*, 445. For the Sharābī madrasa in Baghdad, see Nājī Maʿrūf, *al-Madāris al-Sharābiyya bi-Baghdād, Wāsiṭ wa-Makka* (Baghdad: al-Irshād, 1965), 143–146.

20. Tāj al-Dīn al-Urmawī, *Kitāb al-Ḥāṣil min al-Maḥṣūl*, ed. ʿAbd al-Salām Maḥmūd Abū Nājī, 2 vols. (Benghazi: Jāmiʿat Qāryūnis, 1994), 1:224–225. Urmawī completed this commentary in 614/1218, see *Kashf*, 2:1615–1616; Shams al-Dīn al-Iṣfahānī, *al-Kāshif ʿan al-Maḥṣūl fī ʿilm al-uṣūl*, Süleymaniye, Köprülü 498, fol. 1b, copied 699/1300.

21. For Kātibī's studies with Abharī, see the collection of four handbooks of philosophy by Najm al-Dīn al-Kātibī, Süleymaniye, Köprülü 1618, discussed in Sayyid Muḥammad ʿImādī Ḥāʾirī, *Az nuskhahā-yi Istānbūl: Dastnivīshāʾī dar falsafa, kalām, ʿirfān* (Tehran: Mīrāth-i Maktūb, 1391sh/2012), 63–76, plates 18–29; *PPT*, 100–106, 115–125, 536. For Kātibī's work in Maragha, see *Alqāb*, 3:54–55, 123, 5:489. See also Khaled El-Rouayheb, "al-Kātibī al-Qazwīnī," *EI3*.

22. See *APW*, 130–133, 155–159 (work 11). For commentaries on the *Ishārāt*, see Michael Rapoport, "The Life and Afterlife of the Rational Soul: Chapters VIII–X of Ibn Sīnā's Pointers and Reminders and Their Commentaries" (PhD diss., Yale University, 2018), 203–215, 358–368. On the *Ishārāt*, see Abharī, *Mabdaʾ*, fol. 66b; *Kashf*, 1:94–95, 97; Sarıoğlu and Naṣīr, "Abharī," 1:197, 199, §§6, 24. For questions concerning Abharī's transmission of the *Ishārāt* to

Ṭūsī, see Ayman Shihadeh, *Doubts on Avicenna: A Study and Edition of Sharaf al-Dīn al-Masʿūdī's Commentary on the Ishārāt* (Leiden: Brill, 2016), 13–14. For Abharī's connections to the observatory through the son of one of his pupils, Jamāl al-Dīn Aḥmad al-Qazwīnī, see *Alqāb*, 3:572, §2320.

23. *Āthār*, 378. See Fathalla Kholeif, *A Study on Fakhr al-Dīn al-Rāzī and His Controversies in Transoxiana* (Beirut: Dar el-Machreq, 1966). For further context, see Shihadeh, *Doubts*, 7–43; *FPP*, 226–239, 264–304.

24. The full epistolary exchange does not survive. See Naṣīr al-Dīn al-Ṭūsī, *Ajwibat al-masāʾil al-naṣīriyya*, ed. ʿAbd Allāh Nūrānī (Tehran: Pizhūhishgāh-i ʿUlūm-i Insānī wa-Muṭālaʿāt-i Farhangī, 1383sh/2004), 273–276 (citing Ibn al-Haytham). See Ibn al-Haytham, *Shukūk ʿalā Baṭlamiyūs*, ed. ʿAbd al-Ḥamīd Ṣabra [Abdelhamid Sabra] and Nabīl al-Shihābī (Cairo: Maṭbaʿat Dār al-Kutub, 1971), 24–34; translated in Don Voss, "Ibn Al-Haytham's Doubts concerning Ptolemy: A Translation and Commentary" (PhD diss., University of Chicago, 1985), 42–52. See also *Tadhkira*, 1:48–51 (introduction), 195–222 (translation and edition), 2:427–457 (commentary).

25. *Āthār*, 377 (reading Ibn Surayj); M fols. 49b–50a (W 86–87). For a similar hadith, see Shams al-Dīn al-Dhahabī, *Siyar aʿlām al-nubalāʾ*, ed. Shuʿayb al-Arnāʾūṭ et al., 24 vols. (Beirut: Muʾassasat al-Risāla, 1996), 10:46; Ibn ʿAsākir (d. 571/1176), *Tabyīn kadhib al-muftarī fī-mā nusiba ilā l-imām Abī l-Ḥasan al-Ashʿarī* (Damascus: Maṭbaʿat al-Tawfīq, 1347/1928~), 45–56; Abū Dāwūd al-Sijistānī (d. 275/889), *Sunan*, 2 vols. (Vaduz, Liechtenstein: Jamʿiyyat al-Maknaz al-Islāmī, 2000), "Kitāb al-Malāḥim," 2:715, chap. 38.1, §4293. See also Ella Landau-Tasseron, "The 'Cyclical Reform': A Study of the *Mujaddid* Tradition," *Studia Islamica* 70 (1989): 79–117.

26. See Christopher Melchert, *The Formation of the Sunni Schools of Law, 9th–10th Centuries C.E.* (Leiden: Brill, 1997); Ahmed El Shamsy, *The Canonization of Islamic Law: A Social and Intellectual History* (Cambridge: Cambridge University Press, 2013).

27. See Patricia Crone and Martin Hinds, *God's Caliph: Religious Authority in the First Centuries of Islam* (Cambridge: Cambridge University Press, 1986), 80–96; Muhammad Qasim Zaman, "The Caliphs, the ʿUlamāʾ, and the Law: Defining the Role and Function of the Caliph in the Early ʿAbbāsid Period," *Islamic Law and Society* 4, no. 1 (1997): 1–36.

28. See Michael Cook, *Commanding Right and Forbidding Wrong in Islamic Thought* (Cambridge: Cambridge University Press, 2001), 114–138; Daphna Ephrat, "Religious Leadership and Associations in the Public Sphere of Seljuk Baghdad," in *The Public Sphere in Muslim Societies*, ed. Miriam Hoexter, Shmuel Eisenstadt, and Nehemia Levtzion (Albany: SUNY Press, 2002), 31–48. For Nishapur, see Margaret Malamud, "The Politics of Heresy in Medieval Khurasan: The Karramiyya in Nishapur," *Iranian Studies* 27, no. 1/4 (1994): 37–51. For Isfahan, see Ibn Abī l-Ḥadīd, *Sharḥ*, 4:348; treated in Woods, "A Note."

29. For Shāfiʿī law in the charter of the Niẓāmiyya in Baghdad, see Makdisi, *Colleges*, 302–303. For the charter of the Mustanṣiriyya complex, see *Ḥawādith*, 58–61; Ibn Wāṣil (d. 697/1298), *Mufarrij al-kurūb fī akhbār Banī Ayyūb*, ed. Jamāl al-Dīn al-Shayyāl et al., 5 vols. (Cairo: Maṭbaʿat Jāmiʿat Fuʾād al-Awwal, 1953–1975), 5:316–317. See also Nājī Maʿrūf, *Tārīkh ʿulamāʾ al-Mustanṣiriyya* (Baghdad: al-ʿĀnī, 1959), 1–10.

30. On the *ṣandūq al-sāʿāt*, see *Āthār*, 316–317, compare 189, 604. Description from *Ḥawādith*, 79; Maʿrūf, *Tārīkh*, 267–269. See also Riḍwān Ibn al-Sāʿātī, *ʿIlm al-sāʿāt wa-l-ʿamal bihā*, ed. Muḥammad Aḥmad Duhmān (Damascus: Maktab al-Dirāsāt al-Islāmiyya, 1981); *WTI*, 3:98–99, trans. 3:98–99; Ibn Razzāz al-Jazarī, *al-Jāmiʿ bayn al-ʿilm wa-l-ʿamal al-nāfiʿ fī ṣināʿat*

al-ḥiyal, ed. Aḥmad Yūsuf al-Ḥasan (Aleppo: Maʿhad al-Turāth al-ʿIlmī al-ʿArabī, 1979), 1–78; translated as *The Book of Knowledge of Ingenious Mechanical Devices*, trans. Donald Hill (Dordrecht: D. Reidel 1974), 17–40. See also Donald Hill, *Arabic Water-Clocks* (Aleppo: University of Aleppo, 1981), 69–71; Elly Truitt, *Medieval Robots: Mechanism, Magic, Nature, and Art* (Philadelphia: University of Pennsylvania Press, 2015), 141–153. Before being acquired by the Museum of Fine Arts, Boston, from the collection of Victor Goloubew, the folio of the water clock (figure 2.3) once formed part of Süleymaniye, Aya Sofya 3606.

31. For Sharābī's role in the appointment of Muʿtaṣim, see Ibn Wāṣil, *Mufarrij*, 5:318. On the judiciary, see *Ḥawādith*, 164 (stipulation of Shāfiʿī law), 203, 226 (Nahruqallī), 212 (resignation of Ḥanafī judge); *Alqāb*, 1:87–88 (Nahruqallī), 2:175–176 (Dāmghānī), 2:667 (appointment by Nahruqallī).

32. *Ḥawādith*, 178, 202.

33. On stipends, see Wadād al-Qāḍī, "The Salaries of Judges in Early Islam: The Evidence of the Documentary and Literary Sources," *JNES* 68, no. 1 (2009): 9–30; Makdisi, *Colleges*, 58–59. For the madrasa complex, see *Ḥawādith*, 74, 198; *Alqāb*, 3:96; Maʿrūf, *Madāris*, 271–272.

34. See Jonathan Bloom, *Paper before Print: The History and Impact of Paper in the Islamic World* (New Haven, CT: Yale University Press, 2001), 47–56; Shawkat Toorawa, *Ibn Abī Ṭāhir Ṭayfūr and Arabic Writerly Culture: A Ninth-Century Bookman in Baghdad* (London: Routledge-Curzon, 2005), 56–60.

35. This description builds on the Shāfiʿī jurist Ibn Abī l-Dam al-Hamadhānī (d. 642/1244), *Kitāb Adab al-qaḍāʾ*, ed. Muḥyī Hilāl al-Sarḥān, 2 vols. (Baghdād: al-Irshād, 1984), 1:317–353. See also Karen Bauer, "Debates on Women's Status as Judges and Witnesses in Post-Formative Islamic Law," *JAOS* 130, no. 1 (2010): 1–21.

36. *Āthār*, 140, 426 (Dhū l-Nūn), 322–323 (Muḥāsibī), 324, 329–330 (Junayd), 171–172, 571 (Tustarī), 165–168 (Ḥallāj), 394–395 (Suhrawardī). See Łukasz Piątak, "Between Philosophy, Mysticism and Magic: A Critical Edition of Occult Writings of and Attributed to Shihāb al-Dīn al-Suhrawardī (1156–1191)" (doctoral diss., University of Warsaw, 2018), 20, 23–24, 104 (Arabic), 406. See also *FPP*, 244–263.

37. See Daniel Gimaret, *La doctrine d'al-Ashʿarī* (Paris: Cerf, 1990), 459–467. Compare Johnathan Brown, "Faithful Dissenters: Sunni Skepticism about the Miracles of Saints," *Journal of Sufi Studies* 1, no. 2 (2012): 123–168.

38. *Āthār*, 297–298, 373. See ʿAbd al-Karīm al-Qushayrī (d. 465/1072), *Risāla*, ed. ʿAbd al-Ḥalīm Maḥmūd and Maḥmūd ibn al-Sharīf, 2 vols. (Cairo: Dār al-Kutub al-Ḥadītha, 1966), 2:538.

39. *Āthār*, 269, 295, 363–364, 563, 354, 497. Compare *Enz.*, 46–47, 49, 63–64; *CAL*, 46, 61.

40. *Āthār*, 438; quoted in *Alqāb*, 5:372, §5290. Reading *tajrīd* in light of *Ikhwān*, 3:89, 233; 4:287.

41. See Devin DeWeese, "Cultural Transmission and Exchange in the Mongol Empire: Notes from the Biographical Dictionary of Ibn al-Fuwaṭī," in *Beyond the Legacy of Genghis Khan*, ed. Linda Komaroff (Leiden: Brill, 2006), 11–29.

42. M fol. 1a (W 1). See *Alqāb*, 2:66–67, §1050, 4:261, §3806, 5:372, §5290.

43. For Qazwīnī's transmission of Ibn Mājah, see *Alqāb*, 1:313–314, §444.

44. *Āthār*, 436–437; cited in *Enz.*, 36–37. For more on ancestors in Qazvin, see ʿAbd al-Karīm Ibn Muḥammad al-Rāfiʿī (d. 623/1226), *al-Tadwīn fī akhbār Qazwīn*, ed. ʿAzīz Allāh al-ʿUṭāridī, 4 vols. (Beirut: Dār al-Kutub al-ʿIlmiyya, 1987), 4:181–182. For the reading *kammūnī*, cumin merchant, and not Kamūna, a tribal name, see ʿAbd al-Karīm al-Samʿānī, *al-Ansāb*, ed.

ʿAbd al-Raḥmān ibn Yaḥyā l-Muʿallimī, 12 vols. (Hyderabad: Maṭbaʿat Majlis Dāʾirat al-Maʿārif al-ʿUthmāniyya, 1962–1982), 10:471.

45. *Āthār*, 419, 495.

3: Astral Power

1. *Maʿādin*, 5–6, translated in *Avicennae De congelatione et conglutinatione lapidum, Being Sections of the Kitâb al-Shifâʾ*, ed. and trans. Eric John Holmyard and Desmond Christopher Mandeville (Paris: P. Guethner, 1927), 23–25 (translation); M fol. 6b; B fols. 5b, 100b (W 11, 209). Cited in Shams al-Dīn al-Shahrazūrī (d. 687/1288), *Rasāʾil al-shajara al-ilāhiyya*, ed. Najafqulī Ḥabībī, 3 vols. (Tehran: Muʾassasa-yi Pizhūhishī-yi Ḥikmat wa-Falsafa, 2004–2006), 2:288.

2. See Aristotle, *al-Āthār al-ʿulwiyya = Aristotle's Meteorology in the Arabico-Latin Tradition*, ed. Pieter Schoonheim (Leiden: Brill, 2000), 5; *GAS*, 7:212–215; *CAL*, 108–111.

3. For a defense of astrology, see Ibn Ṭāwūs (d. 664/1266), *Faraj al-Mahmūm fī tārīkh ʿulamāʾ al-nujūm* (Qom: Manshūrāt al-Raḍī, 1944), translated in Zeina Matar, "The *Faraj al-Mahmūm* of Ibn Ṭāwūs: A Thirteenth-Century Work on Astrology and Astrologers" (PhD diss., New York University, 1987); George Saliba, "The Role of the Astrologer in Medieval Islamic Society," *Bulletin d'études orientales* 44 (1992): 45–67. For astrology as divine beneficence, see Robert Morrison, *Islam and Science: The Intellectual Career of Niẓām al-Dīn al-Nīsābūrī* (London: Routledge, 2007), 66–77.

4. M fol. 79a (W 143, *hayʾat al-arḍ*); B fol. 53a (*ʿilm al-hayʾa*). For *ʿilm al-hayʾa* as a branch of the mathematical sciences, see Ibn Sīnā, "Fī aqsām al-ʿulūm al-ʿaqliyya," in *Tisʿ rasāʾil fī-l-ḥikma wa-l-ṭabīʿiyyāt* (Cairo: Maṭbaʿa Hindiyya, 1908), 104–118, at 111.

5. See Kara Richardson, "Avicenna's Conception of the Efficient Cause," *British Journal for the History of Philosophy* 21, no. 2 (2013): 220–239. For secondary causation in Ashʿarī theology, see *GPT*, 124–141; Jon McGinnis, "Occasionalism, Natural Causation and Science in al-Ghazālī," in *Arabic Theology, Arabic Philosophy: From the Many to the One*, ed. James E. Montgomery (Leuven: Peeters, 2006), 441–463.

6. Ibn Sīnā, "Aqsām," 110–111. The term *tasyīr* is translated here as life span. For its technical sense as an arc on the heavenly sphere used for determining the length of life, see Kūshyār ibn Labbān, *al-Madkhal fī ṣināʿat aḥkām al-nujūm = Kūšyār Ibn Labbān's Introduction to Astrology*, ed. and trans. Michio Yano (Tokyo: Bikohsha Co., 1997), ix (editor's introduction), 216–235 (3.20–21). For his critique of astrology, see Ibn Sīnā, *Risāla fī ibṭāl aḥkām al-nujūm = Réfutation de l'astrologie*, ed. and trans. Yahya Michot (Beirut: Albouraq, 2006), 11–14 (astrological qualities associated with planets), 15–16 (Ptolemy), 38–49 (complexity of celestial influences); trans. 68–77, 78–82, 140–142. See also *Ilāhiyyāt*, 2:440; Qazwīnī too suggests that the Ptolemy of the *Almagest* did not author the astrological corpus, *Āthār*, 572; compare Naṣīr al-Dīn Ṭūsī, *Sharḥ-i Thamara-yi Baṭlamiyūs*, ed. Jalīl Akhawān Zanjānī (Tehran: Āyīn-i Maktūb, 1378sh/1999), 2.

7. For references to *munajjimūn*, see M fol. 13a (Mercury), 13b (Venus), 14b (Sun), 16a (Mars), 16b (Jupiter), 17a (Saturn), 18a (names and titles of fixed stars), 22a (twelve signs of the zodiac), 27a (Gamma Leonis). For the section missing between folios 22b and 23a, see V fol. 24a (Virgo), 25a (southern constellations).

8. *Maktūm*, fols. 7a–b, 19a–21a, 23a, 70a–b. See Michael-Sebastian Noble, *Philosophising the Occult: Avicennan Psychology and "The Hidden Secret" of Fakhr al-Dīn al-Rāzī* (Berlin: De Gruyter, 2021), 201–208. See *FPP*, 273n43.

9. Kūshyār, *Madkhal*, 8–11 (defense of astrology), 72–73, 78–79 (dynastic changes), 80–81 (earthquakes, comets), 92–93, 110–111, 146–147 (floods), and 236–259 (propitious timings).

10. *Maktūm*, fol. 2b (conquering without contact). See also *Sittīnī*, 420–426 (49. *ʿilm al-hayʾa*), 427–431 (50. *ʿilm al-aḥkām*), 432–437 (51. *ʿilm al-raml*), 438–447 (52. *ʿilm al-ʿazāʾim*), 448–452 (53. *ʿilm-i ilāhiyyāt*); Amina Steinfels, "Knowledge on Display, Fakhr al-Dīn al-Rāzī's Universal Compendium," in *Light upon Light: Essays in Islamic Thought and History in Honor of Gerhard Böwering*, ed. Jamal J. Elias and Bilal Orfali (Leiden: Brill, 2019), 335–346; Matthew Melvin-Koushki, "Powers of One: The Mathematicalization of the Occult Sciences in the High Persianate Tradition," *Intellectual History of the Islamicate World* 5 (2017): 127–199, at 145–147. For ambivalence toward judicial astrology, see Fakhr al-Dīn al-Rāzī, *Munāẓarāt*, in *A Study on Fakhr al-Dīn al-Rāzī and His Controversies in Transoxiana*, ed. and trans. Fathalla Kholeif (Beirut: Dar el-Machreq, 1966), 32–39 (Arabic), 55–62 (translation); for limits of astrological knowledge and critiques leveled at astrologers, see *Mafātīḥ*, 2:228 (Q. 2:32–33), 3:34 (Q. 2:40), 3:73–74 (Q. 2:50), 4:169 (Q. 2:155), 5:130–133 (Q. 2:188), 6:64 (Q. 4:49), 13:55–58 (Q. 6:76–79), 14:146 (Q. 7:58), 19:21–22 (Q. 13:11), 24:225 (Q. 28:1–6), 29:39 (Q. 54:12), 30:59–62, 132 (Q. 67:5, 71:22–24), 32:82, 194 (Q. 102:113–114). Compare Abū Bakarāt al-Baghdādī (d. ca. 560/1165), *Muʿtabar*, 2:235–236.

11. See Stephen Haw, "The Mongol Empire: The First 'Gunpowder Empire'?," *JRAS* 23 (2013): 441–469; *Mongols*, 88–89, 138.

12. See *The Secret History of the Mongols: A Mongolian Epic Chronicle of the Thirteenth Century*, trans. Igor de Rachewiltz, 3 vols. (Leiden: Brill, 2004–2013), 1:1, §1, 1:224–227, 3:3–8. Paul D. Buell and Judith Kolbas, "The Ethos of State and Society in the Early Mongol Empire: Chinggis Khan to Güyük," *JRAS* 26, no. 1/2 (2016): 43–64, at 49–51; *Mongols*, 64–65, 298. See also John Boyle, "Turkish and Mongol Shamanism in the Middle Ages," *Folklore* 83, no. 3 (1972): 177–193.

13. ʿAṭāʾ Mulk al-Juwaynī, *Tārīkh-i jahān-gushā*, ed. Mīrzā Muḥammad Qazwīnī, 3 vols. (London: Luzac, 1912–1937), 1:18; translated as *The History of the World Conqueror*, trans. John Andrew Boyle, 2 vols. (Cambridge, MA: Harvard University Press, 1958), 1:26. For further context, see Matthew Melvin-Koushki, "Early Modern Islamicate Empire: New Forms of Religiopolitical Legitimacy," in *The Wiley Blackwell History of Islam*, ed. by Armando Salvatore et al. (Chichester, West Sussex: John Wiley & Sons, 2018), 353–375.

14. See Juwaynī, *Tārīkh*, 1:43–44, 85–86, 2:105–106; 3:29–30; trans. 1:59, 109–110, 2:374–375, 567–568.

15. See David Pingree, "Astrology," in *Religion, Learning and Science in the ʿAbbasid Period*, ed. M. J. L. Young et al. (Cambridge: Cambridge University Press, 1990), 290–300. For parallel developments, see Paul Magdalino, "Occult Science and Imperial Power in Byzantine History and Historiography (9th–12th Centuries)," in *The Occult Sciences in Byzantium*, ed. Paul Magdalino and Maria Mavroudi (Geneva: La Pomme d'or, 2006), 119–162.

16. See *NG* 386–387, 390. See references to Ṭabasī's *Shāmil* in *Maktūm*, fols. 128b, 136a–b, 144a–b, 186a–188a. On Ṭabasī's handbook and his encounter with Ghazālī, see *Āthār*, 406–407.

17. See ʿAbd al-Ghāfir al-Fārisī (d. 529/1134), *al-Muntakhab min Kitāb al-Siyāq li-l-tārīkh Nīshābūr*, ed. Muḥammad Kāẓim al-Maḥmūdī (Qom: Jamāʿat al-Mudarrisīn fī l-Ḥawza al-ʿIlmiyya, 1362sh/1983), 61, §119.

18. On Quranic ingestion, see Ṭabasī, *Shāmil*, Princeton University, Islamic MSS, New Series, no. 160, fols. 89b–90a, 98a–b, 106a.

19. See Emily Selove, "Magic as Poetry, Poetry as Magic: A Fragment of Arabic Spells," *Magic, Ritual, and Witchcraft* 15, no. 1 (2020): 33–57, and her ongoing work on Sakkākī's handbook of magic. See also *Kashf*, 2:1762–1768; *Maktūm*, fol. 159b. The hadith is indexed in *Musnad*, 46:452. For Rāzī's associations with the Khwārazm-Shāh, see *FPP*, 268–281, 296–300.

20. See Fakhr al-Dīn al-Rāzī, *al-Sirr al-maktūm*, Süleymaniye, Carullah 1482, fol. 121a, copied 614/1218 in Bukhara. A note in the colophon states that the second chapter (*maqāla*) was compared with a copy in the handwriting of Rāzī, and that the majority of it was transcribed from the author's copy, while a bit of it was copied from the hand of Sirāj al-Dīn al-Khwārazmī known as al-Ṣakkākī [*sic*].

21. This story is made famous by the Timurid historian Ghiyāth al-Dīn Khwāndamīr (d. ca. 942/1535), *Ḥabīb al-siyar*, ed. Jalāl al-Dīn Humāʾī, 4 vols. (Tehran: Khayyām, 1333sh/1954), 3:77, 80–81; translated in *Classical Writings of the Medieval Islamic World: Persian Histories of the Mongol Dynasties*, trans. Wheeler Thackston, 3 vols. (London: I.B. Tauris, 2012), 1:44–46. See ʿAbd al-Ḥayy al-Laknawī (d. 1304/1886), *al-Fawāʾid al-bahiyya fī tarājim al-ḥanafiyya*, ed. Aḥmad al-Zaʿbī (Beirut: Dār al-Arqam, 1998), 382–384, §522; compare Ibn Abī l-Wafāʾ al-Qurashī (d. 775/1373), *al-Jawāhir al-muḍiyya*, ed. ʿAbd al-Fattāḥ Muḥammad al-Ḥulw, 5 vols. (Giza: Hajr, 1993), 3:622–623, §1838. For the use of a magic circle to catch birds and animals, see the collection attributed to Buzurg ibn Shahriyār al-Rāmhurmuzī, *Kitāb ʿAjāʾib al-Hind = Livre des merveilles de l'Inde*, ed. Pieter Antonie Van der Lith, trans. Louis Marcel Devic (Leiden: E.J. Brill, 1883–1886), 104–105. On Sakkākī's tomb, see Muḥammad Ḥaydar Dūghlāt (d. 958/1551), *Tārīkh-i Rashīdī*, ed. ʿAbbās Qulī Ghaffārī Fard (Tehran: Mīrāth-i Maktūb, 2004), 524–525, translated in *Classical Writings*, 1:162. Within his lifetime, Sakkākī's renown for Arabic letters had already reached Yāqūt, *Irshād*, 6:2846, §1255.

22. See Ibn al-Akfānī, *Kitāb al-Irshād al-qāṣid ilā asnā l-maqāṣid*, edited in Jan J. Witkam, *De egyptische arts Ibn al-Akfānī (gest. 749 / 1348) en zijn indeling van de wetenschappen* (Leiden: Ter Lugt Pers, 1989), 51, ll. 675–676 (Arabic text); repeated in Taşköprīzāde (d. 968/1561), *Miftāḥ al-saʿāda wa-miṣbāḥ al-siyāda*, 3 vols. (Beirut: Dār al-Kutub al-ʿIlmiyya, 1985), 1:316. For the authorship of the *Ghāyat*, see *GAS*, 4:294–298; Maribel Fierro, "Bāṭinism in Al-Andalus: Maslama b. Qāsim al-Qurṭubī (d. 353/964), Author of the *Rutbat al-Ḥakīm* and the *Ghāyat al-Ḥakīm (Picatrix)*," *Studia Islamica* 84, no. 2 (1996): 87–112; Godefroid de Callataÿ and Sébastien Moureau, "Again on Maslama Ibn Qāsim al-Qurṭubī, the Ikhwān al-Ṣafāʾ and Ibn Khaldūn: New Evidence from Two Manuscripts of the *Rutbat al-ḥakīm*," *al-Qanṭara* 37, no. 2 (2016): 329–372. Once largely settled, the problem of authorship has been reopened by José Bellver, "Between Traditionalist Theology and Bāṭinism in al-Andalus: Maslama b. al-Qāsim and the Author of *Rutbat al-ḥakīm* and *Ghāyat al-ḥakīm*" (unpublished working paper).

23. For an example, see Andrew Peacock, "A Seljuq Occult Manuscript and Its World: MS Paris persan 174," in *The Seljuqs and Their Successors: Art, Culture and History*, ed. Sheila Canby et al. (Edinburgh: Edinburgh University Press, 2020), 163–179.

24. Juwaynī, *Tārīkh,* 1:86–90, trans. 1:109–115; *Ḥawādith,* 108–109; Khwāndamīr, *Ḥabīb,* 3:77–79, trans. 44–45. See David Cook, "Apocalyptic Incidents during the Mongol Invasions," in *Endzeiten: Eschatologie in den monotheistischen Weltreligionen,* ed. Wolfram Brandes and Felicitas Schmieder (Berlin: De Gruyter, 2008), 293–312, at 297–299; Mansura Haider, "The Revolt of Mahmud Tarabi and the Sarbadar Movement," *Proceedings of the Indian History Congress* 52 (1991): 939–949; *Mongols,* 322.

25. See Hamid Dabashi, "Khwājah Naṣīr al-Dīn al-Ṭūsī: The Philosopher/Vizier and the Intellectual Climate of His Times," in *History of Islamic Philosophy,* ed. Seyyed Hossein Nasr and Oliver Leaman, 2 vols. (New York: Routledge, 1996), 1:527–584; *Tadhkira,* 1:1–23.

26. *Āthār,* 301–302. See Farhad Daftary, *The Assassin Legends: Myths of the Isma*ʿ*ilis* (London: I.B. Tauris, 1994).

27. On astral learning in Alamut, see *Science in the City of Fortune: The Dustūr al-Munajjimīn and Its World,* ed. Eva Orthman and Petra Schmidl (Berlin: EB-Verlag, 2017). For the Mongol invasions as providence, see Juwaynī, *Tārīkh,* 1:13–14, 3:139–140, 278, trans. 1:9, 2:658, 725; George Lane, "Whose Secret Intent?," in *Eurasian Influences on Yuan China,* ed. Morris Rossabi (Singapore: Institute of Southeast Asian Studies, 2013), 1–40, at 2–4; *Mongols,* 24.

28. For Alamut's library, see Juwaynī, *Tārīkh,* 3:270, trans. 2:719. For destruction in Medina and in Baghdad, see *Ḥawādith,* 226–228; Cook, "Incidents," 304–305.

29. Rashīd al-Dīn, *Jāmi*ʿ *al-tawārīkh,* Part 1, *Tārīkh-i mubārak-i Ghāzānī,* ed. Muḥammad Rawshan and Muṣṭafā Mūsawī, 4 vols., 2nd ed. (Tehran: Mirāth-i Maktūb, 2015), 2:891–892 (deliberations about the caliph), 896 (Aries in ascendant), 898–899 (Ṭūsī with caliph and Sulaymān Shāh), 926 (execution of Ḥusām al-Dīn); translated in *Classical Writings,* 3:349–350, 352–354, 363; see also *Mongols,* 129, 254.

30. See *Alqāb,* 2:549–550, §1971 (Dāmghānī). For the impact of the collapse of the Abbasids, see Mona Hassan, *Longing for the Lost Caliphate: A Transregional History* (Princeton, NJ: Princeton University Press, 2016), 20–65.

31. For the conquests of Baghdad and Wasit, see *Ḥawādith,* 237; translated in Hend Gilli-Elewy, "*Al-Ḥawādiṯ al-Ǧāmiʿa:* A Contemporary Account of the Mongol Conquest of Baghdad, 656/1258," *Arabica* 58, no. 5 (2011): 353–371, at 368. See also Naṣīr al-Dīn Ṭūsī, "Kayfiyyat-i Wāqiʿa-yi Baghdād," appendix in Juwaynī, *Tārīkh,* 3:280–292; translated in John Boyle, "The Death of the Last Abbasid Caliph: A Contemporary Muslim Account," *Journal of Semitic Studies* 6, no. 2 (1961), 145–161, at 151–161; Rashīd al-Dīn, *Jāmi*ʿ, 2:903, trans. 355; *Duwal,* 471–475.

32. *Āthār,* 495; see *Ḥawādith,* 229, trans. 361; Ṭūsī, "Kayfiyyat," 290, trans. 159, compare 148–149; see also *Mongols,* 128–130, 134–135, 167.

33. *Ḥawādith,* 238, trans. 369. For Bandanījī, see *Ḥawādith,* 258–259; *Alqāb* 1:87n3, compare 5:489, §5550.

34. Rashīd al-Dīn, *Jāmi*ʿ 2:907–908, trans. 356. See *Observ.,* 189–190. See autobiographical references with an abstract of Ibn al-Fuwaṭī's life, *Alqāb,* 1:13–59, 5:655–690.

35. See *Alqāb,* 3:111, §2297 (arrival of scholars); 4:155–156, §3566 (high-ranking Shia authority; teaching of Rāzī's writings); 4:492, §4292 (example of poem on the heavens); 5:117, §4753 (al-Tūnisī); 4:131, §3513 (student from India); 5:657–666 (Ibn al-Fuwaṭī in Maragha). For more, see *Observ.,* 189–223. See the autobiographical description by a chief astrologer in the observatory, Muʾayyad al-Dīn al-ʿUrḍī (d. ca. 665/1266), *Risāla fī kayfiyyat al-arṣād,* KMSI 4345,

fols. 1b–30b, fol. 2b. See also Hadi Jorati, "Science and Society in Medieval Islam: Nasir al-Din Tusi and the Politics of Patronage" (PhD diss., Yale University, 2014), 159–187. On poetry with astral themes, see Elaheh Kheirandish, "Astronomical Poems from the 'Four Corners' of Persia (ca. 1000–1500 CE)," in *Essays in Islamic Philology, History, and Philosophy*, ed. Alireza Korangy et al. (Berlin: De Gruyter, 2016), 51–90.

36. *Alqāb*, 2:175, §1278. See also John Boyle, "The Longer Introduction to the 'Zij-i Ilkhani' of Nasir-ad-Din Tusi," *Journal of Semitic Studies* 8, no. 2 (1963): 244–254; Yoichi Isahaya, "The *Tārīkh-i Qitā* in the *Zīj-i Īlkhānī:* The Chinese Calendar in Persian," *SCIAMVS: Sources and Commentaries in Exact Sciences* 14 (2013): 149–258; Isahaya, "History and Provenance of the 'Chinese' Calendar in the *Zīj-i Īlkhānī*," *Tarikh-e Elm: Iranian Journal for the History of Science* 8 (2009): 19–44.

37. See *Tansūkh*, 1–2 (dedication) 57, 66, 125, 157–158, 163, 188–194 (references to Alexander and Aristotle).

38. *Ḥawādith*, 250–251, 290–291 (Ibn al-Hudhayl), 252, 265 (Ṭūsī in Wasit). Qazwīnī retained his position of judgeship and professor, see *Ḥawādith*, 299; *Alqāb*, 4:470, 477, §§4241, 4257.

39. According to the title page of the early Persian recension, the work was composed (*waḍaʿahu*) in the months of 659/1260~; the text of this recension records the current year (*dar īn vaqt*) as 658/1259~, F pp. 1, 19. See Karin Rührdanz, "Zakariyyā' al-Qazwīnī on the Inhabitants of the Supralunar World: From the First Persian Version (659/1260–61) to the Second Arabic Redaction (678/1279–80)," in *The Intermediate Worlds of Angels: Islamic Representations of Celestial Beings in Transcultural Contexts*, ed. Sara Kuehn et al. (Beirut: Orient-Institut Beirut, 2019), 385–402, at 387–389.

4: Cosmic Order

1. On the dedication, see B fol. 2a; see Juwaynī's title in *Alqāb*, 2:315–316, §1537.

2. For the Juwaynī brothers, see George Lane, *Early Mongol Rule in Thirteenth-Century Iran: A Persian Renaissance* (London: Routledge, 2003), 177–211; Hend Gilli-Elewy, "The Mongol Court in Baghdad: The Juwaynī Brothers between Local Court and Central Court," in *Court Cultures in the Muslim World: Seventh to Nineteenth Centuries*, ed. Albrecht Fuess and Jan-Peter Hartung (London: Routledge, 2011), 168–181. On Ibn Hubayra, see M fol. 69a (W 124); *Āthār*, 367–368; *Kashf*, 2:1168.

3. For manuscripts derived from an authorial copy (i.e., *khaṭṭ al-muṣannif*), see B fol. 135b (undated); Qazwīnī, *Āthār*, Brit., Or. 3623, fol. 173a (copied 27 Dhū l-Qaʿda 729/22 September 1329), based on a copy finished Dhū l-Ḥijja 674/May 1276. The copyist of Or. 3623 is the same as Y fol. 181b (dated 722/1322); see Persis Berlekamp, "From Iraq to Fars: Tracking Cultural Transformations in the 1322 Qazwīnī *ʿAjāʾib* Manuscript," in *Arab Painting: Text and Image in Illustrated Arabic Manuscripts*, ed. Anna Contadini (Leiden: Brill, 2007), 73–91.

4. For draft and fair copies, see Adam Gacek, *Arabic Manuscripts: A Vademecum for Readers* (Leiden: Brill, 2009), 92–93.

5. For Qazwīnī's transmission of the *ʿAjāʾib* in a *majlis* to the Imami authority Ghiyāth al-Dīn ibn Aḥmad Ibn Ṭāwūs (d. 693/1294), see *Biḥār*, 104:105; *Enz.*, 82. The date of dedication is based on the discussion of the solar apogee, where Qazwīnī gives the current year as 661/1262~; see Qazwīnī, *ʿAjāʾib*, Berl., Or. Sprenger 11, fols. 2a (dedication), 14a (solar apogee);

Qazwīnī, *ʿAjāʾib* (Cairo: Maṭbaʿat al-Taqaddum, 1886), 4, 24. B is lacunose here; see London, Wellcome Collection, MS Arabic 18, fols. 4b, 17a, which records the year as 665/1266~. Compare *Enz.*, 82–83.

6. References to a *maqālat al-buldān* in W 115, 156, 171, which are not in M fols. 64a, 83b, 87a. For the date of the geography, see Qazwīnī, *Kitāb al-ʿAjāʾib al-buldān*, in Hartwig Derenbourg and Evariste Lévi-Provençal, *Les manuscrits arabes de l'escurial*, 3 vols. (Paris: E. Leroux and P. Geuthner, 1884–1928), 3:177–178, §1638; dated 661/1262~ in Michael Casiri, *Bibliotheca Arabico-Hispana Escurialensis*, 2 vols. (Madrid: Antonius Perez de Soto, 1760–1770), 2:5, §1632. See *Kashf*, 1:9, 2:1126.

7. For Persian verse, see F 16, 26 (Sanāʾī), 81 (Niẓāmī), 147, 152 (Kamāl al-Dīn Iṣfahānī responding to Abū Nawās), 161, 172 (Sanāʾī), 179 (Firdawsī), 186, 187, 200, 207 (Isfahānī), 266; P 111, 165, 267–268, 406–407, 418, 457, 471–472, 485, 487, 510, 623.

8. See Barbara Schmitz et al., *Islamic and Indian Manuscripts and Paintings in the Pierpont Morgan Library* (New York: Pierpont Morgan Library, 1996), 9–24; for comparisons with B, see *Ilkh.*, 115–123. See Anna Contadini, *A World of Beasts: A Thirteenth-Century Illustrated Arabic Book on Animals (The Kitāb Naʿt al-Ḥayawān), in the Ibn Bakhtīshūʿ Tradition* (Leiden: Brill, 2012), 4–5.

9. The date 658/1259~ in the early Persian version is noted as the current year in section on the solar apogee: see F 19 (P 118); BnF, Supplément persan 1781, fol. 21b; BnF, Supplément persan 2051, fol. 15a; Berl., Or. 14649, fol. 13a; Princeton, NJ, Princeton University, Garrett no. 82G, fol. 17b; KMSI 860, fol. 28b; Cambridge University, Nn.3.74, fol. 19a. Instead of Juwaynī, this version is dedicated to an otherwise unidentified functionary: ʿĀdil Muʾīd Manṣūr Muẓaffar ʿIzz al-Dīn Fakhr al-Islām Malik al-Ṣudūr Sayyid al-Akābir Shāhpūr ibn ʿUthmān (var. ibn ʿAlīkhān). See Storey, 2:124–126, §188. As for the possibility that Qazwīnī produced this Persian text himself, see the review of Bābāpūr and Ghulāmiyya's edition by ʿAlī Ṣafarī Āq-Qilʿa, "ʿAjāʾib al-makhlūqāt wa-gharāʾib al-mawjūdāt," *Guzārish-i Mīrāth* 54–55 (1391sh/2012): 139–145. The same date (658/1259~) in F also appears in copies of the later redaction produced for Ibrāhīm ʿĀdil Shāh (d. 965/1558), see Brit., Or. 1621, fol. 36a; I.O. Islamic 3243, fol. 45b; New Haven, Beinecke Library, MSS Persian +28, p. 29; Kashmir, Srinagar University, Allama Iqbal Library, MS 205, fol. 41b; KMSI 3343/46, fol. 41b; KMSI 2242, fol. 41b. A third Persian recension, with the date 664/1265~, contains no dedication, but with the same section on different people and crafts as in F, e.g., Manchester, John Rylands Library, Persian MS 3, p. 41 (unpaginated); NLM, MS P 1, fol. 19b; BnF, Supplément persan 330, fol. 20b.

10. M fol. 212b.

11. In addition to folios missing from the star catalogue (M fols. 22b–23a, Gemini through Orion), the sections on jinn and demons (M fols. 164b–168b) and a folio on birds (fol. 189a–b) have been reinserted by a latter hand. See *Ilkh.*, 150, 163. The section on the jinn is integral to the original manuscript, promised in the table of contents, M fol. 8b; see Y fols. 10a, 154a. Compare *Cosmos*, 71.

12. For painting and sculpture, see *Āthār*, 54–55, 344, 454, 457, 531. See also M fols. 82b–83b, 160b. For Ṣūfī, *Kawākib*, Brit., Or. 5323, see Moya Carey, "Mapping the Mnemonic: A Late Thirteenth-Century Copy of al-Ṣūfī's Book of the Constellations," in *Arab Painting*, 65–73. For the Arabic *Materia medica* housed in TPMK, Ahmet III 2147, and the dispersed folios from it, as well as the copy produced in 637/1240 in the Niẓāmiyya madrasa of Baghdad (?), housed in Bodl., Arab. d. 138, fol. 210a, see Ernst Grube, "Materialen zum Dioskurides Arabicus," in *Aus der Welt der islamischen Kunst: Festschrift für Ernst Kühnel*, ed. Richard Ettinghausen (Berlin:

Gebr. Mann, 1959), 163–194, at 172, 179. See also George Saliba and Linda Komaroff, "Illustrated Books May Be Hazardous to Your Health: A New Reading of the Arabic Reception and Rendition of the *Materia Medica* of Dioscorides," *Ars Orientalis* 35 (2008): 6–65. For connections between M and Brit., Or. 2784, see Contadini, *Beasts*, 155–156, compare 41–42 (fol. 97a, on wonder). See also Bīrūnī, *Kitāb al-Tafhīm*, Brit., Add. 7697, dated 685/1286, and Iṣṭakhrī, *al-Masālik wa-l-mamālik*, Leiden, Universiteitsbibliotheek, MS Or. 3101, dated 589/1193.

13. See Ḥarīrī, *Maqāmāt*, BnF, Arabe 5847, dated 634/1237; Juwaynī, *Tārīkh*, BnF, Supplément persan 206; Sheila Blair, "Calligraphers, Illuminators and Painters in the Ilkhanid Scriptorium," in *Beyond the Legacy of Genghis Khan*, ed. Linda Komaroff (Leiden: Brill, 2006) 167–182. For painters, see *Alqāb*, 1:120, 478, §§83, 768, see also 4:274, §3843; 4:508, §3250. For pigments, see Contadini, *Beasts*, 12–16, 165–166 (Paul Garside, "Appendix One: Analyses of Colours, Ink and Gold").

14. M fol. 1a, quoted in *Enz.*, 83. Compare B fol. 1b.

15. M fols. 47a (Hārūt and Mārūt), 65b (escaping by a giant bird), 67b (attacked by a string-legged creature), 83a (Behistun carving), 212a (two-headed human in Yemen). On narrative elements in B, see *Ikh.*, 13–22, 70–77.

16. See *Ikh.*, 79–80, 334 (cat. 332), compare 364–365n146.

17. M fol. 7b (W 13), "wa-li-llāhi fī kulli taḥrīkatin/wa-taskīnatin ababan shāhidu || wa-fī kulli shay'in lahu āyatun/tadullu ʿalā annahu wāḥidu." See broadly al-Qāsim ibn Ibrāhīm, *Kitāb al-Dalīl al-kabīr*, published as *Al-Ḳāsim b. Ibrāhīm on the Proof of God's Existence*, ed. and trans. Binyamin Abrahamov (Leiden: E.J. Brill, 1990), 90; Abū ʿUthmān al-Jāḥiẓ (ascribed), *al-ʿIbar wa-l-iʿtibār*, ed. Ṣābir Idrīs (Cairo: al-ʿArabī li-l-Nashr wa-l-Tawzīʿ, 1994), 29–30 (early Christian writings), 60, 73, 74 (references to marvels); *Ḥay.*, 1:208–209, 2:109–110. Compare *Ikhwān*, 3:482.

18. See *Khalīqa*, 13–23 (first cause, monotheism), 51–61, 100–105 (creation, causation), 524–525 (emerald tablet of Hermes, key to alchemy); compare pseudo-Apollonius, *Kitāb Ṭalāsim Balīnās al-akbar*, BnF, Arabe 2250, fol. 94b. See also Julius Ruska, *Tabula smaragdina: Ein Beitrag zur Geschichte der hermetischen Literatur* (Heidelberg: Winter, 1926), 107–124; *NG*, 378–381. See *Āthār*, 572 (Apollonius).

19. M fol. 2a; Y fol. 3a; paraphrased in F 2 (P 82); missing from V fol. 3a, W 4, and B fol. 2a. See *Ilāhiyyāt*, 2:414–422.

20. Ghazālī, *Kitāb al-Imlā'* in *Iḥyā'*, 16:3083–3085 (49–51). Compare *Manāfiʿ*, fol. 55b–56a; 1:174–175 (Greek); 1:188–189 (English); and *Galeni compendium Timaei Platonis*, ed. Paulus Kraus and Richardus Walzer (London: Warburg Institute, 1951), 5, §c. For further context, see Eric Ormsby, *Theodicy in Islamic Thought: The Dispute over al-Ghazālī's "Best of All Possible Worlds"* (Princeton, NJ: Princeton University Press, 1984), 14, 16–31, 38–91; *GPT*, 225–231; Ayman Shihadeh, *The Teleological Ethics of Fakhr al-Dīn al-Rāzī* (Leiden: Brill, 2006), 160–169; Shams Inati, *The Problem of Evil: Ibn Sīnā's Theodicy* (Piscataway, NJ: Gorgias Press, 2017), 65–101.

21. See M fols. 1a, 5a, 7b, 9a (W 3, 13, 15).

22. See Abharī, *Talkhīṣ al-ḥaqā'iq*, Süleymaniye, Köprülü 1618, fol. 2b. See also *APW*, 296–300.

23. M fols. 1b–2a, citing Q. 50:6 and Q. 10:101; M fols. 98b, 144a (W 202–203, 303). See *Maṭālib*, 7:282; *Muḥaṣṣal*, 40–41; *Mafātīḥ*, 17:76–77 (Q. 10:101); 28:155–156 (Q. 50:6); 30:226–229 (Q. 75:22–23, on *naẓar*).

24. Aristotle, *Metaphysica*, ed. Werner Jaeger (Oxford: Clarendon Press, 1957), 993b19–21; translated in Ibn Rushd (d. 595/1198), *Tafsīr mā baʿd al-ṭabīʿa*, ed. Maurice Bouyges, 4 vols. (Beirut: Imprimerie catholique, 1938–1952), 1:11; compare Ibn Sīnā, *Ilāhiyyāt*, 1:3–5.

25. See Ulrich Rudolph, "Occasionalism," in *The Oxford Handbook of Islamic Theology*, ed. Sabine Schmidtke (Oxford: Oxford University Press, 2016), 347–363; *GPT*, 124–133, 215–234; Shihadeh, *Teleological*, 13–44.

26. M fol. 2a, "riyāḍat al-nafs baʿd taḥsīn al-akhlāq wa-tahdhīb al-nafs." See V fol. 3a (W 4). On "arinī l-ashyāʾ kamā hiya," see *Mafātīḥ*, 1:133; 2:218 (Q. 2:31); 13:47 (Q. 6:75); on "tafakkarū fī khalq allāh," see Abū l-Shaykh al-Iṣfahānī, *Kitāb al-ʿAẓama*, ed. Riḍāʾ Allāh Mubārakfūrī, 5 vols. (Riyadh: Dār al-ʿĀṣima, 1998), 1:209–270; *Iḥyāʾ*, 15:2803, 2820–2821 (59, 76–77). On wonder and pleasure, see Aristotle, *Ars rhetorica*, ed. William Ross (Oxford: Clarendon Press, 1959), 1371a31–1371b12; Aristotle, *Ars rhetorica: The Arabic Version*, ed. Malcolm Lyons, 2 vols. (Cambridge: Pembroke College, Cambridge University, 1982), 1:58.

27. See Fakhr al-Dīn al-Rāzī, *Kitāb al-Nafs wa-l-rūḥ wa-sharḥ quwāhumā*, ed. Muḥammad Ṣaghīr Ḥasan Maʿṣūmī (Islamabad: Islamic Research Institute, 1968), 3–12; translated as *Imām Rāzī's ʿIlm al-akhlāq*, trans. Muḥammad Ṣaghīr Ḥasan Maʿṣūmī (Islamabad: Islamic Research Institute, 1969), 43–53; Naṣīr al-Dīn al-Ṭūsī, *Akhlāq-i nāṣirī*, ed. Mujtabā Mīnuwī and ʿAlī Riḍā Ḥaydarī (Tehran: Intishārāt-i Khwārazmī, 1978), 59–65; translated as *The Nasirean Ethics*, trans. George Michael Wickens (London: Allen and Unwin, 1964), 43–48.

28. Ibn Sīnā, *al-Shifāʾ, al-Manṭiq: 8. al-Khaṭāba*, ed. Muḥammad Salīm Sālim (Cairo: al-Maṭbaʿa al-Amīriyya, 1954), "al-taʿallum ladhīdh bi-sabab mā yatawaqqaʿu mina l-taʿjīb (*var.* taʿajjub)," 103; *Ishārāt*, 341–367 (viii–ix), particularly 364 (ix.20). See Michael Rapoport, "The Life and Afterlife of the Rational Soul: Chapters VIII–X of Ibn Sīnā's Pointers and Reminders and Their Commentaries" (PhD diss., Yale University, 2018), 70–163 (chaps. 3–4); *APW*, 151n13, 293–294, 344–345.

29. See *Iḥyāʾ*, 12:2273–2274 (129–130). Compare *FPP*, 218–219, 491–492.

30. See Caroline Bynum, "Wonder," *American Historical Review* 102, no. 1 (1997): 1–26, at 24.

31. See Fārābī, *Falsafat Arisṭūṭālīs*, ed. Muḥsin Mahdī (Beirut: Dār Majallat Shiʿr, 1961), 61. For further, see Salim Kemal, *The Poetics of Alfarabi and Avicenna* (Leiden: E.J. Brill, 1991), 120–128.

32. M fol. 2b. For *mushāhadāt* as empirical experiences, see Dimitri Gutas, "The Empiricism of Avicenna," *Oriens* 40, no. 2 (2012): 391–436, at 428–430.

33. On this expression as a line of poetry, see Sībawayh (d. ca. 180/796), *al-Kitāb*, ed. ʿAbd al-Salām Muḥammad Hārūn, 5 vols. (Cairo: Dār al-Qalam, 1966–1977), 2:295–296n4. See also Abū l-Qāsim al-Zamakhsharī (d. 538/1144), *al-Mustaqṣā fī amthāl al-ʿarab*, ed. Muḥammad ʿAbd al-Muʿīd Khān, 2 vols. (Hyderabad: Dāʾirat al-Maʿārif al-ʿUthmāniyya, 1962), 2:267.

34. M fol. 80b (W 147). Compare Qudāma ibn Jaʿfar (d. 337/948), *Kitāb al-Kharāj wa-ṣināʿat al-kitāba*, ed. Muḥammad Ḥusayn al-Zubaydī (Baghdad: Dār al-Rashīd, 1981), 132; al-Idrīsī (d. 560/1165), *Nuzhat*, 1:6–7.

35. *Buldān*, 1:12.

36. M fol. 212b; Y fol. 192a; D fol. 426a; missing from W, B, V, and F. Compare *Ishārāt*, 391 (x.31), "fī l-ṭabīʿa ʿajāʾib wa li-l-quwā l-ʿāliya al-faʿʿāla wa-l-quwā l-sāfila al-munfaʿila ijtimāʿāt ʿalā gharāʾib." Both Rāzī and Ṭūsī read this critique as directed to those philosophers who rejected the authenticity of miracles, see Fakhr al-Dīn al-Rāzī, *Sharḥ al-Ishārāt wa-l-*

tanbīhāt, ed. ʿAlī Riḍā Najafzāda, 2 vols. (Tehran: Anjuman-i Āthār wa-Mafākhir-i Farhangī, 2005), 2:664–665; Naṣīr al-Dīn al-Ṭūsī, *Sharḥ al-Ishārāt wa-l-tanbīhāt*, 3 vols. (Qum: Nashr al-Balāgha, 1996), 3:418–419. Compare Kashshī, *Ḥadīqat al-ḥaqāʾiq*, Süleymaniye, Köprülü 864, fol. 205a–b.

37. See section on rational soul, M fols. 144a–150b (W 304–305), missing from F 156. This opens with the opening of the didactic poem in Ibn Sīnā, *Manṭiq al-mashriqiyyīn* (Cairo: al-Maktaba al-Salafiyya, 1910), xxii–xxiii; translated in Arthur Arberry, *Avicenna on Theology* (London: John Murray, 1951), 77–78. See *CAL*, 115.

38. M fol. 2b (W 5). See Ibn Sīnā, *al-Shifāʾ, al-Manṭiq: 6. al-Jadal*, ed. Aḥmad Fuʾād al-Ahwānī (Cairo: al-Hayʾa al-ʿĀmma li-Shuʾūn al-Maṭābiʿ al-Amīriyya, 1965), 189–190; *Afʿāl*, 255–256. For loss of wonder through habituation, see *Iḥyāʾ*, 15:2833 (89). On pleasure and wonder, see Ibn Sīnā, *Shifāʾ, al-Manṭiq: 9. al-Shiʿr*, ed. ʿAbd al-Raḥmān al-Badawī (Cairo: al-Dār al-Miṣriyya li-l-Taʾlīf wa-l-Tarjuma, 1966), 24–25; Ismail Dahiyat, *Avicenna's Commentary on the Poetics of Aristotle: A Critical Study with an Annotated Translation of the Text* (Leiden: E.J. Brill, 1974), 35–36, 62–63; Kemal, *Poetics*, 161–176; Lara Harb, *Arabic Poetics: Aesthetic Experience in Classical Arabic Literature* (Cambridge: Cambridge University Press, 2020), 88–101.

39. See pseudo-Aristotle, *Uthūlūjiyā*, Süleymaniye, Aya Sofya 2457, fols. 113a–b, 143b; translated in *Plotiniana Arabica*, trans. Geoffrey Lewis, in *Plotini Opera, Tomus II: Enneades IV–V*, ed. Paul Henry and Hans-Rudolf Schwyzer (Paris: Desclée de Brouwer, 1959), 139, ll. 24–25 (habituation), 225, ll. 21–26 (theophany). For problems with the modern Arabic editions and the broader reception history, see Dimitri Gutas, "The Text of the Arabic Plotinus: Prolegomena to a Critical Edition," in *The Libraries of the Neoplatonists*, ed. Cristina D'Ancona (Leiden: Brill, 2007), 371–384; and broadly Peter Adamson, *The Arabic Plotinus* (London: Duckworth, 2002).

40. See Plato, *Theaetetus* in *Platonis Opera*, ed. E. A. Duke et al. (Oxford: Clarendon Press, 1995), 155d2–4; Aristotle, *Metaphysica*, 982b12–18. See the commentary of Alexander of Aphrodisias, *Aristotelis Metaphysica commentaria*, ed. Michael Hayduck (Berlin: G. Reimer, 1891), 15, 25–30. This passage of the *Metaphysica* is famously missing from the text available in Ibn Rushd, *Tafsīr*, 1:cxvi–cxxiv, cxxvii–cxxxii (editorial introduction). See ʿAbd al-Laṭīf al-Baghdādī, *Mā baʿd al-ṭabīʿiyya*, ed. Yūnus Ajʿūn (Beirut: Dār al-Kutub al-ʿIlmiyya, 2017), 42.

41. The debate whether God experiences wonder draws on a Quranic variant (Q. 37:12), reading *ʿajibtu* for *ʿajibta*, and several hadith. See, for example, Abū Zakariyyāʾ al-Farrāʾ (d. 207/822), *Maʿānī l-Qurʾān*, ed. Yūsuf Najātī and Muḥammad ʿAlī l-Najjār, 3 vols. (Cairo: al-Hayʾa al-Miṣriyya al-ʿĀmma li-l-Kitāb, 1980) 3:384; Ibn Qutayba (d. 276/889), *Taʾwīl mukhtalif al-ḥadīth*, ed. Muḥammad Muḥyī l-Dīn al-Aṣfar (Beirut: al-Maktab al-Islāmī, 1999), 305–306.

42. *Iḥyāʾ*, 11:1994–2010 (60–76), final section of book 29, "Kitāb al-Kibr wa-ʿujb"; compare 15:2802–2844 (58–100), book 39, "Kitāb al-Tafakkur."

43. See Abū Ḥayyān al-Tawḥīdī, *al-Baṣāʾir wa-l-dhakhāʾir*, ed. Ibrāhīm al-Kīlānī, 4 vols. (Damascus: Maktabat Aṭlas, 1964), 1:555; Tawḥīdī, *al-Ṣadāqa wa-l-ṣadīq*, ed. ʿAlī Mutawallī Ṣalāḥ (Cairo: Makatabat al-Ādāb, 1972), 144; Tawḥīdī, *al-Muqābasāt*, ed. Ḥasan al-Sandūbī (Cairo: al-Maṭbaʿa al-Raḥmāniyya, 1929), 274, §71; Tawḥīdī and Miskawayh, *al-Hawāmil wa-l-shawāmil*, edited and translated as *The Philosopher Responds: An Intellectual Correspondence from the Tenth Century*, ed. Bilal Orfali and Maurice Pomerantz, trans. Sophia Vasalou and James Montgomery, 2 vols. (New York: NYU Press, 2019), 1:84–90, §§16.1–6. See also *Mukhtār min kalām al-ḥukamāʾ al-arbaʿa al-akābir*, edited and translated as *Greek Wisdom Literature in Arabic Translation: A Study of the Graeco-Arabic Gnomologia*, ed. and trans. Dimitri Gutas (New Haven,

CT: American Oriental Society, 1975), 160, §2 (Arabic text), 426–429 (commentary on sources). See Baghdādī, *Mā baʿd*, 42, compare 285.

44. M fols. 2b–3a (W 5). Compare *Cosmos*, 40–41; *CAL*, 69–71.

45. M fols. 6a–7a; cf. *CAL*, 73–75. For habit or custom (*ʿāda*) in theological discussions of miracles, see Harry Wolfson, *Philosophy of the Kalām* (Cambridge, MA: Harvard University Press, 1976), 544–558; *GPT*, 194–201.

46. M fol. 6b (W 12). The Juwaynī recension ascribes this statement to the *ḥukamāʾ*, B fol. 6a, missing from M. Compare with talismans and *nīranjāt* in F 221–226 (P 544–546).

47. *Ishārāt*, 389–390 (x.28–30). See also Ibn Sīnā, "Aqsām al-ʿulūm," 110–111. See also Charles Burnett, "*Nīranj:* A Category of Magic (Almost) Forgotten in the Latin West," in *Natura, scienze e societa medievali: Studi in onore di Agostino Paravicini Bagliani*, ed. Claudio Leonardi, Francesco Santi, and Agostino Paravicini Bagliani (Florence: Edizioni del Galluzzo, 2008), 37–66, at 43–44.

48. M fol. 6a (W 10), B fol. 5a; compare *Ishārāt*, 390 (x.29).

49. Ibn Sīnā *al-Shifāʾ, al-Ṭabīʿiyyāt:* 6. *al-Nafs*, ed. Jūrj Qanawātī [Georges Anawati] and Saʿīd Zāyid (Cairo: al-Hayʾa al-Miṣriyya al-ʿĀmma li-l-Kitāb, 1975), 177; *Ishārāt*, 387–388 (x.26). On this thought experiment, see *Maktūm*, fol. 8a; *Mafātīḥ*, 3:225. For the Latin translation produced in twelfth-century Toledo, see *Liber de anima*, ed. Simone van Riet, 2 vols. (Leiden: E.J. Brill, 1968–1972), book 4, 1:64, ll. 25–30. For Ibn Sīnā's psychology in Latin, with a focus on these passages, see Dag Nikolaus Hasse, *Avicenna's De Anima in the Latin West: The Formation of a Peripatetic Philosophy of the Soul, 1160–1300* (London: Warburg Institute, 2000), 153–173. For the metaphor of the plank in late Latin discourses on witchcraft and the paranormal, see Heinrich Kramer (d. 1505) and Jacobus Sprenger (d. 1495), *Malleus maleficarum*, ed. and trans. Christopher S. Mackay, 2 vols. (Cambridge: Cambridge University Press, 2006) 1:231 (14c–d).

50. Rāzī, *Sharḥ al-Ishārāt*, 2:663–664; *Maktūm*, fol. 211b (*al-ʿulūm al-ghāmiḍa al-sharīfa al-ʿajība al-hāʾila al-nāfiʿa al-ḍārra*). See also Tariq Jaffer, "Fakhr al-Dīn al-Rāzī's Taxonomy of Extraordinary Acts," in *Light upon Light: Essays in Islamic Thought and History in Honor of Gerhard Böwering*, ed. Jamal Elias and Bilal Orfali (Leiden: Brill, 2019), 347–365; *GPT*, 197–198.

51. For etymologies of *ṭilism* and *nīranj*, see *NG*, 362–363. See also Charles Burnett, "Talismans: Magic as Science? Necromancy among the Seven Liberal Arts," in *Magic and Divination in the Middle Ages: Texts and Techniques in the Islamic and Christian Worlds* (Aldershot: Variorum, 1996), 1–15.

52. For *nērang* in a post-Sasanian priestly context, see *Dēnkard*, book III, in *The Complete Text of the Pahlavi Dinkard*, ed. Dhanjishah Meherjibhai Madan, 2 vols. (Bombay: Fort Printing Press, 1911), 1:157–158, 399–400. On resonance with Galenic humoralism, see Richard Payne, *A State of Mixture: Christians, Zoroastrians, and Iranian Political Culture* (Berkeley: University of California Press, 2015), 89. For examples of *nērang* as earthly amulets in both Pahlavi and Pāzand, see Antonio Panaino, "Two Zoroastrian Nērangs and the Invocation to the Stars and the Planets," in *The Spirit of Wisdom = Mēnōg ī xrad: Essays in Memory of Ahmad Tafazzoli*, ed. Touraj Daryaee and Mahmoud Omidsalar (Costa Mesa: Mazda, 2004), 196–218.

53. See Albert de Jong, *The Traditions of the Magi: Zoroastrianism in Greek and Latin Literature* (Leiden: Brill, 1997), 222, 387–413.

54. *Maktūm*, fol. 212a–b (accordance with religious law). See also *Maṭālib*, 1:234–235, 8:179–185; *Sittīnī*, 313–332, 431–458.

55. M fol. 16a. The idea features in the *Sindhind*, the early Arabic translation derived from the Sanskrit astronomical collection *Brāhmasphuṭasiddhānta*. See Abū l-Ḥasan al-Masʿūdī (d. 345/956), *Murūj*, 1:85, §153. On the cosmic or great year, see Antonio Panaino, "Between Astral Cosmology and Astrology: The Mazdean Cycle of 12,000 Years and the Final Renovation of the World," in *The Zoroastrian Flame: Exploring Religion, History and Tradition*, ed. Alan Williams et al. (London: I.B. Tauris, 2016), 113–134.

56. *Āthār*, 234, 572. "Highest level of probability" is a translation of *ghalabat al-ẓann*, a preponderance of evidence, or the highest degree of knowledge that can be obtained when there is no unambiguous certainty. See Fakhr al-Dīn al-Rāzī, *al-Maḥṣūl fī ʿilm al-uṣūl*, ed. Ṭāhā Jābir Fayyāḍ al-ʿAlwānī, 6 vols. (Beirut: Muʾassasat al-Risāla, 1992), 5:98–126. The *Aḥkām* of Jāmāsp was well-known in Qazwīnī's day, see Etan Kohlberg, *A Medieval Muslim Scholar at Work: Ibn Ṭāwūs and His Library* (Leiden: E.J. Brill, 1992), 102–103, §11. For an Arabic copy produced in 679/1281, see BnF, Arabe 2487, fols. 39a–55b, esp. fols. 42a–b (Moses), 44a (Jesus), 45a–46a (Muhammad), 46b–47a (fall of Iran), 49b (rise of Turks); and the Persian redaction copied in 741/1341 with references to Genghis Khan and the Mongol invasions, BnF, Supplément persan 380, fols. 14a–62b, at fols. 40a, 44b–45b. Discussed in Edgar Blochet, "Études sur le gnosticisme musulman," *Rivista degli studi orientali* 4, no. 2 (1911): 267–300, at 278–287. See also *NG*, 183–184, 295–296. See also Daniel Sheffield, "Primary Sources: New Persian," in *The Wiley Blackwell Companion to Zoroastrianism*, 529–542, at 534–535.

57. For Arabic manuscripts with 678/1279~ as the current year in the solar apogee, see: M fol. 16a; Y fol. 14a; Harvard University, Fogg Art Museum, 1972.3, fol. 20b; BnF, Smith-Lesouëf 221, fol. 15a; Freer Gallery of Art, Washington, DC, F1954.36, p. 14. For 665/1266~, see: London, Wellcome Collection, MS Arabic 18, fol. 17a; compare BnF, Arabe 2175, fol. 17a; BnF, Arabe 2177, fol. 13b. For 661/1262~, see: V fol. 17b; Berl., Or. Sprenger 11, fol. 14a; Istanbul, Millet Kütüphanesi, Feyzullah Efendi, 1369, fol. 14a; Munich, Bayerische Staatsbibliothek, Cod. Arab. 463, fol. 16a; BnF, Arabe 2173, fol. 20a; BnF, Arabe 2174, fol. 32a; BnF, Arabe 2176, fol. 22b; BnF, Arabe 2178, fol. 16a. As to whether these dates may be errors, see Karin Rührdanz, "Between Astrology and Anatomy: Updating Qazwīnī's *ʿAjāʾib al-makhlūqāt* in Mid-Sixteenth-Century Iran," *Ars Orientalis* 42 (2012): 56–66, at 60. Modulated in Rührdanz, "Zakariyyāʾ al-Qazwīnī on the Inhabitants of the Supralunar World: From the First Persian Version (659/1260–61) to the Second Arabic Redaction (678/1279–80)," in *The Intermediate Worlds of Angels: Islamic Representations of Celestial Beings in Transcultural Contexts*, ed. Sara Kuehn et al. (Beirut: Orient-Institut Beirut, 2019), 385–402, at 387–393. For the solar apogee in connection with Hermes, see Abū ʿUbayd al-Bakrī (d. 487/1094), *al-Masālik wa-l-mamālik*, ed. André Ferré and Adrian van Leeuwen, 2 vols. (Carthage: al-Dār al-ʿArabiyya li-l-Kitāb, 1992), 1:50–51. On the solar apogee in Qazwīnī's day, see George Saliba, "Solar Observations at the Maragha Observatory before 1275," *Journal for the History of Astronomy* 16, no. 2 (1985): 113–122.

5: Terrestrial Designs

1. See *Observ.*, 211–218; *WTI*, 28–33, trans. 2:28–33.

2. Muʾayyad al-Dīn al-ʿUrḍī, *Risāla fī kayfiyyat al-arṣād*, KMSI 4345, fols. 1b–30b; translated in Hugo Josef Seemann, "Die Instrumente der Sternwarte zu Marâgha nach den Mitteilungen von al-ʿUrḍî," *Sitzungsberichte der Physikalisch-medizinischen Sozietät zu Erlangen* 60

(1928): 15–126; *WTI*, 2:38–52, trans. 2:38–52; S. Mohammad Mozaffari and Georg Zotti, "The Observational Instruments at the Maragha Observatory after AD 1300," *Suhayl* 12 (2013): 45–179.

3. ʿUrḍī, *Risāla*, KMSI 4345, fol. 1b, trans. 89–90. See Moya Carey, "The Gold and Silver Lining: Shams al-Dīn Muḥammad b. Muʾayyad al-ʿUrḍī's Inlaid Celestial Globe (ca. AD 1288) from the Ilkhanid Observatory at Marāgha," *Iran: British Institute of Persian Studies* 47 (2009): 97–108.

4. Paul Kunitzsch, "The Astronomer Abu 'l-Ḥusayn al-Ṣūfī and His *Book on the Constellations*," *ZGAIW* 3 (1986): 56–81; Emilie Savage-Smith, "The Islamic Tradition of Celestial Mapping," *Asian Art* 5, no. 4 (1992): 5–28.

5. See Emilie Savage-Smith, *Islamicate Celestial Globes: Their History, Construction, and Use* (Washington, DC: Smithsonian Institution Press, 1984), 26, 61–95, 219–220 (nos. 4–5); Edward Kennedy, "al-Ṣūfī on the Celestial Globe," *ZGAIW* 5 (1989): 48–93.

6. M fols. 25a–31a; compare Ibn Qutayba, *al-Anwāʾ fī mawāsim al-ʿArab* (Hyderabad: Maṭbaʿat Majlis al-Maʿārif, 1956), 8–89. For lunar mansions, see Savage-Smith, *Globes*, 119–132; *Funūn*, 103–124 (Arabic), 391–409 (translation); analyzed in Yossef Rapoport and Emilie Savage-Smith, *Lost Maps of the Caliphs: Drawing the World in Eleventh-Century Cairo* (Chicago: University of Chicago Press, 2018), 46–52.

7. For the early Arabic Hermetic tradition on lunar mansions, see the influential treatise, pseudo-Aristotle, *Kitāb al-Ustuwwaṭās*, BnF, Arabe 2577, fols. 24a–34a.

8. On geomancy, see *Sittīnī*, 432–437. See also Matthew Melvin-Koushki, "Persianate Geomancy from Ṭūsī to the Millennium: A Preliminary Survey," in *The Occult Sciences in Premodern Islamic Cultures*, ed. Nader El-Bizri and Eva Orthmann (Beirut: Orient-Institut Beirut: 2018), 151–199, at 162–163.

9. See Emilie Savage-Smith and Marion Smith, "Islamic Geomancy and a Thirteenth-Century Divinatory Device: Another Look," in *Magic and Divination in Early Islam*, ed. Emilie Savage-Smith (Aldershot: Ashgate, 2004), 211–276, at 247–251, 258–259 (account of the lunar mansions); Arabic passages on 275 (§7), 276 (§12).

10. M fols. 30b, 37b. The Arabic word *aṭlas* is the elative of a root that means effaced or blotted out. Qazwīnī makes no reference to the story of Atlas, but follows the explanation that the sphere is so called because there are no stars within it. See *Tadhkira*, 1:110 (translation), 111 (Arabic, §4), 2:390–391 (commentary).

11. M fol. 30b. For celestial equivalences to the pedestal and throne, see the treatises ascribed to Ibn Sīnā, "Fī ithbāt al-nubuwwāt," in *Tisʿ rasāʾil* (Cairo: Maṭbaʿa Hindiyya, 1908), 121–132, at 128; and "Risāla fī maʿrifat al-nafs al-nāṭiqa wa-aḥwālihā," in *Aḥwāl al-nafs*, ed. Aḥmad Fuʾād al-Ahwānī (Cairo: Dār Iḥyāʾ al-Kutub al-ʿArabiyya, 1952), 181–192, at 189–190. For limits of human reason, see Binyamin Abrahamov, "The *Bi-lā kayfa* Doctrine and Its Foundations in Islamic Theology," *Arabica* 42, no. 3 (1995): 365–379.

12. M fols. 5a (read with B fol. 4a), 31a–b (citing Q. 74:31), 32a (Rūḥ). See Ibn Sīnā, *Ḥudūd*, edited as *Livre des definition*, ed. and trans. Amélie Marie Goichon (Cairo: L'Institut français d'archéologie orientale du Caire, 1963), 26, §46 (Arabic), 38 (French). Compare Qazwīnī on angels with *Maṭālib*, 7:17–23 (cosmic role), 387 (harnessing angels with talismans); see Tony Street, "Medieval Islamic Doctrine on the Angels: The Writings of Fakhr al-Dīn al-Rāzī," *Parergon* 9, no. 2 (1991): 111–127. On the treatment of angels as separate (i.e., noncorporeal)

intellects (*ʿuqūl mujarrada*), in Avicennan terms, and their role in the movement of the cosmos, see Abharī, *Hidāyat al-ḥikma* (Karachi: Maktabat al-Bushrā, 2011), 109–117. See also, broadly, *Enz.*, 262–350.

13. M fol. 9a. On angelic mediation in the writings of Ghazālī, see *GPT*, 128–141, 150–151, 180–182, 242–244, 248. See also Jon McGinnis, "Occasionalism, Natural Causation and Science in al-Ghazālī," in *Arabic Theology, Arabic Philosophy: From the Many to the One*, ed. James E. Montgomery (Leuven: Peeters, 2006), 441–463.

14. M fol. 38a, citing Q. 2:30; see *Āthār*, 487 (referencing Maʿarrī). However, the interlocutor in the anecdote is identified as Ibn Fāris (d. 395/1004) by Badīʿ al-Zamān al-Hamadhānī, *Rasāʾil* (Istanbul: Maṭbaʿat al-Jawāʾib, 1298/1880~), 180–181. See Geert Jan van Gelder, "Good Times, Bad Times: Opinions on *Fasād az-zamān*, 'the Corruption of Time,'" in *Inḥiṭāṭ—The Decline Paradigm: Its Influence and Persistence in the Writing of Arab Cultural History*, ed. Syrinx von Hees (Würzburg: Ergon, 2017), 111–130, at 116–119.

15. M fol. 51a. Qazwīnī does not identify the pious youth, who, in addition to Moses, may be an allusion to the Arabic cycle on the adventures of Bulūqiyā, also an acquaintance of Khiḍr from the Tribe of Israel in Egypt. See Abū Isḥāq al-Thaʿlabī (d. 427/1035), *ʿArāʾis al-majālis* (Beirut: al-Maktaba al-Thaqāfiyya, n.d.), 315–322; *Alf layla wa-layla*, edition attributed to William Hay Macnaghten, 4 vols. (Calcutta: Baptist Mission Press, 1839–1842), 2:589–621, nights 486–498.

16. On Wasit's abandonment, see Niʿmat Allāh al-Jazāʾirī (d. 1112/1701), *Zahr al-rabīʿ* (Beirut: Muʾassasat al-ʿĀlamiyya li-l-Tajdīd, 2000), 208–209; Fuad Safar, *Wâsit: The Sixth Season's Excavations* (Cairo: Institut français d'archéologie orientale, 1945), 3–4, 9–10. For the case that the unidentified ruins correspond to the Sharābī madrasa, see Nājī Maʿrūf, *al-Madāris al-Sharābiyya bi-Baghdād, Wāsit wa-Makka* (Baghdad: al-Irshād, 1965), 288–301; Ṭāriq al-Janābī, "Studies in Mediaeval Iraqi Architecture," 2 vols. (PhD diss., Edinburgh University, 1975), 1:85–95. See also *Enz.*, 79–80; *Cosmos*, 14.

17. Citing Q. 30:42. Compare Q. 3:137, 6:11, 12:109, 16:36, 22:46, 27:69, 30:9, 35:44, 40:21, 82, 47:10. See *Āthār*, 66.

18. Andrew Robinson, *Writing and Script* (Oxford: Oxford University Press, 2009), 1–52. See also Alain George, *The Rise of Islamic Calligraphy* (London: Saqi Books, 2010), 21–53.

19. See Lara Harb, *Arabic Poetics: Aesthetic Experience in Classical Arabic Literature* (Cambridge: Cambridge University Press, 2020), 25–134; Geert Jan van Gelder, "*Taʿadjdjub*," *EI2;* Wolfhart Heinrichs, "Takhyīl," *EI2; Takhyīl: The Imaginary in Classical Arabic Poetics*, ed. Geert Jan van Gelder and Marlé Hammond (Cambridge: Gibb Memorial Trust, 2008); Taneli Kukkonen, "Ibn Sīnā and the Early History of Thought Experiments," *Journal of the History of Philosophy* 52, no. 3 (2014): 433–459. Compare M fol. 162a (W 358), F 167 (flying camel). On the parallel of conceptualizing the *ʿanqāʾ*, an unseen creature, with conceptualizing God, see *Muʿtabar*, 3:129; compare 2:340.

20. For regional specialties, see Jāḥiẓ (ascribed), *Kitāb al-Tabaṣṣur bi-l-tijāra*, ed. Ḥasan Ḥusnī ʿAbd al-Wahhāb (Cairo: al-Maṭbaʿa al-Raḥmāniyya, 1935), 25–29; ʿAbd al-Malik al-Thaʿālibī (d. 429/1038), *Laṭāʾif al-maʿārif* (Cairo: Dār Iḥyāʾ al-Kutub al-ʿArabiyya, 1960), 233–239. For the Indian Ocean, see *Masālik*, 65–66.

21. References to Balīnās al-Ḥakīm, which feature in Qazwīnī among others, are not to Pliny, but to writings associated with Apollonius of Tyana. See *GAS*, 3:354–355; 4:315–317; compare Adam Bieniek, "Pliny the Elder as Seen by al-Qazwīnī," *Folia orientalia* 37 (2001): 19–40.

None of the parallels adduced by Bieniek suggest direct knowledge or engagement with Pliny's natural history. For the classical background, see James Romm, *The Edges of the Earth in Ancient Thought* (Princeton, NJ: Princeton University Press, 1992), 82–120; Mary Beagon, "Situating Nature's Wonders in Pliny's Natural History," *Bulletin of the Institute of Classical Studies*, suppl. no. 100 (2007): 19–40. For Roman triumphal imagery, see Diana Kleiner, *Roman Sculpture* (New Haven, CT: Yale University Press, 1992), 183–191.

22. See Garth Fowden, *Quṣayr ʿAmra: Art and the Umayyad Elite in Late Antique Syria* (Berkeley: University of California, 2004), 41, 42–43, 197–226, 250–251, 261, 264.

23. See Dimitri Gutas, *Greek Thought / Arabic Culture: The Graeco-Arabic Translation Movement in Baghdad and Early ʿAbbāsid Society (2nd–4th / 8th–10th Centuries)* (London: Routledge, 1998), 20–60. Compare George Saliba, *Islamic Science and the Making of the European Renaissance* (Cambridge, MA: MIT Press, 2007), 1–25; Ahmed Ragab, "'In a Clear Arabic Tongue': Arabic and the Making of a Science-Language Regime," *Isis* 108, no. 3 (2017): 612–620.

24. For Ptolemy in Arabic, see Nadīm, 2:214–216; *GAS*, 6:83–97; 7:41–48; 10:31–57.

25. See *Masālik*, 155, 159, 170.

26. See Irene Pajón Leyra, *Entre ciencia y maravilla: El género literario de la paradoxografía griega* (Zaragoza: Prensas Universitarias de Zaragoza, 2011), 29–50, 121–128, 241–263; Klaus Geus and Colin Guthrie King, "Paradoxography," in *Oxford Handbook of Science and Medicine in the Classical World*, ed. Paul Keyser and John Scarborough (Oxford: Oxford University Press, 2018), 431–444. For accounts in Middle Persian, see Domenico Agostini, "Half-Human and Monstrous Races in Zoroastrian Tradition," *JAOS* 139, no. 4 (2019): 805–818. On paradoxography in early Christian miracle collections, see Scott Fitzgerald Johnson, *The Life and Miracles of Thekla: A Literary Study* (Cambridge, MA: Harvard University Press, 2006), 173–220. For later developments in Syriac, see *Tedmrātā*, edited and translated by Sergey Minov as *The Marvels Found in the Great Cities and in the Seas and on the Islands: A Representative of ʿAǧāʾib Literature in Syriac* (Cambridge: Open Book, 2021), 2–31 (editorial introduction).

27. Ascribed to "another Plutarch," in Nadīm, 2:178. See also *NG*, 403; pseudo-Plutarch, *De fluviorum et montium nominibus*, edited and translated as *Fiumi e monti*, ed. and trans. Estéban Calderón Dorda, Alessandro De Lazzer, and Ezio Pellizer (Naples: M. D'Auria, 2003). See M fols. 81b–98a (W 150–202); compare with Abū l-Qāsim al-Zamakhsharī, *Kitāb al-Jibāl wa-l-amkina wa-l-miyāh*, ed. Matthias Salverda de Grave (Leiden: E.J. Brill, 1856).

28. *Masālik*, 182–183. See also Nadīm, 1:304–305; al-Khaṭīb al-Baghdādī (d. 463/1071), *Tārīkh Madīnat al-Salām*, 17 vols. (Beirut: Dār al-Gharb al-Islāmī, 2001), 16:68–70, §7338; *Irshād*, 6:2779–2781, §1207.

29. Nadīm, 1:467–469. See Shawkat Toorawa, "Proximity, Resemblance, Sidebars and Clusters: Ibn al-Nadīm's Organizational Principles in *Fihrist* 3.3," *Oriens* 38, no. 1/2 (2010): 217–247. See also Sinan Antoon, *The Poetics of the Obscene in Premodern Arabic Poetry: Ibn al-Ḥajjāj and Sukhf* (New York: Palgrave Macmillan, 2014), 11–17.

30. Nadīm, 1:467–469 (Ṣaymarī). For dumbfounding repartees, see Ibn Abī ʿAwn (d. 322/934), *al-Ajwiba al-muskita*, ed. Mayy Aḥmad Yūsuf (Cairo: ʿAyn, 1996). See also Badīʿ al-Zamān al-Hamadhānī, *Maqāmāt*, ed. Muḥammad Muḥyī l-Dīn ʿAbd al-Ḥamīd (Cairo: al-Maktaba al-Azhariyya, 1923), 320–359, translated as *The Maqámát of Badíʿ al-Zamán al-Hamadhání*, trans. William Joseph Prendergast (London: Luzac and Co., 1915), 156–164. Compare Ibn Sīnā, *Risāla fī ibṭāl aḥkām al-nujūm*, ed. and trans. Yahya Michot (Beirut: Albouraq, 2006), 32 (Arabic), 120–121 (translation).

31. M fol. 212a (W 451, talking crow); M fol. 163b (W 364, emotion). The debate between jinn is missing from W, but original to M fol. 168b; B fol. 102a–b; V fols. 196b–197a; F 237–238; see *Kitāb al-Naqā'iḍ*, ed. Anthony Ashley Bevan, 3 vols. (Leiden: E.J. Brill, 1905–1912), 2:691 (l. 44), 695 (l. 65), 702 (ll. 28, 31). The vulgar variation of the story of ʿŪj appears in Gotha, Universität Erfurt, MS A1507, fol. 227b (W 449). For the verses from the talking crow, see Abū Muḥammad al-Sarrāj (d. 500/1106), *Maṣāriʿ al-ʿushshāq*, 2 vols. (Beirut: Dār Ṣādir, 1958), 1:85–87.

32. Ṣaymarī, *Kitāb Aṣl al-uṣūl*, BnF, Arabe 6808, fols. 21b (Saturn), 31b (Mars), 35b–36b (Venus); 73b, 88b (finding treasure and hidden items), 129a–131b (nativities for Ṣaymarī and his family); see also *GAS*, 7:152–153.

33. Hamadhānī, *Maqāmāt*, 346–353 (Arabic), 160–161 (translation).

34. For early testaments to the *Nights*, see Nadīm, 2:321–322; *Murūj*, 2:406, §1316; Nabia Abbott, "A Ninth-Century Fragment of the 'Thousand Nights': New Light on the Early History of the *Arabian Nights*," *JNES* 8, no. 3 (1949): 129–164, at 132–133, 138–144, and plates 15–16. For wonder and the *Nights*, see Roy Mottahedeh, "*ʿAjā'ib* in the Thousand and One Nights," in *The Thousand and One Nights in Arabic Literature and Society*, ed. Richard Hovannisian and Georges Sabagh (Cambridge: Cambridge University Press, 1997), 29–39.

35. Nadīm, 2:331–332; *Kashf*, 2:1126 (*ʿajā'ib al-baḥr*); *CAL*, 119–120. For *khurāfa* and *asmār* and the question of entertainment, see Rina Drory, "Legitimizing Fiction in Classical Arabic Literature," in *Models and Contacts: Arabic Literature and Its Impact on Medieval Jewish Culture* (Leiden: Brill, 2000), 37–47; Shawkat Toorawa, *Ibn Abī Ṭāhir Ṭayfūr and Arabic Writerly Culture: A Ninth-Century Bookman in Baghdad* (London: RoutledgeCurzon, 2005), 46–50, 79–82.

36. For Ibn al-Mundhir al-Harawī, see Ibn ʿAsākir, *Tārīkh madīnat Dimashq*, ed. by Muḥibb al-Dīn al-ʿAmrawī, 80 vols. (Beirut: Dār al-Fikr, 1995–2001), 2:247–248, 56:31–34; Dhahabī, *Siyar aʿlām al-nubalā'*, ed. Shuʿayb al-Arnā'ūṭ et al., 24 vols. (Beirut: Mu'assasat al-Risāla, 1996), 14:221–222, §123. For the title, see *Kashf*, 2:1437. Harawī's collection is listed in the natural history ascribed to ʿImād al-Dīn Ibn al-Athīr al-Jazarī (d. 699/1299), *Tuḥfat al-ʿajā'ib wa-ṭurfat al-gharā'ib*, BnF, Arabe 5863, fol. 2a; see *NG*, 38; cited in the treatise ascribed to Nūr al-Dīn Ibn Shabīb (fl. 732/1332), *Jāmiʿ al-funūn wa-sulwat al-maḥzūn*, ed. and trans. Mary Frost Pierson (PhD diss., Brandeis University, 1976), 217 (Arabic text), 3 (translation); see *GAS*, 15:74–75. For the Islamic *bona fides* of merchants, see Abū ʿUthmān al-Jāḥiẓ, "Madḥ al-tijāra," *Rasā'il*, ed. ʿAbd al-Salām Muḥammad Hārūn, 4 vols. (Cairo: Maktabat al-Khānjī, 1964–1979), 4:253–258; compare al-Khaṭīb al-Baghdādī, *al-Riḥla fī ṭalab al-ḥadīth*, ed. Nūr al-Dīn al-ʿItr (Beirut: Dār al-Kutub al-ʿIlmiyya, 1975), 71–96, esp. 77–82, §§4–5.

37. For Indian Ocean and Pacific island trade, see Robin Donkin, *Beyond Price: Pearls and Pearl-Fishing: Origins to the Age of Discoveries* (Philadelphia: American Philosophical Society, 1998); Robin Donkin, *Dragon's Brain Perfume: An Historical Geography of Camphor* (Leiden: Brill, 1999). See also Bryan Averbuch, "From Siraf to Sumatra: Seafaring and Spices in the Islamicate Indo-Pacific, Ninth–Eleventh Centuries C.E." (PhD diss., Harvard University, 2013).

38. On the Cirebon shipwreck circa 970, see Horst Hubertus Liebner, "The Siren of Cirebon: A Tenth-Century Trading Vessel Lost in the Java Sea" (PhD diss., University of Leeds, 2014); Carolyn Needell, "Cirebon: Islamic Glass from a 10th-Century Shipwreck in the Java Sea," *Journal of Glass Studies* 60 (2018): 69–113; Averbuch, *Siraf*, 261–268.

39. *Alqāb*, 2:559, §1989; *Wāfī*, 8:284, §3706; Ibn al-Dubaythī (d. 637/1239), *al-Mukhtaṣar al-muḥtāj ilayhi min tārīkh . . . al-Dubaythī*, redacted by al-Dhahabī, ed. Muṣṭafā Jawād, 3 vols.

(Baghdad: Maṭbaʿat al-Maʿārif, 1951–1977), 1:225, §449. On the Ribāṭ al-Maʾmūniyya, see Erik Ohlander, *Sufism in an Age of Transition: ʿUmar al-Suhrawardī and the Rise of the Islamic Mystical Brotherhoods* (Leiden: Brill, 2008), 88–89, 112, 297.

40. *Murūj*, 1:113, 125, §§216, 246 (Sīrāfī sailors), 172, §351 (Abū Zayd in Basra), 173, §355 (*akhbār ʿajība* from China in Masʿūdī's lost *Akhbār al-zamān*), 135, 199–200, §§269, 417–418 (Masʿūdī in India). See Nadīm, 1:474–475; S. Maqbul Ahmad, "Travels of Abu 'l Hasan ʿAli b. al Husayn al-Masʿudi," *Islamic Culture* 28, no. 4 (1954): 509–524.

41. ʿUmarī, 2:199–242. A composite edition of ʿUmarī's selections was published as Abū ʿImrān Mūsā ibn Rabāḥ al-Sīrāfī, *al-Ṣaḥīḥ min akhbār al-biḥār wa-ʿajāʾibihā*, ed. Yūsuf al-Hādī (Damascus: Dār Iqraʾ 2006); see Jean-Charles Ducène, "*Comptes rendus:* Abū ʿImrān . . . édité par Yūsuf al-Hādī," *Journal asiatique* 298, no. 2 (2010): 579–584; Jean-Charles Ducène, "Une nouvelle source arabe sur l'océan Indien au Xe siècle: Le *Ṣaḥīḥ min aḫbār al-biḥār wa-ʿağāʾibihā* d'Abū ʿImrān Mūsā ibn Rabāḥ al-Awsī al-Sīrāfī," *Afriques: débats, méthodes et terrains d'histoire* 6 (2015): 1–13.

42. See Andrew Ehrenkreutz, "Kāfūr, Abū l-Misk," *EI2*. For the *ṣaḥīḥ* collections, see Jonathan Brown, *The Canonization of al-Bukhārī and Muslim: The Formation and Function of the Sunnī Ḥadīth Canon* (Leiden: Brill, 2007), 99–153. For Ibn Rabāḥ, see Nadīm, 2:625; Abū l-Qāsim al-Balkhī (d. 319/931) et al., *Faḍl al-iʿtizāl*, ed. Fuʾād Sayyid (Tunis: al-Dār al-Tūnisiyya, 1974), 331; Taqī l-Dīn al-Maqrīzī (d. 845/1442), *al-Muqaffā l-kabīr*, ed. Muḥammad Yaʿlāwī, 8 vols. (Beirut: Dār al-Gharb al-Islāmī, 1991), 1:147, 7:171. See also Abū Ḥayyān al-Tawḥīdī, *Kitāb al-Imtāʿ wa-l-muʾānasa*, ed. Aḥmad Amīn and Aḥmad al-Zayn, 3 vols. (Beirut: al-Maktaba al-ʿAṣriyya, 1953), 1:107–129.

43. Süleymaniye, Aya Sofya 3306, fols. 1a (title page), 91b, 92b (double colophon). On the reading of the date, see the editorial introduction in Buzurg ibn Shahriyār, *Kitāb ʿAjāʾib al-Hind = Livre des merveilles de l'Inde*, ed. Pieter Antonie Van der Lith, trans. Louis Marcel Devic (Leiden: E.J. Brill, 1883–1886), vii–viii; Ducène, "Une nouvelle source," 2. Compare *CAL*, 120–125.

44. Buzurg ibn Shahriyār, *ʿAjāʾib*, 19–29 (*Jazīrat al-nisāʾ*), 29–35 (mermen). See also Albrecht Rosenthal, "The Isle of the Amazons: A Marvel of Travellers," *Journal of the Warburg Institute* 1, no. 3 (1938): 257–259.

45. Buzurg ibn Shahriyār, *ʿAjāʾib*, 36–38 (giant turtle); M fol. 75a–b; *Ḥay.*, 7:106–107 (told of a *saraṭān*). See Michael Jan de Goeje, "La légende de Saint Brendan," in *Actes du huitième Congrès international des orientalistes*, 2 vols. (Leiden: E.J. Brill, 1891–1893), 2:43–76, at 47–48. For Jāḥiẓ, see Gustave E. von Grunebaum, *Medieval Islam: A Study in Cultural Orientation* (Chicago: University of Chicago Press, 1946), 299–301. For the late Arabic translation of the *Physiologus*, see Jan Pieter Niolaas Land, *Anecdota Syriaca*, 4 vols. (Leiden: E.J. Brill, 1862–1875), 4:115–176, at 173–174 (*aspidochelone*). See also Oktor Skjærvø, "Karsāsp," *EIr.*

46. ʿUmarī, 2:201–206, 213–214, 218, 225, 236–238, 241–422.

47. ʿUmarī, 2:204, 207; M fol. 65a–b (W 117–118).

48. For string-legged savages, known in Persian as *davālpā*, see Natalia Tornesello, "From Reality to Legend: Historical Sources of Hellenistic and Islamic Teratology," *Studia Iranica* 31 (2002): 163–192. See also Maurice Pomerantz, "Tales from the Crypt: On Some Uncharted Voyages of Sindbad the Sailor," *Narrative Culture* 2, no. 2 (2015): 250–269.

49. See the mission sent by the vizier Yaḥyā ibn Khālid al-Barmakī (d. 190/805) to India, redacted by Abū Yūsuf al-Kindī in Nadīm, 2:423–425; *Masālik*, 66–68, 71–72; Bruce Lawrence,

Shahrastānī on the Indian Religions (Paris: Mouton, 1976), 18–29, 125–142. For examples of temples and sex, see Aloka Parasher and Usha Naik, "Temple Girls of Medieval Karnataka," *Indian Economic and Social History Review* 23, no. 1 (1986): 63–78; Wendy Doniger, *The Hindus: An Alternative History* (New York: Penguin Press, 2009), 438–444.

50. For parallels with Polyphemus, see von Grunebaum, *Medieval Islam*, 301–304. On the phoenix, see Herodotus, *Historiae = The Persian Wars*, ed. and trans. Alfred Godley, 4 vols. (Cambridge, MA: Harvard University Press, 1920–1925), 1:358–360, book 2.73–74; Philostratus of Athens (d. ca. 244 CE), *The Life of Apollonius of Tyana*, ed. and trans. Christopher Jones (Cambridge, MA: Harvard University Press, 2005), 311, book 3.49. Compare *Mahābhārata*, ed. Vishnu Sitaram Sukthankar et al., 19 vols. (Pune: Bhandarkar Oriental Research Institute, 1933–1971), 1:157–162, Ādi Parva, chaps. 29–30 (Garuda with elephant and tortoise). See also Rudolf Wittkower, "Eagle and Serpent: A Study in the Migration of Symbols," *Journal of the Warburg Institute* 2, no. 4 (1939): 293–325.

51. See M fol. 193a (W 419–420: *ʿanqāʾ*); F 271 (Sīmorg); *Ḥay.*, 7:120–121. See also Hanns-Peter Schmidt, "Simorğ," *EIr;* Guy Ron-Gilboa, "*ʿAnqāʾ Mughrib:* The Poetics of a Mythical Creature," *Journal of Abbasid Studies* 8, no. 1 (2021): 75–103. See also Deborah Black, "Avicenna on the Ontological and Epistemic Status of Fictional Beings," *Documenti e studi sulla tradizione filosofica medievale* 8 (1997): 425–453.

52. See William Eamon, *Science and the Secrets of Nature: Books of Secrets in Medieval and Early Modern Culture* (Princeton, NJ: Princeton University Press, 1994), 39–90; Michelle Karnes, *Medieval Marvels and Fictions in the Latin West and Islamic World* (Chicago: University of Chicago Press, 2022), chap. 1. See also Liana Saif, *The Arabic Influences on Early Modern Occult Philosophy* (New York: Palgrave Macmillan, 2015), 27–94.

53. See David Williams, *Deformed Discourse: The Function of the Monster in Mediaeval Thought and Literature* (Montreal: McGill-Queen's University Press, 1996), 23–103; Claude Lecouteux, *Les monstres dans la pensée medievale europeenne: Essai de presentation* (Paris: Presses de l'Université de Paris-Sorbonne, 1993), 15–81, 116–125. See also Lorraine Daston and Katharine Park, *Wonders and the Order of Nature, 1150–1750* (Cambridge, MA: Zone Books, 1998), 173–214; Surekha Davies, *Renaissance Ethnography and the Invention of the Human: New Worlds, Maps and Monsters* (Cambridge: Cambridge University Press, 2016), 23–46.

54. See Aristotle, *Generation of Animals*, ed. and trans. Arthur Leslie Peck (Cambridge, MA: Harvard University Press, 1942), 416–422; Aristotle, *Generation of Animals: The Arabic Translation Commonly Ascribed to Yahyā ibn al-Biṭrīq*, ed. J. Brugman and Hendrik Joan Drossaart Lulofs (Leiden: E.J. Brill, 1971), 144 (67b5), 150–152 (69b26–70b25), 157 (72a36), 251 (index). Compare Aristotle, *Kitāb al-Ḥayawān*, edited as *The Arabic Version of Aristotle's Historia Animalium*, ed. Lourus S. Filius (Leiden: Brill, 2019), 501 (index); Ibn Sīnā, *al-Shifāʿ, al-Ṭabīʿiyyāt:* 8. *al-Ḥayawān*, ed. ʿAbd al-Ḥalīm Muntaṣir et al. (Cairo: al-Hayʾa al-Miṣriyya al-ʿĀmma li-l-Taʾlīf wa-l-Nashr, 1970), 423–424.

55. See Aristotle, *De Animalibus, Michael Scot's Arabic-Latin Translation: Generation of Animals*, ed. Aafke M. I. van Oppenraaij (Leiden: E.J. Brill, 1992), 182–183, compare 175, 384 (*ʿajab*), 488–489 (*teras*, etc.). William of Moerbeke (d. 1286), *De generatione animalium*, ed. Hendrik Joan Drossaart Lulofs (Leiden: E.J. Brill, 1966), 132–133, 135, 213 (index). For categories of monsters in classical Latin, see Kenneth F. Kitchell, "A Defense of the 'Monstrous' Animals of Pliny, Aelian, and Others," *Preternature: Critical and Historical Studies on the Pre-*

ternatural 4, no. 2 (2015): 125–151. See also Lisa Verner, *The Epistemology of the Monstrous in the Middle Ages* (London: Routledge, 2005), 11–43; Asa Simon Mittman, "Are the 'Monstrous Races' Races?" *postmedieval* 6, no. 1 (2015): 36–51. Also see Silvia Manzo, "Monsters, Laws of Nature, and Teleology in Late Scholastic Textbooks," in *Contingency and Natural Order in Early Modern Science*, ed. Pietro Daniel Omodeo and Rodolfo Garau (Cham, Switzerland: Springer, 2019), 61–92, at 69–72.

56. M fols. 207a–212a (W 448–452); compare M fols. 98a–99b, 165a (W 202–203, 368). The claim of physicians and astrologers is missing from W, but in M fol. 210a; B fol. 134a. See Ibn Sīnā, *Ḥayawān*, 423–424 (hybrid animals). On cranes, see Homer, *Iliad*, ed. and trans. Augustus Taber Murray, revised William Wyatt, 2 vols. (Cambridge, MA: Harvard University Press, 2014), 1:128 (3.2–6); Aristotle, *Historia animalium, Books VII–X*, ed. and trans. David Balme (Cambridge, MA: Harvard University Press, 1991), 130–132 (597a3–9); Aristotle, *Kitāb al-Ḥayawān*, 288.

57. See Jon McGinnis, "Avicenna's Natural Philosophy," in *Interpreting Avicenna: Critical Essays*, ed. Peter Adamson (Cambridge: Cambridge University Press, 2013), 71–90; Catarina Belo, "The Concept of 'Nature' in Aristotle, Avicenna and Averroes," *Kriterion* 56, no. 131 (2015): 45–56.

58. For complexities and modulations, see Michael Gomez, *African Dominion: A New History of Empire in Early and Medieval West Africa* (Princeton, NJ: Princeton University Press, 2018), 54–57; Dahlia E. M. Gubara, "Revisiting Race and Slavery through 'Abd al-Rahman al-Jabarti's *'Aja'ib al-athar*," *Comparative Studies of South Asia, Africa, and the Middle East* 38, no. 2 (2018): 230–245; Supriya Gandhi, "Locating Race in Mughal India," *Renaissance Studies* 75, no. 4 (2022, forthcoming).

59. Louise Marlow, *Hierarchy and Egalitarianism in Islamic Thought* (Cambridge: Cambridge University Press, 1997), 22–28. See also Maysam Al Faruqi, "*Umma:* The Orientalists and the Qurʾānic Concept of Identity," *Journal of Islamic Studies* 16, no. 1 (2005): 1–34.

60. See David Goldenberg, *The Curse of Ham: Race and Slavery in Early Judaism, Christianity, and Islam* (Princeton, NJ: Princeton University Press, 2003), 47, 110–112; compare Haroon Bashir, "Black Excellence and the Curse of Ham: Debating Race and Slavery in the Islamic Tradition," *ReOrient* 5, no. 1 (2019): 92–116. See also David Nirenberg, *Neighboring Faiths: Christianity, Islam, and Judaism in the Middle Ages and Today* (Chicago: Chicago University Press, 2014), 232–264; Geraldine Heng, *The Invention of Race in the European Middle Ages* (Cambridge: Cambridge University Press, 2018), 15–54.

61. For Galen on the centaur, rendered as *al-insān al-farasī*, see *Manāfiʿ*, fol. 40b–41a; 1:125–126 (Greek); 1:155–156 (English). For the afterlives of Ḥunayn ibn Isḥāq's use of *ʿanqāʾ mughrib* for Aristotle's sphinx, see Ron-Gilboa, "*ʿAnqāʾ*," 86. See also Black, "Avicenna."

62. See also Jāḥiẓ, "Risālat al-maʿād wa-l-maʿāsh," *Rasāʾil*, 1:91–142, at 119–120 (the unseen); Masʿūdī equates the juridical category of *khabar al-istifāḍa* with *khabar al-tawātur* in *Murūj*, 2:370–371, §1354.

63. ʿUmarī, "anna l-baḥr lā ḥaraja ʿalā man ḥaddatha ʿan ʿajāʾibihi wa-akhbara ʿan gharāʾibihi," 2:242; this is a play on the proverb "ḥaddith ʿan al-baḥr wa-lā ḥaraja," Jāḥiẓ, *al-Bayān wa-l-tabyīn*, ed. ʿAbd al-Salām Muḥammad Hārūn, 4 vols. (Cairo: Maṭbaʿat Lajnat al-Taʾlīf wa-l-Tarjuma wa-l-Nashr, 1948), 2:113, cited in M fol. 3b (W 7). For more, see Sīrāfī, *Ṣaḥīḥ*, 223n1.

6: Alchemical Bodies

1. Citing Qazwīnī and his title in ʿUmarī, 20:26. Parallels are identified by the editor: 20:25–72 (domesticated and wild animals), 73–112 (birds), 113–146 (insects and reptiles), 147–165 (animals of the sea), 169–275 (plants), 22:101–107 (minerals), 108–224 (stones), 225–240 (oily substances). Others can be adduced.

2. For overviews, see Manfred Ullmann, *Islamic Medicine*, trans. Jean Watt (Edinburgh: University Press, 1978), 21–24, 55–69; Peter Pormann and Emilie Savage-Smith, *Medieval Islamic Medicine* (Edinburgh: Edinburgh University Press, 2007), 41–79.

3. M fols. 98b–99a (W 203–204). On *kāʾināt*, see Ibn Sīnā, *Ḥudūd*, 19, §33 (s.v.ʿ*unṣur*); Ibn Sīnā, *al-Shifāʾ*, *Ṭabīʿiyyat:* 3. *al-Kawn wa-l-fasād*, ed. Maḥmūd Qāsim (Cairo: Dār al-Kitāb al-ʿArabī li-l-Ṭibāʿa wa-l-Nashr, 1969), 79, 199; see *Mulakhkhaṣ*, fols. 204b–205a.

4. M fols. 99a–b (W 204–205: *filizzāt*), 100a–b (W 208: *aḥjār*), 111b–112a (W 242: *al-ajsām al-duhniyya*); compare, *CAL*, 109–111. On *khārṣīnī*, translated here as "Chinese iron," see Fabian Käs, *Die Mineralien in der arabischen Pharmakognosie*, 2 vols. (Wiesbaden: Harrassowitz, 2010), 1:537–539.

5. See *Maʿādin*, 20–22 (formation of minerals), 22–23 (against alchemy), translated in *Avicennae De congelatione et conglutinatione lapidum, Being Sections of the Kitâb al-shifâʾ*, ed. and trans. Eric John Holmyard and Desmond Christopher Mandeville (Paris: P. Guethner, 1927), 33–42 (English). See *Mabāḥith*, 2:215–219 (minerals and alchemy); *Mulakhkhaṣ*, fols. 210b–212a; *Sittīnī*, 313–318 (alchemy), 319–324 (gems and stones). On alchemy, see *Āthār*, 145, 248, 462, 548. For sulfur and mercury as basis for transmutation, see *Khalīqa*, 246–266 (3.4); Jābir ibn Ḥayyān, *Mukhtār rasāʾil Jābir ibn Ḥayyān*, ed. Paul Kraus (Cairo: Maktabat al-Khānjī, 1935), 330–331 (philosopher's stone), compare 538–539. See also Paul Kraus, *Jābir ibn Hayyān: Contribution à l'historire des idées scientifiques dans l'Islam*, 2 vols. (Cairo: Institut français d'archéologie orientale, 1942–1943), 2:1n1, 21n5.

6. See Ibn Sīnā, *al-Qānūn fī l-ṭibb*, 4 vols. (Beirut: Dār Iḥyāʾ al-Turāth al-ʿArabī, 2005), "Kitāb 2. fī l-adwiya l-mufrida," 1:291–515, 2:1–197; "Kitāb 5. fī l-adwiya l-mukarraba," 4:423–605.

7. Pseudo-Aristotle, *Kitāb al-Aḥjār*, translated in Julius Ruska, *Das Steinbuch des Aristoteles, mit literargeschichtlichen Untersuchungen nach der arabischen Handschrift der Bibliothèque nationale* (Heidelberg: C. Winter, 1912), 43–46 (textual history), 80–92 (passages in Qazwīnī), 93–125 (Arabic ed.). Compare Manfred Ullmann, "Der literarische Hintergrund des Steinbuches des Aristoteles," in *Actas, IV Congresso de Estudos Árabes e Islâmicos, Coimbra-Lisboa, 1 a 8 de setembro de 1968, Actas* (Leiden: E.J. Brill, 1971), 291–299. The section on stones is translated in Julius Ruska, *Das Steinbuch aus der Kosmographie des Zakarijâ ibn Muḥammad ibn Maḥmûd al-Ḳazwînî* (Kirchhain: Max Schmersow, 1896). See also ʿ*Uyūn*, 1:69; Abū l-Rayḥān al-Bīrūnī, *al-Jamāhir fī l-jawāhir*, ed. Yūsuf Hādī (Tehran: Mirāth-i Maktūb, 1995), 116.

8. *Āthār*, 269. For name of Hermes Trismegistus in Arabic, see *Ghāyat*, 225. See also Kevin van Bladel, *The Arabic Hermes: From Pagan Sage to Prophet of Science* (Oxford: Oxford University Press, 2009), 121–163.

9. See *Khalīqa*, 1–4. For the *Dhakhīra*, see *NG*, 376–377; *GAS*, 4:103–104. See translation and facsimile ed. of Madrid, Escorial Library, MS n. 947 in Ana Maria Alfonso-Goldfarb and Safa Abou Chahla Jubran, *Livro do tesouro de Alexandre um estudo de hermética árabe na oficina da história da ciência* (Petrópolis, Brazil: Editora Vozes, 1999), fols. 2a–4b (Aristotle's discovery

and translation), 43a–46a (talismans of Apollonius). See also Lucia Raggetti, "Apollonius of Tyana's Great Book of Talismans," *Nuncius* 34, no. 1 (2019): 155–182, at 176–178. For the redaction of the *Physika kai mystika* ascribed to Democritus, see *The Four Books of Pseudo-Democritus*, ed. and trans. Matteo Martelli (Leeds: Maney, 2013). Martelli argues against the ascription to Bolos of Mendes (introduction, 29–48). See also *GAS*, 4:49–50; *NG*, 393–395, 428–429; van Bladel, *The Arabic Hermes*, 50–52.

10. Alexander of Aphrodisias, *Fī mabādiʾ al-kull = Alexander of Aphrodisias on the Cosmos*, ed. and trans. Charles Genequand (Leiden: Brill, 2001), 112–115, ll. 127–129, see also 16–20 (introduction), 164–165 (commentary). On the *philosophos* as wonderworker, see Paul Magdalino and Maria Mavroudi, "Introduction," in *The Occult Sciences in Byzantium*, ed. Magdalino and Mavroudi (Geneva: La Pomme d'or, 2006), 11–38, at 12–14.

11. For an overview of the corpus, see Liana Saif, "A Preliminary Study of the Pseudo-Aristotelian Hermetica: Texts, Context, and Doctrines," *Al-ʿUṣūr al-Wusṭā: Journal of Middle East Medievalists* 29 (2021): 20–80; *NG*, 374–375; *GAS*, 4:102; 7:57. For the vocalization as *Ustūwwaṭāṣ*, see BnF, Arabe 2577, fols. 1a–34a, 1b–2a (micro- and macrocosms), 3a (*ummahāt*), 25a (*Isṭamākhīs*). The opening parallels pseudo-Aristotle, *Kitāb al-Madīṭīs*, focusing on, among other topics, the unique properties of animals, Bodl., Marsh 556, fols. 3a–110b, 4a (discovery of the book), 5b–6a (micro- and macrocosms), 7a (*ummahāt*), 110a–b (conspectus of interrelated titles). The treatise entitled *Isṭamākhīs* is also preserved in Marsh 556, fols. 111a–152a, 111a–b (preparation of the work for Alexander's campaign), 113a–b (*al-ṭibāʿ al-tāmm*, perfect nature). For an overview of BnF, Arabe 2577, see Edgar Blochet, "Études sur le gnosticisme musulman," *Rivista degli studi orientali*, 4, no. 1/2 (1911): 47–79, 267–300, at 62–79, 267–277. For more on Alexander and Aristotle, see *Maktūm*, fol. 105b.

12. Nadīm, 2:442, 445–446, 466; Johann Fück, "The Arabic Literature on Alchemy according to an-Nadīm (A.D. 987): A Translation of the Tenth Discourse of *The Book of the Catalogue* (*al-Fihrist*) with Introduction and Commentary," *Ambix* 4, no. 3/4 (1951): 81–144, at 88, 90–91, 109. See also Martelli, *Four Books*, 57–63 (introduction).

13. See *Masālik*, 159; compare *Āthār*, 268–269. See Alexander Fodor, "The Origins of the Arabic Legends of the Pyramids," *Acta Orientalia* 23, no. 3 (1970): 335–363; *GAS*, 15:50–51; Michael Cooperson, "al-Maʾmūn, the Pyramids and the Hieroglyphs," in *ʿAbbasid Studies II: Occasional Papers of the School of ʿAbbasid Studies*, ed. John Nawas (Leuven: Peeters 2010), 165–190.

14. Nadīm, 2:447, 457; Fück, "Alchemy," 92, 103, 107, 133–134; *Khalīqa*, 2.

15. For talismans of Apollonius, see *Āthār*, 431, 442, 445, 485–486, 495, 524, 553, 572, 594, 605–606. Compare Ibn al-Faqīh, *Kitāb al-Buldān*, ed. Yūsuf Hādī (Beirut: ʿĀlam al-Kutub, 1996), 197, 420, 422, 496, 497, 504, 532, 534, 548, 591; *Buldān*, 1:267, 350, 2:205, 381, 3:103, 4:258, 421, 452, 5:415–416. See also Kraus, *Jābir*, 292–293n13. For *finjān al-sāʿāt* as a clepsydra, see Dozy, 2:283.

16. *Āthār*, 8, 314–315, 417. Compare *Buldān*, 1:460; al-Khaṭīb al-Baghdādī, *Tārīkh Madīnat al-Salām*, 17 vols. (Beirut: Dār al-Gharb al-Islāmī, 2001), 1:382–383.

17. Nadīm, 2:341. See also *NG*, 379–380; *GAS*, 4:90. See pseudo-Apollonius, *Kitāb Ṭalāsim Balīnās al-akbar*, BnF, Arabe 2250, fols. 100b (jinn, demons, ghouls), 106a–b (baths in Edessa), 110b–111a (inducing memory loss), 111a–b (expelling foes), 112b–113a (wells), 115b–116b (magic mirror of Alexandria), 117a–b (inducing sleep), 124b (Persia), 125b (Nubia), 128a–129b (commerce, performed for a merchant in the West), 131b–132a (keeping women chaste).

18. On the use of *charaktēres*, see pseudo-Apollonius, *Apotelesmata Apollonii Tyanensis*, ed. François Nau, in *Patrologia Syriaca*, part 1, vol. 2 (Paris: Firmin-Didot, 1907): 1362–1392, at 1391, l. 3, rendered as *qalafṭīriyyāt*, etc. in BnF, Arabe 2250, fol. 115a–b, 116b–117a. The Arabic is based on an otherwise lost intermediary, see Raggetti, "Apollonius," 163. On *qalafṭīriyyāt*, see Dozy, 2:397; *NG*, 362. Compare Richard Gordon, "*Signa Nova et Inaudita:* The Theory and Practice of Invented Signs (*Charaktêres*) in Graeco-Egyptian Magical Texts," *MHNH: Revista internacional de investigación sobre magia y astrología antiguas* 11 (2011): 15–44; Gideon Bohak, "The Charaktêres in Ancient and Medieval Jewish Magic," *Acta classica universitatis scientiarum Debreceniensis* 47 (2011): 25–44.

19. M fols. 31a–32a, 150a, 150b (W 56–57, 318, 328).

20. See Zosimos, *Muṣḥaf al-ṣuwar*, Istanbul, Arkeoloji Müzeleri, MS 1574, fols. 9a–10b, published as *The Book of Pictures*, ed. and trans. Theodor Abt and Salwa Fuad (Zurich: Living Human Heritage, 2011); pseudo-Apollonius, *Ṭalāsim*, BnF, Arabe 2250, fol. 134a–b. See also Christopher Faraone, *Talismans and Trojan Horses: Guardian Statues in Ancient Greek Myth and Ritual* (Oxford: Oxford University Press, 1992). See also Garth Fowden, *The Egyptian Hermes: A Historical Approach to the Late Pagan Mind* (Princeton, NJ: Princeton University Press, 1993), 1–11, 22–44, 120–126; Gideon Bohak, *Ancient Jewish Magic: A History* (Cambridge: Cambridge University Press, 2008), 41–44, 63–67, 114–123. For ancient Hermetica in Sasanian Iran, see van Bladel, *The Arabic Hermes*, 4–13, 23–63.

21. See *Maktūm*, fols. 4b–5a (definition of talismans), 72a–78a (Hermes on the *muṣḥaf al-qamar*, the stations of the Moon), 78a–b (citing Apollonius on the hours); compare BnF, Arabe 2250, fol. 94b. For further citations to occult Hermetic teachings, see *Maṭālib*, 5:164, 173. See also *Sittīnī*, 325–332, 333–336.

22. Drawn from citations to Apollonius in M: fols. 75b (W 137: *sulaḥfā*), 76a (W 138: *samak*), 76b (W 139–140: *ḍifdiʿ*), 100a (W 208: *usrub*), 106a (W 226: *zujāj*), 114a (W 248: *utrujj*), 117a (W 252: *khirwaʿ*), 118b (W 256: *zaytūn*), 125b (W 267–268: *nārajīl*), 131a (W 278: *jirjīr*), 133a (W 282: *khass*), 134a–b (W 283–284: *diflā*), W 285 (*zaʿfarān*, missing from M), 136a (W 287: *shibitt*), 140a (W 295: *kuzbara*), 143a (W 299: *nānkhwāh*), 170a (W 377: *ḥimār*), 178b (W 393: *khinzīr*), 172b (W 381–382: *baqar*), 174a (W 384: *ḍaʾn*), 174b (W 385: *maʿz*), 176b (W 389: *arnab*), 180a (W 396: *dhiʾb*), 181a (W 397: *sinnawr*), 182a (W 398–399: *ḍabuʿ*), 183b (W 401: *fīl*), 185a (W 404: *kalb*); B fol. 118a (W 411: *khuṭṭāf*); M fol. 190a (W 413, *khuffāsh*), 190a (W 414: *dajāj*), 190b (W 414–415: *rakhama*), 190b (W 415: *zāgh*), 193b–4a (W 421: *ghurāb*), 196b (W 426: *hudhud*), 197a (W 426: *waṭwāṭ, bālwāya*), 198a (W 429: *afʿā*), 200b (W 434: *dīk al-jinn*), 203b (W 440: *ʿankabūt*).

23. M fol. 176b (W 389), citing Imruʾ al-Qays, *Dīwān*, via Abū Saʿīd al-Aṣmaʿī (d. 213/828), ed. Muḥammad Abū l-Faḍl Ibrāhīm (Cairo: Dār al-Maʿārif, 1958), 128, §18; *Enz.*, 176. See Marie-Louise Thomsen, "The Evil Eye in Mesopotamia," *JNES* 51, no. 1 (1992): 19–32; John Boyle, "The Hare in Myth and Reality," *Folklore* 84, no. 4 (1973): 313–326, at 324–325.

24. For literature on *khawāṣṣ*, see Kraus, *Jābir*, 2:61–95; *NG*, 393–416; William Newman, "The Occult and the Manifest among the Alchemists," in *Tradition, Transmission, Transformation: Proceedings of Two Conferences on Pre-modern Science Held at the University of Oklahoma*, ed. F. Jamil Ragep et al. (Leiden: E.J. Brill, 1996), 173–198; Lucia Raggetti, "The 'Science of Properties' and Its Transmission," in *In the Wake of the Compendia: Infrastructural Contexts and the Licensing of Empiricism in Ancient and Medieval Mesopotamia*, ed. J. Cale Johnson (Berlin: De Gruyter, 2015), 159–176. For Qazwīnī on plants, see also Eilhard Wiedemann, "Übersetzung und Besprechung des Abschnittes über die Pflanzen von Qazwînî," *Aufsätze zur arabischen Wis-*

senschaftsgeschichte, 2 vols. (Hildesheim: Olms, 1970), 2:372–407. On agricultural recipes, Qazwīnī frequently cites a *ṣāḥib al-filāḥa,* wrongly identified with Ibn al-ʿAwwām (d. ca. 580/1185) in *CAL,* 54; cf. *Enz.*, 168n138. For agronomic literature, see *GAS,* 4:303–346; *NG,* 427–451; Jaakko Hämeen-Anttila, *The Last Pagans of Iraq, Ibn Waḥshiyya and His Nabatean Agriculture* (Leiden: Brill 2006), 10–33, 52–84; Balīnās (Balyās) al-Ḥakīm, *Kitāb al-Filāḥa,* edited and translated in Concepción Vázquez de Benito, "El manuscrito No. X de la colección Gayangos (fols. 1–98)," *Boletín de la Asociación Española de Orientalistas* 9 (1973): 73–124; 10 (1974): 215–304.

25. For the Galenic treatment of unsayable properties and its afterlives, see Julius Röhr, *Der okkulte Kraftbegriff im Altertum* (Leipzig: Dieterich, 1923), 98–133; Brian Copenhaver, "The Occultist Tradition and Its Critics," in *The Cambridge History of Seventeenth-Century Philosophy,* ed. Daniel Garber and Michael Ayers (Cambridge: Cambridge University Press, 1998), 454–512, at 459, 504n16. See also, Manfred Ullmann, *Wörterbuch zu den griechisch-arabischen Übersetzungen des 9. Jahrhunderts* (Wiesbaden: Harrassowitz, 2002), 305–306.

26. See *Afʿāl,* 261–262, compare 252. See also F 224. On extraordinary phenomena, see Ibn Sīnā, "Risāla fī l-fiʿl wa-l-infiʿāl," first epistle in *Majmūʿ rasāʾil al-shaykh al-raʾīs* (Hyderabad: Dāʾirat al-Maʿārif al-ʿUthmāniyya, 1353/1934~), 9–10; see miracles as "al-muʿjizāt al-mukhālifa li-majrā l-ṭabīʿa," in Ibn Sīnā, "Fī aqsām al-ʿulūm al-ʿaqliyya," in *Tisʿ rasāʾil* (Cairo: Maṭbaʿa Hindiyya, 1908), 114. For amber, see Ibn Sīnā, *Qānūn,* 1:479; Abū Bakr al-Rāzī, *Kitāb al-Khawāṣṣ,* Cairo, Dār al-Kutub, MS Taymūriyya Ṭibb 264, at 1 (paginated); Bīrūnī, *Jamāhir,* 343, 353; M fol. 159b (W 334); John Riddel, "Amber in Ancient Pharmacy: The Transmission of Information about a Single Drug," *Pharmacy in History* 15, no. 1 (1973): 3–17. For further on amber and the vinegar-repelling stone, see *NG,* 112, 125, 397, 408; Hans Daiber, "Qosṭā ibn Lūqā (9. Jh.) über die Einteilung der Wissenschaften," *ZGAIW* 6 (1990): 93–129, at 102 (*mubghiḍ al-khall*).

27. On *ṣarʿ* as jinn possession, see Michael Dols, *Majnūn: The Madman in Medieval Islamic Society* (Oxford: Clarendon, 1992), 211–260. Ibn Sīnā and Abū Bakr al-Rāzī preferred physiological explanations, though other physicians accepted demon possession as a cause; see Abū l-Qāsim al-Zahrāwī (d. ca. 404/1013), *Taṣrīf li-man ʿajiza ʿan al-taʾlīf,* Süleymaniye, Hasan Hüsnü Paşa 1361/1, fols. 61b, 62b. As a malady, *ṣarʿ* is also treated in books of talismanic and astral magic, e.g., *Maktūm,* fols. 166a–170a.

28. See M fol. 121b (W 360: peony); Ibn Sīnā, *Qānūn,* 1:132, 2:90; Abū Bakr al-Rāzī, *Khawāṣṣ,* 24; Abū Bakr al-Rāzī, *al-Ḥāwī fī l-ṭibb,* ed. Muḥammad Muḥammad Ismāʿīl, 23 vols. (Beirut: Dār al-Kutub al-ʿIlmiyya, 2000), 21:104–106 (3232–3234), §623. Compare Bohak, *Jewish Magic,* 41–42. See Galen, *De simplicium medicamentum,* in *Opera omnia,* ed. and trans. Carolus Gottlob Kühn, 20 vols. (Leipzig: C. Cnoblochii, 1821–1833), 11:859–860; discussed in Caroline Petit, "Galen, Pharmacology and the Boundaries of Medicine: A Reassessment," in *Collecting Recipes: Byzantine and Jewish Pharmacology in Dialogue. Science, Technology, and Medicine in Ancient Cultures,* ed. Lennart Lehmhaus and Matteo Martelli (Berlin: De Gruyter, 2017), 51–79, at 63–64. For charlatans, see Abū Bakr al-Rāzī, *Akhlāq al-ṭabīb,* ed. ʿAbd al-Laṭīf Muḥammad al-ʿAbd (Cairo: Dār al-Turāth, 1977), 89–95; compare Galen, *Fī minḥat afḍal al-aṭibbāʾ = Galeni De optimo medico cognoscendo,* ed. Albert Iskandar (Berlin: Akademie-Verlag, 1988), 44.

29. Rāzī, *Khawāṣṣ,* 1–3. Compare *GAS,* 3:285–286; *NG,* 407.

30. See Rāzī, *Khawāṣṣ,* 3, 36–37, 49. On the rain stone, see Ibn al-Faqīh, *Buldān,* 640; Bīrūnī, *Jamāhir,* 357–359; M fol. 104b (W 221); *Tansūkh,* 160–163; *Āthār,* 247; Ruska, *Ḳazwînî,* 19; *Ghāyat,* 397–398. See also *NG,* 121, 132, 136; John Boyle, "Turkish and Mongol Shamanism in the Middle Ages," *Folklore* 83, no. 3 (1972): 177–193, at 185–192; Frantz Grenet and Samra

Azarnouche, "Where Are the Sogdian Magi?," *Bulletin of the Asia Institute* 21 (2007 [2012]): 159–177, at 171–172. For magnets, see Rāzī, *Khawāṣṣ*, at 3–4; Jāḥiẓ, *Ḥayawān*, 4:112; *Murūj*, 2:91, §816; *Ghāyat*, 399–400; Bīrūnī, *Jamāhir*, 349, citing Galen; Bīrūnī, *al-Ṣaydana fī l-ṭibb*, ed. and trans. Hakim Mohammed Said (Karachi: Hamdard Academy, 1973), 349 (Arabic), 307 (English); *Tansūkh*, 133–134; M fols. 2b, 110b (W 5, 339–340), citing pseudo-Aristotle, *Aḥjār*, 108–109, §15 (Arabic), 154–155 (translation); Wiedemann, "Über Magnetismus," *Aufsätze*, 1:28–37; Kraus, *Jābir*, 2:72n1; *NG*, 398–399 (magnet). See also Daryn Lehoux, "Tropes, Facts, and Empiricism," *Perspectives on Science* 11, no. 3 (2003): 326–345.

31. On emeralds, see M fol. 106b (W 227), translated in Ruska, *Ḳazwînî*, 25; Rāzī, *Khawāṣṣ*, 13; Rāzī, *Ḥāwī*, 19:136 (2860), 20:211 (3117), §405; *Ghāyat*, 397; Bīrūnī, *Jamāhir*, 272–273; *Tansūkh*, 60–61; Kraus, *Jābir*, 2:74n1; *NG*, 408. Compare Abū Rayḥān al-Bīrūnī, *Risāla li-l-Būrīnī fī fihrist kutub Muḥammad ibn Zakariyyāʾ al-Rāzī*, ed. Paul Kraus (Paris: Calame, 1936), 29–46.

32. See Rāzī, *Khawāṣṣ*, 50; Alexander of Tralles, *Therapeutica = Alexander von Tralles: Original-Text und Übersetzung nebst einer einleitenden Abhandlung*, ed. and trans. Theodor Puschmann, 2 vols. (Vienna: Braumüller, 1878–1879), 2:473–475; Petit, "Galen," 71–73. Alexander cites the otherwise lost Galenic work *On Medicine according to Homer*, which Ḥunayn ibn Isḥāq says follows Galen's style, but not method (*madhhab*). See *ʿUyūn*, 1:101–102; *GAS*, 3:139, §155; 4:162–164.

33. See Galen, *De simplicium*, 12:289. See M fol. 164a (W 366); Ibn al-Bayṭār, *Jāmiʿ li-l-mufradāt al-adwiya wa l-aghdhiya*, 4 vols. (Beirut: Dār al-Kutub al-ʿIlmiyya, 1992), 1:134.

34. See *Manāfiʿ*, "min muṣannafāt muqaddam al-aṭibbāʾ fāʿil al-ʿajāʾib maẓhar al-gharāʾib Jālīnūs," fol. 1a, copied by ʿAbd Allāh ibn al-Fatḥ ibn ʿAlī l-Bundārī l-Iṣfahānī in 682/1283. Compare Galen, *Fī minḥat*, 60–63.

35. Wonderment is cultivated throughout Galen's *De usu partium*, e.g., *Manāfiʿ*, fols. 11b, 14b, 15a, 16b, 299b–300a; 1:29–30, 40, 42, 47, 2:447–448 (Greek); 1:87, 95, 96, 100, 2:730–731 (English). Compare Ḥunayn ibn Isḥāq, *Risāla . . . fī dhikr mā tarjama min kutub Jālīnūs = Ḥunayn ibn Isḥāq on His Galen Translations*, ed. and trans. John Lamoreaux (Provo, UT: Brigham Young University Press, 2016), 63; *GAS*, 3:106–108. See also Ahmed El Shamsy, "Al-Ghazālī's Teleology and the Galenic Tradition: Reading *The Wisdom in God's Creation* (*al-Ḥikma fī makhlūqāt Allāh*)," in *Islam and Rationality: The Impact of al-Ghazālī—Papers Collected on His 900th Anniversary*, ed. Georges Tamer and Frank Griffel, 2 vols. (Leiden: Brill, 2016), 2:90–112, at 100–110.

36. M fols. 143b–144a (W 302–303). Compare Ghazālī, *al-Ḥikma fī makhlūqāt Allāh*, ed. Muḥammad Rashīd Qabbānī (Beirut: Dār Iḥyāʾ al-ʿUlūm, 1978), 13–14; *Maṭālib*, 1:233–236.

37. See Ibn al-Jazzār, *Kitāb al-Khawāṣṣ = Die Risāla fī l-Ḫawāṣṣ des Ibn al-Ǧazzār: Die arabische Vorlage des Albertus Magnus zugeschriebenen Traktats De mirabilibus mundi*, ed. and trans. Fabian Käs (Wiesbaden: Harrassowitz, 2012), 9–22 (editor's introduction), fol. 158b (ancient knowledge and experience); on Ibn al-Jazzār's otherwise lost *ʿAjāʾib al-buldān*, see Ibn al-Bayṭār, *Jāmiʿ*, 2:473; Abū ʿAbd Allāh al-Zuhrī (fl. 545/1150), *Kitāb al-Jaʿrāfiyya*, ed. Muḥammad Ḥājj Ṣādiq (Damascus: al-Maʿhad al-Faransī, 1968), 83n2, §217. Compare Abū Rayḥān al-Bīrūnī, *al-Āthār al-bāqiya ʿan al-qurūn al-khāliya*, ed. Parwīz Adhkāyī (Tehran: Mīrāth-i Maktūb, 1380sh/2001), 285–286, §77; *Kashf*, 2:1126, 1437. The title does not feature in Bīrūnī, *Risāla*, 29–46.

38. Abū Jaʿfar al-Ḥāsib al-Ṭabarī (attributed), *Tuḥfat al-gharāʾib*, ed. Jalāl Matīnī, 2nd ed. (Tehran: Kitābkhāna, Mūza wa-Markaz-i Asnād-i Majlis-i Shūrā-yi Islāmī, 2012), 35nn1–2 (editor's introduction with list of Qazwīnī's citations), 95–96 (table of contents), 294, 301, 371,

384–385, 398–399, 402–403 (commentary with parallels in Qazwīnī). There are significant differences in wording between the early Persian recension of Qazwīnī's natural history and the edition of the *Tuḥfat*. Compare, for instance, citations in F 60, 82, 85–87, 91, with *Tuḥfat*, 205, 211–212, 215, 223–235.

39. For a feminist critique of virtue ethics, see Zahra Ayubi, *Gendered Morality: Classical Islamic Ethics of the Self, Family, and Society* (New York: Columbia University Press, 2019), 59–60, 70–114, 275–277.

40. For early exegetical and legal discourses, see, for instance, Karen Bauer, *Gender Hierarchy in the Qurʾān: Medieval Interpretations, Modern Responses* (Cambridge: Cambridge University Press, 2015); Aisha Geissinger, *Gender and Muslim Constructions of Exegetical Authority: A Rereading of the Classical genre of Qurʾān Commentary* (Leiden: Brill, 2015); Kecia Ali, *Marriage and Slavery in Early Islam* (Cambridge, MA: Harvard University Press, 2010). On public space, see Marion Katz, *Women in the Mosque: A History of Legal Thought and Social Practice* (New York: Columbia University Press, 2014). For constructions of religious authority, see Asma Sayeed, *Women and the Transmission of Religious Knowledge in Islam* (Cambridge: Cambridge University Press, 2013).

41. Al-Fārābī, "wa-l-madīna al-fāḍila tashbihu l-badan al-tāmm al-ṣaḥīḥ alladhī tataʿāwanu aʿḍāʾuhu kulluhu," *Mabādiʾ ārāʾ ahl al-madīna al-fāḍila = Al-Farabi on the Perfect State,* ed. and trans. Richard Walzer (Oxford: Clarendon Press, 1985), 230, §4. M fol. 144a (W 303–304). See *CAL*, 114.

42. Naṣīr al-Dīn Ṭūsī, *Akhlāq-i nāṣirī*, ed. Mujtabā Minuwī and ʿAlī Riḍā Ḥaydarī (Tehran: Intishārāt-i Khwārazmī, 1978), 37–41. On *siyāsa* as the Greek *politeia*, see Hans Daiber, "Qosṭā," 110–112. See also Christian Lange, *Justice, Punishment, and the Medieval Muslim Imagination* (Cambridge: Cambridge University Press, 2008), 14, 42–50, 54–65, 180–181, 223, 235–236.

43. M fol. 145a (W 305–306). See *Iḥyāʾ*, 8:1440–1441 (96–97); Ibn Sīnā, *al-Mabdaʾ wa-l-maʿād*, ed. ʿAbd Allāh Nūrānī (Tehran: Dānishgāh-i Tihrān, 1984), 109–110. See also Erez Naaman, "Nurture over Nature: Habitus from al-Fārābī through Ibn Khaldūn to ʿAbduh," *JAOS* 137, no. 1 (2017): 1–24.

44. M fol. 2a (W 4: *riyāḍāt* and *tahdhīb*); M fol. 98b (W 202).

45. On perfect nature, see *Khalīqa*, 1, 7, 456–457; *Ghāyat*, 187–194; *Murūj*, 2:88, §811; *Muʿtabar*, 2:391–392; *Maktūm*, fols. 88b–89a, 90b–91a, 212a; *Maṭālib*, 7:142, 400; 8:136–137, 144. See also Michael-Sebastian Noble, *Philosophising the Occult: Avicennan Psychology and 'The Hidden Secret' of Fakhr al-Dīn al-Rāzī* (Berlin: De Gruyter, 2021), 23–28, 76–82, 94–97, 229–249; Łukasz Piątak, "Between Philosophy, Mysticism and Magic. A Critical Edition of Occult Writings of and Attributed to Shihāb al-Dīn al-Suhrawardī (1156–1191)" (doctoral diss., University of Warsaw, 2018), 23, 76, 198–199 (Arabic), 399–400. Compare *Āthār*, 395; *FPP*, 243. See also Bernd Roling, "The Complete Nature of Christ: Sources and Structures of a Christological Theurgy in the Works of Johannes Reuchlin," in *The Metamorphosis of Magic from Late Antiquity to the Early Modern Period*, ed. Jan Bremmer and Jan Veenstra (Leuven: Peeters, 2002), 231–266, at 238–244.

46. M fol. 149b; Y fol. 139b (W 317). The early Persian recension is truncated, featuring only the position of the philosophers and masters of talismans, F 161. For the Platonic background, see Charles Stang, *Our Divine Double* (Cambridge, MA: Harvard University Press, 2016), 20–63.

47. See *Maʿālim*, 116; compare *Maṭālib*, 7:142–143. See also Obuchi Hisashi, "Fakhr al-Dīn al-Rāzī and Occult Science as Philosophy: An Aspect of the Philosophical Theology of Islam at the Beginning of the Thirteenth Century," *Annals of Japan Association for Middle East Studies* 34, no. 1 (2018): 1–33, at 21–22. On *al-nafth fī l-rūʿ*, as angelic inspiration in the heart or mind, see *Iḥyāʾ*, 15:2794, 2815–2816 (50, 71–72); *Mafātīḥ*, 32:194–195 (Q: 113.4). Compare Qazwīnī's Arabic discussion of perfect nature with his Persian treatment of *nīranjāt*, citing Fakhr al-Dīn al-Rāzī, in F 224–225 (P 536).

48. W 322–327. This section, *fī tawalludihi min al-nuṭfa*, is missing from Y and M, though referenced in the index (M fol. 8a), preserved in the recension of 661/1262~, reflected in V fols. 168a–170a; in the Juwaynī version in London, Wellcome Collection, MS Arabic 18, fols. 240a–242a; the Persian recension, F 162–163 (P 437–440). See M fol. 160a–b (W 352–354: reproductive organs); M fols. 162b–163b (W 360–364: divergence of people). For variants and a translation of the section on gestation, see Jan Jaap de Ruiter, "Human Embryology in Zakariyā al-Qazwīnī's *The Marvels of Creation*," *Tijdschrift voor de geschiedenis der geneeskunde, natuurwetenschappen, wiskunde en techniek* 9, no. 3 (1986): 99–117. On *khunthā* in Islamic law, see Indira Falk Gesink, "Intersex Bodies in Premodern Islamic Discourse: Complicating the Binary," *Journal of Middle East Women's Studies* 14, no. 2 (2018): 152–173; in medical discourses, Ahmed Ragab, "One, Two, Or Many Sexes: Sex Differentiation in Medieval Islamicate Medical Thought," *Journal of the History of Sexuality* 24, no. 3 (2015): 428–454. For women as sources of carnal desire and demonic traps, see M fols. 144b–145a, 165a (W 305, 368); see also deceptions (*ḥiyal*) of women in F 228–229 (P 551–553).

49. M fol. 16a (W 25); compare P 456–457; *Āthār*, 232–233. See also Benjamin Isaac, *The Invention of Racism in Classical Antiquity* (Princeton, NJ: Princeton University Press, 2004), 114–124 (on Athens).

50. M fols. 149a–150b (W 317–322); *Āthār*, 573. See also Robert Hoyland, "Physiognomy in Islam," *Jerusalem Studies in Arabic and Islam* 30 (2005): 361–402; Anna Akasoy, "Arabic Physiognomy as a Link between Astrology and Medicine," in *Astro-Medicine: Astrology and Medicine, East and West*, ed. Anna Akasoy et al. (Florence: SISMEL, 2008), 119–141.

51. See pseudo-Aristotle, *Uthūlūjiyā*, Süleymaniye, Aya Sofya 2457, fols. 143a, 144b, 145a, translated in *Plotiniana Arabica*, trans. Geoffrey Lewis, in *Plotini Opera, Tomus II: Enneades IV–V*, ed. Paul Henry and Hans-Rudolf Schwyzer (Paris: Desclée de Brouwer, 1959), 137, ll.13–17, 141, l. 35, 143, l.40 (river). Compare ʿAbd al-Laṭīf al-Baghdādī, *Mā baʿd al-ṭabīʿiyya*, ed. Yūnus Ajʿūn (Beirut: Dār al-Kutub al-ʿIlmiyya, 2017), 283–284. For connections with Kindī's philosophy, see Liana Saif, *The Arabic Influences on Early Modern Occult Philosophy* (New York: Palgrave Macmillan, 2015), 30–36.

52. Süleymaniye, Aya Sofya 2457, fol. 146a (*jinn*), trans. 145, l.43. For further background, see Wendy Helleman, "Plotinus and Magic," *International Journal of the Platonic Tradition* 4 (2010): 114–146; Stang, *Divine Double*, 185–230.

53. See *Ikhwān*, 4:283–463. For the so-called shorter epistle on magic (4:283–312), see *On Magic: An Arabic Critical Edition and English Translation of Epistle 52a*, ed. and trans. Godefroid de Callataÿ and Bruno Halflants (Oxford: Oxford University Press, 2011). For the authorship, see Godefroid de Callataÿ, "Magia en al-Andalus: *Rasāʾil Ijwān al-Ṣafāʾ, Rutbat al-ḥakīm* y *Gāyat al-ḥakīm (Picatrix)*," *al-Qantara* 34, no. 2 (2013): 297–344; see also Guillaume De Vaulx d'Arcy, "The *Epistles of the Brethren of Purity* edited by the Institute of Ismaili Studies," *Mélanges de l'Institut dominicain d'études orientales* 34 (2019): 253–328.

54. See *Āthār,* 571 (Shihāb al-Dīn Suhrawardī). For further, see Cornelis van Lit, *The World of Image in Islamic Philosophy: Ibn Sina, Suhrawardi, Shahrazuri, and Beyond* (Edinburgh: Edinburg University Press, 2017); John Walbridge, *The Wisdom of the Mystic East: Suhrawardi and Platonic Orientalism* (Albany: SUNY Press, 2001).

55. See M fol. 104a (W 220: *ḥajar al-ʿuqāb*); Ruska, *Ḳazwînî,* 18; pseudo-Aristotle, *Aḥjār,* 114, §31 (Arabic), 165 (translation); *Tansūkh,* 145. See also Kraus, *Jābir,* 2:72n2; M fol. 105b (W 224–225: *dahnaj*); Ruska, *Ḳazwînî,* 22–23; pseudo-Aristotle, *Aḥjār,* 103–104, §7 (Arabic), 145–146 (translation); *Tansūkh,* 118–120; M 110a (W 237–238: *mirād*); Ruska, *Ḳazwînî,* 36.

56. M fols. 103a (W 217: *ḥajar al-bāh, ḥajar al-baḥr*), 108a (W 221–232: *furslūs*), 109b–110a (W 236–237: *mās, mānṭas*); Ruska, *Ḳazwînî,* 14–15, 30, 34–36; pseudo-Aristotle, *Aḥjār,* 105–106 (Arabic, §9), 149–150 (translation).

7: Long Divided

1. For the Earth as a grape in water, see Ibn Khaldūn, *al-Muqaddima,* ed. ʿAbd al-Salām al-Shaddādī, 5 vols. (Casablanca: Khizānat Ibn Khaldūn, 2005), 1:71, 84; as egg yolk, see M fol. 79a (W 144). For the Stoic background, see pseudo-Plutarch, *Kitāb al-Ārāʾ al-ṭabīʿiyya = Placita philosophorum,* published as *Aetius Arabus: Die Vorsokratiker in arabischer Überlieferung,* ed. and trans. Hans Daiber (Wiesbaden: Franz Steiner, 1980), 25 (Arabic text), ll. 2–3 (II 2.1–2), 378 (commentary). For the cosmic egg in Arabic geography, see *Masālik,* 4–5; *Murūj,* 2:367–368, §1326; *Nuzhat,* 1:7–9; *Buldān,* 1:16; ʿUmarī, 1:121. On the classical conception of the inhabited quarter, see Ptolemy, *Kitāb al-Jughrāfiyā,* Süleymaniye, Aya Sofya 2610, fol. 4a–4b.

2. Library of Congress, G93.Q3185 1553, fol. 6a (pagination starts at the end of manuscript); compare with image in Bodl., Turk D. 2, fol. 155a; British Library, Add 7894, fol. 167a; Manisa, Turkey, Halk Kütüphanesi, MS 3109, fol. 170b. Sürūrī transmits the account via Ibn ʿAbbās and not, as Qazwīnī does, from Wahb ibn Munabbih. Skeptically ascribed to storytellers (*quṣṣāṣ*) in *Buldān,* 1:23–24; Abū Naṣr al-Maqdisī (fl. 355/966), *Kitāb al-Badʾ wa-l-tārīkh,* ed. and trans. Clément Huart, 6 vols. (Paris: E. Leroux, 1899–1919), 2:247–249.

3. See Arent Jan Wensinck, *The Ocean in the Literature of the Western Semites* (Amsterdam: Johannes Müller, 1918), 3–4, 24; Daniel Prior, "Travels of Mount Qāf: From Legend to 42° 0′ N 79° 51′E," *Oriente Moderno* 89, no. 2 (2009): 425–444. For an overview of the Indic context, see Joseph Schwartzberg, "Cosmographical Mapping," *HOC,* 2.1:332–387.

4. M fol. 80b (W 145). For Qāf as a superstructure, see Abū l-Ḥasan al-Bakrī (d. 289/902?), *al-Anwār fī mawlid al-nabī* (Beirut: Maktabat al-Alfayn, 1986), 9–10; Abū l-Shaykh al-Iṣfahānī, *Kitāb al-ʿAẓama,* ed. Riḍāʾ Allāh Mubārakfūrī, 5 vols. (Riyadh: Dār al-ʿĀṣima, 1998), 4:1411–1412, §931; Thaʿlabī, *ʿArāʾis al-majālis* (Beirut: al-Maktaba al-Thaqāfiyya, n.d.), 4, 145, 320; Abū Jaʿfar al-Ṭabarī, *Tārīkh al-rusul wa-l-mulūk = Annales quos scripsit Abu Djafar Mohammed Ibn Djarir at-Tabari,* ed. Michael Jan de Goeje et al., 3 series (Leiden: E.J. Brill, 1879–1901), I, 49–50.

5. *Āthār,* 14; M fols. 57b, 80b (W 105, 147).

6. For the Earth's shape, see M fol. 79a–b (W 144–145); compare *Buldān,* 1:16; pseudo-Plutarch, *al-Ārāʾ,* 44, ll. 18–23, 45, ll. 1–2 (III 10), 427–428 (commentary with further references). For a demonstration of sphericity, see M fol. 56b (W 101); Ptolemy, *al-Majisṭī,* BnF, Arabe 2482, fol. 4a; Greek text in *Almagest = Syntaxis Mathematica,* ed. Johan Ludvig Heiberg, 2 parts (Leipzig: Teubner, 1898–1903), part 1:16 (book 1.4); see also Yaʿqūb ibn Isḥāq al-Kindī, *Fī l-ṣināʿa*

al-ʿuẓmā, ed. ʿAzmī Ṭāhā al-Sayyid Aḥmad (Cyprus: Dār al-Shabāb, 1987), 145; Ibn Sīnā, *al-Shifāʾ, al-Riyāḍiyyāt: 4 ʿIlm al-Hayʾa*, ed. Muḥammad Riḍā Madwar and Imām Ibrāhīm Aḥmad (Cairo: al-Hayʾa al-Miṣriyya al-ʿĀmma li-l-Kitāb, 1980), 21. For the Earth's circumference, see M fols. 79b–80a (W 142); compare *Funūn*, 133 (Arabic), 417 (translation); *Buldān*, 1:18. See also Raymond Mercier, "Geodesy," *HOC*, 2.1:175–188, at 176–179.

7. The followers of Pythagoras are missing from M; but in F 77; B fol. 53a (W 144). Compare Carl Huffman, *Philolaus of Croton, Pythagorean and Presocratic: A Commentary on the Fragments and Testimonia with Interpretive Essays* (Cambridge: Cambridge University Press, 1993), 231–288. See also pseudo-Plutarch, *Ārāʾ*, 45, ll. 23–24 (III 13.2); Ibn Sīnā, "Risāla fī ʿillat qiyām al-arḍ fī ḥayyizihā," in *Jāmiʿ al-badāʾiʿ*, ed. Muḥyī l-Dīn Ṣabrī al-Kurdī (Cairo: Maṭbaʿat al-Saʿāda, 1335/1917), 152–164, at 163–164; Ibn Sīnā, *al-Shifāʾ, al-Ṭabīʿiyyāt: 2. al-Samāʾ wa-l-ʿālam*, ed. Maḥmūd Qāsim (Cairo: Dār al-Kātib al-ʿArabī, 1969), 58–63. For the stationary Earth in Indic cosmography, see Abū l-Rayḥān al-Bīrūnī, *Taḥqīq mā li-l-hind*, ed. Eduard Sachau (London: Trübner and Co., 1887), 138–139. For geocentrism in scholastic circles, see *Mulakhkhaṣ*, fols. 189b–191b, 193a–194b; *Mabāḥith*, 2:115–119; *Mafātīḥ*, 25:144–145 (Q. 31.10); Najm al-Dīn al-Kātibī al-Qazwīnī, *Ḥikmat al-ʿayn*, ed. Ṣāliḥ Aydın al-Turkī (Cairo: n.p., 2002), 77–78; commentary by ʿAllāma al-Ḥillī (d. 726/1325), *Īḍāḥ al-maqāṣid fī ḥikmat ʿayn al-qawāʾid*, ed. ʿAlīnaqī Munzawī (Tehran: Dānishgāh-i Tihrān, 1959), 242–243; ʿAbd al-Raḥmān al-Ījī (d. 756/1355), *al-Mawāqif fī ʿilm al-kalām* (Beirut: ʿĀlam al-Kutub, n.d.), 218–219.

8. M 57b (W 104). See Bīrūnī, *Taḥdīd nihāyāt al-amākin*, ed. Pavel Georgievich Bulgakov (Cairo: Maʿhad al-Makhṭūṭāt al-ʿArabiyya, 1962), 156; compare *Buldān*, 1:21. On the encircling ocean, see Karen Pinto, *Medieval Islamic Maps: An Exploration* (Chicago: University of Chicago Press, 2016), 79–187. On Atlantic navigation, see Christophe Picard, *L'Océan Atlantique musulman: De la conquête arabe à l'époque almohade—Navigation et mise en valeur des côtes d'al-Andalus et du Maghreb occidental* (Paris: Maisonneuve et Larose, 1997). For wonder tales, see also Christophe Picard, "Récits merveilleux et réalité d'une navigation en Océan Atlantique chez les auteurs musulmans," *Actes des congrès de la Société des historiens médiévistes de l'enseignement supérieur public* 25, no. 1 (1994): 75–87.

9. *Murūj*, 1:137–138, §§273–274. See *Āthār*, 29 (*jazāʾir al-khālidāt*), also 550–551 (Cadiz); M fol. 57b (W 104).

10. See Masʿūdī, *al-Tanbīh wa-l-ishrāf*, ed. Michael Jan de Goeje (Leiden: E.J. Brill, 1894), 69. Greek *stēlē* are often identified as idols, *aṣnām*, rather than pillars, although *asāṭīn* (sg. *usṭūna*) also feature, as in Bīrūnī, *Taḥdīd*, 143. See the translation of pseudo-Aristotle, *De Mundo* (392b17–27, 393b11), known as the *Risālat bayt al-dhahab*, ed. David Alan Brafman (PhD diss., Duke University, 1985), 87, 89–90. See also *Tanbīh*, 171–172, 189; *Murūj*, 1:85, 87, §§153, 158.

11. For the Cadiz lighthouse, see *Nuzhat*, 1:17; Zuhrī, *Kitāb al-Jaʿrāfiyya*, ed. Muḥammad Ḥājj Ṣādiq (Damascus: al-Maʿhad al-Faransī, 1968), 91, §239; Abū Ḥāmid al-Gharnāṭī, *Tuḥfat al-albāb*, ed. Gabriel Ferrand (Paris: Imprimerie nationale, 1925), 69–70 (ascribed to Alexander); *Buldān*, 4:290–291; Abū ʿAbd Allāh al-Ḥimyarī (d. ca. 727/1326), *Kitāb al-Rawḍ al-miʿṭār*, ed. Iḥsān ʿAbbās (Beirut: Maktabat Lubnān, 1975), 448–449. See the *columnae Herculis* in Paulus Orosius (d. ca. 418 CE), *Historiarum adversum paganos*, ed. Karl Friedrich Wilhelm Zangemeister (Leipzig: Teubner, 1889), 5 (§1.2, ll. 20–3); cited in *Carte marine et portulan au XIIe siècle: Le Liber de existencia riveriarum et forma maris nostri Mediterranei (Pise, ca. 1200)*, ed. Patrick Gautier Dalché (Rome: École française de Rome, 1995), 115, ll. 170–171. For transposi-

tion of pillars into idols, see the translation of Orosius, completed in Cordoba before 344/955, *Kitāb Hurūshiyūsh*, ed. Mayte Penelas (Madrid: CSIC, 2001), 21, §15.

12. See Duane Roller, *Through the Pillars of Herakles: Greco-Roman Exploration of the Atlantic* (New York: Routledge, 2006), 23–25, 44–56.

13. Dante, *Inferno*, Canto 26, ll. 108–109. See Earl Rosenthal, "*Plus Ultra, Non plus Ultra*, and the Columnar Device of Emperor Charles V," *Journal of the Warburg and Courtauld Institutes* 34 (1971): 204–228.

14. *Nuzhat*, 1:17; *Āthār*, 29. Bīrūnī, *Taḥdīd*, 156–157.

15. See Aristotle, *De caelo*, ed. and trans. William Keith Chambers Guthrie (Cambridge, MA: Harvard University Press, 1939), book 2, 14, 298a10–16; translated in Aristotle, *Fī l-Samāʾ wa-l-āthār al-ʿulwiyya*, ed. ʿAbd al-Raḥmān al-Badawī (Cairo: Lajnat al-Taʾlīf wa-l-Tarjuma wa-l-Nashr, 1961), 303; Ibn Rushd, *Talkhīṣ al-samāʾ wa-l-ʿālam*, ed. Jamāl al-Dīn al-ʿAlawī (Fez: Jāmiʿat Sīdī Muḥammad ibn ʿAbd Allāh, 1984), 276. On debates about circumnavigation of Africa, see Roller, *Pillars*, 19–20, 23–25, compare 92–93. See also Venetian cosmographer Fra Mauro (d. 1464), in Angelo Cattaneo, *Fra Mauro's Mappa Mundi and Fifteenth-Century Venice* (Turnhout: Brepols, 2011), 117–118, 272–273. Compare *GAS*, 11:349–403 (trans., 2:342–394).

16. *Murūj*, 1:138, §274. See also Abū l-ʿAbbās al-ʿUdhrī (d. 478/1085), *Kitāb Tarṣīʿ al-akhbār wa-tanwīʿ al-āthār* (partial edition), ed. ʿAbd al-ʿAzīz al-Ahwānī (Madrid: Instituto de Estudios Islámicos, 1965), 119; Douglas Morton Dunlop, "The British Isles according to Medieval Arabic Authors," *Islamic Quarterly: A Review of Islamic Culture* 4 (1957): 11–28, at 19–20; Juan Vernet, "Textos árabes de viajes por el Atlántico," *Anuario de Estudios Atlánticos* 1, no. 17 (1971): 401–427, at 403–404.

17. Abū ʿUbayd al-Bakrī, *al-Masālik wa-l-mamālik*, ed. André Ferré and Adrian van Leeuwen, 2 vols. (Carthage: al-Dār al-ʿArabiyya li-l-Kitāb, 1992), 1:203, §§287–288.

18. *Nuzhat*, 2:525.

19. *Nuzhat*, 2:547.

20. A scribal error has the phrase as *maghrūrīn*, i.e., deceived, beguiled, in Ibn al-Wardī, *Kharīdat al-ʿajāʾib wa-farīdat al-gharāʾib*, ed. Anwar Maḥmūd Zanātī (Cairo: Maktabat al-Thaqāfa al-Dīniyya, 2008), 65. However, see Ibn al-Wardī, Leiden, Universiteitsbibliotheek, MS Or. 158, fol. 15b; Idrīsī, *De geographia universali* (Rome: In Typographia Medice, 1592), clime 4, section 1 (p. 16 of unpaginated chapter); compare clime 3, section 1 (p. 2 of unpaginated chapter). For the correct form, see Idrīsī, *Nuzhat*, BnF, Arabe 2221, fols. 91a, 192a. Misread in *GAS*, 13:139 (trans., 4:132); Sezgin, *Piri Reis and the Pre-Columbian Discovery of the American Continent by Muslim Seafarers* (Istanbul: Boyut, 2013), 43.

21. See Vernet, "Textos árabes," 407–409; Picard, *L'Océan Atlantique*, 31–35.

22. On Java, see *Āthār*, 82; M fol. 62 (W 112).

23. See George Tolias, "Isolarii, Fifteenth to Seventeenth Century," *HOC*, 3.1:263–284. See also Simone Pinet, *Archipelagoes: Insular Fictions from Chivalric Romance to the Novel* (Minneapolis: University of Minnesota Press, 2011), ix–xxxv.

24. See *Nuzhat*, 1:218–220.

25. See Rudolf Wittkower, "Marvels of the East: A Study in the History of Monsters," *Journal of the Warburg and Courtauld Institutes* 5 (1942): 159–197, at 161n1, 164.

26. M fols. 62b–63a (W 112–113); see also M fols. 71b–72a, 73a–b (W 129, 132–133); for further on *tinnīn*, see Sara Kuehn, *The Dragon in Medieval East Christian and Islamic Art* (Leiden:

Brill, 2011); Ahmed al-Rawi, "The Religious Connotation of the Islamic Dragon," *Fabula* 53, no. 1/2 (2012): 82–93.

27. ʿUmarī, 4:56; copied in Abū l-ʿAbbās al-Qalqashandī (d. 821/1418), *Ṣubḥ al-aʿshā*, 14 vols. (Cairo: al-Maṭbaʿa al-Amīriyya, 1913–1922), 5:294–295. See also Taqī l-Dīn al-Maqrīzī, *al-Sulūk li-maʿrifat duwal al-mulūk*, ed. Muḥammad ʿAbd al-Qādir ʿAṭāʾ, 8 vols. (Beirut: Dār al-Kutub al-ʿIlmiyya, 1997), 3:261. For further context, see Michael Gomez, *African Dominion: A New History of Empire in Early and Medieval West Africa* (Princeton, NJ: Princeton University Press, 2018), 92–143.

28. See M fols. 57b, 80b (W 105, 147); compare *Buldān*, 1:19; *Āthār*, 18–27.

29. Ibn Khaldūn, *Tārīkh*, 6:226–227; Qalqashandī, *Ṣubḥ*, 5:293–294. Sezgin mistakenly claims that Mansā Mūsā's father, Abū Bakr, undertook the expedition, *GAS*, 13:139–140 (trans., 4:132); Sezgin, *Piri Reis*, 45. As Ibn Khaldūn makes clear, the preceding ruler was the rival Muḥammad ibn Qū; see Nehemia Levtzion, "The Thirteenth- and Fourteenth-Century Kings of Mali," *Journal of African History* 4, no. 3 (1963): 341–353, at 346–347; Gomez, *Dominion*, 3–5, 100–103.

30. As a preview to Iberian New World colonization, see Bartolomé de las Casas, *Historia de las Indias*, ed. Miguel Ángel Medina, vols. 3–5 of *Obras completas*, under the direction of Paulino Castañeda Delgado, 14 vols. (Madrid: Alianza, 1988–1999), 1:429–459.

31. Ibn Khaldūn, "lā yaʿrifūna dīnan wa-lam tablaghhum daʿwatun," *Muqaddima*, 1:89–90.

32. *GAS*, 11:55 (trans., 2:53).

33. See William Brice, "Compasses, Compassi, and *Kanābīṣ*," *Journal of Semitic Studies* 29, no. 1 (1984): 169–178; Svatopluk Soucek, "Islamic Charting in the Mediterranean," *HOC*, 2.1:263–292, at 286n86.

34. On nautical charts, often referred to in modern scholarship as portolan charts, see Evelyn Edson, *The World Map, 1300–1492: The Persistence of Tradition and Transformation* (Baltimore: Johns Hopkins University Press, 2007), 33–59; Tony Campbell, "Portolan Charts from the Late Thirteenth Century to 1500," *HOC*, 1:371–463.

35. For the origins of the nautical chart, see *GAS*, 10:285–337, 11:3–84 (trans., 1:287–340, 2:3–83). Sezgin's idea that Arabic cartography shaped the later development of portolan maps in Latin Christendom is not without controversy; see Sonja Brentjes, "Revisiting Catalan Portolan Charts: Do They Contain Elements of Asian Provenance?," in *The Journey of Maps and Images on the Silk Road*, ed. Philippe Forêt and Andreas Kaplony (Leiden: Brill, 2008), 181–201, at 183–185; Sonja Brentjes, "Medieval Portolan Charts as Documents of Shared Cultural Spaces," in *Acteurs des transferts culturels en Méditerranée médiévale*, ed. Rania Abdellatif et al. (Munich: Oldenbourg, 2012), 135–146; Patrick Gautier Dalché, "Le renouvellement de la perception et de la représentation de l'espace au XIIe siècle," *Renovación intelectual del occidente europeo* (Pamplona: Gobierno de Navarra, 1998), 169–217, at 195–204, 214–216; Dalché, *Carte marine et portulan*, 51–65, compare 10, 126, ll. 546–549.

36. See Patrick Gautier Dalché, "The Reception of Ptolemy's Geography (End of the Fourteenth to Beginning of the Sixteenth Century)," *HOC*, 3.1:285–364. See also Christophe Picard, *La mer des califes: Une histoire de la Méditerranée musulmane (VIIe-XIIe siècle)* (Paris: Éditions du Seuil, 2015), 187–223, 333–344.

37. *Nuzhat*, 1:7–8. See *GAS*, 10:276–294, 11:16–26, 352–353 (trans., 1:288–296, 2:15–26, 345–346). For a study of Idrīsī's influence on cartography in the Latin West, shaped by Sezgin's

theories, see Carsten Drecoll, *Idrísí aus Sizilien: Der Einfluss eines arabischen Wissenschaftlers auf die Entwicklung der europäischen Geographie* (Egelsbach: Hänsel-Hohenhausen, 2000), 19–60. Compare Patrick Gautier Dalché, "Géographie Arabe et Géographie Latine au XIIe Siècle," *Medieval Encounters* 19, no. 4 (2013): 408–433, at 410–411, 422–423. For further context, see *Cartography between Christian Europe and the Arabic-Islamic World, 1100–1500*, ed. Alfred Hiatt (Leiden: Brill, 2021).

38. ʿUmarī, 2:185, 189–192; cited in Tarek Kahlaoui, "The Depiction of the Mediterranean in Islamic Cartography (11th–16th Centuries)" (PhD diss., University of Pennsylvania, 2008), 204. For more, see Jean-Charles Ducène, "Le portulan arabe décrit par al-ʿUmarī," *Comité français de cartographie* 216 (2013): 81–90.

39. See Dionisius Agius, *Classic Ships of Islam: From Mesopotamia to the Indian Ocean* (Leiden: Brill, 2008), 370–372; Giovan Battista Pellegrini, "Terminologia marinara di origine araba in italiano e nelle lingue europee," in *La navigazione mediterranea nell'alto Medioevo: 14–20 aprile 1977*, 2 vols. (Spoleto: Presso la sede del Centro, 1978), 2:797–841.

40. Aḥmad ibn ʿAlī l-Maqrīzī, *Durar al-ʿuqūd al-farīda fī tarājim al-aʿyān al-mufīda*, ed. Maḥmūd al-Jalīlī, 4 vols. (Beirut: Dār al-Gharb al-Islāmī, 2002), 2:406–408; María Jesús Viguera Molíns, "Eco árabe en un viaje genovés a las islas canarias antes de 1340," *Medievalismo* 2 (1992): 257–258. On war galleys, see Agius, *Ships*, 348–351.

41. See Ibn Baṭṭūṭa, *Riḥlat Ibn Baṭṭūṭa = Tuḥfat al-nuẓẓār fī gharāʾib al-amṣār wa-ʿajāʾib al-asfār*, ed. ʿAbd al-Hādī al-Tāzī, 5 vols. (Rabat: Akādīmiyyat al-Mamlaka al-Maghribiyya, 1997), 1:77–78, 150–151.

42. On the Dulcert portolan of 1339, see Ernest-Théodore Hamy, *Angelino Dulcert de Majorque (1339)*, 2nd ed. (Paris: H. Champion, 1903). See also Campbell, "Portolan Charts," 378n65, 410. On the Pizigani portolan of 1367, see Mario Longhena, "La Carta dei Pizigano del 1367 (posseduta dalla Biblioteca Palatina di Parma)," *Archivio storico per le province Parmensi*, 4th ser., 5 (1953): 25–130, at 58–61. For the Catalan atlas, see Campbell, "Portolan Charts," 372, 393n208, 430, 432n417, 434–435, fig. 19.5; Brentjes, "Portolan Charts."

43. For the inscription by the statue in the western margin, see Gerald Roe Crone, "The Pizigano Chart and the 'Pillars of Hercules,'" *Geographical Journal* 109, no. 4–6 (1947): 278–279; Longhena, "La Carta," 54–57; Armando Cortesão, *Esparsos*, 3 vols. (Coimbra: University of Coimbra, 1974–1975), 3:106–108, and fig. 2.

44. On the Tunisian portolan housed in TPMK, Hazine 1823, see *GAS*, 11:31–32 (trans., 2:31–32); Mónica Herrera-Casais, "The 1413–4 Sea Chart of Aḥmad al-Ṭanjī," in *A Shared Legacy: Islamic Science East and West—Homage to Professor J. M. Millàs Vallicrosa*, ed. Emilia Calvo (Barcelona: Universitat de Barcelona, 2008), 283–307. On figure 7.4, TPMK, Hazine 1822, a portolan produced more than a century later by Ḥājj Abū l-Ḥasan during the reign of Süleyman the Magnificent (r. 926–974/1520–1566), which contains full details of the Canaries, see Soucek, "Charting," 276, fig. 14.4.

8: Across the Globe

1. M fol. 87b (W 170). On multiple worlds, see Bīrūnī responding to Ibn Sīnā, *al-Asʾila wa-l-ajwiba*, ed. Sayyid Ḥusayn Naṣr and Mahdī Muḥaqqiq (Tehran: Markaz-i Muṭālaʿāt wa-Hamāhangī-yi Farhangī, 1352sh/1973), 90–98, §§24–32; Alesandro Bausani, "Some

Considerations on Three Problems of the Anti-Aristotelian Controversy between al-Bīrūnī and Ibn Sīnā," in *Akten des siebten Kongresses für Arabistik und Islamwissenschaft, Göttingen*, ed. Albert Dietrich (Göttingen: Vandenhoeck und Ruprecht, 1976), 74–85. See also *Mafātīḥ*, 1:14; *Maṭālib*, 6:193–195; *Mabāḥith*, 2:151–155; Ījī, *al-Mawāqif* (Beirut: ʿĀlam al-Kutub, n.d.), 256. Compare Steven Dick, *Plurality of Worlds: The Extraterrestrial Life Debate from Democritus to Kant* (Cambridge: Cambridge University Press, 1982).

2. M fols. 4b–6a (W 8–9); *Enz.*, 100–101 (diagram 1). Compare *Maʿālim*, 32–33; *Maṭālib*, 4:9–12, 7:7–9; *Muḥaṣṣal*, 65–66, 82; Naṣīr al-Dīn al-Ṭūsī, *Talkhīṣ al-Muḥaṣṣal*, ed. ʿAbd Allāh Nūrānī (Beirut: Dār al-Aḍwāʾ, 1985), 93–95, 142–143; Ījī, *Mawāqif*, 185–186, 263. See also *PPT*, 450–459.

3. See Ījī, *Mawāqif*, 207; commentary al-Sharīf al-Jurjānī (d. 816/1413), *Sharḥ*, ed. Muḥammad Badr al-Dīn, 8 vols. (Cairo: Maṭbaʿat al-Saʿāda, 1907), 7:107. See Abdelhamid I. Sabra, "Science and Philosophy in Medieval Islamic Theology: The Evidence of the Fourteenth Century," *ZGAIW* 9 (1994): 1–42, at 34–41; Robert Morrison, "What Was the Purpose of Astronomy in Ījī's *Kitāb al-Mawāqif fī ʿilm al-kalām*?," in *Politics, Patronage, and the Transmission of Knowledge in 13th–15th Century Tabriz*, ed. Judith Pfeiffer (Leiden: Brill, 2014), 201–229, at 206–218; and, broadly, *PPT*, 133–144, 499–503.

4. See Robert Morrison, "Quṭb al-Dīn al-Shīrāzī's Hypotheses for Celestial Motions," *Journal for the History of Arabic Science* 13, no. 1/2 (2005): 21–140, at 36; Ahmad Dallal, *Islam, Science, and the Challenge of History* (New Haven, CT: Yale University Press, 2010), 68, 82–86.

5. Shīrāzī, *Durrat al-tāj li-ghurrat al-dabbāj*, ed. Muḥammad Mishkāt, reissued ed., 5 vols. (Tehran: Intishārāt-i Ḥikmat, 1369sh/1990), 1:72 [152]. On knowledge acquired through conception, *taṣawwur*, and assent, *taṣdīq*, and the expression *fī nafs al-amr*, see *Ishārāt*, 40–41 (i.4), 170–171 (ix.5). See also Robert Wisnovsky, "Avicennism and Exegetical Practice in the Early Commentaries on the *Ishārāt*," *Oriens* 41 (2013): 349–378, at 374n40, 375–376; Morrison, "Purpose," 216–218.

6. Shīrāzī, *Durrat*, 1:85–86 [165–166].

7. Shīrāzī, *Durrat*, 1:75–76, 101–102 [155–156, 181–182]. See Matthew Melvin-Koushki, "Powers of One: The Mathematicalization of the Occult Sciences in the High Persianate Tradition," *Intellectual History of the Islamicate World* 5 (2017): 127–199, at 147–149.

8. M fol. 184a–b (W 402–403); ʿUmarī, 20:70. Qazwīnī cites Shahmardān Ibn Abī l-Khayr, identified as the author of the *Nuzhatnāma*, for special properties of rhinoceros horn. Compare Shahmardān Ibn Abī l-Khayr, *Nuzhatnāma-yi ʿalāʾī*, ed. Farhang Jahānpūr (Tehran: Muʾassasa-yi Muṭālaʿāt wa-Taḥqīqāt-i Farhangī, 1362sh/1983~), 14–15 (dedication), 22 (Ibn Sīnā), 32 (Jābir ibn Ḥayyān), 54–65 (rhinoceros, but not this story). For the date of the *Nuzhatnāma*, see the editor's introduction, 44–61.

9. Ṭūsī, *ʿAjāʾib-nāma*, ed. Manūchihr Sutūda (Tehran: Nashr-i Kitāb, 1966), 5, 18, 38 (titles of the work), 14 (dedication); on the date, see the editor's introduction, 15–21; *Kashf*, 2:1127. Cf. Storey, 2:121–122; *GAS*, 15:54–55; *CAL*, 127–130.

10. *Āthār*, 201, 202, 415, 600 (Firdawsī), 415 (ʿUnṣurī), 523 (Niẓāmī and Jurjānī), 412 (Niẓām al-Mulk), 234 (Alexander, son of Dārā ibn Bahmān); M fols. 82b–83a (Shīrīn and Farbād); B 57a–b and F 82–83 (Afarīdūn and Bīwārasb). For Alexander as Persian king, see Abū l-Ḥanīfa al-Dīnawarī (d. ca. 282/895), *Akhbār al-ṭiwāl*, ed. ʿUmar Fārūq al-Ṭabbāʿ (Beirut: Dār al-Arqam, 1995), 31–32; Abū l-Qāsim Firdawsī, *Shāh-nāma*, ed. Jalāl Khāliqī Muṭlaq, 8 vols. (New York: Bibliotheca Persica, 1988–2008), 5:523–526, ll. 95–136.

11. See the dedication in F 2 (P 83).

12. See F 3–4 (index on arts and crafts), 171–229 (section on arts and crafts, lacunose at 226); P 90–91, 456–553.

13. F 221–226. The arts after elemental magic (*nīranjāt*) promised in the index, have not survived: natural magic (*sīmiyāʾ*), alchemy, gemstones, perfume, guarding against animals, geometry, riddles, sleight of hand, artifices (*ḥiyal*). See Julius Ruska, "Ḳazwīnīstudien," *Der Islam* 4 (1913): 236–262, at 244–245. See also Karin Rührdanz, "Illustrated Persian *ʿAjāʾib al-Makhlūqāt* Manuscripts and Their Function in Early Modern Times," in *Society and Culture in the Early Modern Middle East: Studies on Iran in the Safavid Period*, ed. Andrew Newman (Leiden: Brill, 2003), 33–47.

14. For parallels, see *Āthār*, 7–8 (social order), 233–235 (ten great figures of Persian history), 571 (trick of Aristotle), 572 (Jāmāsp); P 470–472 (list of ten figures), 488 (social order), 535 (Jāmāsp), 548–549 (Aristotle). For astral and talismanic iconography, see *Cosmos*, 119–149.

15. See Ḥamd Allāh Mustawfī, *Nuzhat al-qulūb*, ed. Mīr Hāshim Muḥaddith, 2 vols. (Tehran: Intishārāt-i Safīr Ardihāl, 1396sh/2017), 1:97–98 (list of sources), 2:1077–1104 (wonders of Iran and other regions), 1285 (index).

16. For illuminated Persian copies during the period, see Karin Rührdanz, "Qazwīnī's *ʿAjāʾib al-Makhlūqāt* in Timurid Manuscripts," in *Iran, questions et connaissances: Actes du IVe Congrès européen des études iraniennes*, 3 vols. (Paris: Association pour l'avancement des études iraniennes, 2002), 2:473–484; Rührdanz, "Populäre Naturkunde illustriert: Text und Bild in persischen *ʿAjāʾib*-Handschriften spätjala'iridischer und frühtimuridischer Zeit," *Studia Iranica* 34 (2005): 231–256, at 233, 243–244; Vivek Gupta, "Remapping the World in a Fifteenth-Century Cosmography: Genres and Networks between Deccan India and Iran," *Iran* 59, no. 2 (2021): 151–168. The bifolium frontispiece (figure 8.2) was removed from TPMK, Revan Köşkü 1660 and purchased by the Metropolitan Museum of Art in New York. See Zeren Tanındı, "Arts of the Book: The Illustrated and Illuminated Manuscripts Listed in ʿAtufi's Inventory," in *TOK*, 1:213–239, at 223, 237n54.

17. See Jean-Charles Ducène, "Les encyclopédies et les sciences naturelles dans le monde arabe médiéval (XIIe–XIVe siècle)," in *Encyclopédire: Formes de l'ambition encyclopédique dans l'Antiquité et au Moyen Âge*, ed. Arnaud Zucker (Turnhout: Brepols, 2013), 201–212; Elias Muhanna, "Why Was the Fourteenth Century a Century of Arabic Encyclopaedism?," in *Encyclopaedism from Antiquity to Renaissance*, ed. Jason König and Greg Woolf (Cambridge: Cambridge University Press, 2013), 343–356.

18. See V fol. f. 250b (Manfalūṭ on the Nile); BnF, Arabe 2175, fols. 8b–9a, 224a (conjoined twins of Egypt), 225b (Jerusalem); BnF, Arabe 2177, fols. 6a (conjoined twins of Egypt), 165a (Christian copyist).

19. For direct citations, see the index in Kamāl al-Dīn al-Damīrī, *Ḥayāt al-ḥayawān al-kubrā*, ed. Ibrāhīm Ṣāliḥ, 4 vols. (Damascus: Dār al-Bashāʾir, 2005) 7:397, 475. For their proximity in the catalogues of the Istanbul archives held now in the Süleymaniye, see, for instance, Aya Sofya 2927–2933 and 2934–2939; Fatih 4158–4170 and 4171–4179; Nuruosmaniye 3009–3019 and 3020–3027. See also the lithograph edition with Qazwīnī in the margins of Damīrī, *Ḥayāt al-ḥayawān al-kubrā* (Cairo: al-Maṭbaʿa al-Maymaniyya, 1311/1893).

20. For the *Akhbār al-zamān*, see Süleymaniye, Aya Sofya 2938, fols. 217b–319a (copied before 894/1489) and Aya Sofya 2939, fols. 287b–425b (copied 892/1487); compare *GAS*, 1:334, 15:43–51. For interpolations on Egypt, see Qazwīnī, *ʿAjāʾib*, BnF, Arabe 2179, fols. 62a–78b.

21. Yāzıcıoğlu Aḥmed Bīcān, *ʿAcāʾib al-makhlūqāt,* BnF, Turc 161, fols. 1b–2a. *GAS,* 15:82; Charles Rieu, *Catalogue of the Turkish Manuscripts in the British Museum* (London: Longmans and Co., 1888), 106–107 (MS Sloane 4088). See Marinos Sariyannis, "*Ajāʾib ve gharāʾib:* Ottoman Collections of Mirabilia and Perceptions of the Supernatural," *Der Islam* 92, no. 2 (2015): 442–467, at 448, 451, 454, 462; Âmil Çelebioğlu, "Ahmed Bîcan," *DIA,* 2:49; *GAS,* 15:82.

22. Ibn al-Wardī, *Kharīda,* 22–24 (dedication, sources), 227, 250–251, 282, 284–285, 289, 293 (Qazwīnī), 423 (date), 424–476 (end of time), 474–480 (*Qilāda*). On the authorship, see editor's introduction, 6–8; *GAS,* 15:79–82. The modern edition is wanting: read *ḥalabiyya* for *jalīla* and *jughrāfiyā* for *Jaʿfar al-anbiyāʾ* (!). Compare Bodl., Selden Superius 69, fols. 1b, 2a, copied 892/1487 in Medina. On the Mamluk emir Shāhīn, evidently Shāhīn al-Arghūn Shāwī, see Ibn Taghrībirdī, *al-Manhal al-ṣāfī wa-l-mustawfī baʿd al-wāfī,* ed. Muḥammad Amīn, 13 vols. (Cairo: al-Hayʾa al-Miṣriyya al-ʿĀmma li-l-Kitāb, 1984–2009), 6:211–212, §1179.

23. See Feray Coşkun, "An Ottoman Preacher's Perception of a Medieval Cosmography: Maḥmūd al-Haṭīb's Translation of *Kharīdat al-ʿAjāʾib wa Farīdat al-Gharāʾib,*" *Al-Masāq* 23, no. 1 (2011): 53–66.

24. ʿAṭūfī, *Kitāb al-Kutub,* TPMK, Török F. 59, fols. 2a, 85a, "al-siyar wa-tawārīkh . . . ʿajāʾib al-makhlūqāt, ṣuwar al-aqālīm," in *TOK,* 2:5–6, 7n4. See Cornell Fleischer and Kaya Şahin, "On the Works of a Historical Nature in the Bayezid II Library Inventory," in *TOK,* 1:569–596, at 570, 582 (§90–92), 590 (§§258–259), 593 (§347). Compare Pasha Khan, *The Broken Spell: Indian Storytelling and the Romance Genre in Persian and Urdu* (Detroit: Wayne State University Press, 2019), 133–161.

25. ʿAṭūfī, *Kitāb al-Kutub,* fols. 4a, 302a, 311a, "al-umūr al-ʿajība." See Noah Gardiner, "Books on Occult Sciences," in *TOK,* 1:735–765, at 745–747. See *Falnama: The Book of Omens,* ed. Massumeh Farhad and Serpil Bagci (Washington, DC: Smithsonian Institution, 2009), 68–75, 295–305.

26. See Ṭūsī, *Tercüme-i Acâʾibü'l-mahlûkât,* trans. Rükneddīn Aḥmed, ed. Bekir Sarıkaya and Günay Kut (Istanbul: Türkiye Yazma Eserler Kurumu Başkanlığı, 2019), 103 (dedication), 116, 188, 190, 202, 231, 237, 249, 258, 269, 271, 305, 311, 340, 351, 384, 442, 470 (talismans and *nīranjāt*).

27. See Pınar Emiralioğlu, "Books on the Wonders of Creation and Geography in ʿAtufi's Inventory," in *TOK,* 1:597–606, at 604–606.

28. Ptolemy, *Kitāb al-Jughrāfiyā,* Süleymaniye, Aya Sofya 2610, fols. 3a (dedication), 7b (merchants), 8a (Islands of the Blessed), 13b (Strait of Hercules), 34b–35a (Iberia), 78b–79a (Mountain of the Moon), 107b–108a (Mecca and Medina), 123a (Yājūj and Mājūj). For the seal, see Zeren Tanındı, "Preliminary List of Manuscripts Stamped with Bayezid II's Seal and Transferred from the Topkapi Palace Inner Treasury to Other Library Collections," in *TOK,* 1:983–1009, at 986, §22. See also Ahmet Karamustafa, "Military, Administrative, and Scholarly Maps and Plans," *HOC,* 2.1:209–227, at 210nn9–10; Karen Pinto, "The Maps Are the Message: Mehmet II's Patronage of an 'Ottoman Cluster,'" *Imago Mundi* 63, no. 2 (2011): 155–179, 157–158; Pınar Emiralioğlu, *Geographical Knowledge and Imperial Culture in the Early Modern Ottoman Empire* (Farnham, Surrey: Ashgate, 2014), 72–74. See Qazwīnī, *ʿAjāʾib,* BnF, Arabe 2179, fol. 63a (Mountain of the Moon), quoting Masʿūdī's *Murūj.*

29. Abū l-Khayr Ṭaşköprīzāde, *Miftāḥ al-saʿāda wa-miṣbāḥ al-siyāda,* 3 vols. (Beirut: Dār al-Kutub al-ʿIlmiyya, 1985), 1:259, 362–363, 2:315, 3:552.

30. See Abū l-Khayr Ṭaşköprīzāde, *Shaqāʾiq al-nuʿmāniyya fī ʿulamāʾ al-dawla al-ʿuthmāniyya* (Beirut: Dār al-Kitāb al-ʿArabī, 1975), 64; *Kashf*, 2:1437. See also Zeynep Buçukcu, "Mahmud bin Kadı-i Manyas'in A'cebü'l-'Üccab Adli Eserinin Transkripsiyon ve Dizini" (MA thesis, Hacettepe University, Ankara, 2017), 3a (dedication), 60a (magic squares), 73a–b, 78b–79a, 88b, 89a, 91b (cyphers and magical diagrams).

31. See Ibn Zunbul al-Maḥallī, *ʿAjāʾib al-makhlūqāt wa-gharāʾib al-mawjūdāt*, Harvard University, Houghton Library, MS Arab 263, fols. 90b–163b. For Ibn Zunbul, *Qānūn al-dunyā*, TPMK, Revan Köşkü 1638, see Rachel Milstein and Bilha Moor, "Wonders of a Changing World: Late Illustrated *ʿAjāʾib* Manuscripts," *Jerusalem Studies in Arabic and Islam* 32 (2006): 6–58.

32. See Palmira Brummett, *Ottoman Seapower and Levantine Diplomacy in the Age of Discovery* (Albany: SUNY Press, 1994); Giancarlo Casale, *The Ottoman Age of Exploration* (Oxford: Oxford University Press, 2010).

33. Edmundo O'Gorman, *La invención de América: El universalismo de la cultura de Occidente* (Mexico City: Fondo de Cultura Económica, 1958), 68–76, 90–92, 122–126, 151–159; Walter Mignolo, *The Darker Side of the Renaissance: Literacy, Territoriality, and Colonization* (Ann Arbor: University of Michigan Press, 1995), 259–269.

34. Alfonso García-Gallo, "Los sistemas de colonización de Canarias y América en los siglos XV y XVI," in *I Coloquio de Historia Canario-Americana* (Las Palmas: Cabildo Insular de Gran Canaria, 1977), 424–442.

35. James J. Parsons, "The Migration of Canary Islanders to the Americas: An Unbroken Current since Columbus," *Americas* 39, no. 4 (1983): 447–481, at 452–454.

36. Surekha Davies, *Renaissance Ethnography and the Invention of the Human: New Worlds, Maps and Monsters* (Cambridge: Cambridge University Press, 2016), 183–216.

37. Mercedes García-Arenal, "Moriscos e indios: Para un estudio comparado de métodos de conquista y evangelización," *Chronica nova* 20 (1992): 153–175, at 162; William Mejías-López, "Hernán Cortés y su intolerancia hacia la religión azteca en el contexto de la situación de los conversos y moriscos," *Bulletin Hispanique* 95, no. 2 (1993): 623–646.

38. See Juan de Ovando (d. 1575), "De los judíos luteranos, moros, conversos, y moriscos," in *Gobernación espiritual y temporal de las Indias*, in *Colección de documentos inéditos de ultramar*, 25 vols. (Madrid: Real Academia de la Historia, 1885–1932), 20:169–174; Esteban Mira Caballos, "Los prohibidos en la emigración a América (1492–1550)," *Estudios de historia social y económica de América* 12 (1995): 37–53. Compare with the Portuguese use of deportment, Geraldo Pieroni, "Outcasts from the Kingdom: The Inquisition and the Banishment of New Christians to Brazil," in *The Jews and the Expansion of Europe to the West, 1450–1800*, ed. Paolo Bernardini and Norman Fiering (New York: Berghahn Books, 2001), 242–251.

39. For the bull *Inter caetera* (1493), see Francisco López de Gómara, *Historia general de las Indias*, 2 vols. (Madrid: Calpe, 1922), 1:49–154. See also James Muldoon, "Papal Responsibility for the Infidel: Another Look at Alexander VI's 'Inter Caetera,'" *Catholic Historical Review* 64, no. 2 (1978): 168–184.

40. See Juan de Ribera (d. 1611), *Catechismo para instrucción de los nuevamente convertidos de moros* (Valencia: Pedro Patricio Mey, 1599), 85–86, 120, 155; Jaime Bleda (d. 1622), *Coronica de los moros de España: Diuida en ocho libros* (Valencia: Felipe Mey, 1618), 52. Compare Karoline Cook, *Forbidden Passages: Muslims and Moriscos in Colonial Spanish America* (Philadelphia: University of Pennsylvania Press, 2016), 31nn103–104; Lino Gómez Canedo, "¿Hombres y bestias? Nuevo examen de un viejo tópico," *Estudios de historia novohispana* 1 (1966): 29–51.

41. Cook, *Forbidden*, 1–9; Michael Gomez, "Muslims in Early America," *Journal of Southern History* 60, no. 4 (1994): 671–710.

42. Piri Reis, *Kitāb-ı Baḥriyye*, Süleymaniye, Aya Sofya 2612, fol. 3a (copied 932/1525~), published as *Kitab-ı Bahriye*, ed. Ertuğrul Zekâi Ökte, trans. Vâhid Çabuk, Tülây Duran (Turkish text) and Robert Bragner (English text), 4 vols. (Istanbul: Historical Research Foundation, 1988); Svatopluk Soucek, *Piri Reis and Turkish Mapmaking after Columbus: The Khalili Portolan Atlas* (London: Nour Foundation, 1992), 49–50; Giancarlo Casale, "Did Alexander the Great Discover America? Debating Space and Time in Renaissance Istanbul," *Renaissance Quarterly* 72, no. 3 (2019): 863–909. See also Emiralioğlu, *Knowledge*, 94–103, 121–124, 132–135.

43. For this interpretation, see Casale, *Ottoman Age*, 23. More broadly, see *Anatolia to Aceh: Ottomans, Turks and Southeast Asia*, ed. Andrew Peacock and Annabel Teh Gallop (Oxford: Oxford University Press, 2015).

44. See Bartolomé de las Casas, *Historia de las Indias*, ed. Miguel Ángel Medina, vols. 3–5 of *Obras completas*, ed. Paulino Castañeda Delgado, 14 vols. (Madrid: Alianza, 1988–1999), 1:582, chap. 45; this is based on his transcription of the ship log, de las Casas, *Diario del primer y tercer viaje de Cristóbal Colón*, ed. Consuelo Varela, vol. 14 of *Obras completas*, 75 (1 de noviembre). See also Abbas Hamdani, "Columbus and the Recovery of Jerusalem," *Journal of the American Oriental Society* 99, no. 1 (1979): 39–48, 43. Sanjay Subrahmanyam, *The Career and Legend of Vasco Da Gama* (Cambridge: Cambridge University Press, 1998), 81, 114–115, 133.

45. Piri Reis, *Baḥriyye*, fol. 41a–41b. On cartographic material that may well have derived from Columbus or his associates, see Paul Kahle, "A Lost Map of Columbus," *Geographical Review* 23, no. 4 (1933): 621–638; Gregory McIntosh, *The Piri Reis Map of 1513* (Athens: University of Georgia Press, 2000), 131–140. See also Casale, "Alexander," 873.

46. On Columbus's belief that he had indeed reached Asia, see de las Casas, *Historia*, 1:552, chap. 33, compare at 398–402, chap. 12; 2:1379, chap. 20; compare de las Casas, *Diario*, 41, 73–74 (30 de octubre), etc. More broadly, see Pauline Moffitt Watts, "Prophecy and Discovery: On the Spiritual Origins of Christopher Columbus's 'Enterprise of the Indies,'" *American Historical Review* 90, no. 1 (1985): 73–102; Valerie Irene Jane Flint, "The Medieval World of Christopher Columbus," *Parergon* 12, no. 2 (1995): 9–27.

47. TPMK, Revan Köşkü 1633 mük, published in facsimile edition as Yusuf Akçura, ed., *Pîrî Reis Haritası* (Ankara: Türk Tarih Kurumu, 1999), transcribed on foldout, §6; Kahle "Lost," 624, §16; Ayşe Afetinan, *The Oldest Map of America, Drawn by Pirî Reis*, trans. Leman Yolaç (Ankara: Türk Tarih Kurumu, 1954), 31, §6; McIntosh, *Piri*, 15–16, §6.

48. Akçura, *Pîrî*, §5; Kahle, "Lost," 624–626, §15; Afetinan, *Map*, 28–32, §5; McIntosh, *Piri*, 70–71, §5.

49. See Piri Reis, *Baḥriyye*, fol. 39b–40a. See Kahle, "Lost," 626nn9–10; Soucek, "Islamic Charting," 270; McIntosh, *Piri*, 72–73, 114, §3.

50. For his theory that Muslims first discovered the Americas, Sezgin purports to identify the Río de la Plata estuary in the Piri Reis map, *GAS*, 13:142–143 (trans., 4:134–135). For a rejection of the identification, originally proposed by Kahle, see McIntosh, *Piri*, 38, 41.

51. Akçura, *Pîrî*, §14; Kahle, "Lost," 636, §1; Afetinan, *Map*, 32–33, §14; McIntosh, *Piri*, 32–33. On *yāpāmūndo*, see Casale, "Alexander," 873, 881. For the sea-beast, see Clara Strijbosch, *The Seafaring Saint: Sources and Analogues of the Twelfth Century*, trans. Thea Summerfield (Dublin: Four Courts, 2000), 51–54, 195–196. Compare the Easter feast with the Nowruz celebration in Buzurg ibn Shahriyār, *ʿAjāʾib*, 36–38.

52. Akçura, *Pîrî*, §19; Kahle, "Lost," 638, §8; Afetinan, *Map*, 33, §19; McIntosh, *Piri*, 33, §19.

53. Akçura, *Pîrî*, §15; Kahle, "Lost," 624, §15; Afetinan, *Map*, 29, §5; McIntosh, *Piri*, 70, §5.

54. Piri Reis, *Baḥriyye*, fols. 41a–42a. See also Soucek, "Islamic Charting," 277n42; McIntosh, *Piri*, 73.

55. Recorded in Bartolomé de las Casas, *Diario*, 191.

56. Bartolomé de las Casas, *Historia*, 1:396–398, chap. 11.

57. See Pierre d'Ailly, *Ymago mundi*, ed. Edmond Buron, 3 vols. (Paris: Maisonneuve frères, 1930), 1:210; see also William Randles, "Classical Models of World Geography and Their Transformation Following the Discovery of America," in *The Classical Tradition and the Americas*, ed. Wolfgang Haase and Meyer Reinhold (Berlin: De Gruyter, 1994), 5–76, at 43–49.

58. See James Romm, "New World and '*novos orbes*': Seneca in the Renaissance Debate over Ancient Knowledge of the Americas," in Haase and Reinhold, *The Classical Tradition and the Americas*, 77–116.

59. Akçura, *Pîrî*, §6; Kahle, "Lost," 624, §16; Afetinan, *Map*, 31, §6; McIntosh, *Piri*, 15, §6. See *Buldān*, 1:25.

60. See *Kashf*, 1:310, 928. On the authorship, see Baki Tezcan, "The Many Lives of the First Non-Western History of the Americas: From the *New Report* to the *History of the West Indies*," *Osmanlı Araştırmaları/Journal of Ottoman Studies* 40 (2012): 1–38, at 33–34.

61. See Thomas Goodrich, *The Ottoman Turks and the New World: A Study of Tarih-i Hind-i garbi and Sixteenth-Century Ottoman Americana* (Wiesbaden: Harrassowitz, 1990), 28.

62. See Goodrich, *Ottoman*, 31–38.

63. *Tārīh-i yeñi dünyā*, Newberry Library, Chicago, MS Ayers 612, fol. 18b; English translation, Goodrich, *Ottoman*, 120. See Casale, "Alexander," 895–896. For the date of MS Ayers 612, see Goodrich, *Ottoman*, 22; compare Tezcan, "Lives," 15n42.

64. M 72a–b (W 130–131). See also Kazue Kobayashi, "The Illustration of the Old Man of the Sea and the Story of Sindbad the Sailor: Its Iconography and Legendary Background," *Senri Ethnological Studies* 55 (2001): 101–119.

65. Ibn al-Wardī, *Kharīda*, 151–152. Qazwīnī quotes an otherwise unidentified Samarqandī, M fol. 58a (W 105–106). See Shams al-Dīn al-Dimashqī (d. 727/1327), *Nukhbat al-dahr fī ʿajāʾib al-barr wa-l-baḥr*, published as *Cosmographie de Chems-ed-Din Abou Abdallah Mohammed ed-Dimichqui*, ed. August Ferdinand Mehren and Christian Martin Fraehn (Saint Petersburg: Académie impériale des sciences, 1866), 134–136. Yūsuf Hādī equates Samarqandī with Saʿīd ibn al-Ḥasan al-Samarqandī, an early authority on the Turks in Ibn al-Faqīh, *Kitāb al-Buldān*, ed. Yūsuf Hādī (Beirut: ʿĀlam al-Kutub, 1996), 6, 34–37, 643.

66. For criticism of Ibn al-Wardī and Masʿūdī, see *Tārīh*, MS Ayers 612, fol. 18a–b; translation, 110–111.

67. *Tārīh*, MS Ayers 612, fol. 17a–b; translation, 104; see Tezcan, "Lives," 6.

68. See BnF, Supplément turc 242, fol. 141a (colophon, copied 990/1582~), published as a facsimile edition, with translation and commentary, as *Book of Felicity*, ed. Mónica Miró and trans. Yorgos Dedes (Barcelona: M. Moleiro Editor, 2007). Suʿūdī draws extensively from the *Kitāb al-Bulhān*, Bodl., Or. 133. See Stefano Carboni, "The 'Book of Surprises' (*Kitab al-bulhan*) of the Bodleian Library," *La Trobe Journal* 91 (2013): 22–34. For Murād and occult learning, see Özgen Felek, "Fears, Hopes, and Dreams: The Talismanic Shirts of Murād III," *Arabica* 64, no. 3/4 (2017): 647–672; Özgen Felek, "(Re)creating Image and Identity Dreams and Visions as a

Means of Murād III's Self-Fashioning," in *Dreams and Visions in Islamic Societies,* ed. Özgen Felek and Alexander Knysh (Albany: SUNY Press, 2012), 249–272.

69. On the "merman," see Goodrich, *Ottoman,* 59–60, 198–204; Tezcan, "Lives," 7–8.

70. E.g., the header "Section on the Sea of Magellan and Its Wonders," has no direct analogue in the European sources; *Tārīh,* MS Ayers 612, fol. 58b; translation, 206; Harvard Art Museums, Acc. 1985.270 (Persian translation), *Tārīkh-i yangī dunyā,* f. 27b. More broadly, see Sanjay Subrahmanyam, "Monsters, Miracles and the World of *'Aja'ib-o-Ghara'ib:* Intersections between the Early Modern Iberian and Indo-Persian Worlds," in *Naturalia, mirabilia y monstrosa en los imperios ibéricos,* ed. Eddy Stols et al. (Leuven: Leuven University Press, 2006), 275–306.

71. *Tārīh,* MS Ayers 612, fol. 46a; BnF, Supplément turc 521, fol. 30b; translation, 173.

72. Piri Reis, *Baḥriyye,* fol. 40a; see McIntosh, *Piri,* 44, §24.

73. Piri Reis, *Baḥriyye,* fol. 40a–b.

74. See Mohammad Hassan Khalil, *Islam and the Fate of Others: The Salvation Question* (Oxford: Oxford University Press, 2012); Richard Foltz, *Animals in Islamic Tradition and Muslim Cultures* (Oxford: Oneworld, 2006), 6.

75. See Lewis Hanke, "Pope Paul III and the American Indians," *Harvard Theological Review* 30, no. 2 (1937): 65–102; and Rolena Adorno, *The Polemics of Possession in Spanish American Narrative* (New Haven, CT: Yale University Press, 2007), 99–124.

76. See, for instance, Abū l-Ḥasan al-Ashʿarī, *Kitāb al-Lumaʿ = The Theology of al-Ashʿarī,* ed. and trans. Richard McCarthy (Beirut: Imprimerie catholique, 1953), 47, 71, §§107, 170. See also Ayman Shihadeh, *Teleological Ethics of Fakhr al-Dīn al-Rāzī* (Leiden: Brill, 2006), 146, 160–169; Gustave von Grunebaum, "Observations on the Muslim Concept of Evil," *Studia Islamica* 31 (1970): 117–134; Eric Ormsby, *Theodicy in Islamic Thought: The Dispute over al-Ghazālī's "Best of All Possible Worlds"* (Princeton, NJ: Princeton University Press, 1984), 16–31; Christian Lange, *Paradise and Hell in Islamic Traditions* (Cambridge: Cambridge University Press, 2016), 4–13.

77. See Fariba Zarinebaf, *Mediterranean Encounters: Trade and Pluralism in Early Modern Galata* (Oakland: University of California Press, 2017), 89–125; Stanford Shaw, *The Jews of the Ottoman Empire and the Turkish Republic* (New York: New York University Press, 1991), 1–108. See also Leigh T. I. Penman, "The Hidden History of the Cosmopolitan Concept: Heavenly Citizenship and the Aporia of World Community," *Journal of the Philosophy of History* 9, no. 2 (2015): 284–305.

78. See Casale, *Ottoman Age,* 160–163.

79. See Francisco López de Gómara, *La conquista de Mexico* (Saragossa: Agustin Millan, 1552), fol. 87a; compare Sebastian Covarrubias (d. 1613), *Tesoro de la Lengua Castellana o Española* (Madrid: Luis Sanchez, 1611), s.v. *tornadiço.* The Italian is explicit: "christiani nuovi," in Francisco López de Gómara, *Historia delle nuove Indie Occidentali,* trans. Agostino di Cravaliz (Venice: Appresso Pietro Bosello, 1560), fol. 215b.

80. *Tārīh,* MS Ayers 612, fol. 77b; BnF, Supplément turc 521, fol. 52b; translation, 253.

81. See Andrew C. Hess, "The Battle of Lepanto and Its Place in Mediterranean History," *Past and Present* 57 (1970): 53–73. For diplomatic exchanges, see Susan Skilliter, *William Harborne and the Trade with Turkey, 1578–1582: A Documentary Study of the First Anglo-Ottoman Relations* (Oxford: Oxford University Press 1977); Jonathan Burton, *Traffic and Turning: Islam and English Drama, 1579–1624* (Newark: University of Delaware Press, 2005), 57–68.

82. On Mīr ʿAlī Beg, see Casale, *Ottoman Age,* 152–179.

9: On the Edge

1. D fol. 213b; NLM, MS Persian 29, fol. 92a–b. See Nūr al-Dīn Jahāngīr, *Tūzuk-i Jahāngīrī = Jahāngīr-nāma,* ed. Muḥammad Hāshim (Tehran: Bunyād-i Farhang-i Īrān, 1359sh/1980~), 87, 114–115; translated as *The Jahangirnama: Memoirs of Jahangir, Emperor of India,* trans. Wheeler Thackston (Oxford: Oxford University Press, 1999), 99, 125; Muʿtamad Khān Mīrza Muḥammad, *Iqbālnāma-yi Jahāngīrī,* 3 vols., ed. ʿAbd al-Ḥayy and Aḥmad ʿAlī (Calcutta: Asiatic Society of Bengal, 1865), 1:35, 57–58.

2. See Corinne Lefèvre, *Pouvoir imperial et élites dans l'Inde moghole de Jahāngīr* (Paris: Les Indes savants, 2017), 109–144; Iqtidar Alam Khan, "The Matchlock Musket in the Mughal Empire: An Instrument of Centralization," *Proceedings of the Indian History Congress* 59 (1998): 341–359.

3. Thomas Roe, *The Embassy of Sir Thomas Roe to the Court of the Great Mogul,* ed. William Foster, 2 vols. (London: Hakluyt Society, 1899), 2:212–214, 224–226, 334. For the mission to the New World, see Michael Brown, *Itinerant Ambassador: The Life of Sir Thomas Roe* (Lexington: University Press of Kentucky, 1970), 12–17. See also Yael Rice, "Lines of Perception: European Prints and the Mughal *Kitābkhāna,*" in *Prints in Translation, 1450–1750: Image, Materiality, Space,* ed. Suzanne Schmidt and Edward Wouk (London: Routledge, 2017), 203–223.

4. Jahāngīr, *Tūzuk,* 5, 123, 199; trans. 24, 133, 206. On the *payrūj,* see Jamāl al-Dīn Injū, *Farhang-i Jahāngīrī,* ed. Raḥīm ʿAfīfī, 2 vols. (Mashhad: Dānishgāh-i Mashhad, 1972–1973), 2:693. For Roe in the Mughal court, see Supriya Gandhi, *The Emperor Who Never Was: Dara Shukoh in Mughal India* (Cambridge, MA: Harvard University Press, 2019), 19–26.

5. See Lee Siegel, *Net of Magic: Wonders and Deceptions in India* (Chicago: University of Chicago Press, 1991); Michael Slouber, *Early Tantric Medicine: Snakebite, Mantras, and Healing in the Garuda Tantras* (Oxford: Oxford University Press, 2017).

6. See ʿAbd al-Razzāq al-Qāshānī (d. ca. 730/1329), *Sharḥ Fuṣūṣ al-ḥikam* (Qom: Intishārāt Bīdār, 1370sh/1991~), 82, 119, 299; compare Ibn Turka (d. 835/1432), *Sharḥ Fuṣūṣ al-ḥikam,* ed. Muḥsin Bīdārfar, 2 vols. (Qom: Intishārāt Bīdār, 1420/1999~), 1:66–67, 252, 2:622; ʿAbd al-Raḥmān Jāmī (d. 898/1492), *Naqd al-nuṣūs fī sharḥ Naqsh al-fuṣūs,* ed. William Chittick and Jalāl al-Dīn Āshtiyānī (Tehran: Anjuman-i Shāhanshāhī-yi Falsafa-yi Īrān, 1977), 29, 39, 65–73. See also William Chittick, "*Waḥdat al-Wujūd* in India," *Ishraq: Islamic Philosophy Yearbook* 3 (2012): 29–40. For the perfect human, see Chittick, *The Sufi Path of Knowledge: Ibn Al-Arabi's Metaphysics of Imagination* (Albany: SUNY Press, 1989), 27–30, 46–47, 366–372.

7. For examples of Persian copies in India, see Rampur, Raza Library, MSS Misc. 50 and Misc. 51, described in Barbara Schmitz and Ziyauddin Desai, *Mughal and Persian Paintings and Illustrated Manuscripts in the Raza Library* (New Delhi: Indira Gandhi National Centre for the Arts, 2006), 83–85 (I.3), 90–95 (II.1), 113 (III.6), 129 (III.27); Salar Jung Museum, Hyderabad, MSS Geo. 4–8. For Heidelberg, Völkerkundemuseum, MS P.St.P IV 8, dated 944/1537~ (?), but likely produced later, see Brijinder Nath Goswami and Anna Libera Dahmen-Dallapiccola, "The Wonders of Creation: An Illustrated Manuscript of Qazwini's *Ajaib-al-Makhluqwat* from Gujarat in the E & J Von Portheim Shiftung [*sic*] Heidelberg," *Roopa-Lekha* 50, no. 1/2 (1978–1979): 16–29. See also *Catalogue of the Arabic and Persian Manuscripts in the Khuda Bakhsh Oriental Public Library,* 42 vols. (Patna: The Library, 1970–2006), 7:193–194, MS 634, fols. 1b–157b; Srinagar, Kashmir University, MS 205, fol. 2b; NLM, MSS P2, P3, P29; New Delhi, National

Museum, MS 57.26.2. For further, see Vivek Gupta, "Wonder Reoriented: Manuscripts and Experience in Islamicate Societies of South Asia (ca. 1450–1600)" (PhD diss., University of London, SOAS, 2020), 268–270. On expanding book culture, see Matthew Melvin-Koushki, "*Taḥqīq* vs. *Taqlīd* in the Renaissances of Western Early Modernity," *Philological Encounters* 3 (2018): 193–249, at 228–230.

8. For Arabic manuscripts, see Rampur, Raza Library, MS A. 4601 (copied 1000/1591~) and MS A. 4600 (copied 979/1571), possessing the same colophon date and cycle of illustrations in Brit., I.O. Islamic 845 (copied 979/1571), I.O. Islamic 1377, and Or. 4701. The earlier cycle behind these copies is discussed in Julie Badiee, "An Islamic Cosmography: The Illustrations of the Sarre Qazwīnī" (PhD diss., University of Michigan, 1978), 331–335. Badiee's identification of the copyist Ibn Kamāl Ḥusayn (var. al-Ḥusayn ibn Kamāl al-Dīn) is not followed here. For more, see Gupta, "Wonder," 116–128. See also the later eighteenth-century Arabic copy, Cambridge, MA, Harvard Art Museums, Acc. 1972.3. For the seal of Amānat Khān, see *Arts of the Islamic World: Wednesday 11 October 2006* (London: Sotheby's, 2006), 35 (Lot 31).

9. For early attitudes on the *farang,* see Rudi Matthee, "Between Aloofness and Fascination: Safavid Views of the West," *Iranian Studies* 31, no. 2 (1998): 219–246; Sanjay Subrahmanyam, "Taking Stock of the Franks: South Asian Views of Europeans and Europe, 1500–1800," *Indian Economic and Social History Review* 42, no. 1 (2005): 69–100.

10. Fakhr al-Dīn Āzarī [Ādharī], *Gharā'ib al-dunyā,* Baltimore, The Walters Art Museum, MS W. 652, fol. 214a (colophon). For further biographical details and manuscripts, see Storey, 2:126–127; A. ʿA. Rajā'ī, "Āẕarī (Āḏarī) Ṭūsī," *EIr.*

11. Ādharī, "Kīmiyā'-i sīmiyā' o-nayranjāt/rahzanānd dar ṭarīq-i najāt//Ay vujūd-at ṭilism-i ganj-i vujūd/nīst ghayr az tū kīmiyā'-i mawjūd," and "ān gharā'ib hama ishārāt-ast/v-ān ʿajā'ib hama ʿibārāt-ast," *Gharā'ib-i dunyā,* Leipzig, Universitätsbibliothek, Vollers 937, fols. 2b (alchemy), 3a (wonders as allegories), 5a–b (definitions); MS W. 652, fols. 3a–b, 6a–b.

12. See Brit., I.O. Islamic 78; Bodl., Ouseley 48 and Elliot 47; BnF, Supplément persan 1148; KMSI 7694. See also *Kashf,* 2:1126. On illuminated adaptations of Qazwīnī's natural history during this period, see Karin Rührdanz, "Qazwīnī's *ʿAjā'ib al-Makhlūqāt* in Timurid Manuscripts," in *Iran, questions et connaissances: Actes du IVe Congrès européen des études iraniennes,* 3 vols. (Paris: Association pour l'avancement des études iraniennes, 2002), 2:473–484; Karin Rührdanz, "Between Astrology and Anatomy: Updating Qazwīnī's *ʿAjā'ib al-makhlūqāt* in Mid-Sixteenth-Century Iran," *Ars Orientalis* 42 (2012): 56–66.

13. Jahāngīr, *Tūzuk,* 323, trans. 319. See Qazwīnī, *ʿAjā'ib,* Brit., Or. 12,220, fol. 271a (copied 909/1503 in Herat); G. M. Meredith-Owens, "A New Illustrated Manuscript of the *ʿAjā'ib al-Makhlūqāt,*" *British Museum Quarterly* 23, no. 1 (1960): 67–68.

14. Freer Gallery of Art, Washington, DC, F1945.9a.

15. Gerard Mercator and Jodocus Hondius, *Atlas sive cosmographicae meditationes de fabrica mundi et fabricati figura,* 4th ed. (Amsterdam: Iudoci Hondij, 1613), "China." The Latin caption and image are drawn from the atlas of the Dutch cartographer Abraham Ortelius (d. 1598), *Theatrum Orbis Terrarum* (Antwerp: Christoffel Plantin, 1584), "Chinae." For more, see Johannes Keuning, "The History of an Atlas: Mercator-Hondius," *Imago Mundi* 4, no. 1 (1947): 37–62. Roe, *Embassy,* 2:413–414, 416–417.

16. See Irfan Habib, "Cartography in Mughal India," *Indian Archive* 28 (1979): 88–105; Sumathi Ramaswamy, "Conceit of the Globe in Mughal Visual Practice," *Comparative Studies in*

Society and History 49, no. 4 (2007): 751–782; Ebba Koch, "The Symbolic Possession of the World: European Cartography in Mughal Allegory and History Painting," *Journal of the Economic and Social History of the Orient* 55, no. 2/3 (2012): 547–580.

17. Description of Mughal wall map of the world drawn from *Catalogue of Historic Documents in Kapad Dwara, Jaipur,* 2 vols. (Jaipur: Jaigarh Public Charitable Trust, 1988–1990), 2:144, painted on cloth, ca. 1620, size 259 × 153 cm. The map is in the estate of Raja Jagat Singh of Jaipur (d. 1997), which remains under litigation. According to the catalogue, it is painted in two hemispheres with longitudes and latitudes and India shown prominently in yellow, Indian cities identified in the Nagari script, a depiction of the Wall of China, and aquatic animals in the Pacific.

18. See Freer Gallery of Art, Washington, DC, F1948.19a; compare Los Angeles County Museum of Art, M.75.4.28.

19. Abū l-Fazl [Faḍl] al-ʿAllāmī (d. 1101/1602), *Āʾīn-i Akbarī*, ed. Heinrich Blochmann, 2 vols. (Calcutta: Baptist Mission Press, 1872–1877), 2:26. See Alexander and the sea of darkness from Qazwīnī's *ʿAjāʾib* in the encyclopedia of Mughal courtier Ṣāliḥ Ṣādiqī (d. 1061/1651), *Shāhid-i ṣādiq*, Brit., Egerton 1016, fol. 331b; a marginal note from the early nineteenth century identifies the story with America. See also Murtaḍā Ḥusayn Bilgrāmī (d. ca. 1210/1795), *Ḥadīqat al-aqālīm*, lithographed ed. (Lucknow: Nawal Kishore, 1296/1879), 503.

20. For conquistadors as deities, see Camilla Townsend, "Burying the White Gods: New Perspectives on the Conquest of Mexico," *American Historical Review* 108, no. 3 (2003): 659–687.

21. See *Āthār*, 143, 214. In the context of Akbar's court, see also Qāḍī Aḥmad Tattavī (d. 996/1588) et al., *Tārīkh-i alfī*, ed. Ghulām-Riḍā Ṭabāṭabāʾī Majd, 8 vols. (Tehran: Intishārāt-i ʿIlmī wa-Farhangī, 1382sh/2003~), 3:1958.

22. ʿAllāmī, *Āʾīn-i Akbarī*, 2:193 (Adam), 196–200 (Alexander), trans. 3:359–360, 365–373; *Tārīkh-i alfī*, 1:380–381 (Alexander's Indian dialogues). See also Gandhi, *Emperor*, 18, 25–26, 28, 57, 137, 203, 216, 269n62. For Ottoman uses, see Giancarlo Casale, "Did Alexander the Great Discover America? Debating Space and Time in Renaissance Istanbul," *Renaissance Quarterly* 72, no. 3 (2019): 863–909, 864n3, 866–868, 871.

23. ʿAllāmī, *Āʾīn-i Akbarī*, 2:226, trans. 3:423 (*ṣulḥ-i kull*). See also *Tārīkh-i alfī*, 3:1524 (injunctions of Aristotle to Alexander). The gnomology ascribed to Aristotle shaped the reception of Alexander in Islamic letters; see *Kitāb al-Aḥwāl wa-l-akhbār al-Iskandariyya*, published as *The Correspondence between Aristotle and Alexander the Great*, ed. and trans. Miklós Maróth (Piliscsaba: The Avicenna Institute of Middle Eastern Studies, 2006), 12–19 (Arabic text). For the Mughal policy of religious tolerance, see Rajeev Kinra, "Handling Diversity with Absolute Civility: The Global Historical Legacy of Mughal *Ṣulḥ-i Kull*," *Medieval History Journal* 16, no. 2 (2013): 251–295; Rajeev Kinra, "Revisiting the History and Historiography of Mughal Pluralism," *ReOrient* 5, no. 2 (2020): 137–182.

24. ʿAbd al-Qādir Badāʾūnī, *Muntakhab al-tawārīkh*, ed. Mawlawī Aḥmad ʿAlī Ṣāḥib and Tawfīq Hāshim Pūr Subḥānī, 3 vols. (Tehran: Anjuman-i Āthār wa-Mafākhir-i Farhangī, 1379–1380sh/2000–2001), 2:177 (court), 2:202–203 (organ from Farangistān).

25. *Razm-nāma* = *Mahābhārata*, ed. Muḥammad Riḍā Jalālī Nāʾīnī and Narayan S. Shukla, 4 vols. (Tehran: Kitābkhāna-yi Ṭahūrī, 1358–1359sh/1979–1980), 1:29 (Garuda, encircling ocean, water of life), 2:7–8 (Qāf, seven climes); compare *Mahābhāratam*, with the Commentary of Nīlakaṇṭha, ed. Ramachandrashastri Kinjawadekar et al., 7 vols. (Pune: Shankar Narhar

Joshi, 1929–1936), 1:76, 78, book 1, Ādi Parvan, chap. 29, ll. 41 (at the ocean), 43, 45, 51 (*amrita*); 3:15, book 6, Bhishma Parvan, chap. 6, ll. 42–53 (Kailash, Meru, and seven climes). For Khiḍr and Alexander, see ʿAllāmī, *Āʾīn-i Akbarī*, 2:224. For wonder in the Persian reception of the *Mahābhārata* and the *Rāmāyana*, see Audrey Truschke, *Culture of Encounters: Sanskrit at the Mughal Court* (New York: Columbia University Press, 2016), 109–110, 137–138, 210–211.

26. *Razm-nāma*, 1:236, 3:105, 141–142, 183–185 (monotheism, gnosticism, liberation), 1:231 (*karṇaprāvaraṇa* = *galīm pūshān*, blanket-clad; *ekapāda* = *yak pā*, one-footed), 4:292–294 (kingdom of women). See broadly, Truschke, *Culture*, 121–133; Bacil Kirtley, "The Ear-Sleepers: Some Permutations of a Traveler's Tale," *Journal of American Folklore* 76, no. 300 (1963): 119–130. For astonishment (*camatkāra*) and the marvelous (*adbhuta*) in Sanskrit aesthetic theory, see David Shulman, *More than Real: A History of the Imagination in South India* (Cambridge, MA: Harvard University Press, 2012), 63, 69, 99, 244, 247–248.

27. For illustrations accompanying the *Razm-nāma*, see Asok Kumar Das, *Paintings of the Razmnama* (Usmanpura, Ahmedabad: Mapin, 2005); Yael Rice, "A Persian Mahabharata: The 1598–1599 *Razmnama*," *Mānoa* 22, no. 1 (2010): 125–131. Several have confused scattered folios of the *Bābur-nāma* in CBL In. 06 with *Wonders and Rarities*, on which see Linda Leach, *Mughal and Other Indian Paintings from the Chester Beatty Library*, 2 vols. (London: Scorpion Cavendish, 1995), 1:117–131. For the *Yog Vashisht*, CBL In. 05, see Leach, *Mughal*, 1:155–195. For the book of the zodiac and talismans, Rampur, Raza Library, Album no. 2, see Schmitz and Desai, *Mughal and Persian Paintings*, 20–27, plates 13–19; identified with selections from the Persian translation produced in the Delhi Sultanate of Rāzī's *al-Sirr al-maktūm* by Francis Richard, "Une page du recueil de talismans et de signes zodiacaux d'Akbar dans la collection Gentil," *Artes Asiatiques* (1986): 116–117; discussed in Yael Rice, "Cosmic Sympathies and Painting at Akbar's Court," *Marg* 68, no. 2 (2016): 88–99.

28. For the early translation of Rāzī's *al-Sirr al-maktūm*, see BnF, Supplément persan 384, fols. 1b–2b. On later observatories, see *Observ.*, 258–305, 358–361. See also Sāmir ʿAkkāsh, *Marṣad Isṭanbūl* (Beirut: al-Markaz al-ʿArabī li-l-Abḥāth wa-Dirāsat al-Siyāsāt, 2017), 51–83.

29. For CBL In. 02 and In. 54, see Leach, *Mughal*, 2:819–889, 891–903. For a third copy, London, Wellcome Collection, MS Persian 373, see Emma Flatt, "The Authorship and Significance of the *Nujūm al-ʿulūm*: A Sixteenth-Century Astrological Encyclopedia from Bijapur," *JOAS* 131, no. 2 (2011): 223–244. See also Flatt, *The Courts of the Deccan Sultanates: Living Well in the Persian Cosmopolis* (Cambridge: Cambridge University Press, 2019), 210–267.

30. *Nujūm al-ʿulūm*, CBL In. 02, fols. 114a–1126b, 221a–226b. Much of the manuscript is out of order, this treatise included. The later section (fols. 221a–226b) is actually the opening of the treatise, on conjuring or summoning spirit beings, which starts with a general description of the practice of preparing the *mandal* and opens with directions for summoning the astral spirit of the Sun.

31. See Manfred Mayrhofer, *Etymologisches Wörterbuch des Altindoarischen*, 3 vols. (Heidelberg: C. Winter, 1986–2001), 2:294 (*máṇḍala-*).

32. See Kazuyo Sakaki, "Yogico-Tantric Traditions in the *Ḥawḍ al-Ḥayāt*," *Journal of the Japanese Association for South Asian Studies* 17 (2005): 135–156, at 148, 152; Carl Ernst, *Refractions of Islam in India: Situating Sufism and Yoga* (New Delhi: Sage, 2016), 208, 227–228 (chart. 10.7), 315 (§3.1).

33. Description drawn from the table of contents in the *Nujūm al-ʿulūm*, fols. 2b–4b. Only a fraction of the topics promised survive: fols. 4b–14a (seven heavens and angelic inhabitants), 14a–58a (the seven planets, constellations, stations, and degrees), 58b–97a (on the *Kitāb-i Vārshīk*, a Sanskrit collection of astrology), 98a–113a (Apollonius, *Mamālik al-mulūk*), 114a–126b, 221a–226b (*Taskhīrāt* of Muḥammad ibn Sirāj al-Dīn Sakkākī), 127a–171a (astrological arithmetic), 171b–221b, 227a–229b, 238a–313a (sections on Indic astrological arts of the *chakr*), 230a–237b (forts and battle formations), 313b–334a (horses), 334a–338b (elephants), 338b–348b (weapons).

34. *Nujūm al-ʿulūm*, "kuntu kanzan makhfiyyan fa-aḥbabtu an uʿrifa fa-khalaqtu l-khalqa," fol. 4b.

35. See Flatt, *Courts*, 224–229, 255; see Sakaki, "Yogico-Tantric Traditions," 139.

36. *Nujūm*, fols. 11a (citing *ʿAjāʾib al-makhlūqāt*), 14a–21b, 54a–58b (northern and southern constellations). The passages overlap directly with the Persian version produced for Ibrāhīm ʿĀdil Shāh; see D fols. 40a–48a, 52b–58b. General parallels have been noted between these illustrations in the *Nujūm* and in Qazwīnī's *Wonders*, though without attention to the actual textual interconnections. See Leach, *Mughal*, 2:826–827; Deborah Hutton, *Art of the Court of Bijapur* (Bloomington: Indiana University Press, 2006), 58; Flatt, *Courts*, 248.

37. See Roy Fischel, "*Ghariban* in the Deccan: Migration, Elite Mobility, and the Making and Unmaking of an Early Modern State," in *Iran and the Deccan: Persianate Art, Culture, and Talent in Circulation, 1400–1700*, ed. Keelan Overton (Bloomington: Indiana University Press, 2020), 127–144.

38. See Sanjay Subrahmanyam, "An Infernal Triangle: The Contest between Mughals, Safavids and Portuguese, 1590–1605," in *Iran and the World in the Safavid Age*, ed. Willem Floor and Edmund Herzig (London: I.B. Tauris, 2012), 103–130.

39. A. A. Kadiri, "Adil Shahi Inscriptions from Panhala," in *Epigraphia Indica, Arabic and Persian Supplement, 1968*, ed. Z. A. Desai (Calcutta: Government of India Press, 1969): 64–79, at 69–71, plate 18a; ʿUmar Khayyām, "dar dil natavān dirakht-i andūh nishānad/hamvāra kitāb-i khurramī bayād khwānad/may bayād khwurd o-kām-i dil bayād rānad/paydāst ki chand dar jahān khwāhī mānad," in *Ṭarabkhāna = Rubāʿiyyāt-i Khayyām*, ed. Jalāl al-Dīn Humāʾī (Tehran: Anjuman-i Āthār-i Millī, 1342sh/1964), 57, §207.

40. D fols. 7b–8a.

41. For copies of the translation produced for Ibrāhīm ʿĀdil Shāh, see Hyderabad, Salar Jung Museum, MS Geo. 6 (Acc. 3948), 3; Srinagar, Kashmir University, MS 205, 2b; New Delhi, National Museum, MS 58.48, fols. 1b–2a (unfoliated), with a forged purchased date of 977/1569~ at the end the colophon; NLM, MS P29, lacunose in the opening. See also the following manuscripts: D fols. 7b–8a, copied before 1061/1650; Brit., Or. 373, copied apparently in 1205/1790~; Brit., I.O. Islamic 3243, fols. 4b–5a, copied in Lahore in 1854; KMSI 2242/36, fol. 2a–b, an illuminated copy produced in India, likely late eighteenth century. See also Berlin, Pergamonmuseum, Museum für Islamische Kunst, I. 9493; and Storey, 2:126 (d).

42. See Kamāl al-Dīn ibn Fakhr al-Dīn Jahrūmī, *Barāhīn-i qāṭiʿa*, KMSI 10678, fol. 2a–b (dedication to Ibrāhīm ʿĀdil Shāh II), an expanded Persian adaptation completed in 994/1586 of Shihāb al-Dīn al-Haythamī (d. 973/1566), *al-Ṣawāʿiq al-muḥriqa*, a polemical Sunni defense of first three caliphs. Compare Firishta, *Gulshān-i Ibrāhīmī = Tārīkh-i Firishta*, 2 vols. (Kanpur: Nawal Kishore, 1301/1883), 1:2 (royal encomia), 2:2 (Ottoman linage), partial

translation as *History of the Rise of the Mahomedan Power in India till the Year A.D. 1612*, trans. John Briggs, 4 vols. (London: Longman and Co., 1829), 3:4.

43. Firishta, *Gulshān*, 2:372–373, trans. 4:439–440.

44. See Subrahmanyam, "Taking Stock," 80–89; Muzaffar Alam and Sanjay Subrahmanyam, "Southeast Asia as Seen from Mughal India: Tahir Muhammad's 'Immaculate Garden' (ca. 1600)," *Archipel* 70 (2005): 209–237.

45. See Wārid, *ʿAjāʾib al-buldān*, section from *Mirʾāt-i wāridāt*, Bodl., Ouseley 213, fols. 6a–b (Java), 7a (Barṭāyīl, Palace Island, reference to *Wonders and Rarities*), 9b–12b (Baygū, in Burma), 12b–19b (Aceh), 22b–30a (*bilād-i farang*).

46. D fols. 6a (date of index to marginalia), 9a (treatise on medicaments), 19a (Sharaf al-Dīn Manerī), 22a (Brethren of Purity), 22b (division of the sciences), 30b (subjugation of planetary forces), 147a–149a (epistle on existence). See Ikhwān al-Ṣafāʾ, *Mujmal al-ḥikma*, ed. Īrāj Afshār (Tehran: Pizhūhishgāh-i ʿUlūm-i Insānī wa-Muṭālaʿāt-i Farhangī, 1996), 26–27. The epistle contains the opening to the commentary of Mīr Sayyid ʿAlī Hamadhānī (d. 786/1384) copied in Khwāja Muḥammad Pārsā (d. 822/1420), *Sharḥ Fuṣūṣ al-ḥikam*, ed. Jalīl Misgar-nizhād (Tehran: Markaz-i Nashr-i Dānishgāhī, 1366sh/1987), 5–11. The marginalia overlaps with several later copies, including: New Haven, Beinecke, Persian MSS +28; Brit., I.O. Islamic 3243, indexed in Hermann Ethé, *Catalogue of Persian Manuscripts in the Library of the India Office* (Oxford: H. Hart, 1903), 369–374, §714; and NLM, MS Persian 29, described by Emile Savage-Smith, *Islamic Medical Manuscripts at the National Library of Medicine*, hosted on www.nlm.nih.gov.

47. D fols. 15a (Socrates), 19b (divinatory passages by ʿAlī), 173a–b (Alexander and philosophers), 175a (Noah's progeny), 186a (people of the cave), 201b–211a (historical digest).

48. D fols. 60b–61a, quoting Dārā Shikoh, *Majmaʿ al-baḥrayn*, ed. Muḥammad Riḍā Jalālī Nāʾīnī (Tehran: Chāp-i Tābān, 1335sh/1956~), 41–45, drawn from the *Bhāgavata Purāṇa*; see Supriya Gandhi, "Mughal Self-Fashioning, Indic Self-Realization: Dārā Shikoh and Persian Textual Cultures in Early Modern South Asia" (PhD diss., Harvard University, 2011), 193–199 (commentary), 323–328 (translation).

49. D fols. 130b–131b (Bābā Lāl), 137b–139a (Gītā), 126a–b, 196a (Urdu passages). For Ṣūfī Sharīf, see Gandhi, *Emperor*, 8, 72, 194–195, 198.

50. See D fols. 99a (*fāl-nāma*), 113b–114b (on physiognomy), 127a–130a (Rūmī), 141a–142a (Naqshbandī path), 183a–185a (properties of women), 186a (people of the cave), 195a (*jafr*), 200a–202a (geomancy), 166b–167a, 212b, 222b, 234a–b, 237a, 242b, 247a, 250a–251b, 253b, 311a, 316a, 346b, 349b, 350a, 351a, 359b, 373b, 389a, 400a (cyphers, talismans, magic squares), 295a, 296a (plague), 294a (defeat an army), 295a (lengthening life), 295b (obtaining love), 363a, 383a (alchemy), 385b, 389b, 410a (increasing sexual potency), 340b (faculty of amazement unique to humans).

51. See Qazwīnī, *ʿAjāʾib*, Manchester, John Rylands Library, Persian MS 3, copied 1041/1632, inspected in Lucknow 1131/1719, first year of reign of Shāhjahān II, with partially illegible seal of Muʿizz al-Dawla dated 1162/1748~, and seal of Iltifāt ʿAlī Khān, noting it was gifted in 1167/1754, first year of the reign of ʿĀlamgīrī II, fol. 4a.

52. Qazwīnī, *ʿAjāʾib*, NLM, MS Persian 29, fol. 173a.

53. See Charles Raikes's note, pasted into the opening flyleaf of Qazwīnī, *ʿAjāʾib*, Brit., I.O. Islamic 3243.

CODA

1. Description drawn from: *Catalogue officiel, publié par ordre de la Commission impériale* (Paris: E. Panis, 1855), 504–506; *Promenades dans l'Exposition universelle de 1855* (Paris: Joel Cherbuliez, 1855), 161–162; *Album de l'Exposition universelle dédié à S. A. I. le prince Napoléon,* 3 vols. (Paris: Bureaux de L'Abeille impériale, 1856–1859), 1:182, 451–452; *Exposition universelle de 1855: Rapports du jury mixte international* (Paris: Imprimerie impériale, 1856), 53, 559, 575–576; T. Bouquillard, *Exposition universelle de 1855: Plan-guide du Palais de l'industrie* (Paris: E. Panis, 26 August 1855).

2. See Priti Joshi, "Miles Apart: The India Display at the Great Exhibition," *Museum History Journal* 9, no. 2 (2016): 136–152; Arthur MacGregor, *Company Curiosities: Nature, Culture and the East India Company, 1600–1874* (London: Reaktion Books, 2018), 192–203.

3. See Benjamin Gastineau, "Produits de l'orient," *L'écho de l'industrie de l'Exposition universelle de 1855,* no. 35 (8 July 1855): 1–2.

4. Qazwīnī, *ʿAjāʾib,* Brit., I.O. Islamic 3243, fol. 1a. See also Hermann Ethé, *Catalogue of Persian Manuscripts in the Library of the India Office* (Oxford: H. Hart, 1903), 36, §714; compare 199, §504, 200–201, §506.

5. See Dag Nikolaus Hasse, *Success and Suppression: Arabic Sciences and Philosophy in the Renaissance* (Cambridge, MA: Harvard University Press, 2016). For further context, see also *The "Arabick" Interest of the Natural Philosophers in Seventeenth-Century England,* ed. G. A. Russell (Leiden: E.J. Brill, 1994); G. J. Toomer, *Eastern Wisedome and Learning: The Study of Arabic in Seventeenth-Century England* (Oxford: Oxford University, 1996).

6. Samuel Bochart, *Hierozoicon: Sive bipertitum opus de animalibus sacrae scripturae,* 2 vols. (London: Tho. Royroft, 1663), 1: "Praefatio ad lectorem," unpaginated, xxiv (division of animals), lii (Arabic sources), lxii (Pliny), lxxviii (list of Arabic sources), 48 (fabulous nature of Qazwīnī's book), 831–832 (hyena), 860 (feral cat), 901 (camel), 934–939 (*unicornis*), 904–905 (*camelopardalis*); 2:312 (Pliny), 775 (crocodile), 811–812 (griffin), 845–846 (*nasnās;* proclivity toward fables), 847–848 (*miʿrāj* and *burāq*), 853 (*ʿanqāʾ*), 858–859 (*insān al-māʾ*). See also Pierre Ageron, "Dans le cabinet de travail du pasteur Samuel Bochart: L'érudit et ses sources arabes," in *Érudition et culture savante: De l'Antiquité à l'époque moderne,* ed. Véronique Sarrazin and François Brizay (Rennes: Presses universitaires de Rennes, 2015), 128–143.

7. See Johann Jakob Hofmann (d. 1706), *Lexici universalis historico-geographico-chronologico-poetico-philologici,* 4 vols. (Basil: Widerhold, 1683), 1:148, 930; 3:230, 339, 691, 759, 770, 4:ix, 12; Filippo Buonanni (d. 1725), *Observationes circa viventia . . . cum Micrographia curiosa* (Roma: D.A. Herculis, 1703), 35–36 (in *Micrographia curiosa*); Johannes Jacobus Scheuchzer (d. 1733), *Physica sacra,* 4 vols. (Augsburg and Ulm: n.p., 1731–1735), 1:825, 957; Friedrich Christian Lesser (d. 1754), *Insecto-theologia* (Frankfurt: M. Blochberger, 1738), 261.

8. See Gottfried Hagen, *Ein osmanischer Geograph bei der Arbeit: Entstehung und Gedankenwelt von Kātib Čelebis Ǧihānnümā* (Berlin: Schwarz, 2003), 251–274, 291–297. See also Pınar Emiralioğlu, "The Ottoman Enlightenment: Geography and Politics in the Seventeenth- and Eighteenth-Century Ottoman Empire," *Medieval History Journal* 22, no. 2 (2019): 298–320.

9. Barthélemy d'Herbelot, *Bibliothèque orientale* (Paris: Compagnie des libraires, 1697), title page and 69. See also Nicholas Dew, *Orientalism in Louis XIV's France* (Oxford: Oxford University Press, 2009), 168–204.

10. See Paulo Lemos Horta, *Marvellous Thieves: Secret Authors of the Arabian Nights* (Cambridge, MA: Harvard University, 2017), 17–87.

11. Jonathan Scott (d. 1829), *The Arabian Nights Entertainments,* 6 vols. (London: Longman, Hurst, Rees, Orme and Brown, 1811), 1:i; Duncan Macdonald, "Concluding Study," in *The Vital Forces of Christianity and Islam: Six Studies by Missionaries to Moslems* (London: Oxford University Press, 1915), 215–239, at 216.

12. See Samuel Friedrich Günther, *Neue arabische Anthologie, oder, auserlesene Sammlung seltener,* 2 vols. (Leipzig: Johnn Friebrid Junius, 1791), 1:180–207; William Ouseley, *The Oriental Collections,* 3 vols. (London: Cadell and Davies, 1797–1800), 1:15–16, 131, 214, 297, 374–378, 2:193, 3:75–77; Johann Jahn, *Arabische Chrestomathie* (Vienna: C.F. Wappler und Beck, 1802), ix–x, 55–58, 61–62, 69, 71–72; Antoine Isaac, Baron Silvestre de Sacy, *Chrestomathie arabe,* 3 vols. (Paris: Imprimerie impériale, 1806), 3:369–508, at 419–420, 423. This section was translated by Antoine Léonard de Chézy (d. 1832), first published as *Extraits du Livre des merveilles de la nature et des singularités des choses créées* (Paris: Imprimerie impériale, 1805).

13. Christian Ludwig Ideler, *Untersuchungen über den Ursprung und die Bedeutung der Sternnamen: Ein Beytrag zur Geschichte des gestirnten Himmels* (Berlin: Weiss, 1809), 375–406 (edition), 409–428 (translation). See Paul Kunitzsch, *Arabische Sternnamen in Europa* (Wiesbaden: Harrassowitz, 1959), 15–21, 25, 97–224.

14. Thomas John Newbold, "Notice of the Ajaib-al-Mukhlukat," *Journal of the Asiatic Society of Bengal* 13, no. 2 (1844): 632–666. On the metropolitan center, colonial peripheries, and indigenous authority, see David Chidester, *Empire of Religion: Imperialism and Comparative Religion* (Chicago: University of Chicago Press, 2014), 5–11.

15. See Patrick Russell, *The Natural History of Aleppo,* 2 vols. (London: G.G. and J. Robinson, 1794), 2:184 (Bochart via Qazwīnī), 234 (Qazwīnī on the ant), 2:98–133. See Scott, *Arabian Nights,* 1:xxvii–xxxi. For further, see Maurits H. van den Boogert, *Aleppo Observed: Ottoman Syria through the Eyes of Two Scottish Doctors, Alexander and Patrick Russell* (Oxford: Oxford University Press, 2010); van den Boogert, "Patrick Russell and the Arabian Nights Manuscripts," in *Scholarship between Europe and the Levant in Honour of Alastair Hamilton,* ed. Jan Loop and Jill Kraye (Leiden: Brill, 2020), 276–298.

16. See Emilie Savage-Smith, "Drug Therapy of Eye Disease in Seventeenth-Century Islamic Medicine: The Influence of the 'New Chemistry' of the Paracelsians," *Pharmacy in History* 29, no. 1 (1987): 3–28; Miri Shefer, "An Ottoman Physician and His Social and Intellectual Milieu: The Case of Salih bin Nasrallah Ibn Sallum," *Studia Islamica* 106, no. 1 (2011): 102–123.

17. See F. Jamil Ragep, "Copernicus and His Islamic Predecessors: Some Historical Remarks," *History of Science* 45, no. 1 (2007): 65–81; F. Jamil Ragep, "Ibn al-Shāṭir and Copernicus: The Uppsala Notes Revisited," *Journal for the History of Astronomy* 47, no. 4 (2016): 395–415; Robert Morrison, "A Scholarly Intermediary between the Ottoman Empire and Renaissance Europe," *Isis* 105, no. 1 (2014): 32–57.

18. See Avner Ben-Zaken, *Cross-Cultural Scientific Exchanges in the Eastern Mediterranean, 1560–1660* (Baltimore: Johns Hopkins University Press, 2010), 47–75, 139–166; Hagen, *Geograph,* 370–371.

19. See Samuel Pye (d. 1772), *The Mosaic Theory of the Solar, or Planetary, System* (London: W. Sandby, 1766); listed in H. W. Hemsworth, *Catalogue of Books in the Library at Freemasons' Hall, London* (London: Printed privately, 1869), 27, 35.

20. See Maurits H. van den Boogert, "Scottish Freemasonry in Ottoman Izmir," in *Ottoman Izmir: Studies in Honour of Alexander H. de Groot*, ed. Maurits H. van den Boogert (Leiden: Nederlands Instituut voor het Nabije Oosten, 2007), 103–121, 117, 119. See also Jessica Harland-Jacobs, *Builders of Empire: Freemasons and British Imperialism, 1717–1927* (Chapel Hill: University of North Carolina Press, 2007), 1–20.

21. For an early account of Freemasons, see ʿAbd al-Laṭīf Shushtarī, *Tuḥfat al-ʿālam*, ed. Ṣamad Muwaḥḥid (Tehran: Ṭahūrī, 1363sh/1984~), 258–259, who notes that in Calcutta there were many Muslims who had joined the group.

22. See Ghulām Ḥusayn Khān Ṭabāṭabāʾī, *Siyar al-mutaʾakhkhirīn*, printed by ʿAbd al-Majīd, 2 vols. (Calcutta: Dār al-Imāra, 1248/1832~), 1:378; *FW*, 7:153, §29; 365, §50. See also Gulfishan Khan, *Indian Muslim Perceptions of the West during the Eighteenth Century* (Karachi: Oxford University Press, 1998), 92–95, 402 (index).

23. See Michael Fisher, *Counterflows to Colonialism: Indian Travellers and Settlers in Britain, 1600–1857* (Delhi: Permanent Black, 2004); Muzaffar Alam and Sanjay Subrahmanyam, *Indo-Persian Travels in the Age of Discoveries, 1400–1800* (Cambridge: Cambridge University Press 2007); Siobhan Lambert-Hurley and Sunil Sharma, *Atiya's Journeys: A Muslim Woman from Colonial Bombay to Edwardian Britain* (New Delhi: Oxford University Press 2010); Naghmeh Sohrabi, *Taken for Wonder: Nineteenth-Century Travel Writing from Iran to Europe* (Oxford: Oxford University Press, 2012); Nile Green, *The Love of Strangers: What Six Muslim Students Learned in Jane Austen's London* (Princeton, NJ: Princeton University Press, 2015).

24. See Mīr Muḥammad Ḥusayn Iṣfahānī, *Risāla dar aḥwāl-i mulk-i farang wa-hindūstān*, Mumbai, K.R. Cama Oriental Institute, MS no. 515 (R. 4.51), fols. 1b–2a; MS no. 625 (R. 1.33), fols. 1b–2a; Ṭabāṭabāʾī, *Siyar*, 1:378. See also the posthumous treatise, Mīr Muḥammad Ḥusayn Iṣfahānī, *Risāla-yi hayʾat-i jadīd-i Angrīzī*, edited by ʿAlī Dihghāhī in *Jashn-nāma-yi Ustād Sayyid Aḥmad Ḥusaynī Ishkavarī*, ed. Rasūl Jaʿfariyān (Tehran: Nashr-i ʿIlm, 2013), 475–493. Compare Aḥmad Ḥusaynī Ardakānī (fl. 1826), *Mirʾāt al-akwān*, ed. ʿAbd Allāh Nūrānī (Tehran: Shirkat-i Intishārāt-i ʿIlmī wa-Farhangī, 1996), 46.

25. Shushtarī, *Tuḥfat*, 238, 245, 284, 287, 300–305. Compare *Tuḥfat*, KMSI 2787, at 57; Mumbai, K.R. Cama Oriental Institute, MS no. 1591 (R. 4.5), fol. 180b. For more, see Juan Cole, "Invisible Occidentalism: Eighteenth-Century Indo-Persian Constructions of the West," *Iranian Studies* 25, no. 3/4 (1992): 3–16; Khan, *Perceptions*, 100–105, 395 (index); Hamid Dabashi, *Reversing the Colonial Gaze: Persian Travelers Abroad* (Cambridge: Cambridge University Press, 2020), 25–46.

26. Shushtarī, *Tuḥfa*, 467–468. See also Mīrzā Abū Ṭālib, *Khulāṣat al-afkār*, KMSI 1306, fols. 268b–271a (unnumbered); Dhū l-Fiqār ʿAlī Mast, *Riyāḍ al-wifāq*, Urdu translation by Sayyid Ḥasan and ʿAṭā Kakvī (Patna: Dāʾira-yi Adab, 1967), 59–60.

27. See *CPC*, 7:262–263, §850.

28. See Roy Wolper, "The Rhetoric of Gunpowder and the Idea of Progress," *Journal of the History of Ideas* 31, no. 4 (1970): 589–598.

29. See Iṣfahānī, *Hayʾat-i jadīd*, 482–485. See Khan, *Perceptions*, 132–138.

30. See Abū Bakr al-Marghīnānī (d. 593/1197), *Hidāya-yi fārsī*, trans. Mīr Muḥammad Ḥusayn, et al., 2 vols. (Calcutta: Chāpkhāna-yi Hindūstānī, 1807), 1:2. See Scott Alan Kugle, "Framed, Blamed and Renamed: The Recasting of Islamic Jurisprudence in Colonial South Asia," *Modern Asian Studies* 35, no. 2 (2001): 257–313.

31. See Ethé, *Persian Manuscripts*, 254, §626; 823, §1503; 1355–1356, §§2518–2519.

32. See *CPC*, 5:122, §814; Iṣfahānī, *Risāla*, 485. See also Ṭabāṭabāʾī, *Siyar*, 1:378, cited in Khan, *Perceptions*, 94n95.

33. See *CPC*, 7:137, §382; 207–209, §604. See also *CPC*, 7:xxiv (index); 8:27–29, §59n1, and xxxix (index); 9:278–279, §1307; 290, §1381; and xviii (index). See also *FW*, 10:197–198, §140; compare *FW*, 15:555, §39; 680, §52; *Diplomatic Correspondence between Mir Nizam Ali Khan and the East India Company, 1780–1798*, ed. Yusuf Husain (Hyderabad: Central Records Office, 1958), 57, and index.

34. See *CPC*, 9:211, §925; 278–279, §1307; 291n1, §1391; *FW*, 15:560, §61; 617, §58; 16:159, §6; 442n15.1.

35. See Iṣfahānī, *Risāla*, 485–487, 493; Khan, *Perceptions*, 285.

36. See Abū l-Khayr Ghayāth al-Dīn, *Majmūʿa-yi Shamsī = A Concise View of the Copernican System of Astronomy, Mejmua Shemsi* (Calcutta: T. Hubbard, 1807), 2 (dedication to the governor general). For debates on heliocentrism, see Hāshim ʿAlī al-Riḍawī (fl. 1235/1819), *Mirʾāt al-bilād*, Brit., Or. 202, fols. 6b–7a. Compare Sumathi Ramaswamy, *Terrestrial Lessons: The Conquest of the World as Globe* (Chicago: University of Chicago Press, 2017), 108–112. In the Ottoman context, see Robert Morrison, "The Reception of Early Modern European Astronomy by Ottoman Religious Scholars," *Archivum Ottomanicum* 21 (2003): 187–195; then later Daniel Stolz, *The Lighthouse and the Observatory: Islam, Science, and Empire in Late Ottoman Egypt* (Cambridge: Cambridge University Press, 2018).

37. See the addition of *yangī dunyā*, the New World, in the following late eighteenth-century manuscripts of Ṣāliḥ Ṣādiqī, *Shāhid-i ṣādiq*, Bodl. Ouseley 292, fol. 26b (copied 1194/1780 in Lucknow), and Ouseley 166, fol. 15a (copied 1196/1782). Compare *The Geographical Works of Sadik Isfahani* (London: J. Murray, 1832), 56–57. The New World is missing from earlier recensions: Brit., I.O. Islamic 1537, fol. 662a (dated 1117/1705) and the edition *Khātima-yi Shāhid-i Ṣādiq: Dar ḍabt-i asmāʾ-i jughrāfiyāʾī*, ed. Mīr Hāshim Muḥaddith (Tehran: Kitābkhāna, Mūza wa-Markaz-i Asnād-i Majlis-i Shūrā-yi Islāmī, 1377sh/1998~), 203. See later marginalia in Ṣādiqī, *Shāhid-i ṣādiq*, Brit., Egerton 1016, fols. 323a (position of the Sun), 327a (theories of lightening), 329a–b (geographical errors of the author), 330b–332b (modern seafaring, the Western Hemisphere, distance from London to the Canaries), 372a (London), 377b (fountain of life), 387a (Persians and *farang* on Alexander the Great), 443b (Wahhabi sack of Karbala in 1802), 445a (correcting *Wonders and Rarities* for location of the Wāqwāq islands), 446a (on Holland). This manuscript bears the Persian seal of English mercenary Richard Rotton (fl. 1805) dated April 14, 1791 (fol. 3b). On Rotton, who served in the Maratha army and took an Indian wife, see Rosie Llewellyn-Jones, *The Great Uprising in India, 1857–58: Untold Stories, Indian and British* (Woodbridge: Boydell Press 2007), 60–61.

38. See the *Illustrated London News*, November 21, 1863, 519, column a. For reference to the Indian press, see Edward Rehatsek, *Prize Essay on the Reciprocal Influence of European and Muhammedan Civilization* (Mumbai: Education Society's Press, 1877), v. For ridicule in the London press, see "Shameful Barbarity," *Punch: Or the London Charivari* 45 (Nov. 28, 1863), 218, where the outcome of the prize is predicted: "The Mahommedan mind will shut up."

39. Sir Sayyid Aḥmad Khān, "Guftigū [Speech]," *Proceedings of the Scientific Society*, no. 1 (Ghazipur: Printed at Syud Ahmud's Private Press, 1864), 27 (Urdu), 24–25 (English). "No religious work will come under the notice of the Society," in *Bye-Laws for the Scientific Society* (Ghazipur: Syud Ahmud's Private Press, 1864), 1, §2.

40. See Forster Arbuthnot, "Life and Labours of Mr. Edward Rehatsek," *JRAS* (July 1892): 581–595; Rehatsek, *Essay*, 103.

41. See the translated redaction of the speech before the Mahomedan Literary Society of Calcutta (November 1870), Sayyid Karāmat-ʿAlī, *On a Question of Mahomedan Law, Involving the Duty of Mahomedans in British India towards the Ruling Power* (Calcutta: Cambrian Press, 1871).

42. Karāmat-ʿAlī, *Risāla* (Calcutta: Maẓhar al-ʿAjāʾib, 1865), 39; translated by ʿUbayd Allāh al-ʿUbaydī and Sayyid Amīr ʿAlī as *Makhaz-i-Uloom, or a Treatise on the Origin of the Sciences* (Calcutta: Baptist Mission Press, 1867), 39.

43. See Michael Crowe, *The Extraterrestrial Life Debate, 1750–1900* (Cambridge: Cambridge University Press, 1986); Joydeep Sen, *Astronomy in India, 1784–1876* (London: Routledge, 2014), 75–110.

44. Karāmat-ʿAlī, *Risāla*, 34, trans. 34–35; see *Biḥār*, 57:121, §10; citing Muḥammad ibn Muḥammad al-Sabzawārī (fl. 583/1176), *Jāmiʿ al-akhbār*, ed. ʿAlāʾ Āl Jaʿfar (Beirut: Muʾassasat Āl al-Bayt, 1993), 347–348.

45. On the essay's reception, see George Cowell, *Life and Letters of Edward Byles Cowell* (London: Macmillan, 1904), 205; Joseph Garcin de Tassy (d. 1878), *Histoire de la littérature hindoui et hindoustanie*, 3 vols., 2nd ed. (Paris: Adolphe Labitte, 1870–1871), 2:162–163, §4. Letters from Charles Trevelyan in Karāmat-ʿAlī, *Risāla*, 130.

46. "Report of the Committee" in Rehatsek, *Reciprocal Influence*, iv. Muir mistakenly attributes authorship to ʿUbaydī. On the later developments, see Ahmad Dallal, *Islam, Science, and the Challenge of History* (New Haven, CT: Yale University Press, 2010), 169–176; Stefano Bigliardi, "The 'Scientific Miracle' of the Qurʾān," in *Islamic Studies Today: Essays in Honor of Andrew Rippin*, ed. Majid Daneshgar and Walid Saleh (Leiden: Brill, 2017), 339–353. For scientism in the colonial context, see Peter Gottschalk, *Religion, Science, and Empire: Classifying Hinduism and Islam in British India* (Oxford: Oxford University Press, 2013), 33–51.

47. *Jahān-numā*, published by Muḥammad Wajāhat ʿAlī Khān (Meerut: Maṭbaʿ Dār al-ʿUlūm, 1864), 16–17 (mountains of Qāf and Adam's peak), 33–39 (talismans), 26 (Alexander's adventures).

48. *Jahān-numā*, "tamāshā-yi bāzīgarān-i farang aur bangāla," 42–55. See the spurious memoirs of Jahāngīr, *Tūzuk-i Jahāngīrī = Memoirs of the Emperor Jahangueir*, trans. David Price (London: Oriental Translation Fund, 1829), 96–104; excerpted in Khāfī Khān (d. 1144/1731), *Muntakhab al-lubāb*, ed. Mawlvī Kabīr al-Dīn Aḥmad and Mawlvī Ghulām Qādir, 2 vols. (Calcutta: Asiatic Society of Bengal, 1869–1873), 1:308–313. See also Storey, 1:559.

49. *Jahān-numā*, 54–55. See Rosie Llewellyn-Jones, *A Fatal Friendship: The Nawabs, the British, and the City of Lucknow* (Delhi: Oxford University Press, 1985).

50. See *Shujāʿ-i Ḥaydarī*, Houghton Library, Cambridge, MA, Harvard University, MS Persian 63; lithograph (N.p.: Maṭbaʿ-i Chasma-yi Nūr, 1281/1865). See also Storey, 2:141, §212.

51. See Qazwīnī, *ʿAjāʾib al-makhlūqāt* (Tehran: Mīrzā ʿAlīqulī Khūʾī, 1264/1848), fols. 168a (irregularly foliated), 192a, 194a. See Urkīda Turābī, *ʿAjāʾib al-makhlūqāt-i Qazwīnī dar taṣāwīr-i chāp-i sangī-yi Mīrzā ʿAlīqulī Khūʾī* (Tehran: Muʾassasa-yi Farhangī-yi Pizhūhishī-yi Chāp wa-Nashr-i Naẓar, 1393sh/2014), 15–26; Ulrich Marzolph, "Mirzâ ʿAli-Qoli Xuʾi: Master of Persian Lithograph Illustration," *Annali* (Istituto Orientale di Napoli) 57, no. 1/2 (1997): 183–202.

52. See Qazwīnī, *ʿAjāʾib al-makhlūqāt* (Lucknow: Nawal Kishore, 1866), 585. See Urdu translation by Taṣadduq Ḥusayn, *ʿAjāʾib* (Lucknow: Nawal Kishore, 1912), 606. See also colophon

and unpaginated announcement opening the Urdu translation by Mawlānā Muḥammad Nadhīr ʿAlī Ṣāḥib Laknawī, Qazwīnī, *ʿAjāʾib al-makhlūqāt* (Lucknow: Maṭbaʿ-i Aḥmad Wāqiʿ, 1893), 532. For further context, see Ulrike Stark, *An Empire of Books: The Naval Kishore Press and the Diffusion of the Printed Word in Colonial India* (Ranikhet, India: Permanent Black, 2007).

53. On the development of print culture in Muslim societies, see Brinkley Messick, "On the Question of Lithography," *Culture and History* 16 (1997): 158–176; Geoffrey Roper, "The Printing Press and Change in the Arab World," in *Agent of Change: Print Culture Studies after Elizabeth L. Eisenstein*, ed. Sabrina Baron, Eric Lindquist, and Eleanor Shevlin (Amherst: University of Massachusetts Press, 2007), 250–267; Nile Green, "Journeymen, Middlemen: Travel, Transculture, and Technology in the Origins of Muslim Printing," *International Journal of Middle East Studies* 41, no. 2 (2009): 203–224; Ami Ayalon, *The Arabic Print Revolution: Cultural Production and Mass Readership* (Cambridge: Cambridge University Press, 2016), 1–14; Ahmed El Shamsy, *Rediscovering the Islamic Classics: How Editors and Print Culture Transformed an Intellectual Tradition* (Princeton, NJ: Princeton University Press, 2020), 63–92.

54. For the occult sciences in South Asian lithographs, see ʿĀrif Nawshāhī, *Kitāb shināsī-yi āthār-i fārsī-yi chāp-shuda dar shibh-i qāra*, 4 vols. (Tehran: Mīrāth-i Maktūb, 2012), 2:1103–1105, 1108–1109, 1119–1127, 1129–1137.

55. Drawn from the early bilingual catechism *A Compendium of Geography = Khulāṣa ʿilm-i arḍ kā Hindūstānī zabān meṇ* (Calcutta: Printed at the School-Book Society's Press, 1824), 59–60, 120–121, 128–129, 132–133. Compare Ramaswamy, *Terrrestrial Lessons*, 37–90.

56. Nawal Kishore, *Fihrist-i kutub-i mawjūd-i maṭbaʿ-i Munshī Nawal Kishūr wāqiʿ-i Laknaw wa-Kānpūr* (Lucknow: Nawal Kishore, 1874), lithograph in Brit., Or. 14119 b.21, 1 (Index), 88–90 (wonders of the world). Amīn al-Dīn Khān al-Harawī, *Maʿlūmāt al-āfāq* (Lucknow: Nawal Kishore, 1873); Urdu translation: *Maṭlaʿ al-ʿajāʾib*, trans. Mahdī ʿAlī Khān (Lucknow: Nawal Kishore, 1873). See also Storey, 2:142, §213. See the Uzbek adaptation drawn from the ʿĀdil Shāh Persian translation of Qazwīnī and Harawī's *Maʿlūmāt al-āfāq*, published as an illustrated lithograph, *Turkī ʿĀjāʾib al-makhlūqāt* (Tashkent: Ṭabaʿ Uldī, 1335/1917), 2–3, housed in Washington, DC, Library of Congress, QB 43.S5 1917. The illustrations are adapted from the Nawal Kishore lithograph copies.

57. Yūsuf Khān Kambalposh, *ʿAjāʾibāt-i farang* (Lucknow: Nawal Kishore, 1873), translated as *Between Worlds: The Travels of Yusuf Khan Kambalposh*, trans. Mushirul Hasan and Nishat Zaidi (New Delhi: Oxford University Press, 2014).

58. For an overview, see Alison Winter, *Mesmerized: Powers of Mind in Victorian Britain* (Chicago: University of Chicago Press, 1998). See also Wouter Hanegraaff, *New Age Religion and Western Culture: Esotericism in the Mirror of Secular Thought* (Leiden: Brill, 1996), 430–435.

59. Nawal Kishore, *Fihrist*, 57–60. For context, see Pasha Khan, *The Broken Spell: Indian Storytelling and the Romance Genre in Persian and Urdu* (Detroit: Wayne State University Press, 2019), 163–208.

60. See Sonja Brentjes, "The Prison of Categories: 'Decline' and Its Company," in *Islamic Philosophy, Science, Culture, and Religion: Studies in Honor of Dimitri Gutas*, ed. Felicitas Opwis and David Reisman (Leiden: Brill, 2012), 131–156.

61. See Muḥammad Daud Rahbar, "Sir Sayyid Aḥmad Khān's Principles of Exegesis," *Muslim World* 46, no. 2 (1956): 104–112, and no. 4 (1956): 104–112, 324–335.

62. Aḥmad Khān, *Tafsīr al-jinn wa-l-jānn ʿalā mā fī l-Qurʾān* (Agra: Mufīd-i ʿĀmm, 1309/1891~), 7. See also Aḥmad Khān, "Malāʾika, jinn, aur shayāṭīn kī ḥaqīqat imām Ghazālī

ke nazdīq," "Jibra'īl o-Mīkā'īl aur firishtōṇ kā wujūd," and "Firishtōṇ aur shayāṭīn kī ḥaqīqat," in *Maqālāt-i Sir Sayyīd*, ed. Muḥammad Ismāʿīl Pānīpatī, 14 vols. (Lahore: Majlis-i Taraqqī-yi Adab, 1961–1965), 3:46–52, 13:157–185.

63. Aḥmad Khān, "Jādū," *Maqālāt*, 4:300–303; Aḥmad Khān, *Ḥaqīqat al-siḥr* (Lahore: Nawal Kishore, 1910), 3; reprinted in *Maqālāt*, 4:306–339, at 308; citing Badr al-Dīn al-ʿAynī (d. 855/1451), *ʿUmdat al-qāriʾ, sharḥ Ṣaḥīḥ al-Bukhārī*, ed. ʿAbd Allāh Maḥmūd Muḥammad ʿUmar, 25 vols. (Beirut: Dār al-Kutub al-ʿIlmiyya, 2001), "Kitāb al-Ṭibb: Bāb al-siḥr," 21:312.

64. Aḥmad Khān, "nafs-i insānī meṇ ēk aisī quwwat-i barqī aur magnāṭīsī mawjūd hai," *Ḥaqīqat*, 3, 11; *Maqālāt* 4:308, 317.

65. Aḥmad Khān, "Mismarīzm wa-huwa ʿilm al-sīmiyāʾ," *Maqālāt*, 4:288–299. See *Kashf*, 1:12–13, 15, 22–23, 2:1020 (cited here). On *sīmiyāʾ* as "natural magic," see Edward Rehatsek, "Magic," *Journal of the Bombay Branch of the Royal Asiatic Society* 14 (1878–1880): 199–218, at 211–212; Richard Burton (d. 1890), *The Book of the Thousand Nights and a Night*, 12 vols. (London: H.S. Nichols, 1894–1897), 1:281n1.

66. Khān cites *Ikhwān*, 4:309.

67. Aḥmad Khān, "Mismarīzm," 4:289–291.

68. See James Esdaile, *Mesmerism in India, and Its Practical Application in Surgery and Medicine* (London: Longman, Brown, Green, and Longmans, 1846), 43–44. See Gyan Prakash, "Science 'Gone Native' in Colonial India," *Representations* 40 (1992): 153–178, at 161–163; and Winter, *Mesmerized*, 187–212.

69. For *al-siḥr al-ḥalāl*, see *Ikhwān*, 4:313–315, 330–331.

70. William Gregory, *Letters to a Candid Inquirer, on Animal Magnetism* (London: Taylor, Walton, and Maberly, 1851), 59–60; translated as *Ṭilism-i farang*, trans. Motī Lāl (Lucknow: Nawal Kishore, 1868), 48–49. See also, Sayyid Muḥammad Taqī, *Risāla-yi taʾthīr al-anẓār* (Kanpur: Nawal Kishore, 1878), 3–4; Muḥammad Najaf ʿAlī Khān, *Zubdat al-gharāʾib* (Delhi: Maṭbaʿ-i Anjum-i Imrūz, 1867); Mīr Imdād Ḥusayn, *Risāla-yi muʿjizāt-i insānī ba-qāʿida-yi ṭāqat-i miqnāṭīsī* (Lucknow: Munshi Nawal Kishore, 1878).

71. See *Report on the Administration of the Bombay Presidency, 1881–2* (Mumbai: Government Central Press, 1882), 287–288; *Report on the Administration of the United Provinces, 1897–8* (Allahabad: Government Press, 1899), 183.

72. Mawlvī Sayyid Niẓām al-Dīn, *ʿAql o-shuʿūr* (Ratlam: Ratan Prakash, 1873), 4, 289–291 (mesmerism), 245 (Galileo, Copernicus), 248 [148] (astrology), 252 (*raml*), 253–256 (talismans), 274–275 (magicians before Jahāngīr). Compare Choudhri Mohammed Naim, "Prize-Winning Adab," in *Moral Conduct and Authority: The Place of Adab in South Asian Islam*, ed. Barbara Daly Metcalf (Berkeley: University of California Press, 1984), 290–314, at 295–299.

73. George Douglas, Eighth Duke of Argyll, *Autobiography and Memoirs*, 2 vols. (London: J. Murray, 1906), 1:328–331.

74. On mesmerism, clairvoyance, and the Theosophical movement, see Joscelyn Godwin, *Theosophical Enlightenment* (Albany: SUNY Press, 1994), 151–187. See also Hanegraaff, *New Age*, 448–455. On radical politics in the context of colonial India, see Gauri Viswanathan, *Outside the Fold: Conversion, Modernity, and Belief* (Princeton, NJ: Princeton University Press, 1998), 177–207; Leela Gandhi, *Affective Communities: Anticolonial Thought, Fin-de-Siècle Radicalism, and the Politics of Friendship* (Durham, NC: Duke University Press, 2006), 118–126.

75. Modernizing Arabic journals, such *al-Muqtaṭaf* and Jurjī Zaydān's (d. 1914) *al-Hilāl*, regularly published rejections of various aspects of spiritualism, as well as reader inquiries

concerning the validity of occult phenomena, e.g., *al-Muqtaṭaf* 2, no. 3 (1877): 8–11; 3, no. 11 (1879), 306–308; 3, no. 12 (1879), 321–323; 4, no. 11 (1879): 27–28; 15, no. 5 (1891): 306–309; 32, no. 2 (1907): 112–116. See also, for example, *al-Hilāl* 13, no. 3 (1904), 173–174, translated in Paul Starkey, "Jinn, Spiritism, and the Talking Table," in Thomas Philipp, *Jurji Zaidan and the Foundations of Arab Nationalism* (Syracuse, NY: Syracuse University Press, 2014), 156–162. Several other articles can be adduced. See also Muḥammad Farīd Wajdī, *ʿAlā ṭalāl al-madhhab al-māddī*, 4 vols., 2nd ed. (Cairo: Dāʾirat Maʿārif al-Qarn al-ʿIshrīn, 1931), vols. 2 and 3. For Ottoman satires on spiritualism, see Palmira Johnson Brummett, *Image and Imperialism in the Ottoman Revolutionary Press, 1908–1911* (Albany: SUNY Press, 2000), 218–220; Özgür Türesay, "Between Science and Religion: Spiritism in the Ottoman Empire (1850s–1910s)," *Studia Islamica* 113, no. 2 (2018): 166–200; Kutluğhan Soyubol, "In Search of Perfection: Neo-Spiritualism, Islamic Mysticism, and Secularism in Turkey," *Modern Intellectual History* 18, no. 1 (2021): 70–94.

76. See Muḥammad Farīd Wajdī, *Dāʾirat al-maʿārif al-qarn al-rābiʿ ʿashar / al-ʿishrīn*, 10 vols., 3rd ed. (Beirut: Dār al-Maʿārif, 1971), 4:364–400 (*rūḥ*), 6:202–203 (*muʿjiza*), 9:293–302 (*maghnāṭīs*), 10:410–420 (*al-nawm al-maghnāṭīsī*); Rūḥ Allāh Khumaynī [Khomeini], *Kashf-i asrār* (Tehran: n.p., n.d.), 45–58. Both are discussed in Alireza Doostdar, *The Iranian Metaphysicals: Explorations in Science, Islam, and the Uncanny* (Princeton, NJ: Princeton University Press, 2018), 117–119. See also Rashīd Riḍā, "al-Sibiritzm aw istiḥḍār al-arwāḥ," *Manār* 29, no. 5 (1928): 362–371; On Barak, *On Time: Technology and Temporality in Modern Egypt* (Berkeley: University of California Press, 2013), 102, 106; Marwa Elshakry, *Reading Darwin in Arabic, 1860–1950* (Chicago: University of Chicago Press, 2013), 283.

77. See Nikki Keddie, *An Islamic Response to Imperialism: Political and Religious Writings of Sayyid Jamāl ad-Dīn "al-Afghānī"* (Berkeley: University of California Press, 1968), 53–84, 130–180.

78. See Muḥsin al-Mulk, *Majmūʿa-yi Lakcharz aur ispīchiz* (Lahore: Nawal Kishore, n.d.), 80, 94, 105; partially translated as *Causes of the Decline of the Mahomedan Nation* (Mumbai: Bombay Gazette Steam Printing Works, 1891), 39, 65.

79. Muḥammad al-Muwayliḥī, *Ḥadīth ʿĪsā ibn Hishām, aw fatra min al-zamān = What Ibn Hisham Told Us, or, A Period of Time*, ed. and trans. Roger Allen, 2 vols. (New York: NYU Press, 2015), 1:390, §§22.9–10.

80. Muwayliḥī, *Ḥadīth*, 1:304–306, §§17.10–12; Rashīd Riḍā, "Bāb: Asʾila wa-ajwiba," *Manār* 5, no. 2 (1902): 51–8, at 57–58; "Fatāwā: Ḥaqīqat al-jinn wa-l-shayāṭīn," *Manār* 7, no. 18 (1904): 702–711, at 706; "Taʿlīq ʿalā ḥadīth Jābir," *Manār* 29, no. 5 (1928): 377–382, at 377; Shaykh Muḥammad ʿAbduh with Rashīd Riḍā, *Tafsīr al-Manār*, 2nd ed., 12 vols. (Cairo: al-Manār, 1947), 8:366. See also Lutz Berger, "Esprits et microbes: L'interprétation des *ğinn*-s dans quelques commentaires coraniques du XXe siècle," *Arabica* 47, no. 3 (2000): 554–562; Elsharky, *Darwin*, 177; Barak, *Time*, 102.

81. Aḥmad Zakī, "al-khizāna al-muṭalsama al-sāḥira," in Philippe de Tarrazi, *Khazāʾin al-kutub al-ʿarabiyya fī l-khāfiqayn*, 4 vols. (Beirut: Dār al-Kutub, 1947–1948), 1:256, §8.

82. Aḥmad Zakī, *Taqrīr muqaddam ilā maqām fakhāmat al-ṣadr al-aʿẓam Ḥusayn Ḥilmī Bāshā bi-shaʾn tanẓīm Dār al-Kutub al-ʿUthmāniyya bi-l-Qusṭanṭīniyya* (Istanbul: Maṭbaʿat Aḥmad Iḥsān, 1325/1907), 14, §48; discussed in El Shamsy, *Classics*, 131–138. For a continuation of these practices of archival occultation and Islamic reform, see Dahlia Gubara, "Al-Azhar and the Orders of Knowledge" (PhD diss., Columbia University, 2014), 269–272.

83. See Aḥmad Zakī, *al-Ḥaḍāra al-islāmiyya* (Cairo: Dār al-Kitāb al-Miṣrī, 2014), 10–13, 19, 39, 99; Aḥmad Zakī, "Iftitāḥ al-jāmiʿa al-miṣriyya," *al-Muqtaṭaf* 34, no. 2 (1909): 138–145, at 141–145. For further on revivalism and civilization during this period in Arabic letters, see Massad, *Desiring Arabs* (Chicago: University of Chicago Press, 2007), 51–98.

84. Qazwīnī, *ʿAjāʾib*, New Delhi, National Museum, MS 58.48, unpaginated.

Acknowledgments

When the global pandemic brought lockdowns and restrictions, I began to see with new eyes what it might have meant for Qazwīnī to fine-tune *Wonders and Rarities* over the course of his life, returning to the strange beauty in natural order, but with everything around him upended. I saw how the solace of the written word could draw in with its unfinished possibilities, through paths that could unfold across years of refining and subtle emendation in a bookish retreat from the unsettled demands of being in the world.

From an attic office overlooking the elms of New Haven, I could conjure distant worlds through keystrokes and a well-staffed research library ready to lend all measure of remote assistance. No token of recognition captures the gratitude I feel for the support I have had over the years. Through it all has been the guidance and encouragement of Supriya Gandhi, my companion in matters worldly and sublime. Her presence is felt throughout these pages.

Countless versions of this book have come and gone. Put together, rearranged, abandoned, forgotten. Some inchoate musings of what could be, others reaching toward completion. This final form could only have been realized through the careful thought and attention of many readers. I am profoundly grateful to Sharmila Sen for first taking on this project nearly a decade ago. In addition to the editorial and production support at Harvard University Press, the unwavering dedication of Emily Marie Silk has been invaluable, her insights and care unparalleled. Special words of appreciation also go to Valerie J. Turner for her help with proofreading and indexing.

Colleagues, friends, and graduate students at Yale University have provided a steady source of inspiration. My thoughts on the organization of information have been refined through conversations with Ayesha Ramachandran, Deborah Coen, Marta Figlerowicz, Lisa Messeri, and Richard Prum, in the course of our collaboration on a Mellon Sawyer Seminar, "The Order of Multitudes: Atlas, Encyclopedia, Museum" (2020–2022).

Discussions with Sara Omar, while she was a postdoctoral fellow with the Council on Middle East Studies, provided further insights into intersections between science, law, and society. During our symposium "Gender, Authenticity, and Islamic Authority" (2019), I benefited from conversations with Kecia Ali, Karen Bauer, Aisha Geissinger, Niloofar Haeri, Joseph Massad, and Asma Sayeed.

Another symposium, "Magic and the Occult in Islam and Beyond" (2017), brought together scholars working on various aspects of the occult sciences. Their scholarship has informed my thinking on marvels and magic. Particular gratitude goes to Tuna Artun, Godefroid de Callataÿ, Alireza Doostdar, Noah Gardiner, Yahya Michot, Michael Noble, Yael Rice, and Farouk Yahya. Over the years, I have benefited from sustained conversations with Emily Selove, Liana Saif, and Nur Sobers-Khan. Their generosity of spirit and willingness to share and collaborate have been of incalculable significance. I hold special words of appreciation for Matthew Melvin-Koushki, a doyen of the field, whose friendship began many years ago in the ancient labyrinths of Fez. I am especially grateful to him for thoroughly reading the whole manuscript. Beyond his remarkable erudition is an infectious passion to think otherwise along routes long forgotten. So too behind many of my observations on science and magic is the work of Emilie Savage-Smith, whose scholarship, advice, and clarity have been a continual guide.

Among the many colleagues at Yale who have provided encouragement and inspiration, special thanks go to Abbas Amanat, Stephen Davis, Maria Doerfler, Zareena Grewal, Valerie Hansen, Christine Hayes, Samuel Hodgkin, Noreen Khawaja, Hwansoo Kim, Nancy Levene, Alan Mikhail, Evren Savci, Eliyahu Stern, Emily Thornbury, Shawkat Toorawa, Kevin van Bladel, and Marjan Wardaki. Laura Nasrallah and Özgen Felek commented on draft chapters, as did Kathryn Lofton, whose support and keen sense of purpose are beyond compare. I am thankful to Frank Griffel for his engagement with this entire project and for his advice over the years. A word of thanks goes to Giancarlo Casale, Pasha Khan, and Sophia Vasalou, who offered valuable feedback as outside readers in a colloquium on the book, which Frank led and which Katie made possible.

In New Haven, I benefited from conversations with Michelle Karnes, and her work on fiction and medieval marvels, while she was a fellow at the Yale Institute of Sacred Music. Michelle kindly provided insightful feedback on an early draft. I would like to thank Alessia Bellusci, who was a Blaustein postdoctoral associate in Judaic Studies, for references on the comparative history of the occult sciences. The Abdallah S. Kamel Center for the Study of Islamic Law and Civilization at the Yale Law School has brought several scholars whose thinking has been of importance, particularly Marion Katz, Moshe Halbertal, Robert Morrison, and Brinkley Messick.

Many themes explored here formed the basis of doctoral seminars that I have led on Islamic intellectual and social history, Arabic and Persian literature, Orientalism, and theory and method in the study of religion. Over the years I have benefited from discussions with graduate students and early career scholars. Particular gratitude goes to Aseel Alfataftah, Alexa Herlands, Sana Jamal, Lillian McCabe, Ahmed Nur, and Elizabeth Price, who engaged with this project at various stages, as well as to Iman AbdoulKarim, Ryan Brizendine, Kayla Dang, Lynna Dhanani, Conor Dube, Ghayde Ghraowi, Samaah Jaffer, Courtney Lesoon, Michael Lessman, Chris Mezger, Alexander Peña, Naila Razzaq, and Ramona Teepe.

This work has only been possible with the assistance of librarians, curators, and museum directors. Special thanks go to: Ingrid Lennon-Pressey at the Beinecke Rare Book and Manuscript Library; Robin Dougherty in the Near East Collection and Middle East Studies, Sterling Library, at Yale; Hirad Dinavari in the Library of Congress; Ursula Sims-Williams in the British Library; Gillian Evison in the Bodleian Library; Esra Müyesseroğlu in the Topkapı Palace Museum and Library; Khatibur Rahman and Nazia Kamal in the National Museum of India; Nawaz Mody in the K. R. Cama Oriental Institute Library; and the staff and curators in the Süleymaniye Library, the Salar Jung Museum, the Raza Library, and the Allama Iqbal Library at the University of Kashmir.

Over the years, friends, colleagues, and mentors have offered encouragement: the late Shahab Ahmed, Daud Ali, Zayde Antrim, Blain Auer, Zahra Ayubi, Bryan Averbuch, Carla Bellamy, Persis Berlekamp, Issam Eido, Ahmed

El Shamsy, Jamal Elias, Mary Gaylord, Behrooz Ghamari-Tabrizi, Luis Girón-Negrón, Vivek Gupta, Jack Hawley, the late Wolfhart Heinrichs, Tracey Hucks, the late Barbara Johnson, Alexander Key, Christian Lange, Mary McWilliams, Roy Mottahedeh, Elias Muhanna, Erez Naaman, Katharine Park, Frances Pritchett, Nasser Rabbat, Sarah Savant, Sunil Sharma, Daniel Sheffield, Kavita Singh, Devin Stewart, Wheeler Thackston, Jamel Velji, and Laura Weinstein.

Material from the book was presented at various occasions, including the South Asia Seminar at Columbia University. I appreciate the support from The University Seminars at Columbia University for assistance in publication. Some of the topics discussed here I have touched upon or referenced in earlier publications. The treatment of sorcery in chapter 3 builds on ideas discussed in "Commanding Demons and Jinn: The Sorcerer in Early Islamic Thought," published in *The Cambridge History of Magic and Witchcraft in the West* (Cambridge, 2015). My discussion of wonder in chapter 4 is informed by arguments made in "The Wiles of Creation: Philosophy, Fiction, and the *ʿAjāʾib* Tradition," which appeared in the journal *Middle Eastern Literatures* (2010). In the course of chapters 3, 4, and 9, I also expand on research first presented in "Cutting Ariadne's Thread, or How to Think Otherwise in the Maze," published as a postscript to the edited volume, *Islamicate Occult Sciences in Theory and Practice* (Leiden, 2020).

Travels to archives in India, Turkey, and Europe have been supported by research grants from the Yale MacMillan Center, a Fulbright-Nehru Senior Research Fellowship, and a New Directions Fellowship from the Andrew W. Mellon Foundation, which allowed me to delve deeper into Urdu literature and the long reception of Qazwīnī in South Asia.

Through these wanderings there has been the comfort of family and friends: Supriya's parents, Rajmohan and Usha Gandhi, models of unflappable perseverance and openhearted generosity; my mother, Cynthia Eastman, a ready reader and my toughest critic; my father, Mehdi Bahraini Zadeh, an impenetrable cypher, who once thought of naming me Sindbad; my family in California, Eveline, Joe, and Alex; my aunt Susie Eastman, a source of constant sagacity; our friends Todd and Jordan Champagne, Charlotte Whitby-Coles and Amin Hajee, Ravindra, Jayashree, and Archana Rao, Mangal Gole, and Siddhartha Singh.

The numerous opportunities to return to India led me back to the verdant hill station of Panchgani, where Supriya's family has long had ties. Not far from where Captain Newbold met his end, for me the Western Ghats have been a source of rejuvenation, where wisps of clouds rise up over the valley across the towering table land, surrounded by mangoes, bananas, and jackfruit, with parrots and langurs among the fronds. The monsoon brings a remarkable peace, if at times sorrow, with its steady rain. Here Supriya and I have tucked away hours of writing and reflection, watching our son, Anoush, once an infant, now a boy, conjure remote realms—of landryds and dyvinors and wonders beyond belief—some inspired by books, others of his own creation. Worldmaking is important work for young and old alike. But now, many miles removed, so much of it seems but an intangible dream from some distant land.

Illustration Credits

Figure I.1: Dragon Island. Bibliothèque de Bordeaux, France, Qazwīnī, *ʿAjāʾib*, MS 1130, fol. 54a.

Figure I.2: Conjoined twins. Salar Jung Museum, Hyderabad, Qazwīnī, *ʿAjāʾib*, MS Geo. 7, p. 19.

Figure I.3: Diagram of the cosmos. Reproduced by kind permission of the Syndics of Cambridge University Library, Cambridge, Qazwīnī, *ʿAjāʾib*, MS Nn.3.74, fol. 13a.

Figure 1.1: The prophet Solomon. Image © British Library Board, London, Qazwīnī, *ʿAjāʾib*, Or. 14140, fol. 100a.

Figure 2.1: Diagrams of Mercury. Image © British Library Board, London, Qazwīnī, *ʿAjāʾib*, Or. 14140, fol. 8a.

Figure 2.2: Aristotle holding an astrolabe. Topkapı Palace Museum, Istanbul, Ibn Fātik, *Mukhtār al-ḥikam*, Ahmet III 3206, fol. 90a.

Figure 2.3: Castle clock. Photograph © Museum of Fine Arts, Boston, Bartlett Collection, Museum purchase with funds from the Francis Bartlett Donation of 1912 and Picture Fund, acc. 14.533.

Figure 4.1: The Sīmurgh. The Morgan Library and Museum, New York, *Manāfiʿ al-ḥayawān*, M. 500, fol. 55b, Purchased by J. Pierpont Morgan, 1912.

Figure 4.2: Frontispiece. Bayerische Staatsbibliothek, Munich, Qazwīnī, *ʿAjāʾib*, Cod. arab. 464, fol. 1a.

Figure 4.3: A giant serpent or dragon. Image © British Library Board, London, Qazwīnī, *ʿAjāʾib*, Or. 14140, fol. 127a.

Figure 5.1: Constellation of Centaurus. Bayerische Staatsbibliothek, Munich, Qazwīnī, *ʿAjāʾib,* Cod. arab. 464, fol. 24b.

Figure 5.2: Dot asterism of the lunar mansion. Image © British Library Board, Qazwīnī, *ʿAjāʾib*, London, Or. 1621, fol. 60a.

Figure 5.3: Instrument for geomancy. Image © The Trustees of the British Museum, London, Inventory number, 1888,0526.1.

Figure 5.4: Merchant ship. Bibliothèque nationale de France, Paris, Ḥarīrī, *Maqāmāt,* MS arabe 5847, fol. 119b.

Figure 5.5: Shipwrecked merchant. Bayerische Staatsbibliothek, Munich, Qazwīnī, *ʿAjāʾib,* Cod. arab. 464, fol. 65b.

Figure 5.6: Dog-headed creatures. Image © British Library Board, London, Qazwīnī, *ʿAjāʾib,* Or. 14140, fol. 41b.

Figure 7.1: Diagram of the world. Library of Congress, Washington, D.C., Qazwīnī, *ʿAjāʾib,* Shelf mark: G93.Q3185 1553, fol. 6a.

Figure 7.2: World map. Beinecke Rare Book and Manuscript Library, Yale University, New Haven, Qazwīnī, *ʿAjāʾib,* Arabic MSS 575.

Figure 7.3: Map with the New World. Bibliothèque de Bordeaux, France, Qazwīnī,*ʿAjāʾib,* MS 1130 fol. 39r.

Figure 7.4: Portolan map. Topkapı Palace Museum, Istanbul, Hazine 1822.

Figure 8.1: Talisman of Jupiter. Reproduced by kind permission of the Syndics of Cambridge University Library, Cambridge, Qazwīnī, *ʿAjāʾib,* MS Nn.3.74, fol. 192a.

Figure 8.2: Detail of a bifolium frontispiece. Image © The Metropolitan Museum of Art, New York, Fletcher Fund, 1934, Qazwīnī, *ʿAjāʾib,* acc. 34.109.

Figure 8.3: The New World. Piri Reis map, Topkapı Palace Museum, Istanbul, Revan Köşkü 1633 mük.

Figure 8.4: Detail of sailors setting a fire. Piri Reis map, Topkapı Palace Museum, Revan Köşkü 1633 mük, detail.

Figure 8.5: Half-fish, half-man. Newberry Library, Chicago, Suʿūdī, *Tārīh-i yeñi dünyā,* MS Ayers 612, fol. 57a.

Figure 9.1: Encounter in the encircling ocean. The Walters Art Museum, Baltimore, Āzarī, *Gharāʾib al-dunyā,* MS W. 652, fol. 78b.

Figure 9.2a and 9.2b: Painiting of Jahāngīr and Shāh ʿAbbās. Freer Gallery of Art, Smithsonian Institution, Washington, D.C., Purchase—Charles Lang Freer Endowment, F1945.9a.

Figure 9.3: Detail from a map of China. Gerard Mercator and Jodocus Hondius, *Atlas sive cosmographicae meditationes de fabrica mundi et fabricati figura*, fourth edition (Amsterdam: Iudoci Hondij, 1613), between pages 334–335.

Figure 9.4: Subjugating Saturn. Image © The Trustees of the Chester Beatty Library, Dublin, *Nujūm al-ʿulūm*, IN 02, fol. 226b.

Figure 9.5: Colophon with painting. National Library of Medicine, Maryland, Qazwīnī, *ʿAjāʾib*, MS Persian 29, fol. 173a.

Figure C.1: Angels. Image © The Board of British Library, London, Qazwīnī, *ʿAjāʾib*, I.O. Islamic 3243, fol. 91a.

Figure C.2: A couple from Portugal. Houghton Library, Harvard University, Cambridge, *Shujāʿ-i Ḥaydarī*, MS Persian 63, fol. 14a.

Figure C.3: Creatures with strange forms. Image © The Board of British Library, London, Qazwīnī, *ʿAjāʾib*, (Nawal Kishore: Lucknow, 1866), 576, Shelf mark: 14759.d.1.

Figure C.4: Modern map. National Museum, New Delhi, Qazwīnī, *ʿAjāʾib*, MS 58.48, unpaginated.

Index

The letter *f* after a page number denotes a figure

Abbasids, 3, 35, 88, 138, 142, 143; administration of, 65; and caliphs, 62, 74, 107, 150, 289; collapse of, 87–89; and exotica, 142; revolution of, 61, 143; translation movement of, 42, 143–145, 160, 183, 192, 208

ʿAbduh, Muḥammad, 327

al-Abharī, Athīr al-Dīn, 38–40, 56–57; on astronomy, 53–54; and Fakhr al-Dīn al-Rāzī, 40, 348n21; and Ibn Sīnā, 50, 57, 351n22; in Mosul, 47–48; on natural philosophy, 48, 55, 110, 348n21, 350n7; and Qazwīnī, 54, 348n18; teachings of, 49–51, 54, 57; and Urmawī, 56; on void, 50, 135–136

Abū l-Fazl ʿAllāmī, 274–275, 286

Abū Ḥāmid al-Gharnāṭī, 98, 105

Abū Ḥanīfa, 60–61

Abū Nuwās, 137

accident (philosophy), 48, 225

Aceh, Sultanate of Sumatra, 246, 259, 286–287

adab, 104–105; literary conventions of, 115; as comportment, 67

Adam, 275, 289

ʿĀdil Shāhī dynasty, 283–284, 285, 293

admiratio, 159

admiration, 6; toward the self, 119. *See also* astonishment

adventures: of Amīr Ḥamza, 320; beyond pillars of Hercules, 208; of Bulūqiyā, 366n15; of sailors from Lisbon, 209–211; of Saint Brendan, 155; of Sindbad, 301. *See also* Alexander the Great

affection, 193. *See also* love

Africa: Aristotle on, 207–208; circumnavigation of, 20, 208, 249, 283, 381n15; coast of, 154, 199, 206, 207, 219, 242, 246; exploration of, 195, 213, 220*f*, 224, 242; European representations of, 160, 214; inhabitants of, 272; and Muslim merchants, 151, 154; Qazwīnī on, 191, 213; settlements in, 151; and slavery, 305. *See also* Mali Empire

agency, individual, 62, 111

agriculture: classification of, 176; Greek sources on, 181; and lunar mansions, 133; Qazwīnī on, 32, 178, 231, 374n24; writings on, 178, 281

Aḥmad Khān, Sir Sayyid, 309–310, 318, 324–325; on magic, 321–322; and Scientific Society, 311

ʿajab, 16, 119, 145; definition of, 117–118; and Latin discourses, 159–160. *See also* astonishment

ʿajāʾib, 119, 145–146; ʿAlī as bearer of, 17; and *ghar* (abode of wonders), 15–16, 345n19; and *gharāʾib* (rarities), 4, 6, 12, 158, 159, 224, 277; and *al-makhlūqāt* (things created), 29, 107, 229, 238; and *mirabilia*, 159; as miracles, 68; and monsters, 160; in nature, 116; and Qazwīnī's geography, 99; and *thaumata*, 184. *See also* curiosities; marvels; wonders and rarities
Akbar, 263; and Alexander the Great, 275–276; and Portuguese traders, 275; and Sanskrit learning, 277–278
akhlāq, 186–188, 191; calque of Greek, 112; refinement of, 111–112, 187. *See also* ethics; moral philosophy
ʿĀlamgīr, 286
Alamut, 86–87
alchemy, 268, 372n5, 385n13; Apollonius on, 173; Aristotle and 170; vs. chemistry, 296, 302, 328; etymology of, 172; Fakhr al-Dīn al-Rāzī on, 169, 176; Ibn al-Nadīm on, 172–173; Ibn Sīnā on, 76–77, 169; and *kīmiyāʾ*, 172; and Latin Christendom, 296; processes of, 75, 168, 185; Qazwīnī on, 3, 85, 166; Ṣaymarī on, 146; Shahmardān on, 229; writings on, 107, 169, 170–171, 180, 189
alchemists, 84, 169, 172, 180, 229
Alexander of Aphrodisias, 171, 206
Alexander of Tralles, 181, 182–183
Alexander the Great, 80, 149, 202, 212, 230, 288, 384n10; adventures of, 92, 195, 238, 249, 252–254, 274–275; and Aristotle, 170, 195–196, 250, 275, 276, 373n11; and boundary markers, 34, 206, 212; as Dhū l-Qarnayn, 196, 223, 251, 252; and encircling ocean, 196, 223–224, 269*f*, 393n19; and Hermetic teachings, 171, 279; and the New World, 250, 252–254, 269, 274–275; as philosopher king, 143, 275–276; romance, 154, 155, 157; and *The Secret of all Secrets*, 250; and wall of Gog and Magog, 10, 202, 223
Alf layla wa-layla, 149. See also *The Thousand and One Nights*
ʿAlī, nephew of the Prophet, 17, 288–289
ʿAlī ʿĀdil Shāh, 279, 282
Aligarh, 318, 326
ʿAlī ibn Yūsuf ibn Tāshufīn, 211
alms, 89–90. *See also* piety and devotion
Álvares Cabral, Pedro, 248
Amānat Khān Shīrāzī, 266
amber, 180, 375n26
ambiguity, 3, 13, 147, 164, 328, 343n2; and wonder, 5. *See also* uncertainty
Americas, 242, 246, 273; Muslim discovery of, 18–19, 204, 328, 388n50; conquest of, 242–243, 307. *See also* New World
amulets, 82, 133, 167, 174, 175, 180–183; animals used for, 176–177; and elemental magic, 122, 124, 363n52, 385n13; Galen on, 182. *See also* magic; elemental magic; talismans
Anas ibn Mālik, 70–71
al-Andalus, 84, 208, 211, 242, 347n6. *See also* Iberian Peninsula
anecdotes: dubious or strange, 137, 154, 210; as entertainment, 240; of the *farang*, 286; as historical method, 22, 333; illustration of, 105; in marginalia, 261, 288; in *Wonders and Rarities*, 28, 97, 109, 147, 183, 253
Angelino Dulcert of Mallorca, 219
angels: and the cosmos, 50, 106, 136–137, 365n12; and elemental magic, 179; holding up the world, 199, 200*f*, 201; as immaterial intellects, 49, 365n12; and Indic cosmology, 290; mastering, 79, 81, 83–84, 85, 280–281; paintings of, 22, 28*f*, 102, 234*f*, 282, 291, 297*f*; theories on, 136, 321; Qazwīnī on, 3, 135, 136–137, 161, 365n12; sizes and shapes of, 136–137; visionary inspiration from, 68, 69, 378n47
animals, 156, 192, 255, 267, 294, 299, 385n13; Aristotle on, 161; in battle, 268; composite or hybrid, 161–162; fables of, 149; holding up world, 199, 201; hunting, 263; in Persian, 158; properties and uses of, 107, 100, 176–178, 373n11; Qazwīnī on, 160; soul of, 175, 257, 322; strange, 14,

120, 162, 240, 299; in talismans and amulets, 176–177; *See also* beasts; birds
animals, legendary, 164, 299, 371n61. *See also* ʿ*anqā*ʾ; dragons; griffins; mermaids and mermen; Rukhkh; Sīmurgh
animals, types of: birds, 157–158, 165, 219, 359n10; cranes, 161, 371n56; crocodiles, 155; elephants, 155–156, 158, 207, 277; giant fish, 248; listed, 156, 255, 267, 294, 299; tortoises, 155, 158, 277; whale, 199
ʿ*anqā*ʾ, 158, 164, 234*f*, 299, 366n19; as sphinx, 371n61
Antilles, 247
antinomianism, 86–87, 307
antinomies, 145. *See also* perplexity and confusion
apocalypticism. *See* eschatology; millenarianism
Apollonius of Tyana (Balīnās): and Abū Bakr al-Rāzī, 181; as author of *The Secret of Creation*, 107, 169, 170–171, 173, 189; as authority on talismans, 173, 174–176, 281–282; not Pliny, 366n21; on shape of the world, 288; source for Qazwīnī, 169, 173; on special proprieties, 176–178
aporia, 22, 58, 118. *See also* perplexity and confusion
Arabic language: calligraphy, 141, 150; interpreters of, 246–247; into Latin 47, 159, 160, 184, 189, 304; prestige of, 141; poetry, 140; and the *nahḍa*, 19; studies of, 298; translation into, 42, 63, 74, 102, 107, 112, 118, 155, 183, 192
Arabic literature: modern scholars of, 18, 328; and Sakkākī, 84, 356n21; text-reuse in, 38; wonders in, 145; and Yāqūt, 36. *See also adab*; translation
Arabian Nights. See *The Thousand and One Nights*
Aral Sea, 33, 35
archipelagoes, 206, 287, 295; in the Far East, 151; Malay Archipelago, 258, 286–287. *See also* Canary Islands; islands
archives: colonial and national, 9, 14; as historical sources, 23, 333; in Istanbul, 17, 233, 385n19; of National Museum, Delhi, 329; of the Salar Jung Museum, 16
Aristotle, 31, 53*f*, 104, 162, 166; Abū Bakr al-Rāzī on, 181–182; as alchemist, 172; and Alexander the Great, 170, 195–196, 250, 275, 276, 373n11; *Ars rhetorica*, 112; and categories, 225; *De caelo*, 207; and ethics, 186; gnomology ascribed to, 393n23; *Kitāb al-Ḥayawān*, 161, 370n54; and Hermes, 131, 170, 171, 173, 194, 196; on hierarchy, 162–163; lapidary, ascribed to, 169–170, 184, 195–196; on legendary animals, 164, 371n61; and mastery of nature, 171, 192, 195; *Metaphysica*, 111, 118; *Meteorologica*, 74; on monstrous births, 160; and natural philosophy, 75, 168, 188; physics of, 162, 169; on pygmies, 161; as source for Qazwīnī, 92, 169, 195; on shape of the world, 288; *Theology*, ascribed to, 118, 193; on wonder, 118, 119–120; on void, 50, 135
artifices and deceptions, 178, 181; of demons, 196; as magic, 314; and Saturn, 148; of thieves, rogues, and spies, 27, 146, 185, 193, 281, 385n13; of women, 291. *See also* automata; devices and instruments
artisans, 91, 98, 104, 120, 130, 185, 264, 294
artistry, 16, 267, 287
al-Ashʿarī, Abū l-Ḥasan, 27, 62, 68, 108; theology of, 59, 62, 136, 354n5
Assam, 314
assassins, 86
assemblies, 39–40, 55, 146; of Akbar, 276; of Ibn Furāt, 154; of Jahāngīr, 274; for the judiciary, 67; of Juwaynī, 97; of Qazwīnī, 71, 99; of Rāzī, 40; of Ṭūsī, 91. *See also* al-Hamadhānī, Badīʿ al-Zamān; al-Ḥarīrī
asterisms, 129, 132, 134. *See also* constellations
astonishment, 4, 6–7, 115; appeal to, 282; and awe, 295; as bemusement, 147; *camatkāra*, 394n26; at creation, 109; criticism of, 321; definition of, 119–120; and evil eye, 123; finger of, 6, 83, 267; God's experience of, 119; Ibn Sīnā on, 117–118; imperial models for, 143; language of, 17, 173, 277;

astonishment (*continued*)
losing sense of, 118; in *New Report*, 255–256; in the *Nights*, 149; as piety, 113; and perplexity, 109, 120, 267; and truth, 114; unique to humans, 4, 291, 343n4. *See also* admiration; awe; perplexity
astral learning: in Alamut, 86, 357n27; ancient, 75, 126; and conjunctions, 162–163; and Fakhr al-Dīn al-Rāzī, 77–78, 83, 123, 125, 278; for harnessing forces, 78, 288; Hellenistic, 77, 133; in India, 126, 133, 279, 281; and Qazwīnī, 77, 127, 129
astrolabes, 17, 31, 51, 53, 87, 125, 231, 281, 351n13
astrologers: authority of, 77; Christian, 160, 250; Hindu, 16, 17, 263, 279; horoscopes of, 16, 78, 80, 91, 162; Ibn al-Nadīm on, 146; at Mongol court, 79–81, 85, 86, 87–88; Ottoman, 241; Qazwīnī on, 77, 371n56, 354n7; tricks of, 146, 148. *See also* horoscopes
astrology: apocalyptic, 126–127; as applied science, 78, 184; vs. astronomy, 77, 296, 328; critique of, 74–75, 302, 354n6; as divine beneficence, 75, 354n3; and Ibn Sīnā, 76–77, 125; judicial, 74–75, 76–78, 87, 91, 127, 126; in Latin Christendom, 159, 298; natal, 76, 91, 132, 147, 162, 190; and omens, 126; prestige of, 100, 224, 278; of Ptolemy, 74, 77; Qazwīnī on, 3, 85, 126; writings on, 78, 147, 281. *See also* astral learning; geomancy; horoscopes
astronomers, 18, 35, 53, 58, 170, 185, 229, 301. *See also* Copernicus; Ptolemy; al-Ṭūsī, Naṣīr al-Dīn
astronomy: Abharī on, 50–51, 53–54; ancient, 204; and astrology, 74, 77, 296; calculations and tables for, 91, 278–279; discipline of, 10, 41, 45, 50–51, 302; Fakhr al-Dīn al-Rāzī on, 78; and heliocentrism, 303, 306; instruments for, 129; mathematical, 51, 76; Ptolemaic, 46, 51, 74, 129, 204; Ptolemaic, critique of, 58, 226, 303; and Qazwīnī, 45, 76, 204, 301; utility of, 44. *See also* astrology; observatories
atelier, 264, 267, 278
Atlantic Ocean: domination, 219, 222, 243, 258, 259–260; exploration, 18, 206, 218, 250, 253, 380n8; limits of, 20, 208; marvels of, 212, 219; Piri Reis on, 244, 247–248; slavery, 242, 244; trade, 259. *See also* encircling ocean/sea of darkness
Atlas, stories of, 135, 201, 365n10
atlases, 272, 274, 300; Catalan, 219, 383n42
ʿAṭṭār, Farīd al-Dīn, 268
ʿAṭūfī, Hayreddīn, 237–239
authenticity: of anecdotes, 253; of miracles, 361n36; problem of, 116, 164, 185; and hadith, 154
authority: of astrology, 77; of caliphate, 3; of Catholic Church, 243; classical, 10, 180, 183, 197, 251, 252; continuity of, 40, 55; of courts and madrasas, 67; divine and scriptural, 11, 43, 110, 227, 311; of experience, 29, 182; and ideology, 5; and knowledge, 40, 46; of men, 186; modern, 18, 21, 325, 328; of piety, 285; of Quran, 60, 110, 136, 311–312; religious and scientific, 44, 136, 204, 224, 225
automata, 63, 64*f*, 173, 296. *See also* devices and instruments
Avesta, 124, 155
Avicenna. *See* Ibn Sīnā
awe: critique of, 18; cultivation of, 7, 98, 114, 273, 295, 320; language of, 262, 277; as piety, 268; in primates, 4, 343n4. *See also* astonishment
awqāf, 90. *See also* endowments
ʿaẓama, 105–106, 110
Āzarī, Fakhr al-Dīn, 267–268, 270
al-Azhar, 235, 325, 327
Azores, 206
Aztecs, 259

Babel, 145
Bābur, 262, 278, 289
Baghdad: Abbasid capital, 3, 36; books in, 42, 150, 151, 172; floods in, 87; judiciary, 65, 71, 90; madrasas, 45, 55, 62, 63, 70, 89; markets, 10, 133; Mongol conquest of,

72, 79, 87–89, 303; rebuilding of, 90; sectarianism in, 61; Qazwīnī buried in, 71; talismanic statue in, 174
al-Baghdādī, ʿAbd al-Laṭīf, 120
Bahādur Shāh, 289
Bahmanids, 267, 285
Balīnās al-Ḥakīm. *See* Apollonius of Tyana
al-Bāqillānī, Abū Bakr, 59
bayān. *See* eloquence; rhetoric
Bāyezīd II, 237
Bayrām Velī, 235
beasts, 3, 158–159, 191; taming, 68, 273. *See also* animals; monsters; savages
Bengali acrobats, 313
Berbers, 210
bestiaries: 5, 100, 100*f*, 102, 103–104, 155, 159
bewilderment, 142, 241. *See also* perplexity and confusion
Bhagavad Gita, 290
bibliomancy, 238; Quranic, 288. *See also* divination
Bīcān, Yāzıcıoğlu Aḥmed, 235–236, 250
Bihzād, 270
Bijapur, 282–283; translation of *Wonders and Rarities*, 261–262, 265, 285, 289, 291
biographical compendiums, 29, 36–37, 70–71, 90, 146, 237, 238
birds: in the New Word, 255; at sea, 219; talking, 147, 162; trapping through magic, 83, 356n21; wonderous, 264. *See also* animals, types of; phoenix; Rukhkh; Sīmurgh
al-Bīrūnī, Abū Rayḥān, 35, 104, 205, 383n1; and criticism of Abū Bakr al-Rāzī, 181–182; on divisions of the world, 38; lapidary of, 170; on wonder and rarity, 184–185; writings in Persian, 101
Black Sea, 222, 242
Bochart, Samuel, 298–299
books: burning, 87; collecting, 92, 128, 237–238; as commodities and prestige objects, 89, 98; companionship of, 29; culture, 29, 59, 249, 256, 391n7; draft and formal copies, 99; illuminated, 98, 102, 104–105, 233, 282; knowledge obtained by, preserved in, 35, 55, 87, 141; in madrasas, 45; printed, 16, 252, 315–316, 318, 326–327; production, 36, 66, 96, 98–99, 233, 293; published by Nawal Kishore, 319–320; a world within, 7, 197, 336. *See also* libraries; printing press
boundaries: communal, 163; geographical and political, 35, 199, 210, 244; of knowledge, 20, 41, 59, 203, 205–207
Brahmins, 126, 263, 265, 275, 277, 283
Brasil, disappearing island of, 219
Brazilian coast, 248
Brethren of Purity, 42, 195, 288
British Empire: colonial officials of, 308, 322, 324; and colonialism, 10, 16, 304, 311; expansion of, 307; and monarchy, 310
British Isles, 222. *See also* England
British Library, 334
British Museum, 15
Buddhists: cosmology of, 201; devotional objects of, 151; Mongol attitudes toward, 79–80
Bulūqiyā, adventures of, 366n15
Bunker, Chang and Eng, 343n1. *See also* twins, conjoined
Buzurg ibn Shahriyār, 154–155

Cadiz lighthouse, 380n11
caliphs and caliphate, 60, 63, 76, 88, 137, 142, 146, 147, 278, 284, 395n42. *See also* Abbasids; Fatimids; Umayyads
calligraphy, 99, 233, 266, 281, 324; Arabic, 141, 150; Qazwīnī known for, 99; symbolic value of, 318; Yāqūt's mastery of, 36
Canary Islands, 20, 206, 211, 214–215, 219, 220*f*, 222, 242–243, 383n44. *See also* Eternal Islands
cannibals and cannibalism, 151, 213, 249, 257, 287. *See also* savages
capital: accumulation of, 5; investments in, 151; mimetic, 273
Caribbean Islands, 246

cartographers, 216–217, 219, 222, 272, 273, 392n15
cartography: based on Columbus, 20, 388n45; and cosmology, 273; of the imagination, 21; models for, 202, 251; New World, 243, 244, 246–248, 382n35; and Ottoman elite, 203, 300; practices of, 201, 215–216, 217, 219; Qazwīnī, 217; rival Christian, 222, 272; and Wall of China, 273; and wonder, 249. *See also* maps; portolan charts
Caspian Sea, 73, 252–254
categories: of Aristotelian logic, 225; of civilization, 19, 296, 310; of extraordinary phenomena, 74, 121–122; of fiction, 114, 333; of modernity, 21–22; medical, 180; of monsters, 158–160, 370n55; of the secular, 311
Catholic Church, 243–244, 257, 259, 298
causation, 76; endless chain of, 50; Ibn Sīnā on, 124; secondary, 76, 137, 354n5;
causes: divine, 76, 120; of decline, 326, 328; hidden, 5, 118; for revolt of 1857, 310; theophany of, 107
celestial realm: and Earth, 75–76, 327; inhabitants of, 50, 135, 190, 327; phenomena in, 75, 95, 134, 135; and Ptolemy, 77; as *ʿulwiyyāt*, 75, 95. *See also* angels; terrestrial realm
celestial bodies, 49, 51, 129, 178, 194; influences and powers of, 75, 77, 78, 121, 147, 180, 189, 193, 196; movement of, 97, 135–136, 138, 178, 228. *See also* astrology; astronomy; planets; stars; talismans
celestial globes, 31, 51, 129, 130–131, 133, 135
cemeteries and tombs, 2, 71, 87, 141, 142, 240, 283, 356n21
centaur, 164, 371n61
Centaurus (constellation), 131*f*
certainty, 48, 164, 276. *See also* ambiguity; uncertainty
Chaghatay, 83, 85
Chaghatai language, 278
chain of being, 49, 50, 112. *See also* hierarchy
Chaldeans, 173, 179
charaktēres, 174, 374n18. *See also* cyphers
charlatans, 146, 181, 375n28. *See also* artifices and deceptions
chemistry, 10, 18, 296, 328. *See also* alchemy
China: and alchemy, 172; and Alexander, 275; geographical knowledge of, 224, 246, 247, 300; and Jesuits, 274; and Maragha, 91; and Mongols, 79; and Portugese, 286; and trade, 154; and wonders, 153, 156
China, Wall of, 10, 271*f*, 272*f*, 273, 393n17
Christians, 67, 79–80, 175, 235; conquests of 255; and esotericism, 304; and exchanges with Jews and Muslims, 216–217; explorations, 215, 222, 286; Maronite, 298; mercenaries, 219; merchants, 214, 216, 219, 263–264, 273; missionaries, 258, 263, 273; in Mosul, 47; as purveyors of discord, 256–257; on divine design, 5
Cirebon shipwreck, 368n38. *See also* shipwrecks
civilization: category of, 19, 296, 310–311; Islamic, 17, 204, 328; and revivalism, 405n83; and superiority, 19, 321
clairvoyance, 323, 325, 403n74
classification and taxonomy: of existence, 4, 49, 112, 225, 288; of extraordinary phenomena, 122; and Porphyrian tree, 48, 228; Qazwīnī's use of, 6, 41, 95, 109, 224–225. *See also* hierarchy
classifications of science and learning: by ʿAṭūfī, 238–239; divided into practical and theoretical, 111; by Ibn al-Akfānī, 84; by Ibn Sīnā, 76, 78, 288; question of precedence in, 225; by Shīrāzī, 228; by Ṭaşköprīzāde, 240–241
climatic determinism, 31–32, 163, 191, 212, 231, 347n8. *See also* dispositions and natures
clocks: in Constantinople, 173; Mustanṣiriyya madrasa, 63; *ṣandūq al-sāʿāt*, 352n30; Swiss, 16; water, 63, 64f, 373n15. *See also* automata; devices and instruments; time
cognition: gradations of, 225; and rational soul, 49, 110, 186, 187; Qazwīnī's theory

of, 192; and wonder, 8, 23, 29, 278. *See also* epistemology; psychology
colonialism: and administrators, 11, 18, 304, 312, 316, 322, 324; vs. anticolonialism, 325; and education, 309–310, 319; European, 8, 22, 296; and experimentation, 323; history of, 21; and Islamic law, 308; and modernity, 21, 303, 304, 321; and museums, 14–15, 17. *See also* Orientalism
colonization: of Canary Islands, 218, 242; of India, 307–308, 309–310; of New World, 248, 382n30; and Ottomans, 255; and slavery, 243
Columbus, 243, 246, 248; accounts of, 20; Arab origin of, 306; and Aristotle, 249–250; misconceptions of, 247, 388n46; preceded by Muslims, 18–20, 204, 247, 328; precedence of, 20, 247, 254, 388n45
commentaries, 56, 57, 58; on Ibn Sīnā, *Pointers*, 58, 122–123, 351n22. *See also* exegesis
commerce, 231, 287; and exploitation, 307, 314, Ottoman monopoly over, 259; and silent trade, 287. *See also* East India Company; Levant Company; merchants
commodities: circulation of, 14, 314; and colonialism, 15, 294–295; exchange of, 243; New World, 247, 255, 264; as rarities, 2; and slavery, 243. *See also* curiosities; exotica; medicaments; spices
compass: drawing tool, 215–216; magnetic, 307, 318. *See also* portolan charts
confusion. *See* perplexity and confusion; uncertainty
conquests: of Baghdad, 72, 79, 87–89, 303; British and European, 1, 305, 307–308; of Granada, 244; of Iberia, 255; of Mexico, 259; Mongol, 79–80; of New World, 242–243, 252, 307; Ottoman, 235, 235–237, 239, 242; of Wasit, 357n31
conquistadors, 243–244, 256–257; as deities, 393n20; and natives, 275
Constantinople, 173, 235–237, 239. *See also* Istanbul
constellations, 131*f*, 143, 209; illustrations of, 22; in judicial astrology, 126, 127; of Ptolemy, 131, 132; Qazwīnī on, 77, 95, 129, 135, 282; Ṣūfī's catalogue of, 130–132; Umayyad representation of, 143. *See also* celestial globes; planets; stars; zodiac
contemplation, 7, 112, 121; of creation, 110–111, 113, 117, 130, 186, 189, 227, 228; as spiritual exercise, 118–119
conversion: and Catholic Church, 243–244, 257, 259; to Islam, 61, 74, 124, 252; Ismaili, 86; and Jesuits, 274; of Mali elite, 213; of Pacific Islanders, 306; Portuguese policies of, 286; Ottoman attitudes on New World, 258
conversos, 246. *See also* Jews
Copernicus, 18, 303, 324
corruption, 284; and decay, 41, 107; and generation, 138, 167, 168–169, 175; of time, 138
Cortés, Hernán, 259
cosmography and cosmology, 144, 225, 240; Aristotelian, 49; classical accounts of, 326; of d'Ailly, 250; Indic, 158, 201, 262, 263–264, 279, 281, 380n7; Islamic, 204, 312; Latin, 250; Ptolemaic, 135, 144, 199; Zoroastrian, 124. *See also* *ʿilm al-hayʾa*; macrocosm and microcosm
cosmopolitanism, 258
cosmos, 135, 187, 197, 289; accounts of, 204, 282; and angels, 136–137, 365n12; diagram of, 9*f*, 202; in egg, 199, 289, 297, 379n1; Fakhr al-Dīn al-Rāzī on, 135, 184; forces in, 75, 125, 190; interconnection of, 75, 92, 118, 130, 166, 168, 171, 193–194; study of, 75–77, 190, 306; Qazwīnī on, 135, 187. *See also* macrocosm and microcosm
creation: best possible, 108; design of, 105, 158, 183; and God, 121; hierarchical order of, 162, 225; human body as microcosm of, 32, 121, 169, 183–184, 187; Qazwīnī on, 102, 117, 189, 225; stories of, 201; wonders of, 4, 106, 109, 159, 238, 278. *See also* contemplation: of creation; design; nature

Creator. *See* God

creatures, 183, 211–212, 277; aquatic, 181, 211, 273; arboreal, 254; dubious, 299; on islands, 211; strange, 95, 160, 164, 210, 264. *See also* animals

creeds, 68, 263, 306

Ctesiphon, 174

curation, 295; of the past, 14, 328. *See also* museums

cures, 125, 177, 180, 261, 281, 288, 291. *See also* medicine; recipes

curiosity: Europeans as objects of, 263, 298, 305, 314–315; expressions of 4, 6, 27, 210, 214, 262, 276, 296; language of, 145; Muslims lack, 302; promoted in *New Report*, 255–256; and wonder, 118, 120

curiosities: cabinet of, 14, 345n18; collection of, 9, 10, 14, 141, 143, 148, 159, 294, 305; as *gharāʾib*, 29. *See also mirabilia*; rarities; wonders and rarities

cyclops, 157, 370n50

cyphers, 84, 124, 141, 174, 179, 241, 291. *See also* scripts

da Gama, Vasco, 283

d'Ailly, Cardinal Pierre, 250

al-Dāmghānī, Abū Ṭālib, 65, 88

al-Damīrī, Kamāl al-Dīn, 235

Dāwūd ibn Khalaf, 60

d'Anghiera, Pietro Martire, 252

daimones, 190, 194

Dante, 207

Dārā Shikoh, 289–290, 297

dāstān, 320, 333. *See also* romance

de Gómara, Francisco López, 252, 259

de las Casas, Bartolomé, 250

Delhi Sultanate, 35, 278, 285, 289

Democritus, 171, 372n9

demons, 81, 102, 106, 159, 196, 291, 321; and angels, 179; and jinn, 326, 359n10; maleficent forces, 194; Persian *dīv*, 159; possession by, 180, 375n27; summoning, 78, 81, 83–84. *See also* jinn

de Sacy, Antoine Isaac, 301

design: of creation, 105, 106, 158, 165, 183; divine, 3, 5, 43, 108, 113, 118, 121, 228, 241, 304; God's secret, 87; as proof of God, 184, 258, 304; Qazwīnī on, 108, 120

devices and instruments, 51, 129; astrological and astronomical, 87, 129, 133; catapults and siege engines, 33, 79; mathematical, 129; mechanical, 63; musical organ, 276; optical, 304, 306, 309, 327; protective, 173; steam engine, 307, 326. *See also* automata; clocks; compass; printing press

d'Herbelot, Barthélemy, 300

Dhū l-Nūn, 68

diagrams, 8, 102, 104, 281; of cosmos, 9*f*, 202, 280; of lunar mansions, 134; magic, 84–85, 290. *See also* magic circle

Dioscorides, 102

dispositions and natures, 163; environmental, 191; as *hayʾa* and *hexis*, 188; humoral and temperamental, 163, 168; innate and natural, 121, 161, 194, 257; psychic, 122, social, 32, 125. *See also* climatic determinism

disputation, method of, 39, 46

divination, 79, 192, 241; astral, 126, 278; Quranic, 288; science of, 133; writings on, 318. *See also* bibliomancy

divine double, 190, 194

doctors and physicians, 84, 85, 104, 120, 162, 181, 183, 184; European, 302, 304, 322, 323, 324–325, 371n56; on natural wonders and unique properties, 102, 167 180, 185; qualifications of, 181, 302. *See also* medicine

dog-headed people, 159, 161*f*, 257, 287

Dorado, El, 263

Dragon Island, 7*f*, 212

dragons: belief in, 12, 23; on ceramics, 151; theories of, 212; three-horned, 155; as *thuʿbān*, 105, 106*f*

dreams and visions: 22, 68, 69, 235, 148; interpretation of, 76, 79, 146, 238, 318. *See also* prophecy

drugs, 179, 193, 276, 287; Apollonius on, 177; Ibn Sīnā on, 169; and intoxicants, 280. *See also* medicaments; suffumigation
Dutch, merchants and colonialism, 242, 263; 304, 307

Earth, 74, 197; circumference of, 205, 379n6; form and shape of, 204, 250; as a grape in water, 379n1; inhabited, 32, 205; orbit of, 205; pillars of, 201, 273, 327; science, 226; in seven climes, 199, 277; sphericity of, 204–205, 207
East: ideas about, 152, 301; and Latin Christendom, 299; marvels of, 264; otherness of, 12; primitive beliefs of, 15; riches in, 2; vs. West, 14; as timeless, 305; and universal spirituality, 304
Easter, 249, 388n50
Eastern Hemisphere, 246, 273, 330
East India Company, 1, 15, 295, 302, 306, 308, 310–311
Eden, garden of, 275
education: colonial, 309, 319; courtly, 208–209, 239; of Qazwīnī, 30, 46–47; scholastic, 25, 38, 44–45, 48, 95, 117; secular, 311. *See also* madrasas
egalitarianism, 304
elements, foundational, 124, 138, 166, 169, 179, 228; and *ummahāt*, 168, 171. *See also* alchemy; humoralism
elemental magic, 122, 179, 185, 190, 238, 281. *See also* amulets; *nīranjāt*; magic; talismans
elixir, 166, 169. *See also* alchemy
Elizabeth I, 260
eloquence: affective power of, 83, 142, 180; Qazwīnī known for, 104, 285; study of, 82; in *Wonders and Rarities*, 110, 115. *See also* rhetoric
Elysian paradise, 206
emanationism, 326; of Ibn Sīnā, 49, 78; and Ismailis, 86; and Freemasonry, 304; and monism, 265; Platonic, 49, 195; of Plotinus, 42. *See also* Neoplatonism
empiricism, 22, 48, 130, 169, 180, 182–184, 226, 323, 325, 327, 361n32
encircling ocean/sea of darkness: Alexander in, 196, 223, 253, 268, 267*f*, 274, 313; exploration of, 151, 208–212, 213–214; impenetrable, 20, 203–206, 212, 215, 217, 222; in Indic cosmology, 277, 290; and inhabitable lands, 8, 199, 202*f*, 205, 223; and western sea, 19, 208, 247, 249, 257. *See also* Atlantic Ocean; oceans; seas
enchantment, 22–23; and eloquence, 83; as historical method, 330; in India, 262; and modernity, 346n32; Orientals and, 12, 295. *See also* magic
encyclopedias, 237, 282, 300; medical, 16, 41; of positive law, 66; of al-ʿUmarī, 153, 165, 166
endowments, 44, 62–63, 66, 90
England, 310, 319
Enoch, 170. *See also* Hermes
entertainment: and consumption, 5, 314; and refinement, 105; for rulers and courts, 144, 146, 240, 313–314; and Ṣaymarī, 146–147; and storytelling, 28, 142, 320. *See also* pleasure
epistemology: ambiguity and, 343n2; sources for, 41; of Ibn Sīnā, 50; and wonder, 118. *See also* cognition; psychology
erotica and eroticism, 149, 232, 233, 291, 315. *See also* love; nudity; sex
eschatology, 69, 106, 239; and astrology, 126–127; and end of time, 236–237, 240
Esdaile, James, 323
esotericism, 86; Christian, 304; Indic, 281. *See also* occult learning and sciences; secrets and secrecy
Estado da Índia, 283. *See also* Portuguese
Eternal Islands, 206–207, 208, 214–215, 217, 218–219, 222, 240. *See also* Canary Islands
ethics, 185–186, 277; study of, 188, 231. *See also* *akhlāq*; moral philosophy
ethnography, 295, 302, 315, 321
ethnos, 163. *See also* genealogy; group identity; race, theories of
Euclid, 42, 46, 47

Euphrates River, 33, 140
Europe and Europeans, 295; civilizational unity of, 13; as curiosities, 305, 314–315; domination, 321; historical accounts of, 202; learning, 12, 308, 310–312, 319; as magicians, 324; rationality of, 12; superiority of, 19, 296, 309. *See also* colonialism; *farang*
evil: existence of, 107–109, 194; forces of, 159; vs. good, 188; question of, 257–258. *See also* theodicy
evil eye, 121, 123
exegesis: biblical, 300; Quranic, 46, 78, 230, 235, 281. *See also* commentaries
exhibitions and fairs, 14, 295, 307; catalogues, 344n9; Great Exhibition of London, 295; Universal Exhibition of Paris, 295, 296–297
existence: diversity of, 258; divisions of, 49, 225; emanation of, 107; of evil, 107–109; Fakhr al-Dīn al-Rāzī on, 184; of God, 106, 110, 265; as interconnected, 193; Qazwīnī on, 130, 162, 288; strangeness of, 130; ultimate reality of, 265
exorcisms, 70, 265, 327
exotica: appetite for, 142, 254, 264; commodification of, 242, 247, 294–295; flora and fauna as, 243, 249; and geography, 20–21, 301; remoteness of, 102, 207, 210.
experience: authority of, 29; direct, 113, 169, 178, 183–185; empirical, 48, 361n32. *See also* empiricism
exploitation, 242, 306, 308, 314. *See also* colonialism; slaves
explorations, 224, 242, 246, 253, 274; of Atlantic, 206; Christian, 214–215, 222; Iberian, 247; and New World, 20, 212, 214

fables: animal, 114, 149; as *khurāfāt*, 114, 148, 155, 182; Oriental proclivity toward, 11–12, 299; *Wonders and Rarities* as collection of, 302
faculties, 49, 111, 322; of astonishment, 291; cognitive, 110, 116; Ibn Sīnā on, 124; of imagination, 26, 49, 122, 142; influence of, 125, 176; of perception, 187; sacred, 50. *See also* cognition; epistemology; psychology
fāl-nāma, 238, 288
falsafa, 55; relationship to *kalām*, 41–43. *See also* philosophy
al-Fārābī, Abū Naṣr, 42, 136, 187
farang, 286–287, 306–309, 392n9; lands of, 266, 274, 276; as objects of curiosity, 263, 298; practices of, 313–315; on religious diversity, 290. *See also* Europe and Europeans
Farazdaq, 147
Fatimids, 86, 206
Ferdinand and Isabel, 243–244
fiction: category of, 114, 333; vs. fact, 11, 299, 321; Oriental origins of, 12–13
filāḥa, 178, 375n25. *See also* agriculture
Firdawsī, 100, 230, 320
Firishta, Muḥammad Qāsim Astarābādī, 286
fiṭra, 121, 257. *See also* dispositions and natures
forces: angelic, 79, 175, 201; celestial, 122; elemental, 75; of evil, 159; harnessing, 23, 78, 124, 189, 290; hidden, theory of, 167; of humors, 179; in nature, 124, 187, 191; physical, 126, 191, 193; and *rūḥāniyyāt*, 176; spiritual, 171, 189; transformative, 185. *See also* occult powers; properties, unique or special; souls; talismans
fountain of life, 139, 227
Frederick II, 47, 160
free will, 62, 111
Freemasonry, 304, 399n21
French imperialism, 304, 307

Galen: as alchemist, 172; on amulets and incantations, 180, 182–183; authority of, 181; climatic determinism of, 32, 191; on divine design, 43, 184; and empiricism, 180–181; and al-Ghazālī, 184; humoralism of, 167, 363n52; on indescribable properties, 179; on legendary animals, 164, 371n61; on seizures, 180; teachings

of, 42, 108; writings of, 163, 183, 349n26, 376n32; as wonderworker, 183
Galland, Antoine, 156, 300–301
Garuda, 158, 227
genealogy: of ʿĀdil Shāhī rulers, 285; appeals to, 263; and descent, 267; after the Flood, 163–164, 231; of Mongols, 34, 76, 289; notions of, 163; of Qazwīnī, 72; of Shia Imams, 86; of Timurids, 267. *See also* group identity; *nisba*; race, theories of
generation and corruption, 138, 167, 168–169, 175
Genghis Khan, 3, 33, 79, 364n56; descendants of, 289; patronage in court of, 34
geocentrism, 11, 50, 380n7
geography: administrative and descriptive, 36, 144–145, 318; classical Arabic and Persian, 19, 205, 208, 227; and colonial education, 309, 319; of Ḥājjī Khalīfa, 300; of Ibn al-Faqīh, 105; of al-Idrīsī, 207, 216; imagined, 21, 301, 346n31; in Latin, 217; mathematical, 50, 51, 74, 76; and natural history, 238; new models for, 306; philosophers on, 204; of Ptolemy, 204, 216; Qazwīnī on, 25, 71, 231, 288; of world, 239, 274; of Yāqūt al-Rūmī, 37–38
geomancy, 78, 133–134, 241, 255, 290, 318, 365n8
gestation: influences on, 162, 190; section on, 378n48
gharīb: definition of, 28–29, 73–74, 121–122; as extraordinary, 68, 98, 117, 162, 180, 191; and the occult sciences, 84, 126; and strange animals, 160
al-Ghazālī, Abū Ḥāmid: on angels, 136; on contemplating wonders, 113, 118, 119, 120–121, 241; and Galen, 108, 184; on habitus, 188; and Ibn Sīnā, 41, 42, 195; and jinn, 27, 82; on optimism, 108; and Persian, 101, 112; on pride, 119; and refinement of the soul, 112, 186, 241; as renewer, 59; and teleological proof of God, 121, 184
Ghaznavids, 289
ghouls, 159
Gibraltar, Strait of, 205
gilding. *See* gold
Gilgamesh, 33
Gīsū-darāz Muḥammad Ḥusaynī, 283
globes, 309; Earth as, 204; maps of, 270; symbol of, 273. *See also* celestial globes
gnosis, 6, 189, 192, 194, 230, 241, 265, 268; as *maʿrifat*, 277
God: and angels, 138; as architect, 304; arguments for, 41, 106, 183–184, 285; attributes and names of, 34, 69, 84, 97, 109, 118, 227, 280; and best possible world, 108; and devil, 257; human knowledge of, 118, 119, 120, 142, 225, 227; and jinn, 160; judgement of, 111; justice of, 107; lands of, 31, 36, 37, 41; and Mongols, 87; as necessary existent, 110, 265; pen of, 281; power of, 4, 111, 137, 149, 159; as prime cause, 49, 76, 106–107, 109, 112, 120, 137; and prophets and revelation, 137, 139, 172; and Quran, 82, 141, 223, 288; and suffering, 107–108, 138, 140; throne and pedestal of, 136, 281; wisdom and omniscience of, 109, 119, 127, 136, 164, 168, 184; and wonderment, 118, 362n41. *See also* creation; design; wisdom
Gog and Magog, 10, 144, 161, 237, 240, 255; wall of, 200*f*, 202, 223, 268, 273
gold: and alchemy, 169, 172, 244; in amulets and talismans, 174; in the Atlantic, 209; automata made of, 63; conjured, 68; a foundational metal, 168; for gilding manuscripts, 102, 148, 266, 267; jewelry made of, 151, 294; from Mali, 213; in metalwork, 133; from New World, 244, 249, 259. *See also* alchemy; metals
good: conceptions of, 107–108, 109, 110; dispositions toward, 122, 125; vs. evil, 124, 188, 193; the One as, 194; in society and life, 112, 187, 276. *See also* evil; optimism; theodicy
graffiti, 29, 141
Granada, 98, 218, 244

Greek learning and philosophy: alchemy in, 171, 172; attitudes on, 43, 306; curricular study of, 40; and divine double, 190; as *falsafa*, 42; influence of, 310–311; monsters in, 145, 158, 160; place of wonder in, 118, 145, 184; translation of, 42, 63, 143–144, 192; Zoroastrian priests in, 124
Gregory, William, 323–324
griffins, 1, 299
grimoires. *See* handbooks of magic
group identity: allegiance to, 37; categories for, 163–164; as collective, 32; of humans, 192; and pride of place, 63. *See also* sectarianism and factionalism; race, theories of
gunpowder, 79, 307, 318

habitus, 32, 188. *See also* dispositions and natures
Habsburgs, 207, 259
hadith, 28–29, 59, 83, 84; authority of, 110; collections, 154, 369n42; partisans of, 60. *See also* Muhammad, Prophet
Hagia Sophia, 17, 237
Ḥājjī Khalīfa, 300, 322
Hamadan, 33, 174
al-Hamadhānī, Badīʿ al-Zamān, 138, 147–149
Ḥanafī law, 60, 62
Ḥanbalī law, 62
handbooks of magic, 79, 81, 82–84, 175, 178, 189, 190, 193, 195, 232, 356n19
happiness and felicity: pursuit of 3, 23, 112; and Venus, 147
al-Harawī, Amīn al-Dīn Khān, 319
al-Ḥarīrī, 104, 152*f*
Ḥasan-i Ṣabbāḥ, 86
Hastings, Warren, 308–309
Hazār afsān, 149. See also *The Thousand and One Nights*
Hebrew, 81, 235, 298, 299
heliocentrism: 18, 205, 303, 311, 309, 400n36. *See also* planetary motion
hell: Odysseus in, 207; perpetuity of, 258; punishments of, 106
Hercules: pillars of, 205–207, 208, 212; statue of, 206, 222; Strait of, 240
Hermes: and Aristotle, 171, 194, 196; Ibn al-Nadīm on, 172; identity of, 170, 372n8; as prophet Idrīs/Enoch, 175; on solar apogee, 364n57; as source of ancient wisdom, 176; teachings of, 68, 170, 190, 281, 288
Hermetica: in Arabic, 92, 170–171, 172, 175, 189, 194, 196, 365n7; and Freemasons, 304, in Sasanian Iran, 374n20. *See also* perfect nature
Herodotus, 10, 157, 212
hierarchy: in orders of existence, 49, 56, 162–163, 167, 186, 225, 258; civilizational and racial, 19, 307. *See also* classification and taxonomy
hieroglyphics and cuneiform, 33, 141. *See also* cyphers; scripts
ḥikma, 43, 49, 109, 110, 226, 228; as natural philosophy *(al-ḥikma al-ṭabīʿiyya)*, 76; as *sophia*, 43, 184. *See also* design; philosophy; wisdom
Hilla, 65, 89
Hindus: astrologers, 16, 17, 279; contacts and exchanges with, 290; and cosmology, 201; and Persian, 235
Hindustan, 11, 265, 276, 278–279, 287. *See also* India
Hippocrates, 180, 191
history: relation to memory, 23, 139, 141, 327, 328; world, 230, 286, 289, 304
histories, written: and collections of wonders, 238–239; local and regional, 37; of Mongol invasions, 80; monumental and universal, 87, 104, 214, 234, 286; Mughal, 289; and Ottomans, 238–239, 251–252. *See also* natural history, writings on
Homer, 157, 161
Hondius, Jodocus, 272–273
horoscopes: calculations used for, 91, 132; casting, 78, 80, 132; during gestation, 162; by Hindu astrologers, 16, 17; by Jāmāsp, 126–127; rejection of, 327. *See also* astrologers; astrology
Hülegü, 3, 34, 79, 87–88, 92

human body: and cosmic order, 167; and divine design, 107, 119, 183; marvels of, 183; as microcosm, 32, 121, 169, 183–184, 187
humans: characteristics of, 190; capacity of, 7; and cognition, 49; diversity of, 163, 191; ranks of, 50; relation to animals, 49, 110, 159, 170, 192; settlement, 231; social disposition of, 32, 188; and wonder, 4, 291. *See also* perfect human
humans and hybridity: 1, 11, 21, 17, 102, 145, 159–162, 164, 211, 257, 287, 288, 299, 319
humor and laughter, 3, 130, 146–147, 209. *See also* pleasure
humoralism, 124, 162–163, 167–168, 179, 363n52. *See also* dispositions and natures; medicine
Ḥunayn ibn Isḥāq, 183–184, 371n61
hypnosis, 323, 325. *See also* mesmerism

Iberian Peninsula, 211, 214, 242, 244. *See also* al-Andalus
Ibn al-Akfānī, 84
Ibn Amīr Ḥājib, Abū l-Ḥasan ʿAlī, 212–213
Ibn al-ʿArabī, 69, 236, 255, 265
Ibn Bakhtīshūʿ, 100, 101*f*, 104
Ibn Baṭṭūṭa, 219
Ibn al-Furāt, Abū l-Fatḥ, 154
Ibn al-Fuwaṭī, Kamāl al-Dīn, 70–71; on Maragha, 90–91
Ibn Ḥanbal, Aḥmad, 60–61
Ibn al-Jazzār, 184, 376n37
Ibn al-Kalbī, Abū l-Mundhir, 145–146, 173
Ibn Khaldūn, 214–217, 218
Ibn Khurdādhbih, 144–145, 172
Ibn Mājah, 71, 353n43
Ibn Manyās, 241
Ibn al-Mundhir al-Harawī, 150, 368n36
Ibn al-Nadīm: on alchemy, 172, 173; and the *Fihrist*, 146; on Hermes, 172; on Ṣaymarī, 148; on tales, 148–149, 154; on talismans, 174
Ibn Rabāḥ, Abū ʿImrān, 153–156, 369n42
Ibn Sīnā: on alchemy, 169; and Aristotle, 42, 50, 111, 160, 188, 193, 205; on angels, 136; on astrology, 76–78, 125; as autodidact, 50; and emanationism, 42, 78, 195; on evil, 107–108; on extraordinary phenomena and influences, 116, 122–123, 125, 180; on God as necessary existent, 110; on *ḥads*, 50, 350n10; on imaginary beings, 158; in Latin Christendom, 40, 47, 263, 299; and madrasa education, 44, 122; and medicine, 41, 166, 169, 191; metaphysics of, 43, 49–50, 111, 195, 229, 265, 323; and meteorology, 73–74; and minerology, 168; natural philosophy of, 40–43, 58, 75; and Persian, 101, 229; and Plotinus, 193; and *Pointers and Reminders*, 57–58, 112, 116, 117, 122–123; on properties of substances, 169, 179, 180; on prophecy, miracles, and magic, 43, 122, 180, 205, 265, 322–323, 325; psychology of, 49–50, 111, 112, 123–124, 192, 195; on Ptolemy, 77, 205; on Ṣaymarī, 146; on skepticism and idiocy, 116; on strange creatures, 160; on void, 50; on wonders and wonderment, 3, 43, 112, 116–117, 120
Ibn al-Wardī, Abū Ḥafṣ, 210, 236–237, 240, 253, 386n22
Ibn Zunbul al-Maḥallī, 241
Ibrāhīm ʿĀdil Shāh I, 284, 285, 359n9
Ibrāhīm ʿĀdil Shāh II, 285–286
idols, 206, 380n10; worshipers of, 150
Idrīs: Hermes as, 175; prophet, 170. *See also* Hermes
al-Idrīsī, Abū ʿAbd Allāh: geography of, 207, 216–217; in Latin Christendom, 216, 382n37; on mariners from Lisbon, 208–209, 211; on western sea and islands, 211–212, 218, 222
Ikhwān al-Ṣafāʾ. *See* Brethren of Purity
Ilkhanids: administration of, 92, 233; and astrology, 91; and historians, 87; foundation of, 88; fragmentation of, 128; and painting, 28*f*, 101*f*, 104–105, 106*f*, 161*f*
illuminated manuscripts, 98; of Āzarī's poem, 270; with erotica, 291; of natural histories, 229; production of, 98; of *Shujāʿ-i Ḥaydarī*, 314–315. See also *Wonders and Rarities*, copies and translations

Illuminationist philosophy, 68, 189, 195. *See also* emanationism
ʿilm al-hayʾa, 76–78, 240, 350n12, 354n4. *See also* cosmography and cosmology
Iltutmish, Abū l-Muẓaffar, 278
imagination. *See* faculties
incantations and spells: for commanding occult forces, 25, 78, 83–84, 176, 280, 282; critique of, 181, 327; Galen on, 182–183; handbooks of, 81–84; and mesmerism, 325; Plotinus on, 193–194; power of, 75, 125, 149, 179; and wonder, 4; in *Wonders and Rarities*, 179, 232–233; Zoroastrian, 124
India: and alchemy, 172; Alexander in, 275–276; Aristotle on, 207; astronomy in, 278–279; cities and villages of, 290; and colonial education, 319; conquests of, 35, 196; and experiments on natives, 323; Masʿūdī in, 153; and metaphysics, 265; Orientalists and colonial officials in, 2, 10, 296–297, 302–303, 308, 310; Persian in, 285; phoenix in, 157; Piri Reis on, 246; and Portuguese, 249, 283, 286–287; and Saturn, 148; Ṣaymarī in, 148; temples in, 156; wonders of, 154, 157, 239, 262, 263–265, 314; *Wonders and Rarities* in, 265–266, 288–293, 296. *See also* colonialism; East India Company; Hindustan; Mughals
Indian Ocean, 140, 212, 242, 246, 258, 260, 366n20
al-insān al-kāmil, 265, 290, 391n6. *See also* perfect human
infidels, 150, 213, 259; Christians as, 256–257. *See also* idols
influences: astral and celestial, 74–75, 77–78, 176, 179, 187; elemental, 124, 171, 185; of emotions, 123; extraordinary, 121–122, 125, 180, 189; external, 168; mesmeric, 14, 323; spiritual, 322. *See also* properties, unique or special
inhabitable lands, 8, 32, 205, 216; limits of, 152, 208, 212; as quarter of the Earth, 32, 199, 205, 207, 251, 379n1
inimitability. *See* Quran
intellects, 50, 136, 192, 194; angels as, 49. *See also* cognition; faculties
interpreters and translators: and Abbasid translation movement, 42, 183; and Alexander, 253; at courts of law, 67; and European expeditions, 246–247, 249, 264; and Mongols, 79; and Orientalists, 13, 308; Ottoman, 259; with remote islanders, 209–210; at royal courts, 47, 187
intuition, as *ḥads*: 48, 50, 350n6, 350n10
Iran: ancient kingdoms of, 142, 149, 173; book arts in, 105, 268; madrasas in, 230; and Mongol invasions, 36, 39, 79, 80, 100, and Mughals, 270; philosophers of, 314. *See also* Persia
Iranian plateau, 30, 33, 73
Iraq: Abbasids in, 3, 35, 143; book arts in, 105; Mongol conquest of, 79, 80, 89–90, 92–93, 98, 289
Isfahan, 61, 351n16, 352n28
al-Iṣfahānī, Abū l-Shaykh, 105–106
al-Iṣfahānī, Shams al-Dīn, 54–57, 351n17
Islam: authority of, 311, 329; and calendar, 255; vs. Christianity, 256, 257–258; as civilization, 17, 19, 204, 328; and conversion, 150–151, 218, 252, 256, 259; Europe indebted to, 306; history of wonder in, 4–5, 12, 20, 22; Orientalists on, 301; and philosophy, 28, 41–42; reform, 13, 312, 321; rise of, 77, 127, 175; rites and tenets of, 257, 259, 312; and science, 13–14, 18–19, 311; Sunnis as guardians of, 242; territory of, 30–31, 33, 34, 36, 37, 47, 239, 284; and Venus, 147. *See also* law and jurisprudence; natural philosophy; piety and devotion; theology
islands: Antilles, 247; Brasil, 219; Caribbean, 246; Dragon Island, 7*f*, 212; escapades on, 156; inhabitants of, 155, 161; inhabited by jinn, 151, 161; Pacific, 160; Palace Island, 287; Sarandīb, 151, 153, 195, 275; Spice Islands, 246, 258; Wāqwāq, 11, 151, 254, 255, 268; in western ocean, 209–212; of Women, 154. *See also* archipelagoes; Canary Islands; Eternal Islands

Ismailis, 61, 86–87. *See also* Fatimids
al-Iṣṭakhrī, Abū Isḥāq, 104
Istanbul: archives and museums of, 17–18, 385; and commerce, 251; as imperial capital, 224, 237, 246, 251, 259, 327; learning in, 303; and Mughals, 275; observatory of, 279; printing press in, 252; translation of Ptolemy in, 239. *See also* Ottomans; Topkapı Palace

Jābir ibn Ḥayyān, 169, 229
Jahāngīr: autobiography of, 262–264; and European magicians, 313–314; with paintings and globes, 270, 271*f*, 272–273; successors of, 286
al-Jāḥiẓ, Abū ʿUthmān, 155, 164, 369n45
Jains, 201, 282
Jai Singh, 279
Jāmāsp, 126–127, 364n56
Jamshīd, 146, 229
Jarīr, 147
Java, 210, 287
Jerusalem, 35, 55, 246
Jesuits, 263, 273, 274–275
Jews, 67, 235; and divine design, 5; and exchanges with Christians, Muslims, 216–217; and Mongol conquests, 79–80; in Mosul, 47; and New World, 243; and occult sciences, 175; and Ottomans, 258
jinn: and Alexander, 196; commanding and summoning, 28*f*, 70, 78–79, 81–85, 185, 279, 324; as *daimones*, 190, 194; as genies, 159; on islands, 151, 161; market of, 156; as microbes, 327; and mountains of Qāf, 196, 199; and Ottomans, 238, 241, 255; as primitive savages, 161, 321; sex with, 149; stories of, 146; Qazwīnī on, 27–28, 102, 109, 121, 147, 225; and talismans, 174; in Urdu romances, 320. *See also* demons
judges, 39, 59, 65–66, 79; Qazwīnī as, 65, 104
judiciary, 63, 66–67, 71, 353n31
jurisprudence. *See* law and jurisprudence
al-Jurjānī, 230, 384n10
justice, 187, 284; divine, 107–108; ideals of, 258, 276
al-Juwaynī, ʿAlāʾ al-Dīn: and the Mongols, 80, 87, 90; and *Wonders and Rarities*, 97–98, 99, 105

Kaʿba, 202, 240
Kailash, Mount, 277
kalām, 42, 56, 349n25. *See also* theology
Kamāl al-Dīn Ibn Yūnus, 45, 46–47
Kambalposh, Yūsuf Khān, 319–320
karāmāt, 68, 123. *See also* miracles
Karāmat-ʿAlī, Sayyid, 311–312
Karrāmīs, 60–61
al-Kashshī, Zayn al-Dīn, 39–40
Kātib Çelebī. *See* Ḥājjī Khalīfa
Khashkhāsh the Sailor, 208
khawāṣṣ, 107, 167, 178–179, 184. *See also* properties, unique or special
Khiḍr, 138–139, 366n15
Khomeini, Ayatollah, 325
Khuʾī, Mīrzā ʿAlīqulī, 315
khunthā, 191, 378n48
khurāfa, 114, 182, 368n35. *See also* fables
Khwārazm-Shāh, 33–35, 39, 78, 82–83, 182, 289
al-Kindī, Abū Yūsuf, 42, 193, 195, 378n51
kingship, 76, 80, 230, 275, 285, 213–214
Kipling, Rudyard, 15
Kishore, Nawal, 315, 319–320
Kufa, 89
Kūshyār ibn Labbān, 53, 78

lapidaries: ascribed to Aristotle, 92, 169, 173, 184, 195–196; of Bīrūnī, 170; Greek, 155, 180, 369n45; of Naṣīr al-Dīn al-Ṭūsī, 92, 100. *See also* stones; metals; minerals
Lancelotto Malocello, 219
Latin writing and literature: Alexander in, 196; on geography and cosmography, 217, 242, 249, 250; monsters and monstrosity in, 5, 143, 159–160, 162, 257; on occult properties, 179; parallels and connections with Arabic learning, 18, 47, 51, 111, 113, 130, 133, 155, 184, 187, 188, 189, 206, 298–299, 302, 304; Ptolemy's geography in, 216

law and jurisprudence: Anglo-Muhammadan, 308; codes of, 318; positive, 56, 60, 62, 66; reason and revelation in, 43, 188, 226; schools of, 59–62, 282; study of, 44–46, 56–57. *See also* judges; judiciary
Lepanto, Battle of, 259
Levant Company, 260
liberalism, 21, 296, 304
libertines, 146, 307
libraries, 14, 16, 31, 37; at Alamut, 87, 357n28; in Maragha, 70; of Mustanṣiriyya madrasa, 70; Ottoman, 17, 237–239, 327–328; public, 62, 234. *See also* books
lithographs, 7, 16; and Aḥmad Khān, 318; of natural histories, 313, 315, 327; of occult sciences in South Asia, 402n54; of *Wonders and Rarities*, 317*f*, 319, 326–327. *See also* books; Kishore, Nawal; printing press
logic: Aristotelian categories of, 225; basis for dialectal argumentation, 44; curricular study of, 39, 40–41, 43, 46, 48, 54, 55, 57, 58, 110; of Porphyry, 42, 48; Qazwīnī's use of, 109, 225
love: divine, 68; as natural magic, 180, 193; potions, 232; stories of, 230; and wonder, 119. *See also* eroticism; sex; sympathy
lunar mansions. *See* Moon
luxuries, 142, 296. *See also* exotica

al-Maʿarrī, Abū ʿAlāʾ, 138
macrocosm and microcosm: in alchemy, 166; in Hermetica, 171; Indic ideas on, 289; as related to human body, 32, 121, 169, 183–184, 187. *See also* cosmos
madrasas: architectural features of, 47, 66; astrolabes and celestial globes in, 31; of Baghdad, 45, 62–63, 71, 90, 98; of Cairo, 55, 235; of Calcutta, 312; education, 38, 44, 45, 82, 95, 104, 122, 349n29; founded by Sharābī, 56, 62, 65–66, 71, 89, 140, 351n19; illuminated manuscripts in, 45, 63, 103, 104; of Istanbul, 17, 240, 241; libraries of, 17, 60, 62, 66–67, 90; modern study on, 349n29; in Nishapur, 81; Persian translations of the Quran in, 230; relation to law and the judiciary, 60, 65; study of philosophy and natural science in, 44, 226. *See also* al-Azhar
magic: Aḥmad Khān on, 321–322; and black cats, 176; and diagrams, 84–85, 174, 280*f*; and eloquence, 83; Fakhr al-Dīn al-Rāzī on, 78, 83, 123, 125, 278; and *farang*, 287, 313–314, 315, 320, 323–324; as highest stage of learning, 194–195; as licit, 323, 403n69; as *mageia*, 124; and medicine, 172, 192; and miracles, 122–123, 125, 192–193, 195, 265, 323; and Mongols, 80; naturalization of, 43, 75, 118, 121–122, 123; 179–180, 193–195, 265, 298; persecution of, 322; practiced in Africa, 213; and poetry, 268; relation to science and religion, 11, 21, 197, 323; as *siḥr*, 86, 180, 193, 323; and sex, 3, 270, 291; Solomonic, 280–281; and superstition, 12, 181, 327; and Tantra, 84, 282; and Venus, 147; and wonder, 6, 238. *See also* elemental magic; handbooks of magic; natural magic
magic circle, 84, 279–280, 282, 356n21, 394n30
magnetism, 320, 322–323, 325. *See also* mesmerism
magnets, 117, 122, 179–180, 182, 215, 375n30. *See also* compass; properties, unique or special
Mahābhārata, 277–278
madhhab, 59. *See also* law
Maḥmūd of Ghazna, 35, 73
Mahomedan Literary Society of Calcutta, 311
Malacca, 10, 286
Malay Archipelago, 258, 286
Mali Empire, 212–214
Mālik ibn Anas, 60–61
Mālikī law, 60, 62
Mamluks, 55, 79, 88–89, 166, 213, 234, 236, 242, 246, 258
al-Maʾmūn, 107, 145
mandal. *See* magic circle
Manicheans, 175
Mansā Mūsā, 212–214, 382n29

maps: of Arabian Peninsula, 240; boundary lines on, 34–35; and Canary Islands, 219, 220*f*; and colonialism, 309; of globe, 270; of Ibn Khaldūn, 217; of al-Idrīsī, 216–217; as imperial endeavor, 222; and missionaries, 262, 274; and Mughals, 270, 271*f*, 272–273, 393n17; and New World, 243, 254, 324; orientation of, 201; of Piri Reis, 244, 245*f*, 246–249, 251, 257; *qibla*, 236; in *Wonders and Rarities*, 8, 102, 200*f*, 202*f*, 203*f*, 217, 329*f*, 330; woodcut, 252. *See also* cartography; portolan charts
al-Maqrīzī, Taqī l-Dīn, 218, 235
Maragha, observatory, 57, 70, 90, 91–92, 128, 130
marginalia, 149, 250–251, 261, 296, 309, 400n37; and ʿĀdil Shāh translation, 261, 262, 288–291, 296; *Wonders and Rarities* added to Damīrī, 385n19
Marinids: and Atlantic exploration, 218–219; and Canary islanders, 215
Maronite Christians, 298
Mars, 51, 85, 148
marvels: as *ʿajāʾib*, 16; categories for, 162; and divine design, 116; as *mirabilia*, 159–160; of Orient, 12, 264; of sea, 151, 153, 219, 248; as *thaumata*, 145, 184. *See also* *ʿajāʾib*; curiosities; miracles; wonders and rarities
masala, 276. *See also* spices
masnavī: adaptation of Qazwīnī, 267; of Rūmī, 268
al-Masʿūdī, Abū l-Ḥasan, 153, 206, 208, 235
materia medica: Ibn Sīnā on, 176; Qazwīnī on, 102, 178. *See also* medicaments
mathematics: and astrology, 91; Euclidean, 42, 46, 47; and instrumentation, 129; and Mongols, 90; in Orientalist critiques, 302; problems of, 46, 47; as *riyāḍiyyāt*, 41; scholastic study of, 44, 76, 77; training in, 87
al-Māturīdī, Abū Manṣūr, 62
Mecca: as center of world, 202; direction of prayer, 30, 54; at end of time, 237; historical accounts of, 113–114; in maps and books, 240, 272, 319; and Ottomans, 242, 258; pilgrimage to, 65, 144, 212, 267; and the Prophet, 72. *See also* Kaʿba; pilgrims and pilgrimage
medicaments, 11, 31, 288; Galen's knowledge of, 183; Ibn Sīnā on, 41; Sakkākī on, 84
medicine: discourses of, 378n48; encyclopedias of, 16, 41; Galenic writings on, 42, 167, 179; Greek sources on, 181; knowledge of, 125; and magic, 192, 318. *See also* doctors and physicians
Medina: earthquake in, 87, 357n28; as holy city, 319; and legal history, 60; and Ottomans, 242; and the Prophet, 70, 72, 240
Mediterranean: as boundary, 206, 215, 222; navigation in, 215; and Ottomans, 242, 259; portolan charts of, 216–217, 220*f*, 243, 249; and Red Sea, 10, 242, 260
Meḥmed I, 238
Meḥmed II, 239
men: anxieties of, 186; biographies of, 37; dominance of, 186; in public life, 67; sexual desire of, 177; traits of, 191
Mercator, Gerardus, 272, 300
mercenaries, 34, 216, 219, 283, 286, 305
merchants, 142, 151, 153, 287; Christian, 214, 216, 219, 263–264, 273; English, 260; Genoese, 215, 218; vessels, 151. *See also* colonialism; commodities; spices
mercury and sulfur, 168–169, 174, 372n5. *See also* alchemy
Mercury, 52*f*, 350n12
mermaids and mermen, 1, 2, 11, 253, 255, 256*f*, 390n69
Meru, Mount, 201, 277
mesmerism, 14, 320, 322–325, 403n74
metals: formation of, 73; foundational, 168; and metallurgy, 231; and metalwork, 30, 51, 134*f*, 263; and New World, 249; precious, 166, 213, 249, 255; Qazwīnī as authority on, 300; talismanic use of, 174; transmutation of, 169. *See also* lapidaries
meteorology, 74–75, 133, 178; Aristotelian, 73–74; and meteorological phenomena, 4, 73, 87, 121, 311
meteors and meteorites, 74, 121, 300

Middle Persian, 42, 53, 143, 155, 158, 178. *See also* Persian
millenarianism, 126, 255. *See also* eschatology
minerals: and alchemy, 168–169, 176; and classification, 95, 162, 167, 192; and colonialism, 2, 15; and divine design, 10; formation of, 73, 138, 168; Qazwīnī as authority on, 166; properties of, 107, 159, 169, 171; study of 45, 49; and talismans, 173; and wonder, 4, 185, 228, 241. *See also* properties, unique or special; talismans
Mīr Muḥammad Ḥusayn, 305–309
mirabilia, 159–160, 255. *See also* curiosities; wonders and rarities
miracles: as *ʿajāʾib*, 16, 17; Christian collections on, 145, 367n26; Ibn Sīnā on, 117, 265, 325; 361n36; and magic, 75, 122–123, 125, 186, 192–193, 195, 197, 265; and modernity, 23, 197, 321–323, 325; as *muʿjizāt*, 50, 68, 123; prophetic, 50, 68, 123–124, 322; as rupture with custom, 121, 363n45, 375n26; of saints, 68–69; varieties of, 68
mirrors: all-seeing, 145–146; iron, 145, 195
Miskawayh, Abū ʿAlī, 120, 188
misogyny, 191
missionaries: Christian, 18, 258, 263, 273; Muslim, 18, 151, 259, 286, 327
modernity, 21–22; and enchantment, 346n32; European, 13; frameworks of, 19, 328; racial ideologies in, 163
Mongols: cause of civilizational decline, 303; descendants of, 233; ideology of, 79–80, 258; invasions of, 3, 33, 35–36, 40, 46, 61, 70, 80, 87–90, 92, 98, 128; and occult learning, 79, 81, 83, 85, 88, 91; prophecies concerning, 127; and Qazwīnī, 2, 3, 33, 54, 72; and theodicy, 108. *See also* Ilkhanids
monotheism: and creation, 184; and Indian sages, 277; and Muʿtazila, 62; and paganism, 175
monsters: of antiquity, 1, 155; in the New World, 257; in Pliny, 143, 157, 160, 212, 299, 302; and racial ideology, 163; of the sea, 155; varying categories of, 158–162. *See also* curiosities; humans and hybridity; jinn; savages
Moon: influence of, 75; and light of the Sun, 48, 130; mansions or stations of, 132–134, 135, 365n6, 374n21; movement of, 133; in water clock, 63. *See also* sublunar realm
moral philosophy, 112, 185–188
Moses, 127, 139–140, 366n15
mosques: calligraphy on, 141; in Damascus, 63; in Istanbul, 17; in Jerusalem, 235; models of, 294; on the Nile, 235; of the Prophet, 87; Spanish word for, 243; varieties of, 30, 283; in Wasit, 66
Mosul, 35–36, 133; Abharī in, 38, 47–48; course of study in, 46–47; and Mongols, 88; near Nineveh, 10; Qazwīnī in, 34, 45
Mountain of the Moon, 213, 240
mountains: and classification, 95; as cosmic supports, 199, 201; formation of, 137, 226; Ibn Sīnā on, 73; in Sanskrit cosmology, 277; and sphericity of the Earth, 205; wonders of, 145, 185. *See also* Qāf, mountains of
Mughals: collapse of, 305, 310; court, 262–264, 266, 278, 291; and Europeans, 263–264, 274, 286–287; history of, 289; and magicians, 313–314; and New World, 264, 272–273; and painting, 267, 273, 278; religious tolerance of, 276; and Safavids, 270
Muhammad, Prophet: authority of, 60, 62, 87, 284; Companions of, 70; customary blessings upon, 110; and descendants, 86; family of, 17, 147, 175, 320; and God, 111, 119; and jinn, 28; in Medina, 72, 87, 240; miracles of, 48, 68; and modern science, 311–312; in occult writings, 175; and renewal, 59; sayings and actions of, 43, 150, 164, 186, 227
Muḥammad ʿĀdil Shāh, 261, 288
Muḥammad ʿAlī Shāh (nawab), 313–314
Muḥammad Shāh, 279, 287
Muḥsin al-Mulk, Sayyid, 326

Muir, William, 312
mujaddid, 59
multiple worlds, possibility of, 223, 311–312, 383n1
munajjim. *See* astrologers
al-Muqaddasī, Abū ʿAbd Allāh, 37
Murād II, 241
Murād III, 251, 254–255, 259–260, 278
Museum of the History of Science and Technology in Islam, 17–18
museums: as imperial enterprise, 14–15; specimens collected for, 10; visiting, 17, 305, 320; *Wonders and Rarities* in 8–9, 22, 234*f*, 328, 329*f*, 330. *See also* British Museum; Salar Jung Museum
Mustanṣiriyya madrasa, 70, 352n29
al-Mustaʿṣim, 65, 89–90
al-Mustawfī, Ḥamd Allāh, 233
al-Mutanabbī, Abū l-Ṭayyib, 137
Muʿtazilī theology and theologians, 62, 68, 154, 155
Müteferriqa, İbrāhīm, 252
al-Muwayliḥī, Muḥammad, 327
mysticism, 201, 236, 281, 283, 288, 289; and mystics, 69, 151, 153, 234, 265, 266, 289, 290; and poetry, 230, 267; and wonder, 266. *See also* piety and devotion

nahḍa, 19. *See also* Arabic language
nasnās, 162, 164, 299. *See also* monsters; savages
nations: competition among, 295; distant, 240; hegemony over, 112; and Ibn al-Nadīm, 172; modern ideas of, 315; savage, 243; strange, 145, 160, 210, 212; as *umam*, 160
natives: of Canary Islands, 215, 242; and conquistadors, 275; conversion of, 243–244, 257; enslavement of, 218, 242, 305; as erotic, 151, 210; experiments on, 323; ignorance and superstitions of, 10, 12, 310; as informants, 11, 283; mannequins posed as, 294; savagery of, 244, 249, 256, 257, 258, 287; skin color of, 210, 315
nativities. *See* astrology; horoscopes
natural history, writings on: in Arabic, 235, 236, 254; classification of, 238, 240, 319–320, 323, 329; common titles for, 15, 238; illustrated, 158, 233, 291; marginal notes in works of, 309; in museums, 8; Orientalists on, 300, 302; in Persian, 224, 229, 233, 254, 319; poorly understood, 18; popularity of, 268; preceding Qazwīnī, 228; Qazwīnī as authority for, 299, 301; reassessments of, 299; Suʿūdī on, 254; in Urdu, 318, 319. See also *Wonders and Rarities*
natural magic, 180, 314, 322, 385n13, 392n11, 403n65
natural philosophy, 25, 38, 44, 46, 130, 226; Aristotelian, 75, 138, 168; developments in, 164, 171, 175; as *al-ḥikma al-ṭabīʿiyya*, 76; and Ibn Sīnā, 73, 83, 117; in Latin, 159, 298; and madrasa education, 25, 38, 39, 42, 44–46, 95; and magic, 195; and meteorology, 73–74; and theology, 225–226, 313; and minerology, 92. *See also* sciences
nature: conception of, 162, 188; design of, 3; as divine scripture, 113; forces in, 92, 124, 185, 187, 191, 194, 323; and nurture, 186; and religion, 227, 325; secrets of, 4, 43, 68, 159, 170–171, 184, 275; and signs of Creator, 40; and wonders, 7, 116–117, 125. *See also* creation; dispositions and natures; perfect nature
navigation, 219, 249; Atlantic, 380n8; in Mediterranean, 215; technologies of, 215–216, 243
naẓar, 95, 110–111. *See also* contemplation
necessary existent, 110, 265
Neoplatonism, 194. *See also* emanationism
Newbold, Thomas John, 1–2, 9–12, 297, 344n12
New Delhi National Museum, 329
New World, 20, 202, 251–252, 254, 274; Alexander's knowledge of, 250, 252–254, 274–275; categories for, 242, 252; exclusion of Muslims from, 243, 259; Iberian conquests of, 214, 251–252; Piri Reis on, 247–251, 254; riches of, 259, 307; Suʿūdī on, 254–255, 258–259; theory of Muslim discovery of, 18–19, 204, 247, 328, 388n50; as *yangī dunyā*, 252, 400n37

Nile River: cartographic representation of, 272; outpost on, 235; and papyrus, 66; source of, 213, 240
nīranjāt: etymology of, 124, 363n51; and Hermes, 133; in Persian version of *Wonders and Rarities*, 190, 231, 378n47; Qazwīnī and Ibn Sīnā on, 122, 179; and Zoroastrian cosmography, 124, 190. *See also* amulets; elemental magic; talismans
nisba, 37, 39, 54, 57, 72. *See also* genealogy
Nishapur, 61, 352n28
Niẓāmī Ganjavī, 100, 230, 290
Niẓāmiyya madrasas: in Baghdad, 45; charter of, 62, 352n29; illuminated manuscripts in, 359n12; in Nishapur, 81
Niẓām al-Mulk, 45, 86
Noah, 163, 231, 288, 304
North Africa, 20, 44, 222, 242, 301
Nowruz celebration, 155, 388n50
nudity: of angels, 291; of lovers, 315; of islanders and other creatures, 156, 161, 218; of natives, 249, 276, 291, 296. *See also* erotica and eroticism
Nujūm al-ʿulūm, 279, 281–282, 395n33

observatories, 278–279; in Maragha, 57, 70, 90, 91–92, 128, 130
occult learning and sciences, 16, 82–84; and cyphers, 141; and divine writing, 281; and Fakhr al-Dīn al-Rāzī, 38–39, 123; and Hermes, 170; in India, 278–280, 402n54; in Latin, 298; and mesmerism, 324; in natural histories, 229, 233; and Ottomans, 237, 238, 254, 389n68; and self-purification, 193; and wonder, 107, 241; in *Wonders and Rarities*, 231–232, 290, 291
occult forces or powers: and electricity, 322; and magic, 194; and mesmerism, 323–324; in physical world, 75; self-discipline, 190; summoning and commanding, 23, 28, 75, 83, 323, 324; and sympathy, 320; and Theosophical Society, 325
occult properties of substances, 8, 175, 179, 184, 192. *See also* proprieties, unique or special
oceans, 20, 152, 209–212, 214, 257; and Alexander the Great, 253, 393n19; boundaries of, 277; conceptions of, 205; creatures in, 157; of darkness, 205, 214; transformations of, 126–127, 139; unknowability of, 206; wonders of, 153–154, 165. *See also* encircling ocean/sea of darkness; seas
optimism, philosophical, 108. *See also* good; evil; theodicy
Orientalism: dragons as mark of, 12; and ideology of superiority, 11, 14–15, 294–296, 303–304, 310–312, 318, 322; legacy of, 21–22; and Qazwīnī, 296–299, 301–303, 305, 318. *See also* colonialism
Ottomans: domains, 224, 235, 237, 242, 246, 257; elite, 199, 203, 239, 244, 258; libraries of, 17, 237–238, 327; losing territory, 303; in Mediterranean, 259; and Mumluks, 236, 242; and natural histories, 229, 233–234; and New World, 251–252, 255–257, 259–260; and occult sciences, 238, 254–255; and religious tolerance, 258; and Safavids, 260; siege of Valencia, 247; and Spanish and Portuguese, 243;
Oxus River, 33, 230

Pacific Islands, 160, 306
painters, 98, 104–105, 232, 233, 360n13; of Herat, 267; Mughal, 263, 273, 278
painting, 8, 22, 264, 270, 293; allegorical, 273; ceiling frescoes, 143; of Herat, 267–268; mural, 273; in presentation copies, 278; and sculpture, 359n12; in Quṣayr ʿAmra, 143; in *Wonders and Rarities*, 102, 104–105, 266, 268; visual realm of, 278
Palace Island, 287
paper: for bookmaking, 66, 98–99; and Quranic ingestion, 82
Paracelsus, 303
paradoxes and puzzles: as basis for ambiguity, 147; and definition of God, 159; in existence 3, 58; natural, 229; as *paradoxa*, 145, 367n26; and squaring the circle, 47. *See also* ambiguity; aporia

patronage, 3, 63, 92, 98, 224, 307; in court of Genghis Khan, 34; of Khwārazm-Shāh, 39; of Mongols, 80, 89–90; networks of, 25, 60; of Persian literature, 285; of Tīmūr, 267
perception. *See* faculties
perfect human, 265, 288, 290, 391n6
perfect nature, 189–190, 196, 377n45, 378n47; mastering, 171; as *natura completa*, 189
perfume and incense, 31, 84, 147, 150, 247, 179, 280, 294, 385n13
perplexity and confusion: with astonishment, 109, 117–118, 120, 267, 268; as *ḥayra*, 6, 118; and pleasure, 4, 286; in Sanskrit, 277; and uncertainty, 225. *See also* astonishment; uncertainty
Persia: ancient kings of, 34, 149, 173, 289; as midway of world, 30, 192; people of, 72, 172, 192, 319. *See also* Iran
Persian: commentaries and translations of Quran, 230; literature, 5, 141, 146, 159, 230, 233, 262, 263–264, 267–268, 284; Orientalist study of, 308; poetics and poetry, 5, 100, 230, 267–268; prestige of, 100–101, 224, 228, 230, 262, 285; Qazwīnī as speaker and reader of, 35, 72; translations of Sanskrit, 277–278
philosophy: attitudes toward, 43, 55; authority of, 25; and education, 30, 38, 44, 45, 75; and *falsafa*, 55; and *ḥikma*, 43, 184, 226; of illumination, 68; and mystical piety, 69; perennial, 14, 304, 305; and *sophia*, 43, 184; and theology, 41–42, 43–45, 48, 56, 57, 106, 227; and theurgy, 171, 194, 373n10; and universal science, 41, 109; of wonder, 5, 117–120. *See also* emanationism; Illuminationist philosophy; moral philosophy; natural philosophy
phoenix, 157–157, 181, 370n50
physicians. *See* doctors and physicians
physiognomy, 76, 192, 238, 281
piety and devotion: of Alexander, 143, 196, 251; disciplines of, 111, 188; education and teachings, 66, 69, 90, 105–106, 114, 137, 150, 240; expressions of, 16, 17, 82, 201, 215; fanatical, 86; and kingship, 285; and occult sciences, 81, 280; and prayer, 3, 81, 82, 280; and reform, 60; sensibilities of, 226, 228, 236, 237; Sufi, 69, 267, 290; and trade, 150; and wonder at creation, 12, 105, 113, 130, 164, 268, 328. *See also* mysticism
pilgrims and pilgrimage, 65, 144, 212, 217, 266, 286
pillars. *See* Hercules: pillars of
Piri Reis: audience for, 258; *Book of Maritime Knowledge*, 246, 247, 249–250, 257; on Columbus, 250, 254; map of, 244, 245*f*, 246, 247–249, 251; sources of, 251
Pizigani, Domenico and Francesco, 219, 222
planetary motion: critique of Ptolemaic theory of, 226, 303, 306, 312; Ptolemaic theory of, 50–51, 52*f*, 58, 129; Pythagorean theory of, 205
planets: governed by angels, 137, 175; Ibn Sīnā on, 77; as ministers, 187; power and influences of, 75, 80, 85, 148, 168, 175–176; prayers to, 290–291; regulate behavior, 147; subjugation of, 78, 288; tracking positions of, 53–54, 90–91; wandering, 91, 95, 135, 227. *See also* astronomy; astrology; stars
plants: cultivation of, 178; and *materia medica*, 102; paintings of, 102; properties of, 107; and rational soul, 192; relation to fungi and minerals, 162; study of, 45; and Sun, 75, 88. *See also* agriculture; *materia medica*; properties, unique or special
Plato: as alchemist, 172; and divine double, 377n46; on habitus, 188; on hierarchy, 162–163; political theory of, 32, 187; and ring of Gyges, 10, 344n12; on shape of the world, 288; and wonder, 118, 119. *See also* emanationism; Neoplatonism
pleasure: and happiness, 23; and horror, 149; of intellection, 112–113; sexual and erotic, 254, 291; and storytelling, 114, 130, 147, 210, 224–225; and utility, 137; and wonder, 4, 6, 109, 254, 255, 264, 286, 314, 361n26
Pliny: confused with Apollonius, 366n21; on monsters and prodigies, 157, 160, 212, 299; *Naturalis historia*, 143, 160; Qazwīnī likened to, 302

Plotinus: emanationism of, 42; on magic and theurgy, 118, 193–194
Plutarch, 145, 182
poetry and poetics, 105, 137, 144; with astral themes, 91, 357n35; early Arabic, 140; Fakhr al-Dīn al-Rāzī and Sakkākī on, 82; and jinn, 146, 149; Persian, 5, 72, 100, 230, 267; and pleasure, 114; poetic adaptation of *Wonders and Rarities*, 267–268; power of, 83, 142; and Ṣaymarī, 148; of ʿUmar Khayyām, 283–284; and wonder, 4, 5; in *Wonders and Rarities*, 3, 29, 37, 106–107, 115, 133, 137, 147; and Yāqūt, 36, 37; and Venus, 147
Polyphemus the Cyclops, 157, 370n50
Porphyry: logic of, 42, 48; on shape of the world, 288
portolan charts, 216, 222, 382n34, 382n35, 383n44; grids accompanying, 217; of Mediterranean, 220–221*f*, 243, 249
Portuguese: along Africa, 219, 249; Atlantic expeditions of, 248; and colonies in India, 283, 286, 287; empire of, 307; and Mughals, 263–264, 272, 275; and Ottomans, 242, 243, 246–247, 260; in paintings, 316*f*; and Piri Reis, 247, 249; policy of deportment, 387n38
preachers, 98, 114, 142, 201, 234–237, 266
prejudice and bigotry, 164, 191, 319. *See also* climatic determinism; dispositions and natures
printing press, 307, 316, 318, 326–327. *See also* books; lithographs
progress, 13, 19, 302, 307; European, 318, 326, 328
properties, unique or special, 176–186; Abū Bakr al-Rāzī on, 176; in agriculture, 178; in amulets and talismans, 167; Apollonius on, 107, 173, 178; and Aristotle, 195–196, 230; of elemental bodies, 122; Hermes on, 171, 195–196; in minerals, plants, and animals, 31, 41, 142, 159, 171, 173; of planetary spheres, 135; Pliny on, 143; Qazwīnī on, 85, 98, 114, 231–232; of the Quran, 82; regional distribution of, 192, 240, 275; and rivers, 145; Shahmardān on, 229; of the Sun, 126; therapeutic use of, 167, 169. *See also* elemental magic; *nīranjāt*
prophecy: Abharī on, 46; Aḥmad Khān on, 325; Fakhr al-Dīn al-Rāzī on, 190; Ibn Sīnā on, 43, 50, 116, 122; and mesmerism, 323; of Moses, 140; of Muhammad, 88; reality of, 227
prophecies and predictions: and Mongols, 85, 190, 323; of renewal 59; of soothsayers, 121
prophets, 184, 192; miracles of, 50, 123–124, 322; tales of, 114, 288
proverbs, 115, 133, 137, 150, 165
providence, 107–109, 162, 357n27; divine, 31–32, 67, 75, 80, 139
psyche, 49–50, 111, 194, 322. *See also* souls
psychology, 43, 49–50, 111, 123, 192, 195
Ptolemy: and *Almagest*, 51, 74, 77, 132, 204; and astrology, 74, 77–78, 126; on climatic determinism, 5, 31, 163, 191, 207; and constellations, 131–132; on the cosmos, 46, 135, 144, 324; critique of, 58, 128, 204, 311–312; on Earth's circumference, 205; geography of, 115, 199, 216, 239–240, 252; on magnets, 182; and Mountain of the Moon, 213, 240; on planetary motion, 51, 58, 129, 226, 303, 350n12; on prime meridian, 207, 208; Qazwīnī on, 132, 213
Punjab Committee of Lahore, 296
pygmies, 144, 161
pyramids: Abbasid excavations of, 172; built by Hermes, 170, 172; of Giza, 145; and hieroglyphics, 141
Pythagoras: as alchemist, 172; Apollonius as follower of, 169 heliocentrism, 205, 380n7

Qāf, mountains of, 199, 236, 268, 277, 313; as superstructure, 201–202, 379n4
Qajars, 307, 315
Qazvin, 25, 30, 33, 72
al-Qazwīnī, Najm al-Dīn al-Kātibī, 57, 59, 351n21
al-Qazwīnī, ʿImād al-Dīn Zakariyyāʾ al-Kammūnī: and Abharī, 38–39, 44, 45, 51,

135–136; and Abū Bakr al-Rāzī, 181–182, 191; on alchemy, 59, 85, 166, 169; and Alexander the Great, 34, 195–196, 223–224, 230, 253–254; and Apollonius, 107, 173–174, 176–178; and Aristotle, 171, 196; on astrology, 3, 74–75, 77, 126–127; biography of, 25, 29–30, 33–36, 44, 62–63, 69, 70–72; and Fakhr al-Dīn al-Rāzī, 40, 58–59, 68, 75, 81, 135–136, 169, 175–176, 190; genealogy of, 70, 71–72, 104, 293; and Ghazālī, 27, 59, 82, 113, 120, 121, 136, 184, 188; and handbook of teachers, 70–71; and Hermes, 170–171, 173, 195–196; and Ibn al-ʿArabī, 69; and Ibn Sīnā, 73–74, 75, 77, 107, 110–112, 116–117, 120–123, 178–180, 192, 323; influence on, 70, 235, 236, 285, 302; as judge and professor, 3, 25, 65–66, 71, 88, 92, 99, 104; and Juwaynī, 80, 93, 97–98, 99; and Kamāl al-Dīn Ibn Yūnus, 46–47; on miracles and magic, 68–69, 122, 179, 189–190, 232, 323; painting and seal of, 292*f*, 293; and perfect nature, 189–190; and Pliny, 302, 366n21; and Ptolemy, 115, 131–132, 135, 137, 202, 204–205, 213; on Sufis and saints, 68–70; and Suhrawardī, 68, 189, 195; and Ṭabasī, 81; and Urmawī, 56, 71, 72; on writing *Wonders and Rarities*, 71, 92–93, 97, 99–100; on wonder and rarity, 6, 73–74, 117–118, 121–123; and Yāqūt, 37–38, 115–116

qibla, 54, 131, 236. *See also* Kaʿba; Mecca

quadrants, 18, 51, 129. *See also* devices and instruments

qualitates occultae, 179. *See also* properties, unique or special; occult properties of substances

Quran: and ancient ruins, 139, 140; authority of, 136, 312; and bibliomancy, 288; and calligraphy, 141; and contemplation of creation, 4, 106, 110, 119; and conversion of jinn, 28; copies of, 16; cosmography of, 321; and Dhū l-Qarnayn, 196, 251; exegesis of, 46, 56, 78, 235, 238, 281; and Gog and Magog, 144, 223; and law, 60, 62; in madrasa education, 44–45, 63, 66; medicinal and occult uses of, 69, 81–82, 84, 175, 280, 356n18; on miracles and magic, 125; and patriarchy, 186; Persian commentaries and translations of, 230; Qazwīnī as interpreter of, 285; as revealed knowledge, 43, 227; scientific inimitability of, 311–313; sura of the Cave, 139, 288; and universalism, 163

al-Qurṭubī, Maslama ibn Qāsim, 84, 189, 195, 279, 356n22

race, theories of, 163, 307, 315. *See also* climatic determinism; dispositions and natures; genealogies; group formations

Raikes, Charles, 296, 310

rarities, 2, 4, 268; valued, 142, 162, 264; in Europe, 305; spectacle of Eastern, 296. *See also* curiosities; *gharīb*; wonders and rarities

Rashīd al-Dīn, 87, 90, 104

rational soul, 110–112, 187, 189, 192, 322, 362n37; as *al-nafs al-nāṭiqa*, 49–50; power of, 125, 192; and superior rank of humans, 192. *See also* faculties; psychology; reason; souls

al-Rāzī, Abū Bakr (Rhazes), 166, 172, 181–182, 184, 195, 375n27

al-Rāzī, Fakhr al-Dīn: on alchemy, 169; on angels, 136, 189; on astral learning, 77–78, 176; authority of, 40; on eloquence, 82; on geomancy, 133; on God, 136, 184; and *The Hidden Secret*, 78, 81–84, 123, 125, 175–176, 190, 278, 356n20, 394n27, 394n28; and Ibn Sīnā, 38, 40–41, 58, 75, 117, 123, 195; influence on Abharī, 48–49, 54, 348n21; influence on Qazwīnī, 38, 135, 190; and jurisprudence, 56–57, 59; and Khwārazm-Shāh, 82; on magic and miracles, 125, 195, 280, 325–326; on metaphysics of the soul, 112, 136; on minerology, 168; on multiple worlds, 135, 223; and Naṣīr al-Dīn al-Ṭūsī, 91; on perfect nature, 189–190; and Persian, 101; philosophy of, 39, 43, 44, 48, 57, 313; as renewer, 59; and Suhrawardī, 68, 189; on virtue ethics, 112, 186; on void, 135–136

reason: faculty of, 49, 110–111, 116, 187; gradation of, 191–192; and Islam, 18–19, 310; and law, 60, 188; limits of, 108, 136, 137, 223, 365n11; modern ideals of, 18; and revelation or tradition, 43, 58, 62, 227, 228, 312; secular and disenchanted, 21, 22; vs. superstition, 13. *See also* faculties; intuition
recipes: agricultural, 374n24; alchemical, 174; based on unique properties, 176–178, 185–186; magical and talismanic, 81, 82, 84, 133, 176, 178–179, 232, 279, 281. *See also* cures; medicine
reconquista, 242–243. *See also* conquests
Red Sea: and Mediterranean, 110, 260; and Ottomans, 242, 246; and Portuguese, 258; trade in, 151
reform, 309; Aligarh movement, 326; Islamic, 13, 312, 321, 404n82
Reformation, 298, 326
reformists, 18–19, 204, 298, 306, 310, 318, 321, 325, 327
Rehatsek, Edward, 310–311
Renaissance: and cabinets of curiosity, 14; and museums, 16; study of Arabic during, 298; *nahḍa* as, 19
resurrection, 42, 237. *See also* eschatology; salvation
rhetoric: handbooks of, 82; in *Wonders and Rarities*, 110, 115, *See also* eloquence
Riḍā, Rashīd, 327
rivers, 95, 137, 140, 216, 267, 295; sacred, 290; strange phenomena in, 185; treatise on, 145–146
Roe, Thomas, 263–264, 273, 391n4
Roger II, 207
romance: of Alexander, 154–155, 157, 275; *dāstān*, 320; oriental proclivity for, 12
Royal Asiatic Society, 1, 15
Royle, John Forbes, 295
Rūḥ, 136. *See also* angels
rūḥāniyyāt, 171, 175–176, 179, 322. *See also* angels
Rukhkh, 156. *See also* Sīmurgh
Rūmī, Jalāl al-Dīn, 268, 290
Rus, 144, 191
Russell, Patrick, 302
Safavids, 282
sailors, 142, 152*f*, 213, 214; captives from Lisbon, 209–211; tales of, 151, 153, 155–156. *See also* ships and boats; shipwrecks
Saint Brendan, 155, 219, 249
saints, 37, 59, 70, 122, 141, 189; and miracles and wonders, 67–69, 89, 121, 123, 192, 265
al-Sakkākī, Abū Yaʿqūb, 82–85; son of, 279, 281–282; tomb of, 356n21
Ṣalāh al-Dīn al-Ayyūbī, 55
Salar Jung Museum of Hyderabad, 16–17, 345n21
Saljuqs, 35, 289
salvation: and self-discipline, 113, 119; promises of, 106, 150–151; pursuit of, 113
al-Samʿānī, Abū l-Karīm, 36
Samanids, 289
Sanāʾī, 100
Sanskrit: aesthetics of wonder in, 5, 394n26; in Arabic translations, 42, 149; and astrology and astronomy, 53, 279, 281–282, 364n55; and cosmology, 289–290; and magic, 84, 265, 280; and the *Mahābhārata*, 158, 277; and Orientalism, 16; in Persian translations, 276–279, 281
Sarandīb, 151, 153, 195, 275
Sasanians, 41, 42, 149–150, 230
Saturn: conjunction with Mars, 85; and India, 148; ritual to subjugate, 280*f*
savages: by disposition, 32, 191–192, 212; as jinn, 161, 321; and the *Mahābhārata*, 277–278; nations *(barbaricae nationes)*, 243; in the New World, 257–258; tales of, 150–151; ways of conceptualizing, 158–159. *See also* beasts; cannibals and cannibalism; dog-headed people; Gog and Magog; monsters; string-legged creatures
al-Ṣaymarī, Abū l-ʿAnbas: on astrology, 147–148; and Hamadhānī, 148; as judge, 146; and obscenity, 146–147; on wonders of the sea, 153
sciences, 7, 30, 42, 231, 234, 291; of ancients, 44; classification of, 226, 228, 240, 288; conjectural, 76; of cosmic structure, 51,

76–78; in Europe, 11–12, 308; extraordinary, 84, 126; language of, 43, 315; in Latin, 298; modern, 11, 311–312, 325; natural, 40, 43, 225, 228, 322, 318; of natural magic, 314, 322; practical, 85; of properties, 240; rational, 43, 306; of the real, 306; and religion, 11, 21, 23, 43, 56, 70, 197, 226–228, 293, 312; of stars, 77, 303; of talismans, 100, 173; universal, 41, 144, 306. *See also* natural philosophy
Scientific Society of Sayyid Aḥmad Khān, 309
Scot, Michael, 160
scribes, 67, 136, 230; and patrons, 318
scripts: Arabic, 141, 150, 230; ancient, 14, 33, 141; illegible, 2, 124, 172; and performance, 22, 296. *See also* cyphers
seals: magical, 84, 280, 291; as marks of ownership, 293
seas: and fountain of life, 139; seven, 248–249; transformation of, 138–139; wonders of, 20, 105, 146, 148, 151, 153–157, 165, 211, 248, *See also* encircling ocean/sea of darkness; oceans
secrets and secrecy: of alchemy, 173; of Alexander, 230, 250; of Aristotle, 171; biblical, 14; of cyphers and scripts, 141, 174; in Freemasonry, 304; of geomancy, 134; of God, 87; of Hermes, 170, 172; of nature and creation, 4, 43, 68, 159, 170–171, 184, 227, 275
sectarianism and factionalism: in Baghdad, 63; in Bijapur, 283; juridical and theological, 59, 60, 61, 62; transcended, 74, 126, 235, 279, 281. *See also* tolerance and religious impartiality
secularism: and education, 311; logic of, 21; and modernity, 8; and religion, 13
Selīm Khān, 246
Seven Sleepers, 288
sex: differentiation of, 191; with jinn, 149; magic, 3, 232, 270, 291; tales of, 232; women's appetite for, 177, 186; in temples, 156, 369n49
Sezgin, Fuat, 17–18, 204, 247, 335, 382n29, 382n35, 388n50
al-Shāfiʿī, Muḥammad ibn Idrīs, 55, 61–62
Shāfiʿī law, 59–60, 62
Shāh ʿAbbās, 270
Shāhjahān, 286
Shāhpūr ibn ʿUthmān, 231, 359n9
Shahmardān Ibn Abī l-Khayr, 229, 384n8
Shahrāzād, 149. See also *The Thousand and One Nights*
Shāhrukh, 266, 267
al-Sharābī, Sharaf al-Dīn, 55, 62, 65; madrasa of, 62, 65–66, 140, 351n19, 366n16
Shia: and ʿAlī, 17; authority of, 61; of Bengal, 311; in Maragha, 91; and Mughals, 263; and Safavids, 260, 267, 282; and Qazwīnī, 71, 86; and Umayyads, 65
ships and boats, 151, 165, 208, 209, 211, 213–215, 220*f*, 269*f*, 274; carracks and caravels, 244, 245*f*, 247; for pilgrims, 286; Ottoman, 242; Portuguese, 247; warships and galleys, 218, 287. *See also* sailors
shipwrecks, 11, 151, 156, 157*f*, 161*f*
al-Shīrāzī, Quṭb al-Dīn, 226–228
Shujāʿ-i Ḥaydarī, 314–315
Shushtarī, ʿAbd al-Laṭīf, 306
silver: as foundational metal, 168; globe made of, 129; map made of, 216; in New World, 259
sīmiyāʾ, 314, 322, 385n13, 392n11. *See also* natural magic
Sīmurgh, 101*f*, 158. *See also ʿanqāʾ*
Sind, 30, 247
Sindbad the Sailor, 156, 301
Sindhind, 364n55
Siraf, 153
al-Sīrāfī, Abū Zayd, 153
slaves: from Africa, 242–243; from the Canaries, 215, 242, 305; among Christians, 256; and colonialism, 14, 294, 305; and Mongol invasions, 89–90; among Muslims, 36, 154, 244, 246, 247, 283; and society, 31; stories concerning, 147, 149, 151, 155, 156, 209
Slavs, 191

Socrates, 288
solar apogee, 126–127, 358n5, 359n9, 364n57
soothsayers, 79, 81, 121, 181. *See also* divination
sophia, 43, 184. *See also* design; philosophy; wisdom
sorcery, 212, 238, 268, 281, 301, 302; arts of, 265, 322; writings on, 318. *See also* divination; enchantment; handbooks of magic; magic
souls: animal, 49, 257, 322; betterment of, 189–190; discipline and purification of, 68, 111–112, 113, 118–119, 122, 167, 186, 188–190, 192–194, 231, 241, 277, 314; emanation of, 189–190; eternal, 304; Ibn Sīnā on, 43, 49–50, 112, 123; as magnetic force, 320, 322; as *rūḥāniyyāt*, 175; turbid, 189; universal, 322; vegetal, 49; virtuous, 189. *See also* psychology; rational soul; *rūḥāniyyāt*
South America, 248
sovereignty: divine, 277, 278; ideals of, 258; and Juwaynī, 97; universal, 80, 224, 262, 283; world, 283
Spain, 219, 250, 258, 307. *See also* al-Andalus; Iberian Peninsula
speech, 142, 194; divine, 82–83, 268; unintelligible, 161. *See also* eloquence; rhetoric
sphericity of Earth, 204–205, 207, 379n6
sphinx, 164, 371n61
Spice Islands, 246, 258
spices, 31, 151, 242, 244, 259, 264, 276, 287, 294; trade, 142, 249. *See also* commodities; *masala*
spirits, 175, 189–190, 196. *See also* souls
spiritualism, 320, 322, 325, 403n75
star catalogues, 102, 129, 130, 302, 359n10. *See also* constellations
stars: as ensouled, 194; fixed, 135–136, 227, 301; force and influence of, 74–77, 133, 189, 191; governed by angels, 137; mastery of, 83–84; names and titles of, 77, 301–302; observing, 53–54; science of, 77, 303; shooting, 73–74, 78, 87. *See also* astrology; astronomy; planets
statues: as automata, 173; as boundary markers, 206, 222; Buddhist, 151, 174, 206, 222; in museums, 16, 296. *See also* talismans
stones, 195–196; rain, 181. *See also* lapidaries; metals; minerals
string-legged creatures, 161, 369n48; known to Pliny, 157
sublunar realm, 95, 137, 138, 175
al-Sūdān, 208
Suez, 10
suffering, 107–109, 151, 156. *See also* evil; good; theodicy
al-Ṣūfī, ʿAbd al-Raḥmān, 130–131
Sufis, 66, 69, 90, 153, 267, 268, 290. *See also* miracles; mysticism; piety and devotion
suffumigation, magical rites of, 84, 179, 280
al-Suhrawardī, Abū Ḥafṣ, 153
al-Suhrawardī, Shihāb al-Dīn, 68, 189, 195, 322, 325
sulfur. *See* mercury and sulfur
Sumatra. *See* Aceh
Sun, 75, 126–127; movement of, 205, 237, worshipping, 215, 218. *See also* solar apogee
Sunnis, 61–62, 71, 86, 154, 285, 286
superstition, 12, 13, 296, 321, 323, 326
Sürūrī, Musliḥuddīn Muṣṭafā, 199, 201
Suʿūdī, Meḥmed ibn Ḥasan, 251–255, 275; collection on New World, 258–259, 300
sympathy: as basis of paranormal causation, 178–180; criticism of, 321; as hidden force, 125, 166, 168; as a historical method, 23, 330; and Plotinus, 193–194; in talismans and *nīranjāt*, 175, 231
Syriac: in Abbasid translation movement, 42, 143, 169, 178, 192; and Alexander, 196; for biblical studies, 298; terminology, 107, 124; and wonders, 145

al-Ṭabarī, Abū Jaʿfar Muḥammad Ibn Ayyūb, 185, 228–229, 376n38
al-Ṭabasī, Abū l-Faḍl Muḥammad, 81–83, 175, 279, 346n1, 355n16

Ṭabāṭabā'ī, Ghulām Ḥusayn Khān, 308–309
talismans: and Abū Bakr al-Rāzī, 181–182, 184; activation of and recipes for, 124, 185; Apollonius as master of, 107, 170, 173–176, 189; on the authority of Aristotle, 169–171, 196; and Bīrūnī, 185; critique of, 302, 321, 326, 327; and Fakhr al-Dīn al-Rāzī, 175–176, 189, 278; forces of, 96, 178, 196; and idols and statues, 205–206; and lunar mansions, 133; in the *Mahābhārata*, 277; and mesmerism, 320, 323–324; in natural histories, 228–230, 238–239, 241, 281–282, 313, 315, 324; rationalization of, 122, 167, 179, 189–190, 228, 320; in *Wonders and Rarities*, 100, 109, 231–232, 235, 270, 290–291, 298, 315. *See also* elemental magic; handbooks of magic
Tantric philosophy and practice, 84, 281, 282
Ṭaşköprīzāde, ʿIṣām al-Dīn, 240–241
al-Tawḥīdī, Abū Ḥayyān, 119–120
Taurus (constellation), 209
technology: and determinism, 23, 307, 318; and military advances, 303; of modernity, 295, 316; superiority of, 307. *See also* automata; printing press
teratologia, 160. *See also* monsters
terrestrial realm: as sublunar, 6, 49, 75–76, 95–96, 175–176; and boundaries of knowledge, 20, 196, 205–206
text reuse, 38, 348n16
thaumata, 113, 184, 145. *See also* marvels
theodicy, 108. *See also* evil; good; suffering
theology, 43–44, 45, 48, 225; Ashʿarī, 59, 62, 68, 81, 108, 136, 354n5; and astrology, 75, 76; and causation, 111, 120, 137; compendiums of, 226; and divine sciences, 41; and factionalism, 59; 61; and free will, 62, 111; and God's existence, 106; and Ibn Sīnā, 41; as *kalām*, 42, 56; Muʿtazilī, 62; and philosophy, 41–42, 43–45, 48, 56, 57, 106, 227. *See also* design; miracles; philosophy
Theosophical Society, 325, 403n74
The Thousand and One Nights, 12, 149, 238, 295, 368n34; source for conceiving of Orient, 301; Urdu editions of, 320
throne of God, 136, 281, 365n11
al-ṭibāʿ al-tāmm, 171, 189, 373n11. *See also* perfect nature
Tigris River, 25, 30, 140
time: and auspicious timings, 147–148; corruption of, 138; cosmic cycles of, 136, 255; devices to measure, 51, 131, 351n13; Qazwīnī on, 134–135, 137; standardized, 23, 307. *See also* clocks
Tīmūr and Timurids, 233, 262, 267, 270, 278
tolerance and religious impartiality: and Freemasonry, 304; in India, 290; of Mongols, 80; of Mughals, 276, 393n23; of Ottomans, 258; strange to Europeans, 13, 263
Topkapı Palace, 17, 19, 246, 345n25
Tordesillas, Treaty of, 249
truth, 113–114; vs. falsehood, 116; *fī nafs al-amr*, 226; knowledge of, 118; universal, 290
al-Ṭūsī, Muḥammad ibn Maḥmūd, 229, 238–239
al-Ṭūsī, Naṣīr al-Dīn, 70, 85–87, 351n22; and Abharī, 57–58; on ethics, 186–187; on geomancy, 133; and Ibn Sīnā, 58; on Maragha, 90–91; metaphysics of, 112; and Mongols, 90; Persian lapidary of, 92, 100; and Ptolemy, 303; record of observations by, 128
twins, conjoined 1, 2, 8*f*, 162, 234

Ulugh Beg, 278
al-ʿUmarī, Shihāb al-Dīn, 153–155, 166; on encircling ocean, 212; on Mali Empire, 213; on portolan charts, 217; on wonders, 154–155, 165
ʿUmar Khayyām, 283
Umayyads, 42, 61, 65, 143, 208, 289
ummahāt, 168, 171. *See also* elements, foundational
uncertainty, 3, 4, 20, 43, 107, 164, 333; and astral forces, 77; cultivation of, 204, 225, 328; and theodicy, 107. *See also* ambiguity

unity of being, 265
Universal Exhibition of Paris, 295, 296–297
unseen, realm of, 27, 164
ʿUnṣurī, 230
al-ʿUrḍī, Muʾayyad al-Dīn, 129, 357n35
Urdu: and the Brethren of Purity, 322, as literary language, 286; lithography 313–320, 324, 330; natural histories, 313–314; storytelling, 320, 333; and superstition, 326; translations, 313–314, 323–324; *Wonders and Rarities* in, 315, 319
al-Urmawī, Tāj al-Dīn, 56–57, 62, 89, 351n19
Uruk, 33

Vedas, 290
Venus, 147

Wahb ibn Munabbih, 137, 201
wājib al-wujūd, 110, 265
waḥdat al-wujūd, 265
Wajdī, Farīd, 325
Wāqwāq islands, 11, 151, 254, 255, 268
Warton, Thomas, 12, 344n15
Wasit, 25, 30, 65, 89, 140, 357n31, 366n16
Western Hemisphere, 19–20, 160, 204, 251; mainland of, 247; subjugation of, 308
wisdom: divine, 107; as *ḥikma*, 109–110, 170, 184, 226, 228; perennial, 14, 304, 305; as *sophia*, 43, 184
witchcraft, 213, 322. *See also* magic
women: arts and wiles of, 141, 291, 378n48; biographies of, 37; chastity of, 174; degradation of, 296; elite, 186–187; kingdom populated by, 277; and marriage, 89; and misogyny, 191; in public life, 67, 187; sexual appetite of, 177, 186; as sources of temptation, 191, 291, 378n48; superstitions of, 182; traits of, 191; of Wāqwāq, 11, 151, 254. *See also* dispositions and natures
wonderment: appeal to, 12, 183, 282; and contemplation, 4, 7, 29, 130, 145, 186, 210; definition of, 6–7, 67, 119–121; and eloquence, 142; God's experience of, 119, 362n41; as historical method, 6, 334; loss of, 118, 121, 362n38. *See also* astonishment; awe; perplexity and confusion
wonders and rarities: as an emotive registrar, 150, 159, 306; in relation to each other, 4, 117–126; as a way to fathom and organize the world, 6–7, 37, 158, 255, 277, 305, 324; writings on, 114, 238
Wonders and Rarities: authority and influence of, 2–3, 13, 166, 235, 266, 267–268, 282, 287, 298–299, 313, 319; classification of, 238–239, 240, 319–320; composition of, 38, 71, 92, 98–97, 127, 293, 358n5, 359n9, 364n57; and earlier natural histories, 185, 228–229, 233; and entertainment, 3, 115, 129–130, 137, 147; as history, 239; misogyny in, 191; and Orientalism, 1–3, 8, 11, 296–297, 298–299, 300, 302; and patronage, 80, 97–98, 283–284; reception of, 6, 8, 17, 98, 100, 141, 197, 224–225, 228, 232–235, 261–262, 265–266, 270, 283, 287, 291–292; structure of and table of contents for, 6, 45, 95, 109, 167, 197, 225, 288; and taxonomy, 45, 75, 95, 109, 130, 167; title of, 16, 29, 107, 229, 238. *See also* natural history, writings on
Wonders and Rarities, copies/translations of: in different redactions, 98–99, 203, 231, 234–235, 358n5, 364n57; in Europe, 301, 305; in first Persian recension, 100–101, 127, 179, 190, 230–232, 359n9, 335; in India, 2, 262, 265, 329, 391n7, 392n8; in lithograph print, 315, 318, 319, 326; with marginalia, 261, 288–291; modern Arabic edition, 297, 335; modern translations of, 335; with paintings and illustrations, 102–103, 115, 203, 232–235, 266, 268, 288, 291, 293, 319, 329, 335–336; of the Persian translation from Bijapur, 261–262, 282–285, 293; in Turkish, 199, 201, 265, 319; in Universal Exhibition of Paris, 296–297, 310; in Urdu, 315, 319; from Wasit, 102, 127, 167, 239, 335

worldmaking, 21, 411
Wüstenfeld, Ferdinand, 297, 335

Yāqūt al-Rūmī: geography of, 36–37; and Qazwīnī, 37–38, 115–116, 348n17; skepticism of, 173, 201
year transfer, 132
Yoga, 281

Ẓāhirīs, 60–61
Zakī Pāshā, Aḥmad, 19, 204, 327–328
Zangid governors, 35
zīj, 53
zodiac, 280, 394n27; signs of, 132–133, 135, 143
Zoroastrians, 67, 175, 235; cosmography of, 124, 179, writings of, 145
Zosimos, 175